# PRE-ALGEBRA

## An Accelerated Course

*Mary P. Dolciani*
*Robert H. Sorgenfrey*
*John A. Graham*

**Editorial Advisers**

*Richard G. Brown*
*Robert B. Kane*

**HOUGHTON MIFFLIN COMPANY · Boston**
At Dallas Geneva, Ill. Lawrenceville, N.J. Palo Alto Toronto

## AUTHORS

**Mary P. Dolciani**  Professor of Mathematical Sciences, Hunter College of the City University of New York

**Robert H. Sorgenfrey**  Professor of Mathematics, University of California, Los Angeles

**John A. Graham**  Mathematics Teacher, Buckingham Browne and Nichols School, Cambridge, Massachusetts

### Editorial Advisers

**Richard G. Brown**  Mathematics Teacher, The Phillips Exeter Academy, Exeter, New Hampshire

**Robert B. Kane**  Director of Teacher Education and Head of the Department of Education, Purdue University

### Teacher Consultants

**John E. Mosby**  Instructional Coordinator, Dixon School, Chicago, Illinois

**William Voligny**  Mathematics Teacher, Olympia Junior High School, Auburn, Washington

DEFGHIJ-VH-93210/8987

# Contents

iv

# 8 Equations and Inequalities

# 9 The Coordinate Plane

# 10 Areas and Volumes

# Reading Mathematics

This page shows many of the metric measures and symbols that are used in this book. Use this page as a reference when you read the book.

## Symbols

| Symbol | | Page |
|---|---|---|
| $\cdot$ | times | 3 |
| $\approx$ | is approximately equal to | 14 |
| $^-1$ | negative one | 46 |
| $\lvert{}^-3\rvert$ | absolute value of $^-3$ | 46 |
| $>$ | is greater than | 47 |
| $<$ | is less than | 47 |
| $\neq$ | is not equal to | 47 |
| $\geq$ | is greater than or equal to | 47 |
| $\leq$ | is less than or equal to | 47 |
| $-b$ | the opposite of $b$ | 59 |
| $0.4\overline{36}$ | 36 repeats without end | 117 |
| $\overleftrightarrow{PQ}$ | line $PQ$ | 164 |
| $\overrightarrow{BA}$ | ray $BA$ | 164 |
| $\overline{PQ}$ | segment $PQ$ | 164 |
| $\parallel$ | is parallel to | 165 |

| Symbol | | Page |
|---|---|---|
| $AB$ | the length of $\overline{AB}$ | 169 |
| $\cong$ | is congruent to | 169 |
| $\angle$ | angle | 173 |
| $60°$ | sixty degrees | 173 |
| $m\angle A$ | measure of angle $A$ | 173 |
| $\perp$ | is perpendicular to | 173 |
| $\triangle ABC$ | triangle $ABC$ | 178 |
| $\pi$ | pi | 188 |
| $1:5$ | 1 to 5 | 210 |
| $\%$ | percent | 227 |
| $(5, 4)$ | ordered pair 5, 4 | 312 |
| $\sqrt{a}$ | positive square root of $a$ | 394 |
| $\sim$ | is similar to | 406 |
| $3!$ | 3 factorial | 453 |
| $P(E)$ | probability of event $E$ | 461 |

## Metric Measures

### Prefixes

| Prefix | kilo | centi | milli |
|---|---|---|---|
| Factor | 1000 | 0.01 | 0.001 |
| Symbol | k | c | m |

### Base Units

Length: **meter** (m)

Mass: **kilogram** (kg)

Capacity: **liter** (L)

Temperature  **Degree Celsius** (°C)

| | | | |
|---|---|---|---|
| **Length** | 1 mm = 0.001 m | 1 cm = 0.01 m | 1 km = 1000 m |
| | 1 m = 1000 mm | 1 m = 100 cm | 1 cm = 10 mm |
| **Mass** | 1 kg = 1000 g | 1 mg = 0.001 g | 1 g = 0.001 kg |
| **Capacity** | 1 mL = 0.001 L | 1 L = 1000 mL | 1 L = 1000 cm$^3$ |
| **Time** | 60 s = 1 min | 60 min = 1 h | 3600 s = 1 h |

Examples of compound units     kilometers per hour: km/h

square centimeters: cm$^2$          cubic meters: m$^3$

# Diagnostic Test of Whole Number and Decimal Skills

This test reviews the skills of addition, subtraction, multiplication, and division necessary to begin Chapter 1. More practice of these skills can be found on pages 480–483.

## Addition

**Add.**

1.  142
    + 237

2.  374
    + 213

3.  103
    19
    + 42

4.  1007
    285
    + 59

5.  246
    9
    1064
    842
    + 83

6.  5.246
    + 6.38

7.  16.439
    + 28.32

8.  3.84
    2.07
    + 9.39

9.  72.8
    6.349
    + 0.76

10. 0.16
    54.3
    119.057
    + 2.0918

11. $32 + 56$

12. $693 + 105$

13. $34 + 17 + 25$

14. $2306 + 19 + 429 + 1443$

15. $18.7 + 5.394$

16. $0.06 + 19.803$

17. $11.882 + 6.49 + 0.083$

18. $583.117 + 72.5 + 3.76824$

19. $35.402 + 17.6 + 5.28 + 0.314$

20. $3.4289 + 5.005 + 31 + 8.57$

## Subtraction

**Subtract.**

1.  864
    − 231

2.  9748
    − 2635

3.  4693
    − 2758

4.  801
    − 543

5.  3004
    − 2567

6.  6.75
    − 3.81

7.  14.90
    − 7.88

8.  388.6
    − 97.86

9.  705.56
    − 314.6

10. 0.986
    − 0.097

11. $768 − 654$

12. $925 − 713$

13. $853 − 432$

14. $2670 − 357$

15. $5323 − 789$

16. $4803 − 567$

**Subtract.**

**17.** $9.5 - 4.8$

**18.** $46.71 - 22.52$

**19.** $0.039 - 0.0271$

**20.** $0.8743 - 0.33591$

**21.** $0.039 - 0.0271$

**22.** $5.04876 - 229$

---

## Multiplication

**Multiply.**

**1.** $\begin{array}{r} 63 \\ \times\ 21 \\ \hline \end{array}$

**2.** $\begin{array}{r} 42 \\ \times\ 34 \\ \hline \end{array}$

**3.** $\begin{array}{r} 131 \\ \times\ 25 \\ \hline \end{array}$

**4.** $\begin{array}{r} 214 \\ \times\ 37 \\ \hline \end{array}$

**5.** $\begin{array}{r} 381 \\ \times\ 206 \\ \hline \end{array}$

**6.** $\begin{array}{r} 473 \\ \times\ 0.3 \\ \hline \end{array}$

**7.** $\begin{array}{r} 846 \\ \times\ 2.5 \\ \hline \end{array}$

**8.** $\begin{array}{r} 127.3 \\ \times\ 6.6 \\ \hline \end{array}$

**9.** $\begin{array}{r} 67.05 \\ \times\ 2.39 \\ \hline \end{array}$

**10.** $\begin{array}{r} 99.7 \\ \times\ 10.06 \\ \hline \end{array}$

**11.** $51 \times 73$

**12.** $92 \times 34$

**13.** $732 \times 24$

**14.** $947 \times 62$

**15.** $4023 \times 570$

**16.** $9108 \times 6027$

**17.** $18.7 \times 16$

**18.** $0.08 \times 58.6$

**19.** $0.75 \times 0.69$

**20.** $27.9 \times 33.3$

**21.** $6.0810 \times 148.3$

**22.** $5.62 \times 83.109$

---

## Division

**Divide.**

**1.** $7\overline{)91}$

**2.** $4\overline{)64}$

**3.** $8\overline{)296}$

**4.** $29\overline{)522}$

**5.** $74\overline{)3478}$

**6.** $607\overline{)6677}$

**7.** $5\overline{)37.60}$

**8.** $81\overline{)108.54}$

**9.** $5.4\overline{)3348}$

**10.** $1.6\overline{)99.68}$

**11.** $2.86\overline{)247.39}$

**12.** $1.13\overline{)1006.83}$

**Divide. Round to the nearest tenth.**

**13.** $56 \div 3$

**14.** $49 \div 8$

**15.** $319 \div 15$

**16.** $628 \div 20$

**17.** $948 \div 48$

**18.** $9963 \div 542$

**19.** $0.851 \div 0.33$

**20.** $0.909 \div 1.35$

**21.** $0.284 \div 7.31$

**22.** $486 \div 0.391$

**23.** $5.005 \div 0.095$

**24.** $43.761 \div 27.515$

# 1

# Introduction to Algebra

All of the many pieces of information handled by computers are stored and processed by means of tiny microchips, such as the one shown at the right. Each microchip is about 2 mm square and is made up of thin layers of silicon crystals. Each layer is treated chemically and etched photographically with different patterns containing tens of thousands of microscopic switches. Information on the microchip is represented by a code made up of a series of "on" or "off" switches.

Although we use words and numbers to communicate with a computer, the machine does not work directly with those words and numbers. A program within the computer automatically translates the information that we use into the special code that the machine understands. In a similar way, when we work with algebra, we translate our words and ideas into the language of mathematics. In this chapter, you will learn many of the symbols that we use to express mathematical ideas.

## Career Note

Consider how intricate the design of a single microchip is. Electrical engineers are involved in designing and testing new electrical equipment such as the microchip. Electrical engineers must therefore be qualified in both mathematics and science. Most, in fact, specialize in a major field such as communications, industrial equipment, or computers.

# 1-1 Mathematical Expressions

A **numerical expression** is simply a name for a number. For example,

4 + 6 is a numerical expression for the number 10.

Since 4 + 6 and 10 name the same number, we can use the equals sign, =, and write

$$4 + 6 = 10.$$

We **simplify the numerical expression** 4 + 6 when we replace it with its simplest name, 10.

**EXAMPLE 1** Simplify each numerical expression.
  **a.** 400 × 4          **b.** 37 − 19          **c.** 5.1 ÷ 3

**Solution**      **a.** 400 × 4 = 1600      **b.** 37 − 19 = 18      **c.** 5.1 ÷ 3 = 1.7

If a computer can do 100 million arithmetic computations in one second, the table below shows how many million computations the computer can do in two, three, and four seconds.

| Number of Seconds | Millions of Computations |
|:---:|:---:|
| 1 | 100 × 1 |
| 2 | 100 × 2 |
| 3 | 100 × 3 |
| 4 | 100 × 4 |

Each of the numerical expressions 100 × 1, 100 × 2, 100 × 3, and 100 × 4 fits the pattern

$$100 \times n$$

where $n$ stands for 1, 2, 3, or 4. A letter, such as $n$, that is used to represent one or more numbers is called a **variable.** The numbers are called the **values of the variable.**

An expression, such as $100 \times n$, that contains a variable is called a **variable expression.** When we write a product that contains a variable, we usually omit the multiplication sign.

$100 \times n$ may be written $100n$.

$x \times y$ may be written $xy$.

In a numerical expression for a product, such as 100 × 4, we must use a multiplication sign to avoid confusion.

A raised dot is also a multiplication sign.

$$100 \times 4 \text{ may be written } 100 \cdot 4.$$

When we replace each variable in a variable expression by one of its values and simplify the resulting numerical expression, we say that we are **evaluating the expression** or **finding the value of the expression.**

When the value of $n$ is 4, the value of $100n$ is 400.

**EXAMPLE 2**   Evaluate each expression when $a = 6$ and $b = 2$.
  **a.** $9 + a$        **b.** $a \div b$        **c.** $3ab$

**Solution**    **a.** Substitute 6 for $a$.      $9 + a = 9 + 6 = 15$
           **b.** Substitute 6 for $a$ and 2 for $b$.      $a \div b = 6 \div 2 = 3$
           **c.** Substitute 6 for $a$ and 2 for $b$.      $3ab = 3 \times 6 \times 2 = 36$

In the expression $9 + a$, 9 and $a$ are called the **terms** of the expression because they are the parts that are separated by the $+$. In an expression such as $3ab$, the number 3 is called the **numerical coefficient** of $ab$.

---

**Reading Mathematics: *Vocabulary***
Look back at this first lesson. Notice the words in heavier type throughout the text. They are important new words and ideas, such as the following:
           numerical expression          simplify a numerical expression
           variable                       value of a variable
           variable expression            evaluate a variable expression
           terms                          numerical coefficient
When you see a word in heavier type, look near it for an explanation or example to help you understand the new word. For an unusual new word, look up the definition in the glossary to help you understand and remember it.

---

## Class Exercises

**Simplify the numerical expression.**

  **1.** $24 + 18 + 32$          **2.** $125 \div 5$          **3.** $3.6 + 5.1$          **4.** $0.25 \times 10$

**Evaluate the variable expression when $x = 4$.**

  **5.** $x + 7$          **6.** $5x$          **7.** $28 \div x$          **8.** $13 - x$

  **9.** What is another way to write $17 \times x$?

  **10.** In the variable expression $12a$, the numerical coefficient is __?__ and the variable is __?__ .

## Written Exercises

**Simplify the numerical expression.**

**A**
1. $16 \cdot 3$
2. $37 + 12$
3. $114 - 9$
4. $918 \div 6$

5. $1.65 + 12.5$
6. $1.05 + 9.7$
7. $0.5 \times 9$
8. $2.53 \div 11$

**Evaluate the expression when $y = 2$.**

9. $y + 23$
10. $6y$
11. $8 \div y$
12. $4 + y + 7$

**Evaluate the expression when $q = 3$.**

13. $36 \div q$
14. $q \div 3$
15. $q \times 9$
16. $q - q$

**Evaluate the expression when $a = 5$.**

17. $a - 3$
18. $a \times a$
19. $42 + a$
20. $a + a$

**Evaluate the expression when $m = 8$.**

21. $6 + 9 + 3 + m$
22. $m \times 3$
23. $m \times m \times m$
24. $m \div m$

**Evaluate the expression when $x = 8$ and $y = 1$.**

**B**
25. $x + y$
26. $x - y - 2$
27. $3x$
28. $x \div y$

**Evaluate the expression when $c = 7.5$ and $d = 3$.**

29. $c \div d$
30. $c + d + 20$
31. $c \times 15$
32. $4cd$

**Evaluate the expression when $s = 12.3$ and $t = 6.15$.**

33. $9st$
34. $s \div t$
35. $s - t - 3$
36. $73.8 \div s$

**C**
37. Find the value of $a$ for which the expressions $2a$ and $2 + a$ have the same value.

38. Find a value of $x$ for which $x \div 7$ and $7 \div x$ are equal.

---

## Review Exercises

**Perform the indicated operation.**

1. $43.8 + 8.07$
2. $51 \times 3.4$
3. $80.47 - 34.54$
4. $6.29 + 0.124$

5. $11.61 \div 43$
6. $6.328 \times 0.729$
7. $32.004 \div 5.08$
8. $2403 - 976.8$

# 1-2 Order of Operations

The expression

$$2 + (6 \times 3)$$

involves both addition and multiplication. The parentheses indicate that the multiplication is to be done first.

$$2 + (6 \times 3) = 2 + 18 = 20$$

For the expression

$$(2 + 6) \times 3$$

the parentheses indicate that the addition is to be done first.

$$(2 + 6) \times 3 = 8 \times 3 = 24$$

Parentheses used to indicate the order of the arithmetic operations are called **grouping symbols.** Operations within grouping symbols are to be done first.

We usually write a product such as $4 \times (3 + 5)$ without the multiplication symbol as $4(3 + 5)$. We may also use parentheses in any one of the following ways to indicate a product such as $4 \times 8$:

$$4(8) \qquad \text{or} \qquad (4)8 \qquad \text{or} \qquad (4)(8).$$

A fraction bar is both a division symbol and a grouping symbol. Recall that $18 \div 2$ may be written as $\frac{18}{2}$. When operation symbols appear above or below the fraction bar, those operations are to be done before the division. For example, to simplify

$$\frac{6 + 15}{3},$$

we add first:

$$\frac{6 + 15}{3} = \frac{21}{3} = 7.$$

*EXAMPLE 1*　Evaluate each expression when $n = 8$.

　　　　**a.** $4(n - 3)$ 　　　**b.** $\dfrac{45 - 15}{n + 7}$

*Solution*　Substitute 8 for $n$ in each expression.

　　　　**a.** $4(n - 3) = 4(8 - 3) = 4(5) = 20$

　　　　**b.** $\dfrac{45 - 15}{n + 7} = \dfrac{45 - 15}{8 + 7} = \dfrac{30}{15} = 2$

If there are no grouping symbols in an expression, we agree to perform the operations in the following order.

> ### *Rule for Order of Operations*
>
> When there are no grouping symbols:
>
> **1.** Perform all multiplications and divisions in order from left to right.
>
> **2.** Perform all additions and subtractions in order from left to right.

**EXAMPLE 2**   Simplify $392 + 637 \div 49$.

**Solution**        $392 + \underbrace{637 \div 49}$

$\underbrace{392 + \quad 13}$

$\qquad\quad 405$

When a product of two numbers or of a number and a variable is written without a multiplication symbol, as in 5(7) or 4$n$, we perform the multiplication before the other operations.

**EXAMPLE 3**   Evaluate the expression when $x = 6$.

**a.** $27 \div 2x$        **b.** $\dfrac{3x - 2}{4}$

**Solution**        Substitute 6 for $x$ in each expression.

**a.** $27 \div 2x = 27 \div 2(6) = 27 \div 12 = 2.25$

**b.** $\dfrac{3x - 2}{4} = \dfrac{3(6) - 2}{4} = \dfrac{18 - 2}{4} = \dfrac{16}{4} = 4$

---

## Class Exercises

**Tell in which order the operations should be performed to simplify the expression.**

**1.** $6 + 14 \times 3$          **2.** $(6 + 14)3$          **3.** $18 - 12 \div 3 + 1$

**4.** $18 - 12 \div (3 + 1)$          **5.** $23 - 9 \div 5 + 2$          **6.** $(9 + 16) \div (4 + 1)$

Evaluate the expression when $m = 4$.

**7.** $(m + 6)2$

**8.** $5m + 8$

**9.** $3(m - 1)$

**10.** $8 \div m + 7$

**11.** $m(2 + m)$

**12.** $\dfrac{3m}{18 - 6}$

## Written Exercises

**Simplify the expression.**

**A**　**1.** $35 - 14 \div 2 + 64$

**2.** $54 \div 6 + 18 \times 2$

**3.** $44 + 17 - 5 \times 2$

**4.** $(45 - 19)(8 + 7)$

**5.** $(12 + 18) \div (19 - 4)$

**6.** $\dfrac{9 + (4 \times 3)}{7}$

**Evaluate the expression when $n = 7$.**

**7.** $(14 + n)6$

**8.** $36 \div (n - 3)$

**9.** $(n + 28) \div 5$

**10.** $(27 - n)3$

**11.** $12(n - 4)$

**12.** $\dfrac{94 - 38}{n}$

**Evaluate the expression when $t = 10$.**

**13.** $5t \div (14 - 9)$

**14.** $\dfrac{25 - t}{10 - 5}$

**15.** $(t + 6 - 9)t$

**16.** $2(t + 5) - t$

**17.** $(t - 4)(t - 4)$

**18.** $50 \div (t + 15) + t$

**Evaluate the expression when $s = 16$.**

**19.** $7(s + 12)$

**20.** $5(s - 4)$

**21.** $(s + 32) \div s$

**22.** $(3s - 6) \div 7$

**23.** $18(6s - 12)$

**24.** $4s \div (s - 8)$

**Evaluate the expression when $b = 6$ and $c = 7$.**

**B**　**25.** $(b + c)c$

**26.** $3c \div (b - 4)$

**27.** $2bc \div (c - b)$

**Evaluate the expression when $m = 4$ and $n = 9$.**

**28.** $5n \div (m + 5)$

**29.** $(n + m) \div (35 - n)$

**30.** $m(n - m) \div 8$

**Evaluate the expression when $a = 3.6$ and $b = 8.2$.**

**31.** $ab \div 2a$

**32.** $(b - a)(3a + 6)$

**33.** $7ab(3b + a)$

**Evaluate the expression when** $e = 5$, $f = 8$, **and** $g = 13$.

**34.** $(e + f)(g + e)$ **35.** $f \div (g - f) + g$ **36.** $f(e + g) - e$

**Copy the expression as shown. Add grouping symbols so that the value of the expression is 24 when** $x = 3$, $y = 7$, **and** $z = 21$.

**C** **37.** $2x \times y - 4 + 2x$ **38.** $y + z \div 4 \times x + x$ **39.** $x \times y + z \div 3 + 1 - z$

## Review Exercises

**Evaluate the expression when** $x = 2$, $y = 1$, **and** $z = 4$.

**1.** $2x + y$ **2.** $3z - 3x$ **3.** $5xy - z$ **4.** $4xyz - xz$

**5.** $6yz \div 4x$ **6.** $12z \times 3y$ **7.** $7z \times 3x$ **8.** $8xy \div z$

---

### ▮▮▮ Calculator Key-In

To use your calculator to simplify an expression with more than one operation, you must keep in mind the order in which you want the operations to be performed. Try to simplify $2(6 + 4)$ by entering the following on your calculator exactly as it is shown.

Although the correct answer is 20, your calculator will perform the operations in the order in which you entered them and will display 16 for the answer.

To obtain the correct answer, you must enter the expressions in the order in which you want them to be performed. Enter the following exactly as it is shown.

Now your calculator should display the correct answer, 20. By entering $=$ after entering the expression in parentheses, you complete the operation inside the parentheses before doing the next operation. Some calculators will complete the operation for you even if you do not enter $=$ between operations. Check to see if your calculator will.

**Use your calculator to simplify the expression.**

**1.** $12(15 + 9)$ **2.** $57 \div (36 - 17)$ **3.** $(437 + 322) \div 46$

**4.** $(108 + 63) \div (9 - 6)$ **5.** $(55 + 8) \times (2 + 94)$ **6.** $(56 - 32) \div (4 + 16) \times 5$

**8** *Chapter 1*

# 1-3 Exponents and Powers of Ten

When two or more numbers are multiplied together, each of the numbers is called a **factor** of the product. For example, in the multiplication

$$3 \times 5 = 15,$$

3 and 5 are the factors of 15.

A product in which each factor is the same is called a **power** of that factor. For example, since

$$2 \times 2 \times 2 \times 2 = 16,$$

16 is called the *fourth power* of 2. We can write this as

$$2^4 = 16.$$

The small numeral 4 is called an **exponent** and represents the number of times 2 is a factor of 16. The number 2 is called the **base.**

**EXAMPLE 1**   Simplify $4^3$.

**Solution**        $4^3 = 4 \times 4 \times 4 = 16 \times 4 = 64$

The second and third powers of a number have special names. The second power is called the **square** of the number and the third power is called the **cube.**

**EXAMPLE 2**   Read, then simplify.

   **a.** $12^2$      **b.** $9^3$

**Solution**       **a.** $12^2$ is read "twelve squared."
                         $12^2 = 12 \times 12 = 144$

   **b.** $9^3$ is read "nine cubed."
           $9^3 = 9 \times 9 \times 9 = 81 \times 9 = 729$

Powers of 10 are important in our number system. Here is a list of the first five powers of 10.

First power:     $10^1$ (exponent usually not written) $= 10$
Second power: $10^2 = 10 \times 10$ $= 100$
Third power:   $10^3 = 10 \times 10 \times 10$ $= 1000$
Fourth power: $10^4 = 10 \times 10 \times 10 \times 10$ $= 10{,}000$
Fifth power:   $10^5 = 10 \times 10 \times 10 \times 10 \times 10$ $= 100{,}000$

If you study the preceding list carefully, you can see that the general rules below apply.

> ## *Rules*
>
> 1. The exponent in a power of 10 is the same as the number of zeros when the number is written out.
> 2. The number of zeros in the product of powers of 10 is the sum of the numbers of zeros in the factors.

**EXAMPLE 3**  Write 10,000 as a power of 10.

**Solution**  Because there are 4 zeros in 10,000, the exponent is 4.

$$10,000 = 10^4$$

**EXAMPLE 4**  Multiply $100 \times 1000$.

**Solution**  Since there are 2 zeros in 100 and 3 zeros in 1000, the product will have $2 + 3$, or 5, zeros.

$$100 \times 1000 = 100,000$$

Can we give an expression such as $7^0$ a meaning? When the powers of any base are listed in order, we may recognize a pattern. Study the example below.

$$7^4 = 7 \times 7 \times 7 \times 7 = 2401$$

$$7^3 = 7 \times 7 \times 7 = 343$$

$$7^2 = 7 \times 7 = 49$$

$$7^1 = 7$$

Notice that in increasing order each power of 7 is seven times the preceding power. Conversely, in decreasing order, each power of 7 is the quotient of the preceding power divided by a factor of 7. That is, $7^3 = 7^4 \div 7$, $7^2 = 7^3 \div 7$, and so on. This decreasing pattern suggests that $7^0$ (read *7 to the zero power*) is $7^1 \div 7$. Study the example below to verify that the expression $7^0 = 1$.

$$7^0 = 7^1 \div 7 = 7 \div 7 = 1$$

In general,

> ## Definition
> For every number $a$ $(a \neq 0)$, $a^0 = 1$.

---

## Class Exercises

**Name the exponent and the base.**

**1.** $8^3$  **2.** $9^7$  **3.** $3^5$  **4.** $6^4$

**Write as a power of 10.**

**5.** 1000  **6.** 100  **7.** 100,000  **8.** 100,000,000

**Tell the number of zeros in the number or product.**

**9.** $10^4$  **10.** $10^8$  **11.** $1000 \times 1000$  **12.** $100 \times 1,000,000$

**Read the following, then simplify each.**

**13.** $7^2$  **14.** $9^2$  **15.** $13^0$  **16.** $6^2$

**17.** $4^3$  **18.** $8^0$  **19.** $2^7$  **20.** $13^2$

---

## Written Exercises

**Use exponents to write each of the following expressions.**

**A**

**1.** $5 \times 5 \times 5 \times 5 \times 5 \times 5$  **2.** $12 \times 12 \times 12 \times 12$

**3.** $8 \times 8 \times 8 \times 8 \times 8 \times 8 \times 8 \times 8 \times 8$  **4.** $20 \times 20 \times 20 \times 20 \times 20$

**5.** $7 \times 7 \times 7 \times 7 \times 7 \times 7 \times 7$  **6.** $9 \times 9 \times 9 \times 9$

**7.** $4 \times 4 \times 4 \times 4 \times 4 \times 4$  **8.** $3 \times 3 \times 3 \times 3 \times 3$

**Write as a power of 10.**

**9.** 10  **10.** 10,000,000  **11.** 1,000,000  **12.** 1,000,000,000

**Multiply.**

**13.** $1000 \times 1000$  **14.** $10 \times 10,000$  **15.** $100 \times 100$  **16.** $100 \times 100,000$

**Simplify.**

| | | | |
|---|---|---|---|
| **17.** $4^2$ | **18.** $11^3$ | **19.** $16^2$ | **20.** $15^2$ |
| **21.** $20^2$ | **22.** $5^4$ | **23.** $15^3$ | **24.** $2^5$ |
| **25.** $80^2$ | **26.** $40^3$ | **27.** $2^8$ | **28.** $12^3$ |
| **29.** $6^4$ | **30.** $5^5$ | **31.** $16^3$ | **32.** $3^5$ |
| **33.** $5^3$ | **34.** $4^6$ | **35.** $13^2$ | **36.** $11^2$ |

**Multiply.**

*EXAMPLE*   $3^4 \times 2^3$

***Solution***   $3^4 \times 2^3 = (3 \times 3 \times 3 \times 3) \times (2 \times 2 \times 2)$
$$= 81 \times 8$$
$$= 648$$

**B**

| | | | |
|---|---|---|---|
| **37.** $2^4 \times 5^2$ | **38.** $1^3 \times 16^2$ | **39.** $70^2 \times 7^3$ | **40.** $3^4 \times 10^5$ |
| **41.** $0^4 \times 15^8$ | **42.** $15^2 \times 10^3$ | **43.** $31^2 \times 1^5$ | **44.** $20^5 \times 3^2$ |
| **45.** $2^8 \times 1^5$ | **46.** $5^3 \times 3^4$ | **47.** $12^3 \times 2^2$ | **48.** $200^3 \times 3^2$ |
| **49.** $2^3 \times 3^2 \times 10^3$ | **50.** $8^2 \times 5^3 \times 1^4$ | **51.** $119^2 \times 2^5 \times 0^8$ | **52.** $50^4 \times 2 \times 1^5$ |
| **53.** $3^3 \times 2^0$ | **54.** $2^5 \times 3^2$ | **55.** $3^2 \times 10^4$ | **56.** $10^5 \times 11^0$ |

**Evaluate when $a = 3$ and $b = 5$.**

**C**

| | | | |
|---|---|---|---|
| **57.** $a^3$ | **58.** $b^3$ | **59.** $a^5 - b^2$ | **60.** $a^3 + b^2$ |
| **61.** $50 - b^2$ | **62.** $20a^2$ | **63.** $(ab)^2$ | **64.** $a^2b^3$ |
| **65.** $a^3b^3$ | **66.** $(ab)^3$ | **67.** $b^2 - a^2$ | **68.** $(a - b)^2$ |

## *Review Exercises*

**Simplify the expression.**

**1.** $48 + 20 \div 4 + 7$

**2.** $72 \div 9 + 3 \times 8$

**3.** $50 + 35 \div 7 + 2$

**4.** $105 - 30 \times 2 \div 5$

**5.** $36 - 24 \div (3 + 1)$

**6.** $60 + 40 \div (2 + 8)$

**7.** $\dfrac{28 + 20}{4} + 12$

**8.** $\dfrac{78 - 18 \div 3}{4} - 8$

# 1-4 The Decimal System

Our number system uses the powers of 10 to express all numbers. This system is called the **decimal system** (from the Latin word *decem,* meaning *ten*). Using the digits 0, 1, 2, . . . , 9, we can write any number. The **value** of each digit depends on the position of the digit in the number. For example, the 2 in 312 means 2 ones, but the 2 in 298 means 2 hundreds. The decimal system is a system with **place value**.

The chart below shows place values for some of the digits of a **decimal number,** or **decimal.**

**Place-Value Chart**

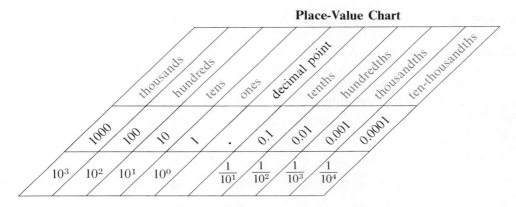

Moving left to right, place values decrease. As illustrated below, each place value is one tenth, or 0.1, of the place value to its left.

$$1000 \times 0.1 = 100 \qquad 10 \times 0.1 = 1 \qquad 0.01 \times 0.1 = 0.001$$

We can see from the place-value chart that values to the left of the decimal point are greater than or equal to 1, while values to the right are less than 1. If a digit is zero, then the product is zero and is usually not written.

When a decimal number is written, the value of the number is the sum of the values of the digits. To illustrate this, we can write decimal numbers in **expanded form,** that is, as a sum of products of each digit and its place value.

**EXAMPLE 1**  Write the decimal in expanded form.

  **a.** 3053         **b.** 16.9         **c.** 0.074

**Solution**   **a.** $3053 = (3 \times 1000) + (5 \times 10) + 3$

  **b.** $16.9 = (1 \times 10) + 6 + (9 \times 0.1)$

  **c.** $0.074 = (7 \times 0.01) + (4 \times 0.001)$

Decimals and whole numbers can be pictured on a number line. The **graph** of a number is the point paired with the number on the number line. The number paired with a point is called the **coordinate** of the point. The graphs of the whole numbers 0 through 8 are shown on the number line below. The coordinates of the points shown are 0, 1, 2, 3, 4, 5, 6, 7, and 8.

Starting with the graph of 0, which is called the **origin,** the graphs of the whole numbers are equally spaced. The greater a number is, the farther to the right its graph is.

The number line can be used to develop a method for **rounding** numbers. For example, we can see that 6.7 is closer to 7 than to 6 by graphing 6.7 on the number line. First we divide the portion of the number line between 6 and 7 into ten equal parts. Then we graph 6.7 on the number line as shown.

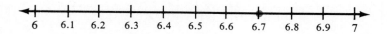

Because 6.7 is closer to 7 than to 6, we say that 6.7 rounded to the nearest whole number is 7. We write 6.7 ≈ 7. The symbol ≈ means *is approximately equal to.*

The rule for rounding decimal numbers can be stated as follows:

## *Rule*

1. Find the decimal place to which you wish to round, and mark it with a caret ( ). Look at the digit to the right.

2. If the digit to the right is 5 or greater, add 1 to the marked digit. If the digit to the right is less than 5, leave the marked digit unchanged.

3. If the marked digit is to the left of the decimal point, replace each digit to the right of the marked place with "0," and drop all digits to the right of the decimal point.
   If the marked digit is to the right of the decimal point, drop all digits to the right of the marked place.

***EXAMPLE 2*** Round 16.0973 to the nearest

                **a.** ten             **b.** tenth            **c.** hundredth

***Solution***    **a.** 16.0973      **b.** 16.0973      **c.** 16.0973
                 ^                 ^                 ^
              20               16.1              16.10

          16.0973 $\approx$ 20      16.0973 $\approx$ 16.1      16.0973 $\approx$ 16.10

Rounding is often used as a quick check on calculations. In general, we **estimate** by rounding to the highest place value of the smaller number for all operations. As a result, our estimated answer should be reasonably close to the exact answer.

***EXAMPLE 3*** Find an estimated answer for each operation.

       **a.**   386      **b.**   2.849      **c.**   32.9      **d.** 12$\overline{)253}$
          + 54         &minus; 0.154       &times; 8.7

***Solution***   **a.**   386   $\longrightarrow$   390      **b.**   2.849   $\longrightarrow$   2.8
             + 54            + 50          &minus; 0.154         &minus; 0.2
                       440                              2.6

                                                      25
      **c.**   32.9   $\longrightarrow$   33      **d.** 12$\overline{)253}$   $\longrightarrow$   10$\overline{)250}$
         &times; 8.7           &times; 9
                     297

## Class Exercises

**Read the number.**

  **1.** 5496       **2.** 0.51       **3.** 28.7       **4.** 0.0317       **5.** 8.026

**Give the value of the digit 6 in the number.**

  **6.** 657.3       **7.** 0.062       **8.** 8.1416       **9.** 6.023       **10.** 408.6

**Round to the place underlined.**

  **11.** 8.0<u>3</u>9       **12.** 2<u>9</u>4.65       **13.** 7<u>5</u>.452       **14.** 1<u>0</u>.988

  **15.** 0.9<u>9</u>6       **16.** 1.35<u>4</u>7       **17.** <u>3</u>19.84       **18.** 0.<u>3</u>828

**Round to the highest place value.**

**19.** 43       **20.** 127.14       **21.** 0.036       **22.** 0.8981

**23.** 1986       **24.** 532.3       **25.** 0.073       **26.** 0.0249

---

## Written Exercises

**Write the decimal in expanded form.**

A    **1.** 38      **2.** 256      **3.** 8091      **4.** 5      **5.** 0.47

     **6.** 28.4      **7.** 0.063      **8.** 9.070      **9.** 0.187      **10.** 5.5

**Write as a decimal.**

**11.** $(5 \times 10) + 4 + (5 \times 0.1) + (7 \times 0.01)$

**12.** $(4 \times 100) + (7 \times 0.01) + (3 \times 0.001) + (2 \times 0.0001)$

**13.** $(9 \times 1000) + 2 + (1 \times 0.1) + (4 \times 0.01) + (6 \times 0.001)$

**14.** $(1 \times 100) + (9 \times 10) + 3 + (2 \times 0.01) + (3 \times 0.001)$

**Write as a decimal.**

**15.** 7 and 43 hundredths      **16.** 11 and 4 tenths

**17.** 19 and 5 thousandths      **18.** 37 ten-thousandths

**19.** 48 ten-thousandths      **20.** 5 and 6 hundredths

**21.** 6 and 25 thousandths      **22.** 94 and 7 ten-thousandths

**Round to the nearest ten.**

**23.** 27.5149      **24.** 82.604      **25.** 293.4      **26.** 70.76

**27.** 648.01      **28.** 108.3      **29.** 159.62344      **30.** 97.23

**Round to the nearest hundredth.**

**31.** 72.459      **32.** 26.804      **33.** 0.0643      **34.** 12.395

**35.** 0.0103      **36.** 8.142      **37.** 18.1657      **38.** 0.70605

**Round to the nearest thousandth.**

**39.** 0.0006      **40.** 12.3568      **41.** 401.0904      **42.** 30.0317

**43.** 250.3407      **44.** 7.0063      **45.** 8.0995      **46.** 0.9996

**Write in expanded form using exponents by writing each number as a sum of multiples of powers of 10.**

*EXAMPLE*   367.04

*Solution*      $(3 \times 10^2) + (6 \times 10^1) + (7 \times 10^0) + \left(4 \times \frac{1}{10^2}\right)$

**47.** 5280 **48.** 64.7 **49.** 183.08 **50.** 0.043

**51.** 0.091 **52.** 12.931 **53.** 7.482 **54.** 0.806

**55.** 204.5 **56.** 0.306 **57.** 38.003 **58.** 10.009

**Select the most reasonable estimated answer.**

**B**  **59.** $89.6 + 13.5$ **a.** 90 **b.** 100 **c.** 70

**60.** $35 + 12 + 26 + 11$ **a.** 70 **b.** 110 **c.** 90

**61.** $65.43 - 8.92$ **a.** 56 **b.** 60 **c.** 90

**62.** $2196 - 924$ **a.** 1300 **b.** 1000 **c.** 3100

**63.** $6.82 \times 4.7$ **a.** 24 **b.** 35 **c.** 28

**64.** $54 \div 2.5$ **a.** 30 **b.** 20 **c.** 18

**65.** $36\overline{)283}$ **a.** 10 **b.** 3 **c.** 7

---

## Review Exercises

**Simplify.**

**1.** $5.7 + (1.3 + 2.4)$ **2.** $(5.7 + 1.3) + 2.4$

**3.** $12 \times (10 \times 8)$ **4.** $(12 \times 10) \times 8$

**5.** $6 \times (4.7 + 5.3)$ **6.** $(6 \times 4.7) + (6 \times 5.3)$

**7.** $3.7 \times (7.9 - 3.9)$ **8.** $(3.7 \times 7.9) - (3.7 \times 3.9)$

### ▮▮▮▮ Challenge

Write 100 using four 5's and any operation symbols you need.

Write 100 using the numbers 1 through 9 and any operation symbols you need.

Write 100 using four 9's and any operation symbols you need.

# 1-5 Basic Properties

Decimals and whole numbers share a number of properties which are used frequently in algebra.

Changing the order of the addends in a sum or the factors in a product does not change the sum or product. For example,

$$9.2 + 4.7 = 4.7 + 9.2 \qquad 3.5 \times 8.4 = 8.4 \times 3.5$$

---

### *Commutative Property*

For all numbers $a$ and $b$,

$$a + b = b + a \qquad \text{and} \qquad a \times b = b \times a$$

---

Changing the grouping of addends in a sum or of factors in a product does not change the sum or product. For example,

| | |
|---|---|
| $5.6 + 0.8 + 11.2$ | $4.1 \times 2.3 \times 7.2$ |
| $(5.6 + 0.8) + 11.2 = 17.6$ | $(4.1 \times 2.3) \times 7.2 = 67.896$ |
| $5.6 + (0.8 + 11.2) = 17.6$ | $4.1 \times (2.3 \times 7.2) = 67.896$ |

---

### *Associative Property*

For all numbers $a$, $b$, and $c$,

$$(a + b) + c = a + (b + c)$$

$$\text{and} \qquad (a \times b) \times c = a \times (b \times c)$$

---

We can use these properties to find the easiest way to add or multiply a long list of numbers.

**EXAMPLE 1**   Use the properties to simplify the expression.

     **a.** $4.8 + 1.1 + 0.2 + 3.9 + 7$        **b.** $2 \times 6 \times 5 \times 3$

**Solution**   One possible way to rearrange the numbers is shown.

     **a.**   $4.8 + 1.1 + 0.2 + 3.9 + 7$        **b.**   $2 \times 6 \times 5 \times 3$

        $(4.8 + 0.2) + (1.1 + 3.9) + 7$          $(2 \times 5) \times (6 \times 3)$

                $5 + 5 + 7$                     $10 \times 18$

                   $10 + 7$                        $180$

                      $17$

The numbers 0 and 1 are called the **identity elements** for addition and multiplication respectively. The word *identity* comes from the Latin word *idem,* which means *the same.* The result of adding 0 to a number or subtracting 0 from a number is the same as the original number. The result of multiplying a number by 1 or dividing a number by 1 is the same as the original number. Study the following examples.

$$4.53 + 0 = 4.53 \qquad 29.7 \times 1 = 29.7$$
$$4.53 - 0 = 4.53 \qquad 29.7 \div 1 = 29.7$$

### Addition and Subtraction Properties of Zero

For every number $a$, $\qquad a + 0 = a \qquad a - 0 = a$

$\qquad\qquad\qquad\qquad\quad 0 + a = a \qquad a - a = 0$

### Multiplication and Division Properties of One

For every number $a$, $\qquad a \times 1 = a \qquad a \div 1 = a$

$\qquad\qquad\qquad\qquad\quad 1 \times a = a \qquad a \div a = 1$

The product of any number and 0 is 0. Similarly, 0 divided by any number is 0.

$$3.784 \times 0 = 0 \qquad 0 \div 3.784 = 0$$

Can we divide by 0? Recall that multiplication and division are inverse operations. Thus, if we were to divide 3.784 by 0, we would have the following.

$$3.784 \div 0 = x \qquad 3.784 = x \times 0$$

We cannot accept the equation at the right because any number times 0 is 0. So it makes no sense to divide by 0.

### Multiplication and Division Properties of Zero

For every number $a$,

$$a \times 0 = 0 \qquad \text{and} \qquad 0 \times a = 0$$

For every number $a$, $a \neq 0$,

$$0 \div a = 0$$

The distributive property is different from the other properties because it involves two operations. The example below illustrates how we may distribute a multiplier over each term in an addition expression.

$$7 \times (8.2 + 1.8) \qquad\qquad (7 \times 8.2) + (7 \times 1.8)$$
$$7 \times 10 \qquad\qquad\qquad 57.4 + 12.6$$
$$70 \qquad\qquad\qquad\qquad 70$$

Therefore $7 \times (8.2 + 1.8) = (7 \times 8.2) + (7 \times 1.8)$.

We may also distribute a multiplier over each term in a subtraction expression.

---

## Distributive Property

For all numbers $a$, $b$, and $c$,

$$a \times (b + c) = (a \times b) + (a \times c)$$
$$a \times (b - c) = (a \times b) - (a \times c)$$

---

**EXAMPLE 2**    Use the distributive property to simplify the expression.
         **a.** $(4 + 6.2)5$          **b.** $(13 \times 2.7) - (13 \times 1.3)$

**Solution**       **a.** $(4 + 6.2)5 = (4 \times 5) + (6.2 \times 5) = 20 + 31 = 51$

               **b.** $(13 \times 2.7) - (13 \times 1.3) = 13(2.7 - 1.3) = 13(1.4) = 18.2$

**EXAMPLE 3**    What value of the variable makes the statement true?
         **a.** $6 + 5 = 5 + m$     **b.** $3.7 \times 4 = 4t$     **c.** $3(14 + 20) = 42 + b$

**Solution**       **a.** $6 + 5 = 5 + m$           **b.** $3.7 \times 4 = 4t$
               $6 + 5 = 5 + 6$, so $m = 6$       $3.7 \times 4 = 4 \times 3.7$, so $t = 3.7$

               **c.** $3(14 + 20) = 42 + b$
               $3(14 + 20) = (3 \times 14) + (3 \times 20) = 42 + 60$, so $b = 60$

---

## Class Exercises

**Name the property illustrated.**

**1.** $7.6 + 0 = 0 + 7.6$

**2.** $(19 \times 3)6.2 = 19(3 \times 6.2)$

**3.** $11.9 \times 1 = 11.9$

**4.** $5(9 + 8.2) = (5 \times 9) + (5 \times 8.2)$

**5.** $138.6 \times 7.4 = 7.4 \times 138.6$

**6.** $6(1.2 + 0.8) = (1.2 + 0.8)6$

Use the properties of addition and multiplication to simplify the expression. Name the property or properties used.

**7.** $0.4 \times 0$  **8.** $3.02 \times 5 \times 2$  **9.** $1 \times 12.87$

**10.** $2.4 + 13 + 2.6$  **11.** $(8 \times 13) + (8 \times 7)$  **12.** $5.93 \times 0$

What value of the variable makes the statement true?

**13.** $25 + 37 = m + 25$  **14.** $(7 \times 6) + (5 \times 6) = (7 + 5)q$

**15.** $(17 + 12) + 8 = b + (12 + 8)$  **16.** $9(w - 20) = (9 \times 35) - (9 \times 20)$

## Written Exercises

Use the properties to simplify the expression. Name the property or properties used.

**A**  **1.** $2.6 + 11.5 + 0.5$  **2.** $0 \times 23.15$

**3.** $4(2.5 + 1.06)$  **4.** $8(40 - 12)$

**5.** $7.24 + 8.97 + 2.76$  **6.** $(22 \times 8) + (22 \times 2)$

**7.** $0.5 \times 2.1 \times 0.2$  **8.** $11.5 + 2.6 + 0.5 + 0.4$

True or false?

**9.** $7.386 + 0 = 0$  **10.** $(3 + 12)6 = (3 \times 6) + (12 \times 6)$

**11.** $(19.7 + 36 + 41.5)0 = 0$  **12.** $15(2 + 7.4) = (15 + 2) \times (15 + 7.4)$

What value of the variable makes the statement true?

**13.** $6 + n = 6$  **14.** $7.02 \times 23 = t \times 7.02$

**15.** $6.4 + t = 3.2 + 6.4$  **16.** $3r = 3$

**17.** $5w = w$  **18.** $2.43 \times 0 = f$

**19.** $(15.9 \times 3)4.2 = g(3 \times 4.2)$  **20.** $(13 - 11.7)8 = (13 \times 8) - (r \times 8)$

**21.** $(2 + 4.8)3 = (2n) + (4.8n)$  **22.** $(3.02 + 4.9)1 = b$

**23.** $5(11.7 + 313) = 58.5 + d$  **24.** $(7 \times 1.2) + (7 \times 3.8) = 7m$

**25.** $(8.31 + 2.73)t = t$  **26.** $2.59 + 7.03 + 18.61 + 3.97 = a + 11$

Use the properties to simplify the expression.

**B**  **27.** $116 \times 3.7 \times 0 \times 4.93 \times 1.47 + 3.88$

**28.** $(78 \times 1) + (1.36 \times 0) + (92 + 0)$

**29.** $(18 + 46) + (12 + 4) + (8 \times 17) + (23 \times 8)$

**30.** $(12 \times 7) + (56 \div 8) + (13 \times 12)$

**31.** $5(81 \div 3) + 5(63 \div 1) + 450$

**32.** $(18 + 9)4 + (12 + 11)4$

**33.** $6(13 + 3) - 4(13 + 3)$

**34.** $1.4(2.61 + 7.39)$

**Find values for *a*, *b*, and *c* that show that the equation is not true for all numbers.**

C  **35.** $(a \times b) + (b \times c) = b(a \times c)$

     **36.** $(a + b)(b + c) = b(a + c)$

     **37.** $(b + c)(a + c) = c(b \times a)$

     **38.** $a + (b \times c) = (a + b) \times (a + c)$

## Self-Test A

**Evaluate the expression when $k = 4$ and $m = 6$.**

**1.** $184 \div k$       **2.** $8 + m + 1$       **3.** $7km$       [1–1]

**Evaluate the expression when $s = 12$ and $t = 18$.**

**4.** $\frac{s}{4} + 6$       **5.** $(t + 2) \div 5$       **6.** $2s - t$       [1–2]

**Simplify.**

**7.** $2^4$     **8.** $8^3$     **9.** $9^1$     **10.** $1000 \times 100$     **11.** $10 \times 10,000$     [1–3]

**Round to the place specified.**

**12.** tens: 84.307     **13.** hundredths: 3.176     **14.** hundreds: 293.84     [1–4]

**Use the properties of addition and multiplication to simplify the expression. Name the property used.**

**15.** $12(15 - 8) + 6 \times 3$       **16.** $(31 \times 4) + (15 \times 4) - 91$     [1–5]

**17.** $7(56 \div 8) - 7(24 \div 6)$       **18.** $9(0.36 \times 4) + 55$

*Self-Test answers and Extra Practice are at the back of the book.*

## 1-6 Equations

A **number sentence** indicates a relationship between two mathematical expressions. A sentence, such as the one below, that indicates that two expressions name the same number is called an **equation.**

$$4 \times 3 = 12$$

The expressions to the left and to the right of the equals sign are called the **sides** of the equation. In the example above, $4 \times 3$ is the left side of the equation, and 12 is the right side of the equation.

A number sentence may be *true* or *false*. For example, $3 + 9 = 12$ is a true equation, but $3 + 9 = 13$ is a false equation.

A number sentence that contains one or more variables is called an **open number sentence,** or simply an **open sentence.** Frequently a set of intended values, called the **replacement set,** for a variable is specified. An open sentence may be true or false when each variable is replaced by one of the values in its replacement set.

When a value of the variable makes an open sentence a true statement, we say that the value is a **solution** of, or **satisfies,** the sentence. We can **solve** an open sentence in one variable by finding all the solutions of the sentence. An open sentence may have one solution, several solutions, or no solutions.

**EXAMPLE 1**   Solve $x + 6 = 13$ for the replacement set $\{5, 6, 7\}$.

**Solution**   Substitute each value in the replacement set for the variable $x$.

| | | |
|---|---|---|
| $5 + 6 = 13$ | $6 + 6 = 13$ | $7 + 6 = 13$ |
| $11 = 13$ | $12 = 13$ | $13 = 13$ |
| false | false | true |

The solution of the equation is 7.

**EXAMPLE 2**   Solve $a - 1.2 = 0.7$ for the replacement set $\{1.8, 1.9, 2.0\}$.

**Solution**   Substitute each value in the replacement set for the variable $a$.

| | | |
|---|---|---|
| $1.8 - 1.2 = 0.7$ | $1.9 - 1.2 = 0.7$ | $2.0 - 1.2 = 0.7$ |
| $0.6 = 0.7$ | $0.7 = 0.7$ | $0.8 = 0.7$ |
| false | true | false |

The solution of the equation is 1.9.

**EXAMPLE 3** The replacement set for $q$ is the set of whole numbers. Find all solutions of

$$2q = 9.$$

**Solution** The replacement set for $q$ is $\{0, 1, 2, 3, \ldots\}$, so $2q$ must be one of the numbers $2 \times 0, 2 \times 1, 2 \times 2, 2 \times 3, \ldots$, or $0, 2, 4, 6, \ldots$. Because 9 is not one of these numbers, the equation $2q = 9$ has no solution in the given replacement set.

Notice that in the solution to Example 2, we used three dots, read *and so on,* to indicate that the list of numbers continues without end.

## Class Exercises

**Tell whether the equation is true or false for the given value of the variable.**

1. $20 - y = 17;\ y = 3$
2. $n \times 7 = 42;\ n = 8$
3. $144 \div r = 46;\ r = 3$
4. $156 + q = 179;\ q = 23$
5. $13.56 + p = 21.87;\ p = 6.31$
6. $8.91 \div c = 2.97;\ c = 3$

**Solve the equation for the given replacement set.**

7. $m + 6 = 72;\ \{50, 60, 70, 80\}$
8. $x \div 12 = 7;\ \{81, 82, 83, 84\}$
9. $r - 23 = 19;\ \{40, 41, 42, 43\}$
10. $b \times 8 = 64;\ \{2, 4, 6, 8\}$
11. $3.69 - n = 1.31;\ \{2.36, 2.37, 2.38\}$
12. $1.91 \times z = 9.55;\ \{4, 5, 6\}$

## Written Exercises

**Tell whether the equation is true or false for the given value of the variable.**

**A**
1. $x + 9 = 35;\ x = 26$
2. $r - 15 = 40;\ r = 55$
3. $10m = 130;\ m = 10$
4. $9 + y = 100;\ y = 91$
5. $44 - q = 11;\ q = 55$
6. $t \times 7 = 84;\ t = 91$
7. $n \div 8 = 104;\ n = 832$
8. $26 \div d = 2;\ d = 52$
9. $x + 6.71 = 10.82;\ x = 4.11$
10. $4.16a = 29.12;\ a = 8$
11. $m - 26 = 59;\ m = 85$
12. $q + 113 = 789;\ q = 901$

**Solve the equation for the given replacement set.**

**13.** $5d = 145$; $\{29, 30, 31\}$  

**14.** $t - 53 = 67$; $\{13, 14, 15\}$

**15.** $b \times 14 = 112$; $\{6, 8, 10\}$  

**16.** $c \div 12 = 228$; $\{17, 19, 21\}$

**17.** $98 - h = 21$; $\{75, 80, 85\}$  

**18.** $38f = 912$; $\{24, 25, 26\}$

**19.** $e \div 19 = 152$; $\{8, 18, 28\}$  

**20.** $46 + n = 99$; $\{50, 52, 54\}$

**21.** $q \div 23 = 66$; $\{1516, 1517, 1518\}$  

**22.** $b \div 41 = 77$; $\{287, 288, 289\}$

**23.** $t + 1.21 = 2.47$; $\{1.25, 1.26, 1.27\}$  

**24.** $y \div 1.2 = 3$; $\{3.3, 3.6, 3.9\}$

**B** **25.** $4n + 7 = 51$; $\{10, 11, 12\}$  

**26.** $58 - 3a = 10$; $\{16, 17, 18\}$

**27.** $4(b - 5) = 28$; $\{10, 11, 12\}$  

**28.** $17(k + 4) = 170$; $\{4, 6, 8\}$

**29.** $(t + 18) \div 3 = 9$; $\{9, 10, 11\}$  

**30.** $56 \div (d + 8) = 4$; $\{5, 6, 7\}$

**31.** $5(2c - 4) = 0$; $\{0, 1, 2\}$  

**32.** $(3f + 7) \div 4 = 4$; $\{3, 4, 5\}$

**Solve the equation. The replacement set is all even whole numbers.**

**33.** $x + 1 = 10$     **34.** $5x = 35$     **35.** $4x = 16$     **36.** $x - 1 = 20$

**Write an equation with the given solution if the replacement set is all whole numbers.**

**C** **37.** 10                **38.** 15                **39.** 100

**40.** Write an equation with no solution if the replacement set is all whole numbers.

**Replace __?__ with $+$, $-$, $\times$, or $\div$ so the equation has the given solution.**

**41.** $x$ __?__ $17$ __?__ $13 = 24$; 20  

**42.** $n$ __?__ $12$ __?__ $8 = 11$; 36

**43.** $14$ __?__ $t$ __?__ $9 = 17$; 27  

**44.** $y$ __?__ $12$ __?__ $6 = 10$; 8

**45.** $d$ __?__ $(16$ __?__ $4)$ __?__ $8 = 8$; 12  

**46.** $27$ __?__ $(q$ __?__ $9)$ __?__ $15 = 45$; 18

**47.** $21$ __?__ $(19$ __?__ $8)$ __?__ $b = 12$; 3  

**48.** $11$ __?__ $(16$ __?__ $a)$ __?__ $4 = 59$; 11

---

## Review Exercises

**Perform the indicated operation.**

**1.** $7.81 + 3.86$     **2.** $11.65 - 8.58$     **3.** $1.18 \times 23$     **4.** $16.74 \div 2.79$

**5.** $2.53 \times 2.9$     **6.** $10.49 + 9.52$     **7.** $13.27 - 10.43$     **8.** $24.84 - 4.14$

# 1-7 Inverse Operations

Addition and subtraction are related operations, as shown by the following facts.

$$5 + 6 = 11$$
$$5 = 11 - 6$$

We say that adding a number and subtracting the same number are **inverse operations.**

The relationship between addition and subtraction holds when we work with variables, as well. Thus we can write the following related equations.

$$n + 6 = 11$$
$$n = 11 - 6$$

We can use this relationship to solve equations that involve addition or subtraction. Throughout the rest of the chapter if no replacement set is given for an open sentence, assume that the solution can be any number.

**EXAMPLE 1** Use the inverse operation to write a related equation and solve for the variable.

    **a.** $x + 9 = 35$         **b.** $y - 12 = 18$

***Solution***    **a.**    $x = 35 - 9$         **b.**    $y = 18 + 12$
                        $x = 26$                       $y = 30$

           The solution is 26.      The solution is 30.

To check each solution, substitute the value of the variable in the original equation.

    **a.** $x + 9 = 35$         **b.** $y - 12 = 18$
        $26 + 9 = 35$ ✓       $30 - 12 = 18$ ✓

Multiplying by a number and dividing by the same number are inverse operations.

$$4 \times 6 = 24$$
$$6 = 24 \div 4$$

We can use this relationship to help solve equations that involve multiplication and division.

**EXAMPLE 2** Use the inverse operation to write a related equation and solve for the variable.

    **a.** $6r = 30$                   **b.** $x \div 7 = 12$

**Solution**  **a.** Recall that $6r$ means $6 \times r$.

$$6 \times r = 30$$
$$r = 30 \div 6$$
$$r = 5$$

The solution is 5.

Check: $6 \times 5 = 30$ ✓

**b.** $x \div 7 = 12$
$$x = 12 \times 7$$
$$x = 84$$
The solution is 84.

Check: $84 \div 7 = 12$ ✓

**EXAMPLE 3**  Use inverse operations to write a related equation and solve for the variable.

**a.** $x + 8.21 = 12.64$  **b.** $2.47a = 12.35$

**Solution**  **a.**
$$x = 12.64 - 8.21$$
$$x = 4.43$$
The solution is 4.43.
Check: $4.43 + 8.21 = 12.64$ ✓

**b.**
$$a = 12.35 \div 2.47$$
$$a = 5$$
The solution is 5.
Check: $(2.47)(5) = 12.35$ ✓

We can use inverse operations to solve equations that contain two operations.

**EXAMPLE 4**  Use inverse operations to solve the equation

$$3n + 25 = 61.$$

**Solution**  First write the related subtraction equation to find the value of $3n$.

$$3n = 61 - 25$$
$$3n = 36$$

Then write the related division equation to find the value of $n$.

$$n = 36 \div 3$$
$$n = 12$$

The solution is 12.

Check: $3(12) + 25 = 61$
$$36 + 25 = 61 ✓$$

## Class Exercises

**Use the inverse operation to state a related equation and solve for $t$.**

**1.** $t + 6 = 15$  **2.** $t - 4 = 7$  **3.** $6t = 48$  **4.** $t \div 8 = 7$

**5.** $t + 24 = 60$  **6.** $13t = 78$  **7.** $t - 26 = 59$  **8.** $t \div 32 = 8$

**9.** $t + 3.19 = 5.26$  **10.** $t - 21.06 = 31.14$  **11.** $3.14t = 12.56$  **12.** $t \div 1.16 = 12$

## Written Exercises

**Use the inverse operation to write a related equation and solve for the variable.**

**A**
**1.** $x + 8 = 15$      **2.** $a + 6 = 11$      **3.** $f + 38 = 74$      **4.** $t + 46 = 91$

**5.** $y - 9 = 14$      **6.** $n - 7 = 9$      **7.** $b - 25 = 32$      **8.** $r - 55 = 87$

**9.** $3c = 27$      **10.** $5g = 45$      **11.** $9m = 108$      **12.** $4d = 88$

**13.** $n \div 6 = 9$      **14.** $s \div 8 = 4$      **15.** $g \div 16 = 7$      **16.** $w \div 9 = 14$

**17.** $a + 17 = 17$      **18.** $j - 54 = 61$      **19.** $14n = 42$      **20.** $e + 26 = 61$

**21.** $h \div 11 = 297$      **22.** $b + 45 = 256$      **23.** $d - 87 = 110$      **24.** $18b = 18$

**25.** $31q = 465$      **26.** $k \div 17 = 527$      **27.** $p + 208 = 358$      **28.** $f - 11 = 523$

**29.** $c + 511 = 536$      **30.** $18r = 414$      **31.** $g - 19 = 401$      **32.** $m \div 4 = 216$

**33.** $x + 1.6 = 31.9$      **34.** $p - 1.9 = 18.4$      **35.** $1.6c = 2.56$      **36.** $m \div 3.27 = 6$

**Use inverse operations to solve.**

**B**
**37.** $3q + 9 = 27$      **38.** $4a + 19 = 39$      **39.** $7d - 12 = 37$

**40.** $5b - 16 = 39$      **41.** $3r + 24 = 63$      **42.** $2s - 30 = 62$

**43.** $9x - 84 = 69$      **44.** $6w + 81 = 486$      **45.** $(z \div 3) + 22 = 30$

**46.** $(l \div 7) - 4 = 4$      **47.** $(c \div 5) - 13 = 11$      **48.** $(v \div 3) + 15 = 36$

**49.** $4s + 7.41 = 9.57$      **50.** $6c + 3.60 = 9.72$      **51.** $7t - 8.43 = 14.67$

**52.** $9c - 8.32 = 7.61$      **53.** $(x \div 3.21) + 9 = 13$      **54.** $(y \div 1.67) + 5 = 12$

**Replace __?__ with $+$, $-$, $\times$, or $\div$ so the equation has the given solution.**

**C**
**55.** $4 \underline{\ ?\ } a \underline{\ ?\ } 5 = 25$; 5      **56.** $7 \underline{\ ?\ } n \underline{\ ?\ } 4 = 14$; 8

**57.** $3 \underline{\ ?\ } c \underline{\ ?\ } 11 = 22$; 11      **58.** $5 \underline{\ ?\ } y \underline{\ ?\ } 3 = 7$; 2

**59.** $24 \underline{\ ?\ } x \underline{\ ?\ } 15 = 18$; 8      **60.** $60 \underline{\ ?\ } m \underline{\ ?\ } 2 = 15$; 2

---

## Review Exercises

**Evaluate the expression when $m = 2$, $n = 1$, and $p = 4$.**

**1.** $2n^2$      **2.** $(m + n)^0$      **3.** $3p^2 + m^2$      **4.** $n^2 + m^3$

**5.** $p^3 - (mn)^2$      **6.** $2p^2 - n^0$      **7.** $4m^3 - n^3$      **8.** $(2mn^2)^0$

# 1-8 A Plan for Solving Problems

What we know about mathematics enables us to solve many problems. Problems, however, are not usually as neatly organized as the information in the expressions with which we have been working. We must sort out and organize the facts of a problem before we begin to solve. A plan such as the one below can be useful in solving many kinds of problems.

---

### *Plan for Solving Word Problems*

1. Read the problem carefully. Make sure that you understand what it says. You may need to read it more than once.

2. Use questions like these in planning the solution:
   What is asked for?
   What facts are given?
   Are enough facts given? If not, what else is needed?
   Are unnecessary facts given? If so, what are they?
   Will a sketch or diagram help?

3. Determine which operation or operations can be used to solve the problem.

4. Carry out the operations carefully.

5. Check your results with the facts given in the problem. Give the answer.

---

**EXAMPLE 1** On Saturday, the Nizel family drove 17 mi from Topsfield to Newton and 22.5 mi from Newton to Harbor Bluffs. The trip took 50 min. On the way back, the Nizels took the same route, but they stopped for lunch after driving 9.5 mi. If they continue on the same route after lunch, how much farther will they have to drive to return to Topsfield?

**Solution**
- The problem asks for the number of miles to return to Topsfield.

- The following facts are given in the problem:
  - 17 mi from Topsfield to Newton
  - 22.5 mi from Newton to Harbor Bluffs
  - drove 9.5 mi back toward Topsfield

*(The solution is continued on the next page.)*

- We have enough facts to solve the problem since we know the distance between the cities and the distance driven toward Topsfield.

- We do not need to know that the trip took 50 min.

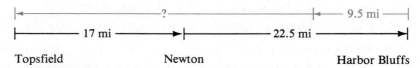

- The sketch shows that we subtract the distance driven back toward Topsfield from the distance between Topsfield and Harbor Bluffs.

$$(17 + 22.5) - 9.5 = 30$$

- Check: If they drive 9.5 mi and 30 mi farther, will the Nizels have driven the distance from Harbor Bluffs to Topsfield?

$$9.5 + 30 = 17 + 22.5 \ \checkmark$$

The Nizels must drive 30 mi to return to Topsfield.

*EXAMPLE 2* Records at the Howard City Weather Bureau show that it rained on a total of 17 days during the months of July through September and on twice as many days during the months of October through December. During October through December, how many days did not have rain?

*Solution*

- The problem asks for the number of days without rain during October, November, and December.

- Given facts: 17 days of rain in July through September
  twice as many days of rain in October through December

- We need to supply these facts:
  31 days in October, 30 days in November, 31 days in December

- To find the number of days without rain, subtract the number of days that did have rain from the total number of days.
  $$(31 + 30 + 31) - (2 \times 17) = 92 - 34 = 58$$

- Check: Are 34 days twice as many as 17?     $34 \div 2 = 17 \ \checkmark$
  Do 58 days without rain and 34 days with rain total the number of days in the three months?
  $58 + 34 = 31 + 30 + 31 \ \checkmark$

During October through December, 58 days did not have rain.

## Class Exercises

**For each problem, answer the following questions.**
**a. What number or numbers does the problem ask for?**
**b. Are enough facts given? If not, what else is needed?**
**c. Are unneeded facts given? If so, what are they?**
**d. What operation or operations would you use to find the answer?**

1. Kevin completed the bicycle race in 2 h 24 min, Lori completed the race in 2 h 13 min, and Helen completed the race in 2 h 54 min. How much faster than Kevin's time was Lori's time?

2. Steve bought 1 lb of Swiss cheese, 12 oz of mild cheddar cheese, and 6 oz of sharp cheddar cheese. How much cheese did he buy in all?

3. Elise bought a record for $5.69, another record for $4.88, and a record cleaning kit for $12.75. How much more than the cost of the records was the cost of the kit?

4. The eighth-grade classes are holding a hobbies and crafts fair on Saturday. Maurice plans to help out at the stamp-collecting booth from 9 A.M. to 11 A.M. and at the model-airplane booth from 2 P.M. to 5 P.M. How many hours does he plan to spend helping?

## Problems

**Solve, using the five-step plan.**

A 1. Mimi is buying weather-stripping tape for some windows. How much should she buy for a window that needs 4.85 m, a window that needs 4.25 m, and a window that needs 2.55 m?

2. Irene Lanata pays $235.40 each month to repay her automobile loan. How much will she pay in one year?

3. Hill School plans to buy 4 computers for each of 12 classrooms. The cost of each computer is $865. What will the total cost be?

4. An 8 mm camera shoots 24 frames of film each second. How many frames will it shoot in 5 min?

5. A package of 2 paintbrushes is on sale for $2.40. How much will 3 packages cost?

6. Roy paid $81.88 for a new jacket and sweater. He then exchanged the sweater, which cost $23.00, for another sweater that cost $19.99. What was the final cost for Roy's jacket and sweater?

**Solve, using the five-step plan.**

7. There are 38 rows with 2 dozen seats each in the Little Theater. An additional 24 people are allowed to stand during a performance. What is the total number of people that can attend a performance?

8. Yesterday it took Jeff Holland 1 h to get to work. This morning, Jeff drove to the train station in 20 min, waited for the train for 7 min, rode the train for 12 min, and then walked for 15 min to get to work. How long did it take Jeff to get to work this morning?

B 9. Tickets for the drama club's performance last weekend cost $2.50 for adults and $2.00 for students. Four hundred twenty adults attended the performance, and 273 students attended. What was the total amount of money collected from tickets for the performance last weekend?

10. Sarah Holness had her car tuned up for $60 and she purchased 4 new tires for $37 each. She gave the cashier 11 twenty-dollar bills. How much change did Sarah receive?

C 11. Joy and David Kramer had $30 to spend on dinner, a movie, and parking. Dinner cost $15.50 and parking cost $4. The Kramers had $2 left after paying for everything. What was the cost of one movie ticket?

12. The museum charges $4.50 per person for a 2 h tour with fewer than 20 people. If 20 or more people take the tour, the charge is $3.75 per person. Of the 23 people in today's tour, 17 had paid $4.50 in advance. How much money will the museum return as a refund?

---

## Review Exercises

**Estimate the answer using rounding.**

1. $267 + 73$    2. $941 - 189$    3. $82.7 + 91.8$    4. $261.6 - 131.9$

5. $24 \times 18$    6. $36 \times 41$    7. $21.3 \times 19.6$    8. $18.7 \times 29.3$

# 1-9 Solving and Checking Problems

The five-step plan shown in the preceding lesson can be used to solve many problems. It is also useful to have methods for checking answers to problems. For example, when planning your solution to a problem, you may find that there is more than one way to proceed. When there is more than one method for solving a problem, you may find it helpful to use one method to obtain an answer and to use the other method to check your results. Study the following example.

**EXAMPLE 1** The Treble Clef celebrated Heritage Day with a two-week sale on the Flexwood turntables. Twelve turntables were sold during the first week of the sale and 17 were sold during the second week of the sale. The sale price for each turntable was $74.90. How much money did the Treble Clef receive for the turntables sold?

**Solution**
- The problem asks for the amount of money received from the sale of the turntables.

- Given facts: 12 turntables sold the first week
  17 more sold the second week
  $74.90 received for each turntable

- There are two ways to solve this problem.

  *Method 1*
  First multiply the sale price by the number sold each week to find the amount of money received each week.

  $$12 \times 74.90 = 898.80 \qquad 17 \times 74.90 = 1273.30$$

  Then add the amounts of money received to find the total amount of money received from the sale of the turntables.

  $$898.80 + 1273.30 = 2172.10$$

  *Method 2*
  First add to find the total number of turntables sold during the sale.

  $$12 + 17 = 29$$

  Then multiply the sale price by the total number sold to find the amount of money received from the sale of the turntables.

  $$29 \times 74.90 = 2172.10$$

- By either method, the Treble Clef received $2172.10 from the sale of the turntables.

Another way to check an answer is to use rounding to find an estimated answer. If the answer and the estimate are close, then the estimate leads us to accept our answer.

**EXAMPLE 2**   Last summer Elka and David drove from Los Angeles to Boston. Along the way they stopped in Albuquerque, Kansas City, Atlanta, and Washington, D.C. They recorded the distance they traveled, as shown below.

| | |
|---|---|
| Los Angeles–Albuquerque | 806 mi |
| Albuquerque–Kansas City | 790 mi |
| Kansas City–Atlanta | 810 mi |
| Atlanta–Washington, D.C. | 630 mi |
| Washington, D.C.–Boston | 437 mi |

How many miles did Elka and David travel?

**Solution**   • The problem asks for the total distance traveled.

• Given facts: traveled 806 mi, 790 mi, 810 mi, 630 mi, 437 mi

• To find the total distance traveled, we add.

$$806 + 790 + 810 + 630 + 437 = 3473$$

• Elka and David traveled 3473 miles on their trip.

To check the answer, we round the distances and add. In this case, round to the nearest hundred miles.

$$800 + 800 + 800 + 600 + 400 = 3400$$

The estimate, 3400 mi, and the answer, 3473 mi, are quite close, so the actual answer seems reasonable.

---

## Class Exercises

**State the operations you would use to solve the problem and the order in which you would use them.**

1. The Dreyer Trucking Company moved 453 cartons one day, and then 485 the next day. On the third day they moved twice as many as on the first two days. What is the total number of cartons moved during those three days?

2. In January, Judy made the following deposits to her savings account: $107.50, $29.35, and $43.20. In February, she deposited twice as much money as in January. How much money did she deposit each month?

**State two methods that you could use to solve the problem.**

3. Three friends went out to dinner. The bill was $41.10, and they left a $6.15 tip. If they divide the total three ways, how much did each person pay?

4. Jay Elder took some clothes to Spotless Drycleaning. He was charged $4.00 for a jacket, $2.50 for a sweater, and $2.25 for a pair of slacks. Jay had coupons that allowed him to deduct $.50 from each item. How much will Jay pay for his drycleaning?

**Estimate the answer to the problem.**

5. Frank is in the check-out line at the grocery store. He has a gallon of milk ($1.83), a bag of flour ($2.15), a box of oatmeal ($1.15), a package of cheese ($1.57), and a dozen eggs ($1.05). How much is the bill?

6. Breda wants to buy a four-door sedan that has a base price of $5624.95, factory-installed options totalling $1213.50, and a destination charge of $183.00. How much does the car cost?

---

## Problems

**Solve. Check to be sure you have answered the question.**

A  1. The August electric bill for $75.80 was twice as much as the July bill. What was the total cost of electricity for July and August?

2. Alvin has $1863.50 in his savings account. His sister Alvis has $756 more in her account. Geoffrey borrowed $257 from each person. How much money do Alvin and Alvis have left in each of their accounts after making the loans?

3. Gregory ordered the following items from the Huntington Gardens catalog: a watering can for $15.80, a trowel for $4.49, and 6 packages of seeds for $.75 each. What is the total cost of the items?

4. Nancy's mobile needs 3 separate pieces of wire that measure 8 cm, 11 cm, and 15 cm. Nina's mobile needs two times the length of wire that Nancy's mobile needs. How much wire is needed for each mobile?

**Solve. Check by using an alternate method.**

5. Each member of the Best Buy Book Club receives 2 bonus points for every book ordered through the club. So far Chris has ordered 3 books in March, 2 in April, and 6 in May. How many bonus points has Chris accumulated so far?

6. The admission ticket to Tyler Amusement Park is $1.25 per person. A total of 815 tickets were sold on Saturday. The attendance decreased by 96 on Sunday. How much money did the park receive from ticket sales in all?

**Solve. Check by estimating the answer to the problem.**

7. Bonnema Brothers recently purchased four beach front lots of land. The areas of the two smaller lots are 2015 ft² and 2248 ft². The areas of the two larger lots are 8730 ft² and 7890 ft². What is the total area of the two smaller lots and the total area of the two larger lots?

8. Lake Tana in Africa has an elevation of 1829 m. Lake Tangra Tso in Tibet is situated 4724 m above sea level. In Europe, Lake Sevan has an elevation of 1915 m. What is the total height of the three lakes?

9. During one game at a bowling tournament the five-member Bright Team scored the following points: 169, 152, 187, 174, and 193. What is the difference between the highest and the lowest scores?

10. Jackie is taking an inventory of the furniture going on sale next week.

| | |
|---|---|
| sofas, 128 | platform beds, 250 |
| love seats, 105 | stereo cabinets, 83 |
| lamps, 216 | bookcases, 45 |

How many items are going on sale?

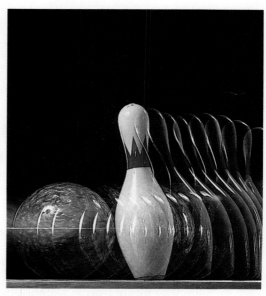

**Solve.**

B  11. The Hillview School Band held a car wash on Friday and Saturday. The charge was $2.25 per car on Friday and $2.50 per car on Saturday. On Friday, 87 cars were washed. On Saturday, 117 cars were washed. What was the total amount of money collected?

12. A photograph is enlarged so that its new dimensions are four times its original dimensions. If the new dimensions are 19.2 cm by 25.6 cm, what were the original dimensions?

13. Today the firm of Beckman and Beckman bought three types of stocks: 4780 shares of utility stocks, 1389 shares of commodity stocks, and 3542 shares of energy-related stocks. This is exactly three times the number of shares the firm bought yesterday. How many shares of stock did the firm buy in the past two days?

14. Yukio bought traveler's checks in the following denominations: five $50 checks, thirty $20 checks, five $10 checks, and twenty $5 checks. What is the total value of the checks bought?

15. A direct dial call from Boston to Australia costs $3.17 for the first minute and $1.19 for each additional minute. A station-to-station operator-assisted call costs $9.45 for the first 3 minutes and $1.19 for each additional minute. How much money would you save by dialing direct for a 5-minute call?

## Self-Test B

**Solve for the given replacement set.**

1. $72 - m = 43$; $\{19, 29, 31\}$       2. $6r = 48$; $\{6, 7, 8\}$     [1–6]

3. $t \div 12 = 11$; $\{23, 24, 25\}$       4. $4d + 16 = 28$; $\{3, 4, 5\}$

**Use inverse operations to solve.**

5. $g - 32 = 12$       6. $7d = 112$       7. $5a + 4 = 49$     [1–7]

**Solve, using the five-step plan.**

8. Laura bought a hammer for $12.95, 5 lb of nails for $5.20, and 8 sheets of plywood for $12 each. What was her total bill?     [1–8]

9. Between the hours of 6 A.M. and 9 P.M., 8 buses that were filled to capacity left the terminal. If the capacity of each bus is the same and 392 tickets were sold, how many passengers were on each bus?

**Solve. Check by estimating.**

10. Jeremy and his roommate share the monthly utility bills evenly. For November the cost of electricity was $87.90, gas was $24.35, heating fuel was $215.80, and water was $36.43. How much did each person pay that month?     [1–9]

*Self-Test answers and Extra Practice are at the back of the book.*

# The Development of Computers

The development of the modern computer began in 1946 with the completion of the ENIAC computer. It weighed 30 tons, contained 18,000 vacuum tubes and 6000 switches, and filled a room 30 feet by 50 feet. Since that time computers have become steadily more compact, powerful, and inexpensive.

Today's large computer systems, called **mainframes,** can process large amounts of data at very fast speeds. **Minicomputers** are smaller and somewhat slower, meeting the needs of colleges and small businesses at lower cost. The smallest of today's computers, such as the computer shown in the photo above, are the **microcomputers.** These computers are often called personal computers because they are inexpensive enough and small enough to go into classrooms and homes. The processing unit of these small computers is the **microprocessor,** a one-quarter-inch-square integrated circuit chip. This tiny chip is more powerful than the ENIAC with its 18,000 vacuum tubes.

The microprocessor controls the microcomputer and performs arithmetic operations. But other parts are needed to make the computer a useful tool. The computer has two kinds of **memory. ROM** (read only memory) permanently stores information needed for the computer to work properly. It cannot be changed by the user. **RAM** (random access memory) is available to the user and can store the user's programs and data. Memory size is measured in **bytes** or K. One K is about 1000 bytes. Each byte can store one character (letter or digit), so an 8K memory can store about 8000 typed characters.

The **keyboard** is used to input programs and data, and the **CRT** screen displays input, results, and graphics. A **disk drive** can be used to read programs and data into the computer from a disk, or to save programs on disk. A **printer** will save output in printed form. A **modem** can connect you to a network of other computers over your telephone line.

As computers have evolved, people have invented programming languages to help users program the computer to solve problems. Some of the more common languages are **BASIC,** which is available on almost all microcomputers, **FORTRAN,** often used for scientific problem solving, and **COBOL,** a business-oriented language. **Pascal** and **Logo** are two languages finding increasing application in education.

The development of computers has opened many new careers. Systems analysts use computers to analyze and solve problems for business and government. Programmers write the programs, or software, that help users apply the computer to their needs. Installation and maintenance of a computer's physical components, or hardware, are done by field engineers.

1. The fastest modern computers can do 100 million arithmetic operations in a second. Estimate how long it would take you to do this many additions. Suppose you are adding two four-digit numbers each time.

2. Each byte of memory will hold one typed character. About how many K of memory would it take to store these two pages? A disk for a microcomputer holds 160 K. About how many pages of this book could you store on one disk?

3. Ask your librarian to help you find out about the Mark I, IBM 360, and UNIVAC 1 computers. Find out about the size of each computer, the number of its components, its purpose, its inventors.

4. The computer language ADA was named after Ada Byron Lovelace (1815–1852). See what you can find out about Ada Lovelace and the computer language.

**Career Activity**

Look in the Help Wanted section of a newspaper and make a list of the job openings for systems analysts and programmers. Include in your list education requirements, what computers or computer languages the candidate should be familiar with, and the salary range.

# Chapter Review

**Match.**

**1.** $22 \times 8$  **2.** $52.6 - 9.95$  **A.** 32  **B.** $10^8$  [1–1]

**3.** $9 \times (3 + 1) - 4$  **4.** $\dfrac{4 + (6 \times 2 \times 5)}{(14 - 12)5}$  **C.** 42.65  **D.** 64  [1–2]

**5.** $4^3$  **6.** 100,000,000  **E.** 6.4  **F.** 176  [1–3]

**True or false?**

**7.** 21.09 to the nearest whole number is 20.  [1–4]

**8.** 124.4 to the nearest hundred is 100.

**9.** 83.415 to the nearest hundredth is 83.42.

**10.** 0.959 to the nearest tenth is 1.0.

**11.** $(17.2 + 1.8)4 = (1.8 + 17.2)4$ illustrates the commutative property.  [1–5]

**12.** $(8 + 7.9)2.3 = (8 \times 2.3) + (7.9 \times 2.3)$ illustrates the distributive property of multiplication with respect to subtraction.

**13.** $1.3(7 + 4) = (7 + 4)1.3$ illustrates the associative property.

**Is the equation true or false for the given value of the variable?**

**14.** $9y = 108;\ y = 12$  **15.** $k \div 4 = 28;\ k = 7$  [1–6]

**True or false?**

**16.** If $k + 5 = 140$, $k = 140 - 5$.  **17.** If $p \div 21 = 14$, $p = 21 - 14$.  [1–7]

**18.** If $9b = 162$, $b = 162 \div 9$.  **19.** If $t - 87 = 87$, $t = 87 - 87$.

**Write the letter of the correct answer.**

**20.** Julia Carmona hired 3 people to landscape her yard. They each  [1–8]
received the same hourly rate and it took them 5 h to do the job. If
her bill was $60, how much did each person earn an hour?
**a.** $6  **b.** $12  **c.** $3  **d.** $4

**21.** To raise money for a local charity, the 26 students of the eighth-  [1–9]
grade class participated in a bike-a-thon. Each of the sponsors
agreed to pay the students $.35 for each mile they rode their bicy-
cles. If 20 students ride 20 mi each and the rest of the students ride
30 mi each, for how many miles will the students be paid?
**a.** $203  **b.** 50 mi  **c.** $17.50  **d.** 580 mi

# Chapter Test

Evaluate the expression when $a = 4$ and $b = 12$.

**1.** $91 + a$  　　　**2.** $27 - b - a$  　　**3.** $5b$  　　　　　**4.** $36 \div a$  　　　　[1–1]

Evaluate the expression when $m = 14$ and $n = 16$.

**5.** $n - 4 \times 3$  　'**6.** $3 \times \frac{m}{7}$  　　　**7.** $2mn - 9$  　　**8.** $(m + n) \div 3$  　　[1–2]

Evaluate.

**9.** $3^4$  　　　　　　**10.** $10^3$  　　　　　　**11.** $2^5$  　　　　　　　　[1–3]

Write as a single power of 10.

**12.** $10^4 \times 10^7$  　　**13.** $10^6 \times 10^6$  　　　**14.** $10{,}000$

Round to the place specified.

**15.** tenths: 7.49  　　**16.** tens: 423.6  　　　**17.** hundredths: 4.283  　　[1–4]

What value of the variable makes the statement true?

**18.** $2.4 + (r + 9.5) = 2.4 + (9.5 + 7.8)$  　　**19.** $19.2k = k$  　　　　[1–5]

**20.** $62d - 19d = (62 - 19)4$  　　　　　　　**21.** $1(3.4 + 1.3) = a$

Solve for the given replacement set.

**22.** $d - 9 = 27$; $\{3, 18, 35\}$  　　　　　**23.** $14r = 70$; $\{3, 4, 5\}$  　　　[1–6]

**24.** $9(x + 4) = 63$; $\{1, 2, 3\}$  　　　　　**25.** $3k + 1 = 13$; $\{4, 5, 6\}$

Use inverse operations to solve.

**26.** $6g = 72$  　　　　**27.** $b \div 9 = 44$  　　　**28.** $3f - 1 = 53$  　　　　[1–7]

Solve, using the five-step plan.

**29.** The tickets for the theater cost $7.50 each.  Miles bought 4 of them　　[1–8]
and gave the cashier a fifty dollar bill.  What was the cost of the
tickets?

Solve and check your answer.

**30.** For the trip, Kari bought 3 stocking caps for $6.75 each and 3 scarfs　　[1–9]
for $8.50 each.  How much money did she spend?

# Cumulative Review

## Exercises

**Simplify.**

**1.** $28 + 781$

**2.** $630 - 52.1$

**3.** $65.1 \div 21$

**4.** $1.2 \times 3.64$

**5.** $48 + 303.9$

**6.** $0.042 \times 0.8$

**7.** $3(14 - 5)$

**8.** $(6 + 3) \div 9$

**9.** $4 \times 5 + 5$

**10.** $16 - 2 \times 3 - 1$

**11.** $(8 + 2) \div (12 - 7)$

**12.** $(14 + 36) \times (54 - 8)$

**Evaluate the expression when $a = 2$, $b = 5$, and $c = 3$.**

**13.** $5b + 18$

**14.** $9c - 12$

**15.** $3b + c \div 2$

**16.** $4ab - 8$

**17.** $30 \div (a + c)$

**18.** $a(12 - c)b$

**Simplify.**

**19.** $3^3$

**20.** $4^3$

**21.** $2^6$

**22.** $6^4$

**23.** $5^6$

**24.** $11^4$

**Select the most reasonable estimated answer.**

**25.** $43.6 - 2.79$    **a.** 41    **b.** 20    **c.** 45

**26.** $9.6 \times 53.66$    **a.** 500    **b.** 450    **c.** 540

**27.** $22.7 + 18.9 + 7.38$    **a.** 49    **b.** 37    **c.** 110

**28.** $165.7 + 38.21 + 6.44$    **a.** 1300    **b.** 210    **c.** 246

**True or false?**

**29.** $16 \div 8 + 2 = 16 \div (8 + 2)$

**30.** $99.5 - (6 + 7) = (99.5 - 6) + 7$

**31.** $(43 \times 0) + (43 \times 1) = 0$

**32.** $11.89 + (426 \div 2) = (11.89 + 426) \div 2$

**Find the solution of the equation for the given replacement set.**

**33.** $k + 16 = 23$; $\{5, 6, 7\}$

**34.** $58 - d = 31$; $\{39, 38, 37\}$

**35.** $7x - 1 = 20$; $\{1, 2, 3\}$

**36.** $8(y - 3) = 32$; $\{7, 8, 9\}$

**37.** $5a = 2a + 57$; $\{19, 20, 21\}$

**38.** $(c + 6) \div 6 = 1$; $\{0, 1, 2\}$

**Use inverse operations to solve.**

**39.** $18 + g = 20$

**40.** $a \div 9 = 18$

**41.** $5z = 65$

**42.** $20d + 8 = 68$

**43.** $7f - 1 = 13$

**44.** $3h + 2 = 26$

## Problems

---

**Problem Solving Reminders**

Here are some reminders that may help you solve some of the problems on this page.

- Determine which facts are necessary to solve the problem.
- Supply additional information if needed.
- Check by using rounding to find an estimated answer.
- If more than one method can be used to solve a problem, use one method to solve and the other to check.

---

**Solve.**

1. Merry and Sandy rented an apartment for $645 each month and shared the rent equally. After 4 months, Tess moved in and the rent was divided three ways. How much was Merry's rent for the year?

2. Hungarian paprika costs $1.30 for 2 oz, $4.00 for $\frac{1}{2}$ lb, and $6.00 for a pound. What is the cost of each ounce if you buy a pound? How much do you save per ounce if you buy a pound?

3. Mal ordered a set of 6 steak knives for $35.00. Additional costs included $4.95 for shipping, $1.25 for a gift box, and $1.75 for tax. What was the total cost of the order?

4. "I can save $19.50 if I buy a half dozen glasses on sale," said Ellis. How much is saved on each glass? If Ellis pays $30.00 for 6 glasses on sale, what was the original price of each glass?

5. House numerals that are 4 in. high cost $4.00 each. Numerals that are 7 in. high cost $10.00 each. How much will it cost to buy numerals that are 4 in. high for your house if your address is 16332 Long Meadow Road?

6. In 1918, a sheet of 100 airmail stamps was mistakenly printed with an airplane upside down. A stamp collector bought the sheet for $.24 per stamp and later sold the sheet for $15,000. How much did the collector make on his lucky buy?

7. A membership to the Science Center costs $53 per year and includes a subscription to a monthly magazine. If the magazine costs $3 per issue, what are the annual dues for membership alone?

8. Dale took advantage of the gas company's offer to make average monthly payments. The payments were based on the average of the two highest and the two lowest bills for the past 12 months. If these bills were $135.50, $142.71, $68.29, and $56.30, what is Dale's average monthly payment?

# 2

# Positive and Negative Numbers

Lightning, as seen in the photograph, is a dramatic, electrical reaction that is usually associated with thunderclouds. It occurs as a result of a sudden, powerful exchange between the positive and negative centers within a cloud, between several clouds, or between a cloud, the air, and the ground. The long flash of light we see is part of the interaction. The positive and negative charges move through the atmosphere so rapidly that a tremendous amount of heat is generated, which warms up the surrounding air so quickly that a thunderous explosion results.

In this chapter you will study operations with positive and negative numbers.

## Career Note

Earthquakes are similar to lightning in that they are sudden, dramatic natural events. Geologists study earthquakes as part of their study of the earth. By studying the structure and history of the rocks beneath the earth's surface, geologists may be able to predict future earthquakes. Geologists can also specialize in locating oil and other raw materials.

# 2-1 The Integers

When we measure temperature, we use a scale that has 0 as a reference point. We use *positive numbers* to indicate temperatures above 0°, and we use *negative numbers* to indicate temperatures below 0°.

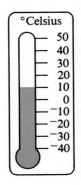

We often have occasion to measure quantities on different sides of a zero reference point, such as distances above and below sea level, time before and after a rocket launch, increases and decreases in stock prices, and deposits and withdrawals in a bank account. Positive and negative numbers help us to measure these quantities.

We may graph both positive and negative numbers on a horizontal number line by extending the number line to the *left* of the origin as shown below. Like the positive whole numbers, the negative whole numbers are equally spaced, but they are positioned to the left of 0.

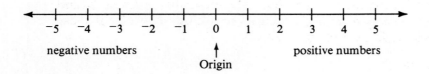

We read ⁻1 as *negative one*. We may read 1 as *positive one,* or simply *one*. For emphasis, we may use the symbol ⁺1 for positive one.

Any pair of numbers, such as 3 and ⁻3, that are the same distance from the origin but in opposite directions are called **opposites.** The opposite of 3 is ⁻3 and the opposite of ⁻3 is 3. The opposite of 0 is 0.

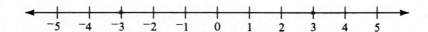

We use the symbol |⁻3|, read *the absolute value of ⁻3,* to represent the distance between ⁻3 and 0. Because ⁻3 is 3 units from the origin, |⁻3| = 3. In general, |n| (read *the absolute value of n*) represents the distance between the number *n* and the origin.

The whole numbers, 0, 1, 2, 3, . . . , together with their opposites, 0, ⁻1, ⁻2, ⁻3, . . . , form the set of numbers called the **integers:**

$$\ldots, {}^{-}3, {}^{-}2, {}^{-}1, 0, 1, 2, 3, \ldots.$$

The **positive integers** are the numbers 1, 2, 3, . . . , and the **negative integers** are the numbers ⁻1, ⁻2, ⁻3, . . . . Although 0 is an integer, it is neither positive nor negative.

**EXAMPLE 1**   Express as an integer.

     **a.** $|^-5|$          **b.** $|0|$

**Solution**   **a.** $|^-5|$ represents the distance between 0 and the number $^-5$. Thus $|^-5| = 5$.

           **b.** $|0|$ represents the distance between 0 and 0. Thus $|0| = 0$.

**EXAMPLE 2**   Arrange $^-3, 1, ^-4, 0, ^-1$ in order from least to greatest.

**Solution**   We can graph the numbers on a number line, with 1 at the right of 0.

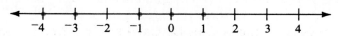

Reading the coordinates of the points from left to right will order the numbers from least to greatest. $^-4, ^-3, ^-1, 0, 1$

Positive and negative numbers may be compared using inequality symbols as well as on the number line. The inequality symbols $>$ and $<$ are used to compare mathematical expressions. The use of these symbols in inequalities is shown below.

$$8 > 6 \qquad\qquad\qquad\qquad 6 < 8$$
Eight is greater than six.          Six is less than eight.

To avoid confusing these symbols, think of them as arrowheads whose small ends point toward the smaller numbers.

We can indicate that one number is between two others by combining two inequalities. We know that $5 < 6$ and $6 < 8$; thus we can write

$$5 < 6 < 8 \qquad \text{or} \qquad 8 > 6 > 5.$$

Other inequality symbols that we use are shown below with their meanings.

      $\neq$     *is not equal to*
      $\geq$     *is greater than or equal to*
      $\leq$     *is less than or equal to*

We can use inequality symbols to write open sentences. The open sentence $n \leq 6$ means $n < 6$ or $n = 6$.

**EXAMPLE 3**   Replace $\underline{\;?\;}$ with $<$ or $>$.

     **a.** $14 \underline{\;?\;} 2$      **b.** $1 \underline{\;?\;} 11$      **c.** $8 \underline{\;?\;} 3 \underline{\;?\;} 0$

**Solution**   **a.** $14 > 2$        **b.** $1 < 11$        **c.** $8 > 3 > 0$

## Class Exercises

**Name an integer that represents each of the following.**

1. 15 s before blastoff of a rocket

2. A gain of 6 yd in a football play

3. A withdrawal of 90 dollars from a bank account

4. An elevation of 350 ft below sea level

5. The opposite of 80          6. The opposite of $^-2$

7. The absolute value of $^-14$          8. The absolute value of 27

9. Name two integers, each of which is 12 units from 0.

10. If $|n| = 15$, then $n = \underline{\ ?\ }$ or $n = \underline{\ ?\ }$.

**Replace __?__ with > or < to make a true statement.**

11. a. $^-6 \underline{\ ?\ } ^-2$    b. $|^-6| \underline{\ ?\ } |^-2|$       12. a. $4 \underline{\ ?\ } ^-5$    b. $|4| \underline{\ ?\ } |^-5|$

**True or false?**

13. $7 > 7$                    14. $18 \le 20$                    15. $15 > 5$

16. $14 > 6 > 2$              17. $20 < 18 < 16$              18. $33 \ge 24 \ge 11$

---

## Written Exercises

**Graph the integers in each exercise on the same number line.**

**A**  1. $0, 1, ^-1, 3, ^-3$       2. $0, 2, ^-2, 5, ^-5$       3. $6, 0, ^-4, ^-9, 7$       4. $^-1, 3, ^-8, 4, ^-6$

**Graph the number and its opposite on the same number line.**

5. 3          6. 10          7. $^-7$          8. $^-2$          9. 0          10. $^-4$

**Replace __?__ with =, >, or < to make a true statement.**

| | | |
|---|---|---|
| 11. $21 \underline{\ ?\ } 14$ | 12. $18 \underline{\ ?\ } 35$ | 13. $76 \underline{\ ?\ } 67$ |
| 14. $104 \underline{\ ?\ } 104$ | 15. $265 \underline{\ ?\ } 256$ | 16. $390 \underline{\ ?\ } 309$ |
| 17. $17 + 82 \underline{\ ?\ } 93$ | 18. $47 - 31 \underline{\ ?\ } 61$ | 19. $25 \div 5 \underline{\ ?\ } 10$ |
| 20. $26 \times 4 \underline{\ ?\ } 52$ | 21. $19 \times 11 \underline{\ ?\ } 208$ | 22. $84 \div 3 \underline{\ ?\ } 24$ |
| 23. $^-3 \underline{\ ?\ } ^-4$ | 24. $^-2 \underline{\ ?\ } 1$ | 25. $7 \underline{\ ?\ } ^-8$ |
| 26. $0 \underline{\ ?\ } ^-2$ | 27. $^-11 \underline{\ ?\ } 0$ | 28. $^-7 \underline{\ ?\ } 10$ |

Use > or < to write a true statement with the given numbers.

**29.** 18, 46, 32

**30.** 29, 5, 31

**31.** 103, 130, 310

**32.** 256, 652, 526

**33.** 986, 689, 698

**34.** 717, 177, 771

Express as an integer.

**35.** $|^-3|$   **36.** $|^-6|$   **37.** $|0|$   **38.** $|12|$   **39.** $|9|$   **40.** $|^-7|$   **41.** $|^-8|$   **42.** $|^-1|$

Write the numbers in order from least to greatest.

**43.** 6, $^-15$, 0, $^-2$

**44.** $^-3$, 1, 0, $^-7$

**45.** $^-12$, 7, $^-8$, 1, $^-1$

**46.** 0, 2, $^-5$, $^-9$, 10

**47.** $^-10$, 4, 14, $^-14$, 8

**48.** 3, 9, $^-13$, 11, $^-15$

**49.** $^-6.4$, 0.6, 3.1, $^-2.7$

**50.** 7.1, $^-0.9$, $^-3.6$, $^-9.4$

For Exercises 51–62, (a) list the integers that can replace *n* to make the statement true, and (b) graph the integers on a number line.

**B**   **51.** $|n| = 6$

**52.** $|n| = 3$

**53.** $|n| = 4$

**54.** $|n| = 5$

**55.** $|n| = 0$

**56.** $|n| = 14$

**57.** $|n| < 2$

**58.** $|n| < 5$

**59.** $|n| \leq 4$

**60.** $|n| \leq 7$

**61.** $2 < |n| < 8$

**62.** $0 < |n| < 3$

Complete with the word *positive* or *negative*.

**C**   **63.** If an integer is equal to its absolute value, then the integer must be a ___?___ integer or 0.

**64.** If an integer is equal to the opposite of its absolute value, then the integer must be a ___?___ integer or 0.

**65.** Explain why there is no number that can replace *n* to make the equation $|n| = ^-3$ true.

## Review Exercises

Graph the number on a number line.

**1.** 4

**2.** 6

**3.** 3

**4.** 0

**5.** 2.4

**6.** 1.3

**7.** $3\frac{1}{2}$

**8.** $1\frac{1}{3}$

**9.** 3.7

**10.** 4.5

## 2-2 Decimals on the Number Line

The graphs of the *positive decimal* 2.5 and its opposite, the *negative decimal* ‾2.5, are shown on the number line below. We graph 2.5 by locating the point that is 2.5 units to the *right* of 0, and we graph ‾2.5 by locating the point that is 2.5 units to the *left* of 0.

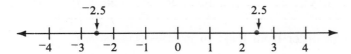

The positive decimals together with the negative decimals and 0 form the set of *decimal numbers*. The set of decimal numbers includes all of the whole numbers and all of the integers.

**EXAMPLE 1**   Write the following numbers in order from least to greatest.

$$‾2, 4.1, 0.2, ‾2.6, ‾1.34$$

**Solution**   We can graph the given numbers on a number line.

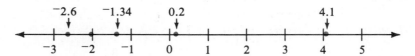

Reading the coordinates from left to right will give the numbers in order from least to greatest.

$$‾2.6, ‾2, ‾1.34, 0.2, 4.1$$

We have been representing decimals by their graphs, that is, by dots on a number line. We can also use directed line segments or arrows to illustrate decimals. Arrows that point to the *left* (the negative direction) represent negative numbers. Arrows that point to the *right* (the positive direction) represent positive numbers.

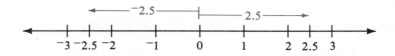

Notice in the diagram above that both the arrow representing ‾2.5 and the arrow representing 2.5 have length 2.5.

An arrow representing a number may have any point on the number line as its starting point, as long as it has length and direction indicated by that number. The length of the arrow is the absolute value of the number that the arrow represents. The direction of the arrow is determined by the sign of the number.

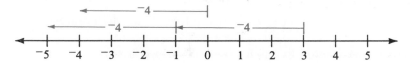

On the number line above, each arrow represents the decimal number ⁻4, for each has length 4 and points to the left.

**EXAMPLE 2**  What number is represented by the arrow above the number line below?

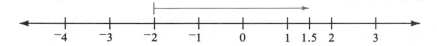

**Solution**  The starting point of the arrow is ⁻2 and the endpoint is 1.5. The arrow points to the right and is 3.5 units long. Thus, the arrow represents the positive decimal number 3.5.

**EXAMPLE 3**  An arrow representing the number ⁻7 has starting point 3. What is its endpoint?

**Solution**  Draw a number line. Starting at 3, draw an arrow 7 units long in the negative direction (left). The endpoint of the arrow is ⁻4.

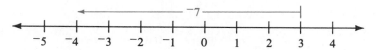

## Class Exercises

**Name a decimal number that represents each of the following.**

1. The opposite of 8.71

2. The opposite of ⁻10.16

3. A discount of fifty-nine cents

4. A rise in body temperature of 0.6°C

5. The absolute value of ⁻67.5

6. The absolute value of 9.07

**7.** Name the letter written above the graph of the given number.

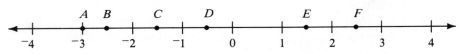

    **a.** ⁻0.5           **b.** ⁻1.5           **c.** 1.5           **d.** ⁻2.5

**8.** State the numbers in Exercise 7 in order from least to greatest.

**Name the number represented by each arrow described below.**

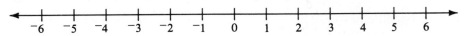

  **9.** Starting point at 0, endpoint at ⁻2.5     **10.** Starting point at ⁻1, endpoint at 3

  **11.** Starting point at 2, endpoint at ⁻4     **12.** Starting point at ⁻0.5, endpoint at 5

## Written Exercises

**Graph the numbers in each exercise on the same number line.**

**A**   **1.** ⁻2, ⁻3.5, 0     **2.** 3.2, ⁻4, ⁻3.2     **3.** ⁻1.5, ⁻7, ⁻3.25     **4.** ⁻0.9, ⁻1, ⁻4.1

**Graph the number and its opposite on the same number line.**

  **5.** 2.25           **6.** 1.9           **7.** ⁻0.5           **8.** ⁻3.1

  **9.** ⁻4.2           **10.** 0.75          **11.** 0.3           **12.** 5.5

**Write the decimal number that is equal to each of the following.**

  **13.** |⁻2.36|     **14.** |1.921|     **15.** |⁻16|     **16.** |⁻100|     **17.** |3.03|     **18.** |⁻0.2|

**Replace __?__ with < or > to make a true statement.**

  **19.** ⁻2.93 __?__ 1.1         **20.** 4 __?__ ⁻0.5          **21.** ⁻8.1 __?__ 2.3

  **22.** ⁻1.95 __?__ ⁻1.96       **23.** ⁻5.01 __?__ ⁻4.99      **24.** ⁻2.99 __?__ ⁻2.98

  **25.** 0.1 __?__ ⁻18.25        **26.** 12.2 __?__ ⁻13.3      **27.** ⁻3.7 __?__ 3.07

**Write the numbers in order from least to greatest.**

  **28.** ⁻2.72, ⁻3, 0.03, ⁻3.5, 0.2         **29.** 6.3, ⁻8, ⁻7.6, ⁻1.75, 6.03

  **30.** 0, 2.99, ⁻10, ⁻0.1, ⁻0.01         **31.** ⁻100.5, ⁻2, 3.11, ⁻2.1, ⁻46.8

  **32.** ⁻0.5, ⁻0.05, ⁻5, ⁻50, 500         **33.** ⁻0.3, 30.3, ⁻0.33, ⁻3.3, 33

**Draw an arrow to represent each decimal number described below.**

**34.** The number 3, with starting point $^-1$

**35.** The number 2.5, with starting point $^-0.5$

**36.** The number $^-5$, with starting point 1.5

**37.** The number $^-3$, with starting point $^-0.5$

**38.** The number 5.5, with starting point $^-3$

**39.** The number $^-4$, with endpoint $^-2$

**40.** The number $^-2$, with endpoint 5

**List the decimal numbers that can replace $x$ to make the statement true.**

**B**  **41.** $|x| = 4.1$        **42.** $|x| = 0.001$        **43.** $|x| = 26.3$

**44.** $|x| = 0$        **45.** $|x| = |^-1.19|$        **46.** $|x| = |^-2.2|$

**Copy and complete the chart so that the two arrows represent the same decimal number.**

|     | Arrow 1 | | Arrow 2 | |
| --- | --- | --- | --- | --- |
|     | **Starting Point** | **Endpoint** | **Starting Point** | **Endpoint** |
| **47.** | $^-2.5$ | $^-7$ | 0 | ? |
| **48.** | ? | 4 | 1.5 | $^-1.5$ |
| **49.** | $^-0.5$ | 8.5 | ? | $^-3$ |
| **50.** | 4 | $^-6.25$ | ? | $^-2$ |

**For Exercises 51–54, (a) list the _integers_ that can replace $n$ to make the statement true, and (b) show their graphs on a number line.**

**51.** $|n| < 4.3$      **52.** $|n| < 2.99$      **53.** $|n| \le 5.001$      **54.** $|n| \le 0.08$

---

## Review Exercises

**Use the properties to simplify the expression. Name the property or properties used.**

**1.** $2(4.5 + 1.07)$      **2.** $3.7 + 12.5 + 0.5$      **3.** $1 \times 27.18$

**4.** $8.69 + 4.78 + 2.31$      **5.** $7(30 - 11.1)$      **6.** $(44 \times 0.3) + (44 \times 0.7)$

**7.** $6.25 \times 43 \times 4$      **8.** $0.2 \times 3.7 \times 0.5$      **9.** $17.6 \times 283 \times 0$

## 2-3 Adding Positive and Negative Numbers

We can use arrows on the number line, as shown below, to add two positive numbers or to add two negative numbers. We draw a solid arrow with starting point 0 to represent the first addend. We draw another solid arrow with *starting point at the endpoint of the first arrow* to represent the second addend. To represent the sum, we draw a dashed arrow from the starting point of the first arrow to the endpoint of the second arrow.

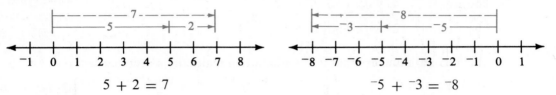

$$5 + 2 = 7 \qquad\qquad\qquad ^-5 + {}^-3 = {}^-8$$

In each case, if we add the absolute values of the addends, we obtain the absolute value of the sum. The sum has the same sign as the addends.

> ### Rules
>
> The sum of two positive numbers is positive.
>
> The sum of two negative numbers is negative.

**EXAMPLE 1**  Find the sum.  **a.** $2.5 + 4.3$  **b.** $^-7 + {}^-1.5$

**Solution**  **a.** Since the addends are positive, the sum is positive.
$$2.5 + 4.3 = 6.8$$

**b.** Since the addends are negative, the sum is negative.
$$^-7 + {}^-1.5 = {}^-8.5$$

We can also use arrows on the number line to add a positive and a negative number. The sum of a positive and a negative number may be positive, negative, or zero, as shown in the following illustrations.

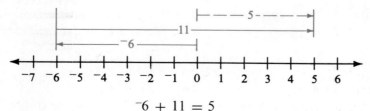

$$^-6 + 11 = 5$$

The positive number 11 has greater absolute value than the negative number $^-6$. Thus the sum is positive.

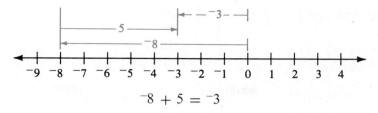

$$^-8 + 5 = ^-3$$

The negative number $^-8$ has greater absolute value than the positive number 5. Thus the sum is negative.

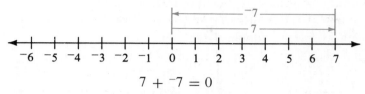

$$7 + ^-7 = 0$$

The positive number 7 and the negative number $^-7$ are opposites and thus have the same absolute value. The sum is zero.

Notice that, in each case, the absolute value of the sum is the *difference* of the absolute values of the addends. The sum has the same sign as the addend with the greater absolute value.

## Rules

The sum of a positive number and a negative number is

1. positive if the positive number has the greater absolute value.
2. negative if the negative number has the greater absolute value.
3. zero if the numbers have the same absolute value.

**EXAMPLE 2** Find the sum.

    **a.** $3.5 + ^-10.5$     **b.** $^-7.6 + 12.2$     **c.** $^-4.8 + 0$     **d.** $^-6.7 + 6.7$

**Solution**

**a.** The negative addend has the greater absolute value, so the sum is negative.
$$3.5 + ^-10.5 = ^-7$$

**b.** The positive addend has the greater absolute value, so the sum is positive.
$$^-7.6 + 12.2 = 4.6$$

**c.** Think of adding 0 as *moving no units* on the number line. Thus, $^-4.8 + 0 = ^-4.8$.

**d.** The numbers have the same absolute value, so the sum is zero.
$$^-6.7 + 6.7 = 0$$

As shown in Example 2, part (c), the addition property of zero holds for the positive and negative decimals. All of the properties for positive decimals hold for negative decimals as well.

**EXAMPLE 3**  Sally Wright bought some stock in ABC Computer Company. The stock went down $2.50 per share in the first week, went up $3.00 in the second week, and went down $1.25 in the third week. If Sally paid $30.50 per share for the stock, did she gain or lose money?

**Solution**
- The question asks if Sally gained or lost money.

- Given information: Sally paid $30.50 per share
  price went down $2.50, went up $3.00, went down $1.25

- First, find the new price per share. Express the given information as a sum of positive and negative decimals.
$$30.50 + {}^-2.50 + 3 + {}^-1.25 = (30.50 + 3) + ({}^-2.50 + {}^-1.25)$$
$$= 33.50 + {}^-3.75$$
$$= 29.75$$
To find whether Sally gained or lost money, compare the new price per share to the price paid per share.
$$29.75 < 30.50$$

- Because the new price per share is less than the price paid per share, Sally lost money.

---

**Problem Solving Reminder**

Many problems involve *more than one step*, but the steps may not always involve operations. In Example 3, we used addition in the first step, but in the second step we compared the answer to the first step with an amount given in the problem.

---

## Class Exercises

**State the addition fact illustrated by the diagram.**

1.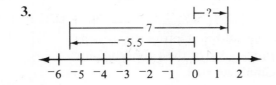

2.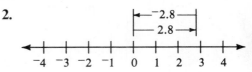

3.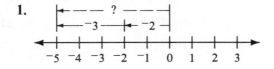

4.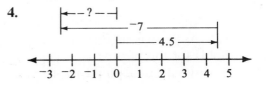

**Without computing the exact sum, state whether the sum is positive, negative, or 0.**

**5.** $-3.4 + {}^-2.6$     **6.** $25.7 + {}^-8.6 + {}^-25.7$     **7.** $2.37 + {}^-9.99$     **8.** ${}^-6.8 + 11.5$

## Written Exercises

**Find the sum by using arrows on a number line.**

**A**  **1.** ${}^-3 + {}^-8$         **2.** $10.7 + {}^-10.7$         **3.** ${}^-12.5 + 22$         **4.** $2.9 + {}^-6.9$

**Find the sum.**

**5.** ${}^-2 + {}^-17$         **6.** ${}^-8 + {}^-9$         **7.** $8.3 + {}^-21.3$         **8.** ${}^-4.6 + 38.6$

**9.** ${}^-0.1 + {}^-0.2$         **10.** ${}^-1.82 + {}^-3.68$         **11.** $16.5 + {}^-16.5$         **12.** ${}^-8.7 + 3.4$

**13.** $16.9 + {}^-0.7$         **14.** ${}^-51.3 + 51.3$         **15.** $12.37 + {}^-8.2$         **16.** ${}^-85 + {}^-41$

**17.** $81.9 + {}^-81.9$         **18.** $32.8 + {}^-36$         **19.** $0 + {}^-0.12$         **20.** ${}^-7.9 + 0$

**21.** ${}^-4.2 + {}^-6.5 + 17$         **22.** $7.1 + {}^-9 + 2.3$         **23.** $5 + {}^-16.9 + 1.1$

**24.** ${}^-8.6 + {}^-17.1 + {}^-4.3$         **25.** ${}^-3.3 + {}^-7.25 + 3.3$         **26.** $0.98 + {}^-13.4 + {}^-0.98$

**What value of the variable makes the statement true?**

**27.** ${}^-8 + x = 4$         **28.** $x + 4 = {}^-9$         **29.** ${}^-41 + x = {}^-53$

**30.** ${}^-19 + x = 0$         **31.** $18.5 + x = 0$         **32.** ${}^-6 + x = {}^-1$

**B**  **33.** ${}^-4.3 + x = {}^-6.7$         **34.** $x + {}^-5.6 = {}^-37$         **35.** $x + 18.6 = {}^-1.2$

**36.** ${}^-0.66 + x = 0.10$         **37.** $12.9 + x = {}^-13$         **38.** $x + 20.2 = {}^-5.1$

**Replace  ?  with =, >, or < to make a true statement.**

**39.** $({}^-25.3 + {}^-8.8)$ __?__ $({}^-12.4 + {}^-19.7)$         **40.** $(6.24 + {}^-15.9)$ __?__ $({}^-6.24 + 15.9)$

**41.** $({}^-34.9 + 27.5)$ __?__ $(9.7 + {}^-18.4)$         **42.** $(14.4 + {}^-18.6)$ __?__ $({}^-3.2 + {}^-0.98)$

**C**  **43. a.** $|{}^-3 + {}^-19|$ __?__ $|{}^-3| + |{}^-19|$         **b.** $|{}^-4.6 + 4.6|$ __?__ $|{}^-4.6| + |4.6|$

**c.** $|{}^-8.7 + 12.6|$ __?__ $|{}^-8.7| + |12.6|$         **d.** $|{}^-18.6 + 4.9|$ __?__ $|{}^-18.6| + |4.9|$

**e.** $|53.5 + 3.7|$ __?__ $|53.5| + |3.7|$         **f.** $|{}^-4 + 13.75|$ __?__ $|{}^-4| + |13.75|$

**g.** On the basis of your answers to parts (a)–(f), write a general rule for $|x + y|$ __?__ $|x| + |y|$ that holds for all numbers $x$ and $y$. Explain why this rule is true.

## Problems

**Solve Problems 1–3 by first expressing the given data as a sum of positive and negative numbers. Then, compute the sum of the numbers and answer the questions.**

**A**  **1.** The temperature in Lynn at 7:00 A.M. was ⁻7°C. By 12:00 noon, the temperature had increased by 13°C, but it then decreased by 3°C between noon and 5:00 P.M. What was the temperature reading at 5:00 P.M.?

**2.** From the Andersons' farm, Bonnie drove 20.4 km due east to Fairvale. From Fairvale, she drove 33.7 km due west to Ward City. How far was she then from the farm and in what direction?

**3.** The Tigers football team gains 3.5 yd on a first down, loses 11 yd on the second down, and gains 2 yd on the third down. Do the Tigers gain or lose total yardage in these three plays? What is the total number of yards gained or lost?

**B**  **4.** For a summer job, Tom plans to clean the Wilsons' house. He estimates that each week he will spend $1.25 and $3.80 on cleaning supplies. How much should he charge the Wilsons if he wishes to make a profit of $6.50 each week?

**5.** Carl purchased stock in the Dependable Equipment Company. The price per share of the stock fell by $4.30 in the first month, rose by $2.50 in the second month, and rose by $2.60 in the third month. Carl sold the stock for $22.00 per share at the end of the third month. Did he gain or lose money?

## Review Exercises

**Evaluate the expression when $s = 2.7$ and $t = 8.4$.**

**1.** $(25 - s)7$

**2.** $102 \div (t + 12)$

**3.** $11(7s - 13)$

**4.** $(t + 6 - 4.4)t$

**5.** $(s + 4)(s + 4)$

**6.** $10st \div (t - 4.2)$

**7.** $st \div 2s$

**8.** $(s + t) \div (t + 13.8)$

**9.** $(t - s)(5t + 8)$

## 2-4 Subtracting Positive and Negative Numbers

You know that $7.5 - 3 = 4.5$. In the preceding lesson, you learned that $7.5 + {}^-3 = 4.5$. Thus, $7.5 - 3 = 7.5 + {}^-3$. This example suggests the following general rule.

> ### Rule
> For any numbers $a$ and $b$,
> $$a - b = a + \text{(the opposite of } b)$$
> or
> $$a - b = a + (-b)$$

Note the lowered position of the minus sign in the expression $(-b)$, above. We use an *unraised* minus sign to mean *the opposite of.* For example,

$-3 = {}^-3$, read *the opposite of three equals negative three*

$-({}^-5) = 5$, read *the opposite of negative five equals five*

Because the numerals $-3$ and ${}^-3$ name the same number, one may be used in place of the other. From now on, we will use an unraised minus sign to denote subtraction, a negative number, and the opposite of a number.

**EXAMPLE 1**  Find the difference.

  **a.** $5 - 13$         **b.** $12.4 - (-8)$         **c.** $-10.9 - (-3.4)$

**Solution**   **a.** $5 - 13 = 5 + (-13) = -8$

  **b.** $12.4 - (-8) = 12.4 + 8 = 20.4$

  **c.** $-10.9 - (-3.4) = -10.9 + 3.4 = -7.5$

It is important to read a variable expression such as $-n$ as *the opposite of n* because $n$ may denote a negative number, a positive number, or 0.

**EXAMPLE 2**  Evaluate the expression when $m = -5.2$.

  **a.** $m - 14$         **b.** $-m - 14$

**Solution**   **a.** $m - 14 = -5.2 - 14 = -5.2 + (-14) = -19.2$

  **b.** $-m - 14 = -(-5.2) - 14 = 5.2 - 14 = 5.2 + (-14) = -8.8$

**Reading Mathematics:** *Using Examples*
The worked-out examples in each lesson show you how the general statements in the lesson can be applied to specific situations. If you need help as you work on the exercises, look back at the examples for models to follow or for ideas on how to begin your solutions.

## Class Exercises

**Complete.**

**1.** $6 - 12 = 6 + \underline{\ ?\ }$

**2.** $-10 - 8 = -10 + \underline{\ ?\ }$

**3.** $6.7 - (-1.5) = 6.7 + \underline{\ ?\ }$

**4.** $-26.01 - (-8.2) = -26.01 + \underline{\ ?\ }$

**Without computing the exact difference, state whether the difference is positive, negative, or 0.**

**5.** $-4.2 - 4.2$
**6.** $4.2 - (-4.2)$
**7.** $4.2 - 4.2$
**8.** $-4.2 - (-4.2)$

**Find the difference.**

**9.** $3 - (-6)$
**10.** $-2 - (-3.5)$
**11.** $-1.8 - 5.8$
**12.** $2.25 - 4$

## Written Exercises

**Write the difference as a sum.**

**A**  **1.** $7 - 19$
**2.** $-21 - 42$
**3.** $6.2 - (-8.3)$
**4.** $-2.9 - (-11.6)$

**5–8.** Find each difference in Exercises 1–4 above.

**Find the difference.**

**9.** $4 - 10$
**10.** $25 - 34$
**11.** $-3 - 24$
**12.** $-12 - 5$

**13.** $9 - (-33)$
**14.** $14 - (-46)$
**15.** $-2 - (-17)$
**16.** $-6 - (-5)$

**17.** $0 - 43$
**18.** $0 - 101$
**19.** $0 - (-20)$
**20.** $0 - (-14)$

**21.** $44 - 0$
**22.** $-16 - 0$
**23.** $6.9 - 8$
**24.** $12 - 20.5$

**25.** $4.3 - 2.1$
**26.** $23.4 - 6.8$
**27.** $-16.1 - 8.5$
**28.** $-0.4 - 8.9$

**29.** $-19 - 5.6$
**30.** $-41.1 - 2.9$
**31.** $0 - (-3.37)$
**32.** $0 - 12.8$

**33.** $12.4 - (-12.4)$
**34.** $-18.1 - (-25)$
**35.** $-52.9 - (-11.6)$

**36.** $(6 - 9.7) - 8.8$
**37.** $(-2.5 - 8.1) - (-12.4)$
**38.** $(0 - 8.3) - (-24.1)$

Evaluate the expression when $a = -4.5$ and $b = -6.2$.

**39.** $-a$      **40.** $-b$      **41.** $-|b|$      **42.** $-|a|$

**43.** $a - b$      **44.** $b - a$      **45.** $-a - b$      **46.** $-b - a$

**47.** $a - (-b)$      **48.** $b - (-a)$      **49.** $-b - (-a)$      **50.** $-a - (-b)$

**What value of the variable makes the statement true?**

**B**    **51.** $2 - d = -6$      **52.** $d - 5 = -13$      **53.** $-3 - d = -11$

     **54.** $8 - d = 13$      **55.** $-7 - d = 12$      **56.** $d - (-6) = -7$

     **57.** $-d - 4 = 14$      **58.** $-d - 5 = -9$      **59.** $7 - (-d) = 14$

     **60.** $d - (-8) = 17$      **61.** $-8 - (-d) = -16$      **62.** $-4 - (-d) = 10$

**C**    **63.** Replace $\underline{\ ?\ }$ with $=$, $>$, or $<$ to make a true statement.
     **a.** $|13.6 - 8.9| \underline{\ ?\ } |13.6| - |8.9|$      **b.** $|-8.9 - (-13.6)| \underline{\ ?\ } |-8.9| - |-13.6|$
     **c.** $|8.9 - 13.6| \underline{\ ?\ } |8.9| - |13.6|$      **d.** $|-8.9 - 13.6| \underline{\ ?\ } |-8.9| - |13.6|$
     **e.** Based on your answers to parts (a)–(d), write a general rule for
     $|x - y| \underline{\ ?\ } |x| - |y|$, where $x$ and $y$ are any decimal numbers.

---

## Problems

**Solve Problems 1–7 by first expressing the given data as a difference of positive and negative numbers. Then, compute the difference of the numbers and answer the question.**

**A**    **1.** On a winter day, the temperature dropped from $-3°C$ to $-11°C$. Find the change in temperature.

     **2.** Find the difference in the ages of two people if one was born in 27 B.C. and the other was born in 16 A.D.

     **3.** The elevation of the highest point in a region is 1226 m above sea level. If the difference between the highest point and lowest point in the region is 1455 m, find the elevation of the lowest point.

     **4.** Two stages of a rocket burn for a total of 114.5 s. If the first stage burns for 86.8 s, how long does the second stage burn?

     **5.** Jan Miller purchased 140 shares of stock in the ABC Company at a price of $18.75 per share. During the next three days, the value declined by $1.00, $1.75, and $1.50. What was the value of a share of ABC stock at the end of three days?

6. A parachutist jumped from an airplane flying at an altitude of 1100 m, dropped 200 m in the first 25 s, and then dropped 350 m in the next 35 s. What was the altitude of the parachutist 60 s after jumping?

7. In Summit City, 78 cm of snow fell on Sunday. The snow melted approximately 5.8 cm on Monday, approximately 7.5 cm on Tuesday, and approximately 12 cm on Wednesday. Approximately how much snow remained?

**B**   8. Donna receives an allowance every 2 weeks that includes $20 for school lunches. During the past 4 weeks, she spent $7.50, $8.25, $5.25, and $8.75 on lunches. How much did Donna have left from the money allowed for lunches for the 4 weeks?

9. Eric Chung had $65.10 in his checking account on June 1. He wrote two checks in June, one for $42.99. Eric forgot to write down the amount of the other check. At the end of the month, he received a notice that his account was overdrawn by $22.11. What was the amount of Eric's second check?

---

## Self-Test A

**Replace __?__ with =, >, or < to make a true statement.**

1. 4 __?__ 7    **2.** 2 __?__ 1    **3.** $^-8$ __?__ 9    **4.** $|^-7|$ __?__ 7    **5.** $|0|$ __?__ 0      [2–1]

**Write the numbers in order from least to greatest.**

6. 0, 5.4, $^-4.52$, $^-0.25$, $^-54$        **7.** $^-3.79$, 37, $^-7.3$, $^-0.37$, $^-0.09$      [2–2]

**Find the sum or difference.**

8. $^-9.3 + 42.3$       **9.** $17.8 + {}^-17.8$       **10.** $8.76 + {}^-10.2$      [2–3]

11. $8 - (-27)$       **12.** $-5.1 - (-5.1)$       **13.** $0 - 36$      [2–4]

**Evaluate the expression when $a = -6.4$ and $b = -5.2$.**

14. $-b - a$       **15.** $a - (-b)$       **16.** $b - |a|$

---

*Self-Test answers and Extra Practice are at the back of the book.*

## 2-5 Multiplying Positive and Negative Numbers

To find a product such as $5(-2)$, we can think of the product as the sum of five identical addends.

$$5(-2) = -2 + (-2) + (-2) + (-2) + (-2)$$

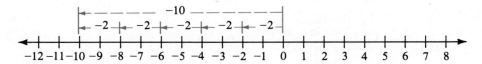

The diagram shows that $-2 + (-2) + (-2) + (-2) + (-2) = -10$. Thus $5(-2) = -10$.

To find the product $-7(2)$, we may use the commutative property of multiplication to write

$$-7(2) = 2(-7).$$

We know that $2(-7) = -7 + (-7) = -14$, so $-7(2) = -14$.

Notice that in the two examples above, the product is the opposite of the product of the absolute values of the numbers. The examples suggest the following rule.

### Rule

The product of a positive number and a negative number is a negative number.

We may use other properties that we have learned for addition and multiplication of positive decimals to determine what a product of negative numbers, such as $-5(-3)$, must be. The multiplication property of zero states that the product of any number and 0 is 0. Thus,

$$-5(0) = 0.$$

Since we know that $3 + (-3) = 0$, we may write

$$-5[3 + (-3)] = 0.$$

By the distributive property, we may write the following.

$$-5(3) + (-5)(-3) = 0$$

$$-15 + (-5)(-3) = 0$$

But we know that $-15 + 15 = 0$, so $-5(-3)$ must equal 15.

Notice that the product of $-5(-3)$ is the product of the absolute values of the factors. The example suggests the following rule.

> ## Rule
>
> The product of two negative numbers is a positive number.

**EXAMPLE 1**  Find the product.
   **a.** $-4.5(8.6)$     **b.** $5.32(-1)$     **c.** $-1(-14.7)$     **d.** $-9.2(-3.1)$

**Solution**  **a.** One number is negative and one number is positive, so the product is negative.
$$-4.5(8.6) = -38.7$$

**b.** One number is positive and one number is negative, so the product is negative.
$$5.32(-1) = -5.32$$

**c.** Both numbers are negative, so the product is positive.
$$-1(-14.7) = 14.7$$

**d.** Both numbers are negative, so the product is positive.
$$-9.2(-3.1) = 28.52$$

Notice in Example 1, parts (b) and (c), that when one of the factors is $-1$, the product is the opposite of the other factor.

> ## Rule
>
> The product of $-1$ and any number equals the opposite of that number.

We may use the rules for products of positive and negative numbers to multiply any number of positive and negative numbers.

**EXAMPLE 2**  Find the product.
   **a.** $-3.7(2.5)(-4.8)$          **b.** $-11.1(-7)(6.5)(-3.2)$

**Solution**  **a.** $-3.7(2.5)(-4.8) = [-3.7(2.5)](-4.8) = (-9.25)(-4.8) = 44.4$

**b.** $-11.1(-7)(6.5)(-3.2) = [-11.1(-7)][(6.5)(-3.2)]$
$$= 77.7(-20.8) = -1616.16$$

Example 2, on the preceding page, illustrates the following rules.

## Rules

For a product with no zero factors:
1. if the number of negative factors is odd, the product is negative.
2. if the number of negative factors is even, the product is positive.

## Class Exercises

**Without computing the exact product, state whether the product is positive, negative, or 0.**

**1.** $-3(-4.2)$      **2.** $2.1(-0.8)$      **3.** $-5(1.6)(-7)$      **4.** $-4.5(3.7)(0)$

**Find the product.**

**5.** $3(-16)$      **6.** $-7(-12)$      **7.** $13(-1)(-5)$      **8.** $-2(-8)(-5)(0)$

## Written Exercises

**Find the product.**

**A**
**1.** $-3(-9)$      **2.** $4(-6)$      **3.** $-8(7)$      **4.** $-7(-11)$

**5.** $2(-8)(-6)$      **6.** $-3(5)(-6)$      **7.** $-2(0)(-12)$      **8.** $14(-1)(0)$

**9.** $-1.5(8)$      **10.** $0.6(-9)$      **11.** $-3.4(-1.5)$      **12.** $-0.4(-0.7)$

**13.** $2.9(-1)$      **14.** $-1(7.84)$      **15.** $-8.8(-1.75)$      **16.** $-1.11(70)$

**17.** $-20(0.25)$      **18.** $12(-1.2)$      **19.** $-15(-30.6)$      **20.** $0.24(-100)$

**21.** $3.9(-17.1)$      **22.** $-5.6(80.1)$      **23.** $-13.7(0)$      **24.** $16.7(0)$

**25.** $-1.8(-1.9)$      **26.** $-0.125(-8.1)$      **27.** $-5.4(20.6)$      **28.** $-8.9(30.9)$

**29.** $10.1(3.75)$      **30.** $4.25(20.4)$      **31.** $-3.72(-16.5)$      **32.** $0.78(-42)$

**33.** $-1.7(-0.2)(-3.1)$      **34.** $-9(-2.7)(-80)$      **35.** $3.25(-17)(0)$

**36.** $-1.21(0)(-1.1)$      **37.** $9.4(-3.5)(-11)$      **38.** $18(-5.75)(6.2)$

**Simplify the expression.**

**B** **39.** $(-1.2 - 6.5)(-1.2 + 6.5)$

**40.** $18 + 3 - (-12 - 7)$

**41.** $(-7)(2.4)(0)(-9.3)(-1) + (-8.2)$

**42.** $(14.4 - 200)(14.4 + 200)$

**43.** $-4[27 - (-9)] + (-4)(-2 - 9)$

**44.** $(-20 + 12)7 + (-2 + 20)7$

**Find the integer $n$ that will make the statement true.**

**45.** $-3(n) = 6$

**46.** $4(n) = -28$

**47.** $-9(-n) = -18$

**48.** $3(n) = 6(-3)$

**49.** $-n(-7) = -14(0)$

**50.** $-5(n) = 25(2)$

**C** **51.** $-3(-1.5)(-n) = -18(-1)(-0.5)$

**52.** $-1.2(30)(-n) = -9(-0.4)(-100)$

**53.** $-1.5(n)(-0.8) = -12(-1.5)(-2)$

**54.** $-0.6(-n)(-1.9) = -18(-1.3)(0)$

**55.** $n(-3.7 + 61.4) = 0$

**56.** $-n(-1.1 - 30.6) = 0$

## Review Exercises

**Use the inverse operation to solve for the variable.**

**1.** $f + 27 = 83$

**2.** $n - 13 = 54$

**3.** $g + 364 = 518$

**4.** $w - 216 = 435$

**5.** $6c = 78$

**6.** $x \div 17 = 6$

**7.** $14j = 364$

**8.** $y \div 37 = 142$

 **Calculator Key-In**

Use your calculator to solve this problem: One day you tell a secret to a friend. The next day your friend tells your secret to two other friends. On the third day, each of the friends who was told your secret the day before tells it to two other friends. If this pattern continues from day to day, how many people will be told your secret on the fourteenth day?

 **Challenge**

Using the first nine counting numbers, fill in the boxes so you get the same sum when you add vertically, horizontally, or diagonally. Can you do this with any nine consecutive counting numbers?

# 2-6 Dividing Positive and Negative Numbers

Recall that multiplication and division are inverse operations for positive numbers. For example, because we know that $4 \times 8 = 32$, we also know that $8 = 32 \div 4$. We can use the relationship between multiplication and division to find quotients of positive and negative numbers. Consider the following examples.

$$4 \times (-8) = -32 \qquad -8 = -32 \div 4$$
$$-4 \times 8 = -32 \qquad 8 = -32 \div -4$$
$$-4 \times (-8) = 32 \qquad -8 = 32 \div -4$$

Notice that in the examples above, the quotient of two numbers with differing signs is the opposite of the quotient of the absolute values of the numbers. The quotient of two numbers with the same sign is the quotient of the absolute values of the numbers.

The examples suggest the following rules for dividing positive and negative numbers.

> ## Rules
>
> The quotient of two positive or two negative numbers is positive.
>
> The quotient of a positive number and a negative number is negative.

By the multiplication property of zero, we know that $-4 \times 0 = 0$ and thus $0 = 0 \div (-4)$. Remember that we cannot divide by 0.

**EXAMPLE** Find the quotient.

a. $-3.06 \div 0.9$     b. $36.8 \div (-2.3)$     c. $-4.046 \div (-1.7)$

**Solution**

a. Since $-3.06$ is negative and $0.9$ is positive, the quotient will be negative.
$$-3.06 \div 0.9 = -3.4$$

b. Since $36.8$ is positive and $-2.3$ is negative, the quotient will be negative.
$$36.8 \div -2.3 = -16$$

c. Since $-4.046$ and $-1.7$ are both negative, the quotient will be positive.
$$-4.046 \div (-1.7) = 2.38$$

## Class Exercises

**Without computing the exact quotient, state whether the quotient is positive, negative, or 0.**

**1.** $-3.6 \div (-40)$     **2.** $-0.216 \div 400$     **3.** $0 \div (-17.5)$     **4.** $850 \div (-0.05)$

**Find the quotient.**

**5.** $-28 \div 7$           **6.** $33 \div (-1)$          **7.** $0 \div -50$

**8.** $-51 \div (-3)$       **9.** $-22 \div 4$          **10.** $-75 \div 15$

---

## Written Exercises

**Find the quotient.**

**A**

**1.** $-18 \div 3$              **2.** $25 \div (-5)$              **3.** $-21 \div (-7)$

**4.** $-54 \div (-18)$        **5.** $0 \div (-7)$               **6.** $0 \div (-24)$

**7.** $144 \div (-12)$         **8.** $-100 \div 25$           **9.** $22.5 \div (-3)$

**10.** $-42 \div 4$            **11.** $-3.6 \div (-1)$         **12.** $-1.01 \div (-1)$

**13.** $-1.75 \div 0.05$       **14.** $-69.3 \div 3.3$         **15.** $-32.86 \div 6.2$

**16.** $-17.05 \div (-1.1)$     **17.** $-0.48 \div (-0.06)$     **18.** $0.06 \div (-0.3)$

**19.** $0 \div (-14.7)$         **20.** $0 \div (-0.25)$        **21.** $-0.9 \div 1.8$

**22.** $-0.042 \div (-0.6)$     **23.** $-38 \div 4$            **24.** $-45 \div (-6)$

**25.** $9.9 \div (-4.5)$        **26.** $46.2 \div (-6)$        **27.** $-13.8 \div (-1)$

**28.** $0.003 \div (-1)$       **29.** $-9.27 \div (-60)$     **30.** $0.25 \div (-40)$

**31.** $13.23 \div (-2.1)$      **32.** $-2.6 \div 0.52$       **33.** $-14.57 \div (-3.1)$

**34.** $-0.53 \div (-0.1)$      **35.** $-18.5 \div 10$       **36.** $3.84 \div (-9.6)$

**Evaluate the expression when $a = -8$ and $b = 2.5$.**

**37.** $a \div b$          **38.** $b \div a$         **39.** $2b \div (-a)$       **40.** $-3a \div b$

**41.** $(a - b) \div b$     **42.** $2ab \div a$       **43.** $-9a \div 3ab$      **44.** $-5b \div (-2a)$

**Use inverse operations to solve for the variable.**

**B**   **45.** $-3n = 6$            **46.** $d \div 5 = -7$        **47.** $b \div (-8) = 9$

**48.** $-15x = -30$       **49.** $-2y - 8 = 8$       **50.** $4c + 12 = -36$

**C**   **51.** Explain why $\left|\dfrac{x}{y}\right| = \dfrac{|x|}{|y|}$ for all decimal numbers for which it is possible to find the quotient $\dfrac{x}{y}$.

**52.** Replace ___?___ with $=$, $>$, or $<$ to make a true statement.

  **a.** $0.10$ ___?___ $2.5$

  $\dfrac{0.10}{0.5}$ ___?___ $\dfrac{2.5}{0.5}$

  $\dfrac{0.10}{-0.5}$ ___?___ $\dfrac{2.5}{-0.5}$

  **b.** $-0.21$ ___?___ $0$

  $\dfrac{-0.21}{70}$ ___?___ $\dfrac{0}{70}$

  $\dfrac{-0.21}{-70}$ ___?___ $\dfrac{0}{-70}$

  **c.** $-3.6$ ___?___ $-1.5$

  $\dfrac{-3.6}{3}$ ___?___ $\dfrac{-1.5}{3}$

  $\dfrac{-3.6}{-3}$ ___?___ $\dfrac{-1.5}{-3}$

Use your answers to parts (a)–(c) to answer parts (d) and (e).

  **d.** If $x < y$ and if $k$ is a positive number, then $\dfrac{x}{k}$ ___?___ $\dfrac{y}{k}$.

  **e.** If $x < y$ and if $j$ is a negative number, then $\dfrac{x}{j}$ ___?___ $\dfrac{y}{j}$.

Write examples similar to those in parts (a)–(c) to answer parts (f) and (g).

  **f.** If $x > y$ and if $k$ is a positive number, then $\dfrac{x}{k}$ ___?___ $\dfrac{y}{k}$.

  **g.** If $x > y$ and if $j$ is a negative number, then $\dfrac{x}{j}$ ___?___ $\dfrac{y}{j}$.

---

## Review Exercises

**Evaluate the expression when $x = 3$, $y = 7$, and $z = 4$.**

**1.** $y^2$

**2.** $5x^2$

**3.** $(7z)^2$

**4.** $z^0$

**5.** $7z^2$

**6.** $(6x)^2$

**7.** $(yz)^0$

**8.** $xy^0$

**9.** $z^3 - x^3$

**10.** $8x^3y$

---

### ▮▮▮ Calculator Key-In

Does your calculator have a change-sign key? The key may look like this: ⊬. If you press this key after entering a number or doing a calculation, the sign of the number displayed on your calculator will change. For example, if you enter 116 ⊬, your calculator will change 116 to $-116$.

**Solve with a calculator that has a change-sign key, if possible.**

**1.** $20.7 + (-19.6)$

**2.** $-55.59 + 438.2$

**3.** $-0.86 + (-27.341)$

**4.** $-426.38 - (-25.004)$

**5.** $-83.5(-61.09)$

**6.** $6.8(-4.17)(-1.61)$

# 2-7 Using Positive Exponents

In Chapter 1, exponents were introduced. Recall that in the expression $3^5$ (called a *power*), 3 is called the *base* and 5 is called the *exponent*.

If a product contains powers of the same base, the product may be written as a single power of that base. For example, $13^2 \times 13^3$ can be written as a single power of 13.

$$13^2 \times 13^3 = (13 \times 13) \times (13 \times 13 \times 13)$$
$$= 13 \times 13 \times 13 \times 13 \times 13$$
$$= 13^5$$

Notice that the exponent in the product is the sum of the exponents in the factors, that is, $2 + 3 = 5$.

In general,

> ## Rule
>
> For every number $a$ ($a \neq 0$) and all whole numbers $m$ and $n$,
> $$a^m \times a^n = a^{m+n}$$

Notice that the bases must be the same.

**EXAMPLE 1**   Write $15^3 \times 15^4$ as a single power of 15.

**Solution**   $15^3 \times 15^4 = 15^{3+4} = 15^7$

**EXAMPLE 2**   Evaluate the expression if $n = 3$.

    **a.** $n^2$      **b.** $4n^2$      **c.** $(4n)^2$      **d.** $n^2 \times n^2$

**Solution**   Replace $n$ with 3 in each expression and simplify.

    **a.** $n^2 = 3^2 = 3 \times 3 = 9$

    **b.** $4n^2 = 4(3^2) = 4 \times 9 = 36$

    **c.** $(4n)^2 = (4 \times 3)^2 = 12^2 = 12 \times 12 = 144$

    **d.** $n^2 \times n^2 = n^{2+2} = n^4 = 3^4 = 3 \times 3 \times 3 \times 3 = 81$

Notice in parts (b) and (c) of Example 2 how grouping symbols change the values of expressions that have the same numbers.

---

**Reading Mathematics: *Study Helps***
Look back at this lesson. Notice that the information in the blue box on page 70 summarizes important ideas from the lesson. The box gives a definition that is applied in the examples. Throughout the book, boxes are used to help you identify important definitions, rules, properties, facts, and formulas. Use them as reminders when you do the exercises and when you review the lesson.

---

## Class Exercises

**Read each expression.**

**1.** $4^5$      **2.** $9^1$      **3.** $15^2$      **4.** $3^7$      **5.** $10^3$      **6.** $2^8$

**Write using exponents.**

**7.** 9 to the third power      **8.** 15 cubed      **9.** 4 squared

**10.** 6 to the fifth power      **11.** 216 is the third power of 6.

**Express the number as a power of 3.**

**12.** 9      **13.** 27      **14.** 3      **15.** 243      **16.** 1

**Simplify the expression.**

**17.** $8^2$      **18.** $2^3$      **19.** $1^{11}$      **20.** $18^0$      **21.** $83^1$

---

## Written Exercises

**Simplify the expression.**

**A**    **1.** $2^6$      **2.** $5^4$      **3.** $10^2$      **4.** $6^3$      **5.** $14^1$

     **6.** $3^2 + 5^2$      **7.** $(3 \times 5)^2$      **8.** $2^4 + 3^2$      **9.** $(5 + 12)^0$      **10.** $(5 + 12)^1$

**Which is greater?**

**11.** $2^3$ or $3^2$      **12.** $5^2$ or $2^5$      **13.** $9 \times 2$ or $9^2$

**14.** $3 \times 10$ or $10^3$      **15.** $(16 \times 4)^2$ or $2 \times 16 \times 4$      **16.** $(10 + 2)^0$ or $10 + 2$

**Write as a single power of the given base.**

**17.** $2^3 \times 2^4$    **18.** $3^2 \times 3^5$    **19.** $10 \times 10^4$    **20.** $5^5 \times 5^6$    **21.** $n^3 \times n^8$

**Evaluate the expression when $m = 5$, $n = 3$, and $p = 2$.**

**22.** $p^2$    **23.** $4m^2$    **24.** $(9n)^2$    **25.** $9n^2$    **26.** $n^0$

**27.** $(8n)^2$    **28.** $np^0$    **29.** $(mn)^0$    **30.** $m^3 - n^3$    **31.** $5m^3n$

**B**  **32.** $(7n)^n$    **33.** $(3m)^p$    **34.** $6^2 \times 6^n$    **35.** $(8 + m)^{n-3}$    **36.** $(15^n)^{p-1}$

**37.** $p^n m^n$    **38.** $(7 + n)^n$    **39.** $np^m$    **40.** $(p^m p^n) + 4$    **41.** $m^p + n^n$

**42.** $(m^p)^n$    **43.** $(m - n)^p$    **44.** $\dfrac{m^n}{m^p}$    **45.** $\dfrac{3p^n}{6p^m}$    **46.** $\dfrac{(m-1)^{n+1}}{p}$

**C**  **47.** Find a value of $n$ such that $(5 + 2)^n = 5^n + 2^n$.

**48.** Is the equation true?

    **a.** $4^4 \div 4^3 = 4^1$        **b.** $5^3 \div 5^1 = 5^2$        **c.** $2^7 \div 2^4 = 2^3$

    **d.** Using your answers from parts (a)–(c), state a general rule to describe what appears to be true for division of powers of the same base.

---

## Review Exercises

**Solve using inverse operations.**

**1.** $8 + x = 6$    **2.** $y + 4 = 1$    **3.** $n + 6 = 8$    **4.** $y + 10 = 6$

**5.** $-7 + a = 5$    **6.** $t + 3 = -8$    **7.** $-2 + c = -1$    **8.** $-13 + x = 0$

---

### ▌▌▌▌ Calculator Key-In

**Use a calculator to simplify the expressions.**

**1.** $15^2 - 13^2$    and    $(15 + 13)(15 - 13)$

**2.** $47^2 - 21^2$    and    $(47 + 21)(47 - 21)$

**3.** $82^2 - 59^2$    and    $(82 + 59)(82 - 59)$

**4.** $104^2 - 76^2$    and    $(104 + 76)(104 - 76)$

Do you recognize a pattern?
Write two expressions that will result in the same pattern.

## 2-8 Negative Integers as Exponents

You know by the rule of exponents that you learned for multiplying powers of the same base that

$$10^1 \times 10^2 = 10^{1+2} = 10^3.$$

Since we want to apply the same rule to negative exponents, we must have

$$10^1 \times 10^{-1} = 10^{1+(-1)} = 10^0 = 1$$
$$10^2 \times 10^{-2} = 10^{2+(-2)} = 10^0 = 1$$

and so on.  We know that

$$10^1 \times \frac{1}{10} = 10 \times 0.1 = 1 \text{ and } 10^2 \times \frac{1}{10^2} = 100 \times 0.01 = 1,$$

so $10^{-1}$ should equal $\frac{1}{10}$ and $10^{-2}$ should equal $\frac{1}{10^2}$.  These examples suggest the following general rule.

---

### Rule

For all numbers $a(a \neq 0)$, $m$, and $n$,

$$a^{-m} = \frac{1}{a^m}$$

---

**EXAMPLE**  Write the expression without exponents.

  **a.** $5^{-2}$  **b.** $(-3)^{-2}$  **c.** $(-4)^{-1}(-4)^{-2}$

**Solution**

**a.** $5^{-2} = \frac{1}{5^2} = \frac{1}{5 \times 5} = \frac{1}{25}$

**b.** $(-3)^{-2} = \frac{1}{(-3)^2} = \frac{1}{(-3)(-3)} = \frac{1}{9}$

**c.** $(-4)^{-1} \times (-4)^{-2} = (-4)^{-1+(-2)} = (-4)^{-3}$
$$= \frac{1}{(-4)^3} = \frac{1}{(-4)(-4)(-4)} = \frac{1}{-64}$$

---

## Class Exercises

**Use the rules for exponents to state the expression without exponents.**

**1.** $3^{-4}$  **2.** $(-6)^{-2}$  **3.** $10^4 \times 10^{-4}$  **4.** $3^5 \times 3^{-7}$  **5.** $(-2)^3(-2)^{-1}$

**Use exponents to state as a power of 2.**

**6.** 8      **7.** $\frac{1}{8}$      **8.** 64      **9.** $\frac{1}{64}$      **10.** $\frac{1}{512}$

---

## Written Exercises

**Write the expression without exponents.**

**A**   **1.** $(-2)^{-5}$    **2.** $3^{-3}$    **3.** $10^{-3}$    **4.** $(-3)^{-5}$    **5.** $1^{-4}$

**6.** $(-1)^{-6}$    **7.** $(-5)^{-2}$    **8.** $4^{-5}$    **9.** $2^{-6}$    **10.** $(-4)^{-2}$

**11.** $7^4 \times 7^{-6}$      **12.** $10^3 \times 10^{-2}$      **13.** $5^{10} \times 5^{-10}$

**14.** $6^{-23} \times 6^{23}$      **15.** $3^{-3} \times 3^0$      **16.** $2^{-3} \times 2^{-4}$

**17.** $(-4)^{-2} \times (-4)^{-2}$      **18.** $(-7)^{-1} \times (-7)^{-1}$      **19.** $(-2)^{-6} \times (-2)^3$

**20.** $(-8)^{-2} \times (-8)^0$      **21.** $6^{-1} \times 6^3 \times 6^{-2}$      **22.** $9^{-5} \times 9^{-1} \times 9^7$

**What value of the variable makes the statement true?**

**23.** $5^n = \frac{1}{125}$      **24.** $4^{-n} = \frac{1}{256}$      **25.** $3^{-n} = \frac{1}{243}$

**26.** $4^2 \times 4^{-2} = 4^n$      **27.** $7^3 \times 7^{-5} = 7^n$      **28.** $9^{-4} \times 9^3 = \frac{1}{9^n}$

**B**   **29.** $3^7 \times 3^n = 3^5$      **30.** $2^{-3} \times 2^n = 2^{-11}$

**31.** $(2)^{-5} \times (2)^n = 8$      **32.** $(-10)^3 \times (-10)^{-n} = -10$

**33.** $144 \times 12^{-2} = 12^n$      **34.** $5^{-3} \times 25 = 5^{-n}$

**35.** $4^n \times 4^{-3} = \frac{1}{16}$      **36.** $6^{-n} \times 6^3 = \frac{1}{216}$

**37.** $9^{-7} \times 9^{-n} = \frac{1}{729}$      **38.** $8^{-4} \times 8^{-n} = \frac{1}{64}$

**39.** $(-5)^{-n} \times (-5)^{-3} = 1$      **40.** $(-3)^{-n} \times (-3)^{-8} = \frac{1}{-243}$

**Simplify. Write the expression with nonnegative exponents.**

**41.** $x^{-5}$      **42.** $n^{-9}$      **43.** $a^{-3} \times a^{-2}$

**44.** $b^7 \times b^{-7}$      **45.** $w^{-10} \times w^3 \times w^{-1}$      **46.** $v^4 \times v^{-12} \times v^3$

**C**   **47.** Explain why $a^m = (-a)^m$ if $m$ is any even integer.

**48.** Explain why $(-a)^n = -1(a)^n$ if $n$ is any odd integer.

## Self-Test B

**Simplify.**

**1.** $4.2(-11.3)$     **2.** $-6.7(20.4)$     **3.** $7.5(-4.2)(-12)$     [2-5]

**4.** $121 \div (-11)$     **5.** $-68.2 \div 2.2$     **6.** $-0.56 \div (-0.07)$     [2-6]

**Evaluate the expression when $a = 2$, $b = 5$, and $c = 3$.**

**7.** $a^3$     **8.** $(bc)^2$     **9.** $(2c)^4$     [2-7]

**Write the expression without exponents.**

**10.** $4^{-2}$     **11.** $(-6)^{-3}$     **12.** $7^5 \times 7^{-8}$     **13.** $(-9)^{-2} \times (-9)^0$     [2-8]

*Self-Test answers and Extra Practice are at the back of the book.*

---

## Challenge

We use the symbol $[x]$ (read *the greatest integer in x*) to represent the greatest integer less than or equal to $x$.

*EXAMPLE*    **a.** $[5.4]$     **b.** $[^-3.2]$

*Solution*    **a.** There is no integer equal to 5.4, so we must find the greatest integer that is less than 5.4.

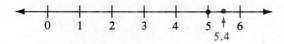

As shown on the number line, the greatest integer that is less than 5.4 is 5. Thus the greatest integer in 5.4 is 5.

**b.** There is no integer equal to $^-3.2$, so we must find the greatest integer that is less than $^-3.2$.

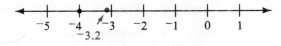

As shown on the number line, the greatest integer that is less than $^-3.2$ is $^-4$. Thus the greatest integer in $^-3.2$ is $^-4$.

**Find the value.**

**1.** $[6.2]$     **2.** $[1.23]$     **3.** $[3]$     **4.** $[45]$

**5.** $[^-12]$     **6.** $[^-1]$     **7.** $[^-4.89]$     **8.** $[^-0.36]$

# Scientific Notation

Scientists frequently deal with data that range from very small to very large magnitudes. For example, when Saturn is closest to Earth, it is about 1,630,000,000 km away. The diameter of a hydrogen atom is approximately $\frac{1}{100,000,000}$ cm. To cope with numbers such as these, a method for writing numbers, called **scientific notation,** has been adopted.

Scientific notation makes use of positive exponents to write large numbers and negative exponents to write small numbers. For example,

$$4800 = 4.8 \times 1000 = 4.8 \times 10^3$$

$$0.000507 = 5.07 \times \frac{1}{10,000} = 5.07 \times 10^{-4}$$

## *Rule*

To express any positive number in scientific notation, write it as the product of a power of ten and a number between 1 and 10.

In addition to being a convenient method for expressing very large or very small numbers, scientific notation provides an exact gauge of the precision of a measurement, based on the smallest unit of calibration on the measuring instrument. Each digit in a number that specifies the degree of precision of measurement is called a **significant digit.**

Zeros that appear to the right of nonzero digits, and to the right of the decimal point, are significant. For example,

0.50 has two significant digits,

40,521 has five significant digits.

The zeros in a measurement such as 41,500 km, however, may be misleading since it is unclear whether the number is rounded to the nearest hundred or is an exact measurement. Scientific notation provides a means of avoiding this confusion. For example, when we write 40,500 as $4.05 \times 10^4$, it means that the measurement is precise to three significant digits. When we write 40,500 as $4.050 \times 10^4$, it means that the measurement is precise to four significant digits.

In general, to write a number in scientific notation, shift the decimal point to just after the first nonzero digit. Then multiply by $10^n$, when $n$ is the number of places the decimal point was shifted. As an example,

$$3\underbrace{165} = 3.165 \times 10^3.$$

Note that 7.46 is written as 7.46 since $10^0 = 1$. Also, 1,000,000 is usually written simply as $10^6$ rather than $1 \times 10^6$.

**Write the number in scientific notation.**

1. 5798          2. 30,090          3. 8,915,673          4. 2,175,000,000

5. 1.75          6. 0.003          7. 0.0501          8. 0.0333

**Write the number in decimal form.**

9. $3.79 \times 10^3$          10. $4.86 \times 10^4$          11. $3.01 \times 10^5$          12. $6 \times 10^9$

13. $5.6 \times 10^{-2}$          14. $7.09 \times 10^{-3}$          15. $3.99 \times 10^{-8}$          16. $2.0111 \times 10^{-6}$

17. The diameter of a red blood cell is about 0.00074 cm. Write this number in scientific notation with two significant digits.

18. An atom of gold is about 0.0000000025 m in diameter. Write this number in scientific notation with two significant digits.

19. The radius of Earth's orbit is 150,000,000,000 m. Write this number in scientific notation with two significant digits.

20. A communications satellite was orbited at an altitude of 625,000 m. Write this number in scientific notation with three significant digits.

# Chapter Review

**Complete. Use =, >, or < to make a true statement.**

1. $2 \underline{\;?\;} 11$

2. $5 \underline{\;?\;} 3$

3. $3 \underline{\;?\;} {}^-4$      [2–1]

4. $0 \underline{\;?\;} {}^-1$

5. $|{}^-9| \underline{\;?\;} 9$

6. $|2| \underline{\;?\;} {}^-2$

7. ${}^-8.7 \underline{\;?\;} {}^-0.87$

8. ${}^-42 \underline{\;?\;} 2.4$

9. $3.05 \underline{\;?\;} -3.55$      [2–2]

10. $0.4 \underline{\;?\;} {}^-4.3$

11. $|{}^-5.6| \underline{\;?\;} {}^-5.6$

12. $|4.93| \underline{\;?\;} |{}^-4.93|$

**True or false?**

13. $0 + {}^-14.2 = 0$

14. $16.8 + {}^-16.8 = 33.6$      [2–3]

15. ${}^-13.2 + {}^-7.8 = {}^-21$

16. $7.6 + {}^-10.5 = 2.9$

17. ${}^-33 + 20.2 = {}^-12.8$

18. $19.5 + {}^-14.3 = {}^-5.2$

19. $37.2 - (-9.6) = 25.6$

20. $-5.8 - (-5.8) = 11.6$      [2–4]

21. $-12.2 - 13.1 = -25.3$

22. $0 - (-0.5) = -0.05$

23. If $a = -7$, $-a = -7$

24. If $b = -2.4$, $-|b| = -2.4$

25. $-40(0.33) = -1.42$

26. $1.2(-6.2) = -7.4$      [2–5]

27. $-17(-24.2) = 411.4$

28. $7(-8.3)0 = -58.1$

29. $-5(2.8)(-20) = 280$

30. $-12(-1)(-8.6) = 103.2$

31. $75.5 \div (-5) = -15.1$

32. $-0.006 \div (-1) = 0.006$      [2–6]

33. $-115.2 \div (-2.4) = 48$

34. $-5.04 \div 3.6 = 1.4$

35. $0 \div (-19.8) = 0$

36. $-6.21 \div (-0.23) = -27$

**What value of the variable makes the statement true? Write the letter of the correct answer.**

37. $2^n = 32$      **a.** 16    **b.** $-5$    **c.** 5    **d.** 30      [2–7]

38. $5^6 \times 5^2 = 5^n$      **a.** 4    **b.** 8    **c.** 12    **d.** 36

39. $7^n = 1$      **a.** 7    **b.** $-6$    **c.** 6    **d.** 0

40. $3^n = \frac{1}{81}$      **a.** 4    **b.** 81    **c.** $-4$    **d.** 9      [2–8]

41. $4^{-n} = \frac{1}{64}$      **a.** 16    **b.** $-16$    **c.** 4    **d.** 3

42. $7^{-6} \times 7^6 = 7^n$      **a.** $-36$    **b.** 0    **c.** 12    **d.** $-12$

# Chapter Test

Replace __?__ with =, >, or < to make a true statement.

**1.** 7 __?__ 10      **2.** 6 __?__ 1      **3.** 0 __?__ ⁻6      **4.** ⁻2 __?__ ⁻3      **[2–1]**

Express as an integer.

**5.** |⁻3|      **6.** |7|      **7.** |⁻12|      **8.** |0|

Write the numbers in order from least to greatest.

**9.** ⁻6.5, ⁻56, 6.05, ⁻556, ⁻0.6      **10.** 3.02, ⁻3.2, ⁻23, 0.32, ⁻333      **[2–2]**

Find the sum.

**11.** ⁻8.4 + 36.8      **12.** ⁻6.3 + ⁻0.12      **13.** 13.2 + ⁻13.2      **[2–3]**

**14.** 14.6 + 23.1      **15.** 0 + ⁻11.5      **16.** 0.89 + ⁻16.1 + ⁻0.94

Find the difference.

**17.** 26.5 − 8.3      **18.** −4.3 − 20.6      **19.** 0 − 13.6      **[2–4]**

**20.** −14.2 − (−9.5)      **21.** 41 − (−11.67)      **22.** −6.4 − (−6.4)

Evaluate the expression when $a = -5$ and $b = -3.6$.

**23.** $-a - b$      **24.** $-a - (-b)$      **25.** $-|b|$

Find the product.

**26.** 12(−6.37)      **27.** −30(0.45)      **28.** −0.37(−20.8)      **[2–5]**

**29.** −1(−14.27)      **30.** 5.4(−8.2)(−3)      **31.** −6.11(−9)(−5.5)

Find the quotient.

**32.** 69.3 ÷ (−3)      **33.** −18 ÷ (−2.5)      **34.** −19.2 ÷ 10      **[2–6]**

**35.** −0.004 ÷ (−1)      **36.** 0 ÷ (−15)      **37.** −0.08 ÷ (−0.2)

Write the expression without exponents.

**38.** $5^3$      **39.** $20^0$      **40.** $4^2 \times 4^3$      **41.** $2^6$      **[2–7]**

**42.** $6^{-2}$      **43.** $(-5)^{-3}$      **44.** $8^{-9} \times 8^7$      **45.** $(-3) \times (-3)^{-2}$      **[2–8]**

# Cumulative Review (Chapters 1 and 2)

## Exercises

**Use inverse operations to solve.**

**1.** $9x = 54$

**2.** $17 + a = 20$

**3.** $a \div 6 = 30$

**4.** $35 - b = 7$

**5.** $15z = 45$

**6.** $w \div 12 = 96$

**7.** $18 + p = 35$

**8.** $n - 5 = 19$

**True or false?**

**9.** $46.7 \times 1.0 = 46.7$

**10.** $87.91 \times 0 = 87.91$

**11.** $4(8 + 2) = 4 \times 8 + 4 \times 2$

**12.** $9 \times 7 - 9 \times 3 = 9 \times (7 - 3)$

**13.** $84 - (2 + 3) = (84 - 2) + 3$

**14.** $63 \times (9 \times 0) = (63 \times 9) \times 0$

**15.** $21(6 \times 3) = (21 \times 6) + (21 \times 3)$

**16.** $(68 - 7) = (7 - 68)$

**17.** $(658 + 15)7 = (7 \times 658) + (7 \times 15)$

**Find the sum.**

**18.** $^-8 + 4$

**19.** $^-3.21 + {^-2.97}$

**20.** $^-4.1 + 6.6 + {^-1.9}$

**Find the difference.**

**21.** $6 - 14$

**22.** $11 - (-13)$

**23.** $1.6 - (-2.7)$

**Find the product.**

**24.** $-7(-11)$

**25.** $-0.2(0.6)$

**26.** $3.7(-2.4)$

**Find the quotient.**

**27.** $-27 \div 3$

**28.** $-66.3 \div (-3)$

**29.** $-0.96 \div 0.6$

**Round to the place specified.**

**30.** tenths: 68.461

**31.** thousandths: 4.00891

**32.** tens: 188.72

**33.** hundreds: 7740.68

**34.** hundredths: 37.5505

**35.** tenths: 909.09

**Simplify the expression.**

**36.** $75.8 + (6.7 + 3.3)^2$

**37.** $100.6 + (3^3 \div 9)$

**38.** $(4^3 - 6) \times (81 - 5^2)$

**39.** $10^3 \times 10^7$

**40.** $2^4 \times 3^2$

**41.** $(2^2)^3$

**Express as an integer.**

**42.** $|-2|$

**43.** $|6|$

**44.** $|0|$

**45.** $|-100|$

**46.** $|-15|$

**Write the integers in order from least to greatest.**

**47.** 12, 15, 0, $-7$, 6, $-20$    **48.** 0, $-5$, 6, $-6$, 5    **49.** 7, 9, $-3$, $-5$, 0, 1

---

## Problems

> **Problem Solving Reminders**
> Here are some reminders that may help you solve some of the problems on this page.
> • Determine which facts are necessary to solve the problem.
> • Determine which operations are needed to solve the problem.
> • Supply additional information if necessary.

**Solve.**

1. At Angler's Supply Company, deluxe waders cost $69.95 a pair. A similar product can be purchased for $10.49 less at Go Fish Discount. What is the price at Go Fish Discount?

2. Alice's Mountain Goat Cheese is currently being promoted through supermarket taste demonstrations. An average of 200 samples are distributed between 11:00 A.M. and 5:00 P.M. each day. To the nearest whole number, find the average number of samples given out each hour. If the daily cost of samples averages $50, what is the average cost per sample?

3. Cabin Fever Ski Area runs a mountain slide ride during the summer months. A single ride costs $3.75 and a day's pass costs $14.00. How many times would you have to ride the slide to make it less expensive to buy a day's pass?

4. Approximately 3.3 Calories are burned per hour per kilogram of body mass by walking. If a student with mass 58 kg takes 20 min to walk to school, how many Calories are burned?

5. The Gourmet Luncheonette is giving out coupons that let customers buy one large roast beef sandwich at the regular price of $2.45 and receive one free. Deli Delights is giving out coupons that let customers deduct $.75 from the regular price of $1.95 for every large roast beef sandwich. Is it less expensive to buy two large roast beef sandwiches at The Gourmet Luncheonette or at Deli Delights?

6. Because of a grain shortage, the price of a loaf of bread increased by 2 cents. When grain again became plentiful, the price decreased by 4 cents, but later it again rose by 5 cents because of inflation. If the original price was 86 cents per loaf, what was the final price?

# 3

# Rational Numbers

Hummingbirds, like the Violet-Capped Woodnymph Hummingbird shown at the right, can fly forward, vertically, and even backward. Perhaps their most unusual feat is their ability to hover apparently motionless in the air while sipping nectar from a flower. This ability comes from their specialized wing structure and its unique movement. The hummingbird pictured lives in the forests of Brazil. It is only $4\frac{1}{2}$ in. long and has a wing beat of 33 beats per second. To photograph the hummingbird so that its wings appear motionless, an electronic flash time of about $\frac{1}{1000}$ of a second was used. Exposure times of $\frac{1}{100,000}$ of a second have been used to study the precise motion of some hummingbirds' wings and to learn exactly how the wings enable hummingbirds to remain at a fixed point with great ease. In this chapter on rational numbers, you will learn about fractions and decimals and their relationship to each other.

## Career Note

Understanding the relationship between animals and their environments is the job of the ecologist. Ecologists study the influence that factors such as temperature, humidity, rainfall, and altitude have on the environment. They monitor levels of pollutants and predict their long term effects on the life cycles of plants and animals.

# 3-1 Factors and Divisibility

A **multiple** of a whole number is the product of that number and any whole number. You can find the multiples of a given whole number by multiplying that number by 0, 1, 2, 3, 4, and so on. For example, the first four multiples of 7 are 0, 7, 14, and 21, since:

$$0 \cdot 7 = 0 \qquad 1 \cdot 7 = 7 \qquad 2 \cdot 7 = 14 \qquad 3 \cdot 7 = 21$$

Any multiple of 2 is called an **even number.** A whole number that is not an even number is called an **odd number.** Notice that since

$$0 = 0 \cdot 2,$$

0 is a multiple of 2 and is therefore an even number.

In general, any number is a multiple of each of its factors. For example, 21 is a multiple of 7 and of 3.

You know that 60 can be written as the product of 5 and 12. Whenever a number, such as 60, can be written as the product of two whole numbers, such as 5 and 12, these two numbers are called **whole number factors** of the first number. A number is said to be **divisible** by its whole number factors. Thus, 60 is divisible by the factors 5 and 12.

To find out if a smaller whole number is a factor of a larger whole number, we divide the larger number by the smaller. If the remainder is 0, then the smaller number is a factor of the larger number. If the remainder is *not* 0, then the smaller number is *not* a factor of the larger number.

**EXAMPLE 1**  State whether or not the smaller number is a factor of the larger.

**a.** 6; 138                           **b.** 8; 154

**Solution**

**a.** Divide 138 by 6.

$$138 \div 6 = 23 \text{ R } 0$$

Since the remainder is 0, 6 is a factor of 138.

**b.** Divide 154 by 8.

$$154 \div 8 = 19 \text{ R } 2$$

Since the remainder is not 0, 8 is not a factor of 154.

**EXAMPLE 2**  Find all the factors of 24.

**Solution**  Try each whole number as a divisor, starting with 1.

| | |
|---|---|
| $24 \div 1 = 24$ | Thus, 1 and 24 are factors. |
| $24 \div 2 = 12$ | Thus, 2 and 12 are factors. |
| $24 \div 3 = 8$ | Thus, 3 and 8 are factors. |

$24 \div 4 = 6$    Thus, 4 and 6 are factors.
$24 \div 5 = 4 \text{ R } 4$    Thus, 5 is not a factor.

Since $24 \div 6 = 4$, the factors begin to repeat, and we do not have to try any whole number greater than 5 as a divisor. Thus, the factors of 24 are 1, 2, 3, 4, 6, 8, 12, and 24.

Sometimes it may be possible to find the factors of a number by an inspection of the digits of the number. For example, let us

consider multiples of 2: 0, 2, 4, 6, 8, 10, 12, 14, 16, . . .
consider multiples of 5: 0, 5, 10, 15, 20, . . .
consider multiples of 10: 0, 10, 20, 30, 40, . . .

From the patterns we see in the last digits of the sets of multiples above, we can devise the following tests for divisibility.

> Divisibility by 2: A whole number has 2 as a factor if its last digit has 2 as a factor.
>
> Divisibility by 5: A whole number has 5 as a factor if its last digit is 5 or 0.
>
> Divisibility by 10: A whole number has 10 as a factor if its last digit is 0.

Suppose we want to check whether a number, such as 712, is divisible by 4. Since any multiple of 100 is divisible by 4, we know that 700 is divisible by 4. To test 712, then, we simply look at the last two digits. Since 12 is a multiple of 4, the number 712 is also a multiple of 4 ($712 \div 4 = 178$). A similar inspection shows that 950 is not a multiple of 4, since the number represented by the last two digits, 50, is not a multiple of 4 ($950 \div 4$ gives 237 R 2). This suggests the following test for divisibility.

> Divisibility by 4: A whole number has 4 as a factor if its last two digits represent a multiple of 4.

**EXAMPLE 3**  Test each number for divisibility by 2, 4, 5, and 10.

**a.** 35 **b.** 150 **c.** 7736 **d.** 920

**Solution**

**a.** 35: Since the last digit is 5 and is odd, 2 and 10 are not factors. Since 35 is not a multiple of 4, 4 is not a factor. Since the last digit is 5, 5 is a factor.

**b.** 150: Since 0 is the last digit, 2, 5, and 10 are factors. 4 is not a factor, since 50 is not a multiple of 4.

**c.** 7736: 2 is a factor, since the last digit, 6, is even. 4 is a factor, since 36 is a multiple of 4. 5 and 10 are not factors, since the last digit is not 5 or 0.

**d.** 920: 2, 5, and 10 are all factors, since the last digit is 0. 4 is a factor, since the last two digits, 20, represent a multiple of 4.

Rules for recognizing numbers divisible by 3 or 9 are a bit more difficult to discover. The rules relate to the sum of the digits of the number. Study the following numbers.

|  | Divisible by 3 | Divisible by 9 | Sum of digits |
|---|---|---|---|
| 393 | yes | no | 15 |
|  | $393 \div 3 = 131$ | $393 \div 9$ gives 43 R 6 |  |
| 394 | no | no | 16 |
|  | $394 \div 3$ gives 131 R 1 | $394 \div 9$ gives 43 R 7 |  |
| 395 | no | no | 17 |
|  | $395 \div 3$ gives 131 R 2 | $395 \div 9$ gives 43 R 8 |  |
| 396 | yes | yes | 18 |
|  | $396 \div 3 = 132$ | $396 \div 9 = 44$ |  |

Notice that when the sum of the digits of the number is divisible by 3 (15 or 18), the number (393 or 396) is divisible by 3. Notice also that when the sum of the digits is divisible by 9 (18), the number (396) is divisible by 9. This illustrates the following tests.

> Divisibility by 3: A whole number has 3 as a factor if the sum of the digits of the number is a multiple of 3.
>
> Divisibility by 9: A whole number has 9 as a factor if the sum of the digits of the number is a multiple of 9.

**EXAMPLE 4**  Test each number for divisibility by 3 and 9.

**a.** 714        **b.** 6291        **c.** 4813

**Solution**   **a.** 7 + 1 + 4 = 12. Since 12 is a multiple of 3 but not a multiple of 9, 714 is divisible by 3, but not by 9.

**b.** 6 + 2 + 9 + 1 = 18. Since 18 is a multiple of 3 and of 9, 6291 is divisible by 3 and by 9.

**c.** 4 + 8 + 1 + 3 = 16. Since 16 is not a multiple of 3 or of 9, 4813 is not divisible by 3 or by 9.

## Class Exercises

**State all the factors of each number.**

**1.** 6          **2.** 10          **3.** 20          **4.** 18          **5.** 30

**6.** What number is a factor of every whole number?

**Write the first five multiples of each number.**

**7.** 9                **8.** 14                **9.** 15                **10.** 18

**Test each number for divisibility by 2.**

**11.** 130                **12.** 4681                **13.** 105                **14.** 3576

**Test each number for divisibility by 4.**

**15.** 8310                **16.** 712                **17.** 86,222                **18.** 5732

**Test each number for divisibility by 5 and 10.**

**19.** 8325                **20.** 7602                **21.** 870                **22.** 6395

**Test each number for divisibility by 3 and 9.**

**23.** 175                **24.** 288                **25.** 651                **26.** 8766

## Written Exercises

**List all the factors of each number.**

**A**   **1.** 42                **2.** 45                **3.** 32                **4.** 40                **5.** 56

**6.** 31                **7.** 84                **8.** 51                **9.** 41                **10.** 112

**State which of the numbers 2, 3, 4, 5, 9, and 10 are factors of the given number. Use the tests for divisibility.**

**11.** 132      **12.** 150      **13.** 195      **14.** 4280      **15.** 567

**16.** 8155      **17.** 43,260      **18.** 720      **19.** 1147      **20.** 78,921

**For each number, determine whether (a) 2 is a factor, (b) 3 is a factor, and (c) 6 is a factor. What appears to be true in order for 6 to be a factor?**

**21.** 1316      **22.** 2,817,000      **23.** 31,027,302

**24.** 1224      **25.** 2,147,640      **26.** 36,111,114

**Supply the missing digit of the first number if it is known to have the other two numbers as factors.**

**B**   **27.** 35?; 2, 3      **28.** 876?; 2, 5      **29.** 910?; 3, 5

    **30.** 472?; 3, 4      **31.** 61?2; 4, 9      **32.** 47?2; 4, 9

**33.** Any multiple of 1000 is divisible by 8. Use this fact to devise a test for divisibility by 8.

**34.** The total number of pages in a book must be a multiple of 32. If the book consists of 10 chapters, each 24 pages long, and 8 pages of introductory material, how many blank pages will be left?

**C**   **35.** Devise a test for divisibility by 25.

**36.** A **perfect number** is one that is the sum of all of its factors except itself. The smallest perfect number is 6, since $6 = 1 + 2 + 3$. Find the next perfect number.

---

## Review Exercises

**Complete with $<$, $>$, or $=$.**

**1.** 80 __?__ 800      **2.** 136 __?__ 119      **3.** 21.6 __?__ 2.29      **4.** 0.87 __?__ 0.0941

**5.** 3.081 __?__ 3.101      **6.** 48.88 __?__ 49.17      **7.** $3^3$ __?__ $(2 + 1)^2$      **8.** $2^3 + 1$ __?__ $(2 + 1)^3$

# 3-2 Prime Numbers and Composite Numbers

Consider the list of counting numbers and their factors given at the right. Notice that each of the numbers 2, 3, 5, 7, and 11 has *exactly* two factors: 1 and the number itself. A number with this property is called a **prime number.** A counting number that has more than two factors is called a **composite number.** In the list 4, 6, 8, 9, 10, and 12 are all composite numbers. Since 1 has exactly one factor, it is neither prime nor composite.

| Number | Factors |
|--------|---------|
| 1 | 1 |
| 2 | 1, 2 |
| 3 | 1, 3 |
| 4 | 1, 2, 4 |
| 5 | 1, 5 |
| 6 | 1, 2, 3, 6 |
| 7 | 1, 7 |
| 8 | 1, 2, 4, 8 |
| 9 | 1, 3, 9 |
| 10 | 1, 2, 5, 10 |
| 11 | 1, 11 |
| 12 | 1, 2, 3, 4, 6, 12 |

About 230 B.C. Eratosthenes, a Greek mathematician, suggested a way to find prime numbers in a list of all the counting numbers up to a certain number. Eratosthenes first crossed out all multiples of 2, except 2 itself. Next he crossed out all multiples of the next remaining number, 3, except 3 itself. He continued crossing out multiples of each successive remaining number except the number itself. The numbers remaining at the end of this process are the primes.

```
     1    2    3    4    5    6    7    8    9
10   11   12   13   14   15   16   17   18   19
20   21   22   23   24   25   26   27   28   29
30   31   32   33   34   35   36   37   . . .
```

The method just described is called the **Sieve of Eratosthenes,** because it picks out the prime numbers as a strainer, or sieve, picks out solid particles from a liquid.

Every counting number greater than 1 has at least one prime factor, which may be the number itself. You can factor a number into prime factors by using either of the following methods.

**Inverted short division**

$$2\overline{)42}$$
$$3\overline{)21}$$
$$7$$

**Factor tree**

```
      42
     /  \
    2    21
        /  \
       3    7
```

Another factor tree for the number 42 is shown at the right. Notice that the prime factors of 42 are the same in either factor tree except for their order. Every whole number is similar to 42 in this respect. This fact is expressed in the following theorem.

$$
\begin{array}{ccc}
 & 42 & \\
3 & & 14 \\
 & 7 & 2
\end{array}
$$

### Fundamental Theorem of Arithmetic

Every whole number greater than 1 can be written as a product of prime factors in exactly one way, except for the order of the factors.

When we write 42 as $2 \cdot 3 \cdot 7$, this product of prime factors is called the **prime factorization** of 42.

*EXAMPLE*   Give the prime factorization of 60.

**Solution**

Method 1

$$
\begin{array}{r}
2\,)\overline{60} \\
2\,)\overline{30} \\
3\,)\overline{15} \\
5
\end{array}
$$

Method 2

$$
\begin{array}{ccc}
 & 60 & \\
2 & & 30 \\
 & 2 & 15 \\
 & 3 & 5
\end{array}
$$

Using either method, we find that the prime factorization of 60 is $2 \cdot 2 \cdot 3 \cdot 5$, or $2^2 \cdot 3 \cdot 5$.

## Class Exercises

**State whether each number is prime or composite.**

**1.** 7          **2.** 9          **3.** 15          **4.** 23          **5.** 22          **6.** 19

**Name the prime factors of each number.**

**7.** 21          **8.** 10          **9.** 18          **10.** 26          **11.** 30          **12.** 70

**Name the number whose prime factorization is given.**

**13.** $3^2 \cdot 5$          **14.** $2^2 \cdot 3^2$          **15.** $2^3 \cdot 3$

**16.** $2 \cdot 7^2$          **17.** $3^2 \cdot 11$          **18.** $2^2 \cdot 5^2$

## Written Exercises

**State whether each number is prime or composite.**

**A**   **1.** 39          **2.** 41         **3.** 51       **4.** 111       **5.** 124       **6.** 321

      **7.** 641      **8.** 753      **9.** 894    **10.** 1164   **11.** 2061   **12.** 3001

**Give the prime factorization of each whole number.**

**13.** 12       **14.** 50       **15.** 24       **16.** 28      **17.** 39      **18.** 56

**19.** 66       **20.** 51       **21.** 54       **22.** 63      **23.** 84      **24.** 90

**25.** 196     **26.** 360     **27.** 308     **28.** 693    **29.** 114    **30.** 1150

**B**  **31.** Explain why 2 is the only even prime number.

     **32.** Write the prime factorizations of the square numbers 16, 36, 81, and 144 by using exponents. What do you think must be true of the exponents in the prime factorization of a square number?

     **33.** Explain why the sum of two prime numbers greater than 2 can never be a prime number.

     **34.** Explain how you know that each of the following numbers must be composite: 111; 111,111; 111,111,111; . . .

     **35.** List all the possible digits that can be the last digit of a prime number that is greater than 10.

**C**  **36.** Choose a six-digit number, such as 652,652, the last three digits of which are a repeat of the first three digits. Show that 7, 11, and 13 are all factors of the number you chose.

     **37.** Since 7, 11, and 13 are factors of any number of the type defined in Exercise 36, what is the largest composite number that is always a factor of such a number? What is the other factor?

     **38.** Give an example to show that the Fundamental Theorem of Arithmetic would be false if 1 were defined to be a prime number.

---

## Review Exercises

**Simplify.**

    **1.** $-4 + 6$      **2.** $8 + (-12)$     **3.** $3 - 4$       **4.** $10 - (-15)$

    **5.** $-12 \cdot (-20)$   **6.** $7 \cdot (-3)$      **7.** $-48 \div (-24)$   **8.** $-72 \div 8$

## 3-3 Positive and Negative Fractions

In any fraction the number above the fraction bar is called the **numerator** and the number below the bar is called the **denominator.** Since the fraction bar indicates division,

$$\frac{3}{5} \text{ means } 3 \div 5.$$

A study of fractions on a number line shows some important properties of fractions.

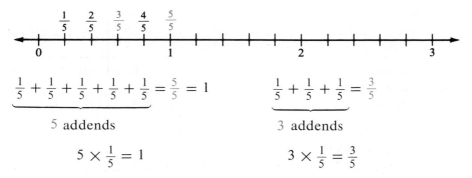

$$\frac{1}{5} + \frac{1}{5} + \frac{1}{5} + \frac{1}{5} + \frac{1}{5} = \frac{5}{5} = 1 \qquad \qquad \frac{1}{5} + \frac{1}{5} + \frac{1}{5} = \frac{3}{5}$$

$$\underbrace{\phantom{\frac{1}{5} + \frac{1}{5} + \frac{1}{5} + \frac{1}{5} + \frac{1}{5}}}_{5 \text{ addends}} \qquad \qquad \underbrace{\phantom{\frac{1}{5} + \frac{1}{5} + \frac{1}{5}}}_{3 \text{ addends}}$$

$$5 \times \frac{1}{5} = 1 \qquad \qquad \qquad 3 \times \frac{1}{5} = \frac{3}{5}$$

We can state these properties in general terms that apply to all positive fractions.

---

### *Properties*

For all whole numbers $a$ and $b$ ($a > 0, b > 0$),

$$\underbrace{\frac{1}{b} + \frac{1}{b} + \cdots + \frac{1}{b}}_{b \text{ addends}} = \frac{b}{b} = 1 \qquad \qquad \underbrace{\frac{1}{b} + \frac{1}{b} + \cdots + \frac{1}{b}}_{a \text{ addends}} = \frac{a}{b}$$

$$b \times \frac{1}{b} = 1 \qquad \qquad \qquad a \times \frac{1}{b} = \frac{a}{b}$$

$$1 \div b = \frac{1}{b} \qquad \qquad \qquad a \div b = \frac{a}{b}$$

---

Just as the negative integers are the opposites of the positive integers, the negative fractions are the opposites of the positive fractions. For every fraction $\frac{a}{b}$ there is a fraction denoted by $-\frac{a}{b}$ that is said to be the **opposite** of $\frac{a}{b}$. On the number line the graphs of $\frac{a}{b}$ and $-\frac{a}{b}$ are on opposite sides of 0 and at equal distances from 0. For example, $\frac{3}{5}$ and $-\frac{3}{5}$ are opposites. They are shown on the number line below.

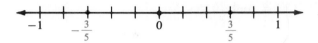

Properties similar to those for positive fractions apply to negative fractions.

$$-\frac{1}{5} + \left(-\frac{1}{5}\right) + \left(-\frac{1}{5}\right) + \left(-\frac{1}{5}\right) + \left(-\frac{1}{5}\right) = -1 \qquad -\frac{1}{5} + \left(-\frac{1}{5}\right) + \left(-\frac{1}{5}\right) = -\frac{3}{5}$$

$$\underbrace{\phantom{-\frac{1}{5} + \left(-\frac{1}{5}\right) + \left(-\frac{1}{5}\right) + \left(-\frac{1}{5}\right) + \left(-\frac{1}{5}\right)}}_{\text{5 addends}} \qquad \underbrace{\phantom{-\frac{1}{5} + \left(-\frac{1}{5}\right) + \left(-\frac{1}{5}\right)}}_{\text{3 addends}}$$

$$5 \times \left(-\frac{1}{5}\right) = -1 \qquad\qquad 3 \times \left(-\frac{1}{5}\right) = -\frac{3}{5}$$

In an earlier section you learned that the quotient of two numbers of opposite sign is negative. Using this rule we can write the following:

$$1 \div (-4) = \frac{1}{-4} = -\frac{1}{4} \qquad\qquad (-1) \div 4 = \frac{-1}{4} = -\frac{1}{4}$$

Therefore, we can write the opposite of $\frac{1}{4}$ as $-\frac{1}{4}$, or as $\frac{-1}{4}$, or as $\frac{1}{-4}$. In general, for $b \neq 0$, $-\frac{a}{b} = \frac{-a}{b} = \frac{a}{-b}$.

*EXAMPLE 1*  Express $-\frac{3}{8}$ in two other ways.

*Solution*  $\qquad -\frac{3}{8} = \frac{-3}{8} = \frac{3}{-8}$

*EXAMPLE 2*  Complete.

a. $\left(-\frac{1}{3}\right) + \left(-\frac{1}{3}\right) + \left(-\frac{1}{3}\right) = \underline{\ ?\ }$     b. $3 \times \underline{\ ?\ } = -1$

c. $2 \times \left(-\frac{1}{5}\right) = \underline{\ ?\ }$     d. $-\frac{2}{5} = \underline{\ ?\ } \div 5$

*Solution*  a. $\left(-\frac{1}{3}\right) + \left(-\frac{1}{3}\right) + \left(-\frac{1}{3}\right) = -1$     b. $3 \times \left(-\frac{1}{3}\right) = -1$

c. $2 \times \left(-\frac{1}{5}\right) = -\frac{2}{5}$     d. $-\frac{2}{5} = \frac{-2}{5} = -2 \div 5$

When we work with fractions having numerators and denominators that are integers, we are working with a new set of numbers called *rational numbers.* Any number that can be represented by a fraction $\frac{a}{b}$, where $a$ and $b$ are integers and $b$ is not 0, is a **rational number.** Notice that the integers themselves are rational numbers. For example, $-9$ can be written as $\frac{-9}{1}$, 0 as $\frac{0}{1}$, and 26 as $\frac{26}{1}$.

## Class Exercises

**Name the rational numbers whose graphs are shown.**

**1.**

**2.**

**3.**

**4.**

**Complete.**

**5.** $\frac{1}{7} + \frac{1}{7} = \underline{\ ?\ }$

**6.** $2 \times \frac{1}{7} = \underline{\ ?\ }$

**7.** $2 \div 7 = \underline{\ ?\ }$

**8.** $\left(-\frac{1}{8}\right) + \left(-\frac{1}{8}\right) + \left(-\frac{1}{8}\right) = \underline{\ ?\ }$

**9.** $3 \times \left(-\frac{1}{8}\right) = \underline{\ ?\ }$

**10.** $3 \div (-8) = \underline{\ ?\ }$

**11.** $4 \times \underline{\ ?\ } = -1$

**12.** $5 \times \underline{\ ?\ } = -\frac{5}{8}$

**13.** $\frac{2}{9} = 2 \div \underline{\ ?\ }$

## Written Exercises

**Graph each set of rational numbers on a number line.**

**A**  **1.** $-1, -\frac{1}{3}, 0, \frac{1}{3}, 1$

**2.** $-1, -\frac{1}{4}, 0, \frac{1}{4}, 1$

**3.** $0, \frac{1}{5}, \frac{5}{5}, \frac{6}{5}$

**4.** $-\frac{4}{3}, -\frac{3}{3}, -\frac{2}{3}, 0$

**5.** $-\frac{5}{4}, -\frac{3}{4}, \frac{3}{4}, \frac{5}{4}$

**6.** $-\frac{5}{3}, -\frac{4}{3}, \frac{4}{3}, \frac{5}{3}$

**Express in two other ways.**

**7.** $-\frac{1}{2}$

**8.** $\frac{-1}{3}$

**9.** $\frac{-1}{11}$

**10.** $\frac{-1}{6}$

**11.** $\frac{2}{-9}$

**12.** $\frac{-9}{10}$

**13.** $-\frac{13}{6}$

**14.** $\frac{5}{-8}$

**15.** $-\frac{3}{4}$

**16.** $\frac{5}{-9}$

**Complete.**

17. $\underline{\ ?\ } \times \left(-\frac{1}{6}\right) = -1$

18. $4 \times \underline{\ ?\ } = \frac{4}{7}$

19. $3 \times \underline{\ ?\ } = \frac{3}{4}$

20. $5 \div 6 = \underline{\ ?\ }$

21. $2 \div \underline{\ ?\ } = \frac{2}{3}$

22. $4 \times \underline{\ ?\ } = -\frac{4}{9}$

**Evaluate the expression when $a = 4$, $b = -3$, and $c = 5$.**

**B** 23. $\frac{a+c}{b}$

24. $\frac{a-c}{a}$

25. $\frac{2b-1}{8}$

26. $\frac{2c-b}{a}$

27. $\frac{a^2-c^2}{a+c}$

**What value of the variable makes the statement true?**

28. $x \times \frac{1}{5} = 1$

29. $3y = -1$

30. $\frac{1}{b} = -\frac{1}{8}$

31. $-\frac{3}{4} = \frac{c}{4}$

32. $-\frac{3}{5} = d \div 5$

33. $\frac{-5}{9} = \frac{5}{m}$

**Write the expression as a positive or a negative fraction.**

*EXAMPLE*   **a.** $3(7)^{-1}$        **b.** $(-5)^{-1}$

*Solution*    **a.** $3(7)^{-1} = 3 \times \frac{1}{7} = \frac{3}{7}$    **b.** $(-5)^{-1} = \frac{1}{-5} = -\frac{1}{5}$

**C** 34. $2(3)^{-1}$    35. $7(6)^{-1}$    36. $5^{-1} + 5^{-1}$    37. $7^{-1} + 7^{-1} + 7^{-1}$

38. $(-2)^{-1}$    39. $(-9)^{-1}$    40. $3(-5)^{-1}$    41. $2(-7)^{-1}$

42. $(-3)^{-1} + (-3)^{-1}$        43. $(-8)^{-1} + (-8)^{-1} + (-8)^{-1}$

---

## Review Exercises

**Complete.**

1. Factors of 72: 1, 2, $\underline{\ ?\ }$, $\underline{\ ?\ }$, 6, $\underline{\ ?\ }$, 9, 12, $\underline{\ ?\ }$, 24, $\underline{\ ?\ }$, 72

2. What is the prime factorization of 72?

3. Factors of 90: $\underline{\ ?\ }$, 2, 3, $\underline{\ ?\ }$, $\underline{\ ?\ }$, 9, $\underline{\ ?\ }$, $\underline{\ ?\ }$, 18, $\underline{\ ?\ }$, $\underline{\ ?\ }$, 90

4. What is the prime factorization of 90?

5. $18x^2y = 3 \times \underline{\ ?\ } \times x \times x \times \underline{\ ?\ } \times y$

6. $24ab^3 = 4 \times \underline{\ ?\ } \times \underline{\ ?\ } \times \underline{\ ?\ } \times b \times b$

# 3-4 Equivalent Fractions

The number line shows the graphs of several fractions.

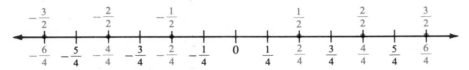

Since $-\frac{1}{2}$ and $-\frac{2}{4}$ have the same graph, they are two names for the same number. Fractions that represent the same number are called **equivalent** fractions. Thus,

$$-\frac{3}{2} \text{ is equivalent to } -\frac{6}{4}; \frac{3}{2} \text{ is equivalent to } \frac{6}{4}.$$

Notice that $\frac{3}{2} = \frac{3 \times 2}{2 \times 2} = \frac{6}{4}$ and $\frac{6}{4} = \frac{6 \div 2}{4 \div 2} = \frac{3}{2}$.

We may state the following general rule.

## Rule

For all numbers $a$, $b$, and $c$ ($b \neq 0$, $c \neq 0$),

$$\frac{a}{b} = \frac{a \times c}{b \times c} \quad \text{and} \quad \frac{a}{b} = \frac{a \div c}{b \div c}.$$

**EXAMPLE 1** Write as an equivalent fraction with a denominator of 12.

   a. $\frac{5}{6}$       b. $-\frac{32}{48}$

**Solution**    a. Since $6 \times 2 = 12$, we write $\frac{5}{6} = \frac{5 \times 2}{6 \times 2} = \frac{10}{12}$.

   b. *Method 1*
   Since $48 \div 4 = 12$, we write

   $$-\frac{32}{48} = \frac{-32}{48} = \frac{-32 \div 4}{48 \div 4} = \frac{-8}{12}, \text{ or } -\frac{8}{12}.$$

   *Method 2*
   $$-\frac{32}{48} = -\frac{32 \div 4}{48 \div 4} = -\frac{8}{12}$$

Notice that we usually write our answers with the minus sign in front of the fraction.

Sometimes we simply show the results of dividing a numerator and denominator by the same number. In Example 1, we can think of dividing by 4 as we write

$$-\frac{\overset{8}{\cancel{32}}}{\underset{12}{\cancel{48}}} = -\frac{8}{12}.$$

A fraction is in **lowest terms** when the numerator and denominator have no common factor other than 1. To write a fraction in lowest terms, we can divide numerator and denominator by a common factor as many times as needed until they have no common factor other than 1.

$$\frac{45}{75} = \frac{45 \div 5}{75 \div 5} = \frac{9 \div 3}{15 \div 3} = \frac{3}{5}$$

Another way to write a fraction in lowest terms is to use the greatest common factor method. The **greatest common factor** (GCF) of two numbers is the greatest whole number that is a factor of each number. To find the GCF of two numbers, we write the prime factorizations of the two numbers, and find the greatest power of each prime factor that occurs in *both* factorizations. The product of these powers is the GCF.

*EXAMPLE 2*  Find the GCF of 84 and 120.

*Solution*  $84 = 2 \times 2 \times 3 \times 7 = 2^2 \times 3 \times 7$
$120 = 2 \times 2 \times 2 \times 3 \times 5 = 2^3 \times 3 \times 5$
$GCF = 2^2 \times 3 = 12$

To write a fraction in lowest terms, we divide the numerator and the denominator by their GCF.

*EXAMPLE 3*  Write $\frac{72}{80}$ in lowest terms.

*Solution*  $72 = 2 \times 2 \times 2 \times 3 \times 3 = 2^3 \times 3^2$
$80 = 2 \times 2 \times 2 \times 2 \times 5 = 2^4 \times 5$
$GCF = 2^3 = 8$

$\frac{72}{80} = \frac{72 \div 8}{80 \div 8} = \frac{9}{10}$

When two numbers have no common factor other than 1, they are said to be **relatively prime.** For example, 9 and 10 are relatively prime. A fraction is in lowest terms when its numerator and denominator are relatively prime.

A **common fraction** is a fraction whose numerator and denominator are both integers; for example, $\frac{2}{5}$ or $-\frac{12}{7}$. A **proper fraction** is a positive fraction whose numerator is less than its denominator, or the opposite of such a fraction; for example, $\frac{3}{8}$ or $-\frac{7}{9}$. A fraction that is not a proper fraction, such as $\frac{9}{4}$ or $-\frac{9}{4}$, is called an **improper fraction.** A number, such as $1\frac{1}{4}$, consisting of a whole number plus a fraction, is called a **mixed number.** Mixed numbers may be written as improper fractions, and improper fractions may be written as mixed numbers. A mixed number is in simple form if the fractional part is in lowest terms.

$$7\frac{15}{27} \text{ in simple form is } 7\frac{5}{9}.$$

**EXAMPLE 4**  Write as a mixed number in simple form.

a. $\frac{12}{8}$         b. $-\frac{12}{8}$

**Solution**    a. $\frac{12}{8} = 12 \div 8$, which gives 1 R4.

Therefore $\frac{12}{8} = 1\frac{4}{8} = 1\frac{1}{2}$.

b. $-\frac{12}{8}$ is *the opposite of* $\frac{12}{8}$, so $-\frac{12}{8} = -1\frac{1}{2}$.

**EXAMPLE 5**  Write $7\frac{2}{5}$ as an improper fraction.

**Solution**    $7\frac{2}{5} = \frac{7}{1} + \frac{2}{5} = \frac{7 \times 5}{1 \times 5} + \frac{2}{5} = \frac{(7 \times 5) + 2}{5} = \frac{37}{5}$

## Class Exercises

**For each fraction, state the GCF of the numerator and denominator. Then state an equivalent fraction in lowest terms.**

1. $\frac{4}{24}$         2. $\frac{5}{20}$         3. $\frac{-12}{16}$         4. $-\frac{8}{12}$         5. $\frac{-9}{12}$

**State an equivalent improper fraction.**

6. $3\frac{2}{5}$         7. $-4\frac{1}{8}$         8. $-7\frac{2}{3}$         9. $9\frac{3}{4}$         10. $-5\frac{1}{6}$

**State an equivalent mixed number in simple form.**

11. $\frac{13}{5}$         12. $-\frac{14}{4}$         13. $\frac{-24}{5}$         14. $\frac{19}{3}$         15. $\frac{-32}{10}$

## Written Exercises

**Complete.**

**A**
**1.** $\frac{2}{3} = \frac{?}{6}$    **2.** $\frac{5}{7} = \frac{?}{21}$    **3.** $1 = \frac{5}{?}$    **4.** $4 = \frac{?}{2}$    **5.** $-\frac{14}{32} = -\frac{?}{16}$

**6.** $\frac{42}{-54} = \frac{?}{-9}$    **7.** $\frac{18}{27} = \frac{2}{?}$    **8.** $-7 = -\frac{?}{3}$    **9.** $\frac{-7}{25} = \frac{-35}{?}$    **10.** $-\frac{3}{4} = -\frac{?}{16}$

**Write as a proper fraction in lowest terms or as a mixed number in simple form.**

**11.** $\frac{21}{35}$    **12.** $\frac{24}{40}$    **13.** $\frac{11}{4}$    **14.** $\frac{25}{7}$    **15.** $-\frac{9}{81}$

**16.** $\frac{54}{81}$    **17.** $-\frac{17}{3}$    **18.** $-\frac{56}{72}$    **19.** $\frac{-34}{85}$    **20.** $-\frac{29}{5}$

**21.** $\frac{125}{12}$    **22.** $\frac{49}{63}$    **23.** $-\frac{79}{13}$    **24.** $\frac{32}{10}$    **25.** $\frac{-300}{7}$

**Write as an improper fraction.**

**26.** $3\frac{1}{4}$    **27.** $2\frac{1}{8}$    **28.** $5\frac{3}{7}$    **29.** $6\frac{3}{16}$    **30.** $5\frac{3}{8}$

**31.** $-4\frac{3}{8}$    **32.** $-3\frac{7}{8}$    **33.** $-16\frac{1}{3}$    **34.** $-7\frac{2}{9}$    **35.** $-8\frac{3}{10}$

**What value of the variable makes the statement true?**

**B**
**36.** $\frac{b}{4} = \frac{6}{12}$    **37.** $\frac{x}{16} = \frac{9}{48}$    **38.** $\frac{a}{6} = \frac{10}{60}$    **39.** $\frac{d}{5} = \frac{-3}{15}$

**40.** $\frac{1}{x} = \frac{4}{16}$    **41.** $\frac{0}{6} = \frac{n}{12}$    **42.** $\frac{5}{-1} = \frac{20}{n}$    **43.** $\frac{1}{3} = \frac{8}{n}$

**Simplify by writing an equivalent fraction in which the numerator and denominator have no common factors.**

**EXAMPLE**    $\dfrac{15x^2y}{20xy^2} = \dfrac{3 \times \overset{1}{\cancel{5}} \times \overset{1}{\cancel{x}} \times x \times \overset{1}{\cancel{y}}}{4 \times \underset{1}{\cancel{5}} \times \underset{1}{\cancel{x}} \times \underset{1}{\cancel{y}} \times y} = \dfrac{3x}{4y}$

**44.** $\frac{2a}{6a}$    **45.** $\frac{4b}{20}$    **46.** $\frac{c}{c^2}$    **47.** $\frac{d^2e}{de^2}$

**48.** $\frac{3x}{15y}$    **49.** $\frac{6n^2}{18n}$    **50.** $\frac{4h}{6hk}$    **51.** $\frac{uv^2}{3u^2v}$

## Review Exercises

**Simplify.**

**1.** $7^2$    **2.** $11^3$    **3.** $6^4$    **4.** $2^8$    **5.** $1^5$

**6.** $5^2 \times 3^2$    **7.** $4^3 \times 9^2$    **8.** $2^4 \times 6^2$    **9.** $3^3 \times 3^2$    **10.** $8^2 \times 2^4$

# 3-5 Least Common Denominator

In calculations and comparisons, we work with more than one fraction. It is sometimes necessary to replace fractions with equivalent fractions so that all have the same denominator, called a *common denominator*. For example, in the addition $\frac{1}{6} + \frac{3}{4}$ we may write $\frac{1}{6}$ as $\frac{2}{12}$ and $\frac{3}{4}$ as $\frac{9}{12}$, using 12 as a common denominator. We may also use 24, 36, 48, or any other multiple of both denominators as a common denominator.

The **least common denominator** (LCD) is the most convenient denominator to use. The LCD is the least common multiple (LCM) of the denominators. The LCD is found using the prime factorizations of the denominators. For each prime factor of any of the denominators, find the highest power of that factor that occurs in *any* prime factorization. The product of these powers is the LCD.

**EXAMPLE 1**    Write equivalent fractions with the LCD: $\frac{5}{48}$, $\frac{11}{120}$.

**Solution**    The LCD is the least common multiple of 48 and 120. Use prime factorization to find the LCM.

Prime factorization of 48:   $2 \times 2 \times 2 \times 2 \times 3$, or $2^4 \times 3$
Prime factorization of 120: $2 \times 2 \times 2 \times 3 \times 5$, or $2^3 \times 3 \times 5$

The LCM is the product of the highest powers of each factor. The LCM $= 2^4 \times 3 \times 5 = 240$, so the LCD $= 240$.

$$\frac{5}{48} = \frac{5 \times 5}{48 \times 5} = \frac{25}{240} \qquad\qquad \frac{11}{120} = \frac{11 \times 2}{120 \times 2} = \frac{22}{240}$$

**EXAMPLE 2**    Replace  $\underline{\phantom{?}?\phantom{?}}$  with $<$, $>$, or $=$ to make a true statement.

    **a.** $\frac{5}{6}\ \underline{\phantom{?}?\phantom{?}}\ \frac{6}{7}$          **b.** $-\frac{5}{8}\ \underline{\phantom{?}?\phantom{?}}\ -\frac{9}{14}$

**Solution**    First rewrite each pair of fractions as equivalent fractions with the LCD. Then compare the fractions.

**a.** The LCD is the LCM of 6 and 7, or 42.

$$\frac{5}{6} = \frac{5 \times 7}{6 \times 7} = \frac{35}{42} \qquad\qquad \frac{6}{7} = \frac{6 \times 6}{7 \times 6} = \frac{36}{42}$$

$$\frac{35}{42} < \frac{36}{42}, \text{ so } \frac{5}{6} < \frac{6}{7}.$$

**b.** The LCD is the LCM of 8 and 14, or 56.

$$-\frac{5}{8} = -\frac{5 \times 7}{8 \times 7} = -\frac{35}{56} \qquad\qquad -\frac{9}{14} = -\frac{9 \times 4}{14 \times 4} = -\frac{36}{56}$$

$$-\frac{35}{56} > -\frac{36}{56}, \text{ so } -\frac{5}{8} > -\frac{9}{14}.$$

When fractions have variables in their denominators, we may obtain a common denominator by finding a common multiple of the denominators.

**EXAMPLE 3**  Write as equivalent fractions with a common denominator: $\frac{2}{a}, \frac{3}{b}$.

**Solution**  $\qquad \frac{2}{a} = \frac{2 \times b}{a \times b} = \frac{2b}{ab} \qquad\qquad \frac{3}{b} = \frac{3 \times a}{b \times a} = \frac{3a}{ab}$

## Class Exercises

**State the LCM of the pair of numbers.**

**1.** 6, 18      **2.** 11, 4      **3.** 10, 8      **4.** 15, 12      **5.** 32, 48

**State the LCD of the pair of fractions.**

**6.** $\frac{1}{2}, \frac{3}{4}$      **7.** $\frac{5}{6}, \frac{1}{2}$      **8.** $\frac{3}{4}, -\frac{1}{3}$      **9.** $-\frac{2}{9}, \frac{1}{6}$      **10.** $\frac{1}{16}, \frac{5}{12}$

## Written Exercises

**Write the fractions as equivalent fractions with the least common denominator (LCD).**

**A**  **1.** $\frac{1}{3}, \frac{1}{12}$      **2.** $\frac{1}{4}, \frac{1}{12}$      **3.** $\frac{3}{4}, \frac{5}{8}$      **4.** $\frac{3}{8}, \frac{3}{16}$

**5.** $\frac{2}{9}, -\frac{1}{27}$      **6.** $-\frac{1}{7}, \frac{1}{49}$      **7.** $\frac{10}{21}, \frac{2}{49}$      **8.** $\frac{7}{18}, \frac{5}{36}$

**9.** $-\frac{2}{3}, \frac{7}{30}$      **10.** $\frac{5}{12}, -\frac{6}{11}$      **11.** $\frac{4}{75}, \frac{7}{100}$      **12.** $\frac{9}{56}, \frac{4}{63}$

**13.** $-\frac{3}{7}, -\frac{7}{112}$      **14.** $-\frac{4}{17}, -\frac{9}{16}$      **15.** $\frac{5}{42}, \frac{5}{49}$      **16.** $\frac{5}{84}, \frac{7}{12}$

**B**  **17.** $\frac{7}{8}, \frac{5}{16}, \frac{21}{40}$      **18.** $\frac{1}{4}, \frac{1}{9}, \frac{1}{5}$      **19.** $\frac{3}{7}, \frac{7}{4}, \frac{4}{9}$      **20.** $\frac{11}{18}, \frac{1}{54}, \frac{2}{27}$

**21.** $\frac{1}{65}, \frac{3}{5}, \frac{9}{26}$      **22.** $-\frac{7}{8}, \frac{9}{28}, \frac{2}{49}$      **23.** $-\frac{5}{6}, \frac{2}{9}, -\frac{7}{8}$      **24.** $\frac{11}{12}, \frac{1}{72}, -\frac{7}{8}$

**25.** $\frac{a}{3}, \frac{b}{6}$      **26.** $\frac{m}{25}, \frac{n}{15}$      **27.** $\frac{h}{25}, \frac{h}{100}, \frac{h}{125}$      **28.** $\frac{a}{2}, \frac{b}{3}, \frac{c}{4}$

**Write the fractions as equivalent fractions with a common denominator.**

**29.** $\frac{1}{c}, \frac{2}{3c}$      **30.** $\frac{1}{x}, \frac{1}{y}$      **31.** $\frac{3}{x}, \frac{1}{y}, \frac{5}{z}$      **32.** $\frac{-1}{r}, \frac{2}{rs}, \frac{r}{s}$

**Replace __?__ with $<$, $>$, or $=$ to make a true statement.**

**33.** $\frac{1}{3}$ __?__ $\frac{3}{6}$

**34.** $\frac{3}{4}$ __?__ $\frac{5}{8}$

**35.** $-\frac{2}{3}$ __?__ $-\frac{7}{12}$

**36.** $-\frac{5}{8}$ __?__ $-\frac{5}{16}$

**37.** $-\frac{3}{5}$ __?__ $-\frac{5}{7}$

**38.** $-\frac{2}{4}$ __?__ $-\frac{3}{5}$

**C**  **39.** Let $\frac{a}{b}$ and $\frac{c}{d}$ be fractions with $b > 0$ and $d > 0$. Write $\frac{a}{b}$ and $\frac{c}{d}$ as equivalent fractions with a common denominator. How do $\frac{a}{b}$ and $\frac{c}{d}$ compare when $ad < bc$? when $ad = bc$? when $ad > bc$? Give two examples to illustrate each of your conclusions.

## Self-Test A

**State which of the numbers 2, 3, 4, 5, 9, and 10 are factors of the given number. Use the tests for divisibility.**

**1.** 756 **2.** 7821 **3.** 11,340 **4.** 34,447 [3–1]

**Find out whether each number is prime or composite. If it is composite, give its prime factorization.**

**5.** 108 **6.** 79 **7.** 87 **8.** 109 [3–2]

**Complete.**

**9.** $3 \times$ __?__ $= -1$ **10.** $7 \div 9 =$ __?__ **11.** $-\frac{2}{3} = \frac{-2}{3} =$ __?__ [3–3]

**Write as a proper fraction in lowest terms or as a mixed number in simple form.**

**12.** $\frac{16}{64}$ **13.** $-\frac{72}{30}$ **14.** $\frac{32}{42}$ **15.** $\frac{-71}{48}$ **16.** $\frac{68}{16}$ [3–4]

**Write as an improper fraction.**

**17.** $2\frac{1}{5}$ **18.** $-3\frac{2}{3}$ **19.** $6\frac{4}{15}$ **20.** $1\frac{7}{12}$ **21.** $-8\frac{5}{8}$

**Write as equivalent fractions with the least common denominator.**

**22.** $\frac{7}{10}, \frac{7}{8}$ **23.** $-\frac{10}{49}, \frac{2}{21}$ **24.** $\frac{3}{50}, \frac{6}{225}$ **25.** $-\frac{8}{15}, \frac{-1}{30}$ [3–5]

*Self-Test answers and Extra Practice are at the back of the book.*

## 3-6 Adding and Subtracting Common Fractions

You have added fractions having a common denominator by adding the numerators and writing the result with the same denominator. We can illustrate the reasoning for this rule using the sum $\frac{2}{13} + \frac{5}{13}$. The reason for each statement at left below is given on the same line at right. All of the properties for addition, subtraction, multiplication, and division of positive and negative numbers hold for the rational numbers.

$\frac{2}{13} + \frac{5}{13} = \left(2 \times \frac{1}{13}\right) \times \left(5 \times \frac{1}{13}\right)$    By the rule $\frac{a}{b} = a \times \frac{1}{b}$

$= (2 + 5) \times \frac{1}{13}$    By the distributive property

$= (7) \times \frac{1}{13}$    Substitution of 7 for $2 + 5$

$= \frac{7}{13}$    By the rule $a \times \frac{1}{b} = \frac{a}{b}$

The methods for adding and subtracting positive fractions apply as well to adding and subtracting negative fractions.

---

### Rules

For all numbers $a$, $b$, and $c(c \neq 0)$,

$$\frac{a}{c} + \frac{b}{c} = \frac{a + b}{c} \quad \text{and} \quad \frac{a}{c} - \frac{b}{c} = \frac{a - b}{c}$$

---

**EXAMPLE 1**    Add or subtract. Write the answers in lowest terms.

         **a.** $-\frac{3}{8} + \frac{1}{8}$          **b.** $\frac{23}{60} - \left(-\frac{17}{60}\right)$

**Solution**      **a.** $-\frac{3}{8} + \frac{1}{8} = \frac{-3}{8} + \frac{1}{8}$    **b.** $\frac{23}{60} - \left(-\frac{17}{60}\right) = \frac{23}{60} + \frac{17}{60}$

                       $= \frac{-3 + 1}{8}$                  $= \frac{23 + 17}{60}$

                       $= \frac{-2}{8}$                      $= \frac{40}{60}$

                       $= \frac{-1}{4}$, or $-\frac{1}{4}$            $= \frac{2}{3}$

When two fractions have different denominators, write the fractions as equivalent fractions with a common denominator before adding or subtracting.

**EXAMPLE 2** Add or subtract. Write the answer in lowest terms.

$$\textbf{a. } \frac{1}{4} - \frac{7}{12} \qquad\qquad \textbf{b. } -\frac{1}{10} + \left(-\frac{5}{6}\right)$$

***Solution***

$$\textbf{a. } \frac{1}{4} - \frac{7}{12} = \frac{3}{12} - \frac{7}{12} \qquad \textbf{b. } -\frac{1}{10} + \left(-\frac{5}{6}\right) = -\frac{3}{30} + \left(-\frac{25}{30}\right)$$

$$= \frac{3-7}{12} \qquad\qquad\qquad = \frac{-3}{30} + \frac{-25}{30}$$

$$= \frac{-4}{12} \qquad\qquad\qquad = \frac{-3 + (-25)}{30}$$

$$= \frac{-1}{3}, \text{ or } -\frac{1}{3} \qquad\qquad = \frac{-28}{30}$$

$$= \frac{-14}{15}, \text{ or } -\frac{14}{15}$$

## Class Exercises

**Add or subtract. Write the answer as a whole number or as a common fraction in lowest terms.**

**1.** $\frac{3}{16} + \frac{8}{16}$      **2.** $\frac{7}{10} - \frac{3}{10}$      **3.** $\frac{7}{11} + \frac{8}{11}$      **4.** $\frac{4}{6} - \frac{11}{6}$

**5.** $-\frac{8}{5} + \frac{4}{5}$      **6.** $\frac{12}{5} - \frac{16}{5}$      **7.** $-\frac{14}{3} + \frac{20}{3}$      **8.** $-\frac{9}{4} - \frac{13}{4}$

**9.** $-\frac{1}{7} + \left(-\frac{3}{7}\right)$    **10.** $\frac{11}{15} - \left(-\frac{2}{15}\right)$    **11.** $-\frac{4}{21} - \left(-\frac{2}{21}\right)$    **12.** $-\frac{7}{10} - \left(-\frac{3}{10}\right)$

## Written Exercises

**Add or subtract. Write the answer as a whole number or as a common fraction in lowest terms.**

**A**   **1.** $\frac{2}{15} + \frac{8}{15}$      **2.** $\frac{9}{20} - \frac{7}{20}$      **3.** $-\frac{8}{17} + \frac{4}{17}$      **4.** $\frac{17}{8} - \frac{9}{8}$

**5.** $\frac{3}{4} + \left(-\frac{3}{4}\right)$      **6.** $-\frac{11}{15} - \frac{16}{15}$      **7.** $-\frac{9}{10} + \left(-\frac{13}{10}\right)$      **8.** $-\frac{7}{16} - \left(-\frac{23}{16}\right)$

**9.** $\frac{3}{7} + \frac{1}{3}$      **10.** $\frac{5}{4} - \frac{1}{2}$      **11.** $-\frac{7}{6} + \frac{14}{3}$      **12.** $-\frac{17}{7} - \frac{7}{3}$

**13.** $\frac{15}{4} + \left(-\frac{33}{8}\right)$  **14.** $-\frac{17}{5} - \frac{11}{3}$  **15.** $-\frac{23}{10} + \left(-\frac{14}{5}\right)$  **16.** $-\frac{21}{4} - \left(-\frac{32}{5}\right)$

**17.** $\frac{11}{12} + \frac{12}{13}$  **18.** $-\frac{5}{6} + \frac{7}{20}$  **19.** $-\frac{5}{8} + \frac{1}{18}$  **20.** $-\frac{2}{15} + \frac{6}{21}$

**B 21.** $\frac{1}{3} + \frac{1}{6} + \frac{1}{12}$  **22.** $\frac{2}{15} + \frac{1}{5} + \frac{2}{3}$  **23.** $\frac{7}{10} + \left(-\frac{1}{2}\right) + \frac{2}{5}$

**24.** $\frac{11}{12} + \frac{2}{7} + \left(-\frac{1}{2}\right)$  **25.** $-\frac{5}{6} + \frac{2}{9} + \left(-\frac{1}{3}\right)$  **26.** $\frac{8}{14} - \left(-\frac{1}{7}\right) + \frac{3}{2}$

**27.** $\frac{3}{2} + \frac{7}{4} + \left(-\frac{5}{3}\right)$  **28.** $-\frac{7}{3} + \frac{2}{3} + \frac{9}{5}$  **29.** $\frac{9}{15} + \left(-\frac{2}{3}\right) - \left(-\frac{1}{6}\right)$

**C 30.** Show that $\frac{a}{c} + \frac{b}{c} = \frac{a+b}{c}$ $(c \neq 0)$ by applying the rule or property that justifies each statement.

$$\frac{a}{b} + \frac{b}{c} = \left(a \times \frac{1}{c}\right) + \left(b \times \frac{1}{c}\right) \qquad \text{Why?}$$

$$= (a + b) \times \frac{1}{c} \qquad \text{Why?}$$

$$= \frac{a+b}{c} \qquad \text{Why?}$$

**31.** As in Exercise 30, show that $\frac{a}{c} - \frac{b}{c} = \frac{a-b}{c}$ $(c \neq 0)$.

## Review Exercises

**Write each fraction as a mixed number.**

**1.** $\frac{11}{8}$  **2.** $\frac{15}{11}$  **3.** $\frac{23}{7}$  **4.** $\frac{31}{6}$

**5.** $\frac{71}{12}$  **6.** $\frac{83}{9}$  **7.** $\frac{121}{13}$  **8.** $\frac{169}{16}$

■■■ **Challenge**

**Write a fraction whose value is between the given fractions.**

**1.** $\frac{2}{5} < \underline{\ ?\ } < \frac{3}{5}$  **2.** $-\frac{2}{3} < \underline{\ ?\ } < -\frac{1}{3}$  **3.** $\frac{1}{8} < \underline{\ ?\ } < \frac{1}{4}$

**4.** $-\frac{2}{7} < \underline{\ ?\ } < -\frac{3}{14}$  **5.** $\frac{1}{6} < \underline{\ ?\ } < \frac{1}{4}$  **6.** $-\frac{5}{6} < \underline{\ ?\ } < -\frac{7}{9}$

# 3-7 Adding and Subtracting Mixed Numbers

The rules for adding and subtracting fractions also apply to mixed numbers. Before mixed numbers are added or subtracted, convert them to improper fractions.

**EXAMPLE 1**  Add or subtract. Write the answer as a proper fraction in lowest terms or as a mixed number in simple form.

$$\textbf{a. } 3\frac{7}{8} + 2\frac{3}{8} \qquad\qquad\qquad \textbf{b. } 2\frac{3}{5} - 1\frac{4}{5}$$

**Solution**

$$\textbf{a. } 3\frac{7}{8} + 2\frac{3}{8} = \frac{31}{8} + \frac{19}{8} \qquad \textbf{b. } 2\frac{3}{5} - 1\frac{4}{5} = \frac{13}{5} - \frac{9}{5}$$

$$= \frac{50}{8} \qquad\qquad\qquad\qquad = \frac{4}{5}$$

$$= 6\frac{2}{8} = 6\frac{1}{4}$$

It is sometimes necessary to write mixed numbers as improper fractions having a common denominator before adding or subtracting.

**EXAMPLE 2**  Add or subtract. Write the answer as a proper fraction in lowest terms or as a mixed number in simple form.

$$\textbf{a. } 5\frac{1}{4} + \left(-2\frac{1}{3}\right) \qquad\qquad \textbf{b. } \frac{5}{6} - 2\frac{3}{8}$$

**Solution**

$$\textbf{a. } 5\frac{1}{4} + \left(-2\frac{1}{3}\right) = \frac{21}{4} + \left(-\frac{7}{3}\right) \qquad \textbf{b. } \frac{5}{6} - 2\frac{3}{8} = \frac{5}{6} - \frac{19}{8}$$

$$= \frac{63}{12} + \left(-\frac{28}{12}\right) \qquad\qquad = \frac{20}{24} - \frac{57}{24}$$

$$= \frac{35}{12} \qquad\qquad\qquad\qquad = -\frac{37}{24}$$

$$= 2\frac{11}{12} \qquad\qquad\qquad\qquad = -1\frac{13}{24}$$

Mixed numbers can be added or subtracted using a vertical format.

**EXAMPLE 3**  Add or subtract. Write the answer as a proper fraction in lowest terms or as a mixed number in simple form.

$$\textbf{a. } 2\frac{3}{10} + 4\frac{1}{10} \qquad\qquad \textbf{b. } 4\frac{7}{8} - 2\frac{1}{5}$$

**Solution**     **a.**  $2\frac{3}{10}$

$$+4\frac{1}{10}$$
$$6\frac{4}{10} = 6\frac{2}{5}$$

**b.**  $4\frac{7}{8} = \phantom{-}4\frac{35}{40}$

$$-2\frac{1}{5} = -2\frac{8}{40}$$
$$2\frac{27}{40}$$

---

## Class Exercises

**Add or subtract. Write the answer as a proper fraction in lowest terms or as a mixed number in simple form.**

1. $1\frac{1}{3} + 3\frac{1}{3}$ 

2. $2\frac{7}{10} - 1\frac{5}{10}$ 

3. $4\frac{13}{15} + 5\frac{4}{15}$ 

4. $5\frac{5}{12} - 4\frac{7}{12}$

5. $-6\frac{1}{4} + 2\frac{3}{4}$ 

6. $5\frac{3}{8} - 6\frac{7}{8}$ 

7. $-8\frac{7}{11} + 9\frac{8}{11}$ 

8. $-6\frac{3}{5} - 8\frac{4}{5}$

9. $-1\frac{5}{6} + \left(-3\frac{1}{6}\right)$ 

10. $2\frac{4}{7} - \left(-7\frac{6}{7}\right)$ 

11. $-10\frac{1}{9} - \left(-8\frac{7}{9}\right)$ 

12. $-3\frac{7}{10} - \left(-1\frac{3}{10}\right)$

---

## Written Exercises

**Add or subtract. Write the answer as a proper fraction in lowest terms or as a mixed number in simple form.**

**A**  1. $5\frac{2}{5} + 3\frac{2}{5}$ 

2. $9\frac{7}{10} - 6\frac{3}{10}$ 

3. $-6\frac{4}{11} + 7\frac{3}{11}$ 

4. $-8\frac{3}{4} - 4\frac{1}{4}$

5. $1\frac{5}{6} + 4\frac{5}{6}$ 

6. $-3\frac{3}{8} - 9\frac{5}{8}$ 

7. $-10\frac{3}{7} + \left(-9\frac{5}{7}\right)$ 

8. $-2\frac{4}{15} - \left(-7\frac{7}{15}\right)$

9. $2\frac{1}{2} + 1\frac{1}{3}$ 

10. $5\frac{3}{5} - 2\frac{1}{4}$ 

11. $-2\frac{1}{6} + 4\frac{2}{3}$ 

12. $-5\frac{3}{7} - 2\frac{1}{3}$

13. $10\frac{3}{8} + \left(-12\frac{7}{12}\right)$ 

14. $-5\frac{2}{5} - 6\frac{5}{6}$ 

15. $-1\frac{9}{10} + \left(-1\frac{2}{3}\right)$ 

16. $-3\frac{7}{9} - \left(-5\frac{11}{15}\right)$

17. $5 + \frac{3}{4}$ 

18. $7 - \frac{1}{5}$ 

19. $-3 - \frac{7}{8}$ 

20. $-2 - \frac{7}{10}$

21. $5\frac{2}{3} + 6$ 

22. $3\frac{9}{10} - 7$ 

23. $-10\frac{1}{5} + (-9)$ 

24. $-7 - \left(-8\frac{1}{9}\right)$

**B**  25. $3\frac{1}{2} + 2\frac{1}{3} + 4\frac{1}{12}$ 

26. $8\frac{3}{5} + 2\frac{1}{2} - 11\frac{1}{10}$ 

27. $-4\frac{1}{3} + 2\frac{1}{4} - 6\frac{1}{6}$

28. $10\frac{1}{8} + \left(-6\frac{3}{4}\right) - 9\frac{11}{24}$ 

29. $-4\frac{1}{7} + 3\frac{1}{2} + \left(-7\frac{9}{14}\right)$ 

30. $-8\frac{9}{10} + 6\frac{2}{5} - 3\frac{1}{2}$

**31.** $-1\frac{2}{3} - \left(-2\frac{1}{5}\right) + 3\frac{7}{15}$  **32.** $-6\frac{7}{10} - \left(-3\frac{2}{5}\right) - \left(-3\frac{3}{10}\right)$  **33.** $3 + 2\frac{1}{3} - 5\frac{1}{5}$

**34.** $4\frac{7}{8} + (-3) - 2\frac{1}{4}$  **35.** $1\frac{1}{5} + \frac{3}{10} - 1\frac{1}{4}$  **36.** $-4\frac{9}{16} + \frac{3}{32} + \frac{1}{2}$

---

## Problems

**Solve.**

**A** **1.** Joan Kent bought $15\frac{3}{4}$ yd of drapery material for $63 at a sale. If she used all except $1\frac{1}{16}$ yd, how much material did she actually use?

**2.** Carl is 6 ft tall. If he grew $1\frac{1}{8}$ in. during the past year and $\frac{3}{4}$ in. the year before, how tall was he one year ago?

**3.** On Monday Kim jogged $1\frac{1}{2}$ mi in $\frac{1}{4}$ h. On Wednesday she jogged $2\frac{1}{3}$ mi in $\frac{1}{3}$ h. How much farther did Kim jog on Wednesday?

**4.** The gas tank of a popular compact car holds $15\frac{2}{5}$ gal of gasoline. How much gas has been used if $10\frac{1}{3}$ gal remain in the tank?

**B** **5.** Last year, total rainfall for April and May was $7\frac{1}{4}$ in. This year 3 in. of rain fell in April and $2\frac{5}{8}$ in. fell in May. How much less rain fell this year than last year during April and May?

**6.** Each share of stock in Unified Electronics had a value of $34\frac{3}{4}$ on Monday. The value of the stock declined by $1\frac{5}{8}$ on Tuesday and by $\frac{7}{8}$ on Wednesday. What was the value of the stock after Wednesday?

**7.** A 512-page book has pages that are 7 in. wide and 9 in. high. The printed area measures $5\frac{3}{8}$ in. by $7\frac{3}{4}$ in. The left margin is $\frac{5}{16}$ in. and the top margin is $\frac{9}{16}$ in. How wide are the margins at the right and at the bottom of the page?

---

## Review Exercises

**Complete.**

**1.** $-\frac{2}{7} = \frac{-2}{7} = \underline{\quad?\quad}$  **2.** $-\frac{4}{11} = \frac{4}{-11} = \underline{\quad?\quad}$  **3.** $-\frac{5}{9} = \frac{-5}{9} = \underline{\quad?\quad}$

**4.** $\frac{7}{-20} = -\frac{7}{20} = \underline{\quad?\quad}$  **5.** $7 \times \underline{\quad?\quad} = -1$  **6.** $6 \times \left(-\frac{1}{7}\right) = \underline{\quad?\quad}$

**7.** $-\frac{4}{11} = \underline{\quad?\quad} \div (-11)$  **8.** $\frac{-1}{4} = 1 \div \underline{\quad?\quad}$  **9.** $3 \div \underline{\quad?\quad} = 1$

## 3-8 Multiplying Fractions

To develop a method for multiplying fractions, we begin by showing that $\frac{7}{3} \times 3 = 7$.

$$\frac{7}{3} \times 3 = \left(7 \times \frac{1}{3}\right) \times 3 \qquad \text{By the rule: } \frac{a}{b} = a \times \frac{1}{b}$$

$$= 7 \times \left(\frac{1}{3} \times 3\right) \qquad \text{By the associative property}$$

$$= 7 \times \left(3 \times \frac{1}{3}\right) \qquad \text{By the commutative property}$$

$$= 7 \times 1 \qquad \text{By the rule: } b \times \frac{1}{b} = 1$$

$$= 7 \qquad \text{By the multiplication property of one}$$

In a similar way, it can be shown that $-\frac{7}{3} \times 3 = -7$. It is possible to prove the following general rule for all fractions.

---

### *Rule*

For all numbers $a$ and $b$ ($b \neq 0$),

$$\frac{a}{b} \times b = a$$

---

The rule tells us how to find the product of a fraction and a whole number. Using the rule on page 92, we can arrive at another rule for finding the product of two fractions such as $\frac{1}{5} \times \frac{1}{2}$. We begin by showing that $10 \times (\frac{1}{5} \times \frac{1}{2}) = 1$.

$$10 \times \left(\frac{1}{5} \times \frac{1}{2}\right) = (2 \times 5) \times \left(\frac{1}{5} \times \frac{1}{2}\right) \qquad \text{Substitution of } 2 \times 5 \text{ for } 10$$

$$= \left(2 \times \frac{1}{2}\right) \times \left(5 \times \frac{1}{5}\right) \qquad \begin{array}{l}\text{By the associative and commutative} \\ \text{properties}\end{array}$$

$$= 1 \times 1 \qquad \text{By the rule: } b \times \frac{1}{b} = 1$$

$$= 1 \qquad \text{By the multiplication property of one}$$

We have shown that $10 \times (\frac{1}{5} \times \frac{1}{2}) = 1$ and we know that $10 \times \frac{1}{10} = 1$, so we conclude that $\frac{1}{5} \times \frac{1}{2} = \frac{1}{10}$. Similarly, we may prove the general rule shown on the following page.

> ## Rule
>
> For all numbers $a$ and $b$ ($a \neq 0$, $b \neq 0$),
>
> $$\frac{1}{a} \times \frac{1}{b} = \frac{1}{ab}$$

**EXAMPLE 1**  Multiply $\frac{1}{-3} \times \frac{1}{4}$.

**Solution**  $\frac{1}{-3} \times \frac{1}{4} = \frac{1}{-3 \times 4} = \frac{1}{-12}$, or $-\frac{1}{12}$

The preceding rules are used to prove the following rule for multiplying two fractions.

> ## Rule
>
> For all numbers $a$, $b$, $c$, and $d$ ($b \neq 0$, $d \neq 0$),
>
> $$\frac{a}{b} \times \frac{c}{d} = \frac{ac}{bd}$$

**EXAMPLE 2**  Multiply. Write the answers to parts (a) and (b) as proper fractions in lowest terms or as mixed numbers in simple form.

  a. $\frac{5}{6} \times \frac{7}{3}$          b. $\frac{3}{2} \times \left(-\frac{5}{7}\right)$          c. $-\frac{7}{a} \times \left(-\frac{4}{b}\right)$

**Solution**  a. $\frac{5}{6} \times \frac{7}{3} = \frac{5 \times 7}{6 \times 3} = \frac{35}{18}$, or $1\frac{17}{18}$

  b. $\frac{3}{2} \times \left(-\frac{5}{7}\right) = \frac{3}{2} \times \frac{-5}{7} = \frac{3 \times (-5)}{2 \times 7} = \frac{-15}{14}$, or $-1\frac{1}{14}$

  c. $-\frac{7}{a} \times \left(-\frac{4}{b}\right) = \frac{-7}{a} \times \frac{-4}{b} = \frac{-7 \times (-4)}{a \times b} = \frac{28}{ab}$

Sometimes it is easier to divide by common factors of the numerator and denominator before multiplying.

**EXAMPLE 3**  Multiply $\frac{-3}{5} \times \frac{15}{16} \times \frac{-2}{3}$.

**Solution**  $\frac{-3}{5} \times \frac{15}{16} \times \frac{-2}{3} = \frac{\overset{-1}{\cancel{-3}}}{\underset{1}{\cancel{5}}} \times \frac{\overset{3}{\cancel{15}}}{\underset{8}{\cancel{16}}} \times \frac{\overset{-1}{\cancel{-2}}}{\underset{1}{\cancel{3}}} = \frac{-1 \times 3 \times (-1)}{1 \times 8 \times 1} = \frac{3}{8}$

**Reading Mathematics:** *Reading Fractions*

Usually you read across a line from left to right, and you read a page from top to bottom. You have learned many special ways to read things in mathematics. For example, you read a fraction from top to bottom. You read first the numerator and then the denominator of each fraction before continuing to the next word or symbol.

To multiply mixed numbers, first write each mixed number as an improper fraction.

**EXAMPLE 4** Multiply. Write the answer as a proper fraction in lowest terms or as a mixed number in simple form.

$$\textbf{a. } 3\tfrac{1}{2} \times \left(-\tfrac{1}{4}\right) \qquad\qquad \textbf{b. } 5\tfrac{1}{4} \times 1\tfrac{3}{7}$$

**Solution**

$$\textbf{a. } 3\tfrac{1}{2} \times \left(-\tfrac{1}{4}\right) = \tfrac{7}{2} \times \tfrac{-1}{4} = \tfrac{-7}{8}, \text{ or } -\tfrac{7}{8}$$

$$\textbf{b. } 5\tfrac{1}{4} \times 1\tfrac{3}{7} = \overset{3}{\underset{2}{\cancel{\tfrac{21}{4}}}} \times \overset{5}{\underset{1}{\cancel{\tfrac{10}{7}}}} = \tfrac{15}{2}, \text{ or } 7\tfrac{1}{2}$$

---

## Class Exercises

**Multiply.**

**1.** $\tfrac{5}{6} \times 6$     **2.** $-\tfrac{3}{5} \times (-5)$     **3.** $-4 \times \tfrac{7}{4}$     **4.** $9 \times \left(-\tfrac{8}{9}\right)$

**5.** $\tfrac{1}{8} \times \tfrac{1}{3}$     **6.** $-\tfrac{1}{4} \times \tfrac{1}{5}$     **7.** $\tfrac{1}{15} \times \left(-\tfrac{1}{2}\right)$     **8.** $-\tfrac{1}{10} \times \left(-\tfrac{1}{10}\right)$

**9.** $\tfrac{5}{8} \times \tfrac{3}{11}$     **10.** $\tfrac{2}{5} \times \left(-\tfrac{3}{7}\right)$     **11.** $1\tfrac{1}{2} \times 2$     **12.** $-3 \times 1\tfrac{1}{5}$

---

## Written Exercises

**Multiply. Write the answer as a proper fraction in lowest terms or as a mixed number in simple form.**

**A**   **1.** $\tfrac{7}{12} \times 12$     **2.** $-6 \times \tfrac{3}{6}$     **3.** $3 \times \left(-\tfrac{2}{3}\right)$     **4.** $-14 \times \left(-\tfrac{5}{14}\right)$

**5.** $-\tfrac{1}{4} \times \left(-\tfrac{1}{7}\right)$     **6.** $\tfrac{1}{8} \times \left(-\tfrac{1}{20}\right)$     **7.** $-\tfrac{1}{10} \times \tfrac{1}{6}$     **8.** $-\tfrac{1}{15} \times \tfrac{1}{2}$

**9.** $\tfrac{2}{3} \times \tfrac{5}{9}$     **10.** $\tfrac{6}{7} \times \tfrac{2}{5}$     **11.** $\tfrac{3}{8} \times \left(-\tfrac{2}{3}\right)$     **12.** $\tfrac{5}{6} \times \left(-\tfrac{2}{5}\right)$

**13.** $-\tfrac{1}{3} \times \tfrac{2}{9}$     **14.** $-\tfrac{3}{8} \times \tfrac{2}{9}$     **15.** $\tfrac{-5}{8} \times (-1)$     **16.** $\tfrac{-2}{5} \times \left(-\tfrac{15}{16}\right)$

**Multiply.**

**17.** $-\frac{1}{4} \times 0$

**18.** $-\frac{3}{4} \times 0$

**19.** $\frac{5}{16} \times \frac{30}{40}$

**20.** $\frac{9}{8} \times \frac{24}{27}$

**B** **21.** $3\frac{1}{4} \times \frac{4}{13}$

**22.** $5\frac{2}{5} \times \frac{5}{9}$

**23.** $4\frac{1}{4} \times 10\frac{1}{3}$

**24.** $10\frac{1}{2} \times 2\frac{3}{4}$

**25.** $-4\frac{2}{7} \times 5\frac{1}{6}$

**26.** $3\frac{1}{8} \times \left(-4\frac{1}{5}\right)$

**27.** $-2\frac{1}{3} \times \left(-1\frac{4}{9}\right)$

**28.** $-6\frac{1}{4} \times \left(-5\frac{2}{5}\right)$

**29.** $\frac{1}{4} \times \frac{1}{3} \times \frac{1}{2}$

**30.** $\frac{5}{16} \times \frac{1}{2} \times \frac{1}{5}$

**31.** $\frac{3}{4} \times \frac{1}{6} \times \frac{1}{9}$

**32.** $-\frac{5}{8} \times \left(-\frac{3}{25}\right) \times \left(-\frac{1}{9}\right)$

**33.** $-2\frac{1}{4} \times 6\frac{1}{2} \times \frac{12}{39}$

**34.** $-5\frac{1}{8} \times \frac{24}{25} \times 10\frac{1}{2}$

**Multiply. Simplify the answer.**

$$EXAMPLE \quad \frac{3}{5} \times 15a = \frac{3}{\underset{1}{\cancel{5}}} \times \frac{\overset{3a}{\cancel{15a}}}{1} = \frac{3 \times 3a}{1 \times 1} = 9a$$

**35.** $\frac{2}{3} \times 9n$

**36.** $5 \times \frac{3x}{10}$

**37.** $-3 \times \frac{y}{6}$

**38.** $-5 \times \left(\frac{-7a}{10}\right)$

**39.** $\frac{2r}{5} \times \frac{1}{r}$

**40.** $\frac{-6}{s} \times \frac{s}{2}$

**41.** $\frac{3c}{5} \times \frac{5}{c}$

**42.** $\frac{2m}{7} \times \frac{14}{m}$

**C** **43.** Let $\frac{a}{b}$ be any fraction ($b \neq 0$). Show that $\frac{a}{b} \times b = a$ by supplying the rule or property that justifies each statement.

$\frac{a}{b} \times b = \left(a \times \frac{1}{b}\right) \times b$     Why?

$\quad\quad = a \times \left(\frac{1}{b} \times b\right)$     Why?

$\quad\quad = a \times \left(b \times \frac{1}{b}\right)$     Why?

$\quad\quad = a \times 1$     Why?

$\quad\quad = a$     Why?

**44.** Let $\frac{a}{b}$ and $\frac{c}{d}$ represent any two fractions ($b \neq 0, d \neq 0$). Show as in Exercise 43 that $\frac{a}{b} \times \frac{c}{d} = \frac{ac}{bd}$.

## Review Exercises

**Evaluate the expression when $x = -8$ and $y = 12$.**

**1.** $5 \times y$

**2.** $7x \div 4$

**3.** $3x - y$

**4.** $x + 8y$

**5.** $2(x + y)$

**6.** $x \div (y - 6)$

**7.** $(2y + 4) \div (x - 1)$

**8.** $(x \div 4) \times (y \div 3)$

# 3-9 Dividing Fractions

To develop a method for dividing fractions, recall that multiplication and division are inverse operations. If $2 \times n = 10$, then $n = 10 \div 2$. Similarly, if $\frac{7}{5} \times x = \frac{2}{3}$, then $x = \frac{2}{3} \div \frac{7}{5}$. We can show by substitution that the multiplication equation is true when the value of $x$ is $(\frac{2}{3} \times \frac{5}{7})$.

$$\frac{7}{5} \times \left(\frac{2}{3} \times \frac{5}{7}\right) = \frac{\overset{1}{\cancel{7}} \times 2 \times \overset{1}{\cancel{5}}}{\cancel{5} \times 3 \times \cancel{7}} = \frac{2}{3}$$

Therefore, the related division equation must also be true when the value of $x$ is $(\frac{2}{3} \times \frac{5}{7})$. That is,

$$\frac{2}{3} \times \frac{5}{7} = \frac{2}{3} \div \frac{7}{5}.$$

Two numbers, like $\frac{5}{7}$ and $\frac{7}{5}$, whose product is 1 are called **reciprocals.** The reciprocal of $\frac{c}{d}$ is $\frac{d}{c}$ because $\frac{c}{d} \times \frac{d}{c} = 1$. Every nonzero rational number has exactly one reciprocal.

We may state the following general rule.

---

## Rule

For all numbers $a$, $b$, $c$, and $d$ $(b \neq 0, c \neq 0, d \neq 0)$,

$$\frac{a}{b} \div \frac{c}{d} = \frac{a}{b} \times \frac{d}{c}$$

To divide by a fraction, multiply by its reciprocal.

---

**EXAMPLE 1**   Name the reciprocal, if any.

    **a.** $\frac{2}{3}$      **b.** $-\frac{5}{8}$      **c.** 2      **d.** 0

**Solution**

    **a.** The reciprocal of $\frac{2}{3}$ is $\frac{3}{2}$ since $\frac{2}{3} \times \frac{3}{2} = 1$.

    **b.** The reciprocal of $-\frac{5}{8}$ is $-\frac{8}{5}$ since $-\frac{5}{8} \times \left(-\frac{8}{5}\right) = 1$.

    **c.** The reciprocal of 2 is $\frac{1}{2}$ since $2 \times \frac{1}{2} = 1$.

    **d.** The equation $0 \times n = 1$ has no solution since 0 times any number is 0. Therefore, 0 has no reciprocal.

**EXAMPLE 2**   Divide.  Write the answer as a proper fraction in lowest terms or as a mixed number in simple form.

$$\text{a. } -\frac{4}{3} \div \frac{5}{8} \qquad \text{b. } 2\frac{1}{3} \div 1\frac{3}{8} \qquad \text{c. } \frac{\frac{2}{5}}{\frac{1}{4}}$$

**Solution**

a. $-\dfrac{4}{3} \div \dfrac{5}{8} = -\dfrac{4}{3} \times \dfrac{8}{5} = -\dfrac{32}{15}$, or $-2\dfrac{2}{15}$

b. $2\dfrac{1}{3} \div 1\dfrac{3}{8} = \dfrac{7}{3} \div \dfrac{11}{8} = \dfrac{7}{3} \times \dfrac{8}{11} = \dfrac{56}{33}$, or $1\dfrac{23}{33}$

c. $\dfrac{\frac{2}{5}}{\frac{1}{4}} = \dfrac{2}{5} \div \dfrac{1}{4} = \dfrac{2}{5} \times \dfrac{4}{1} = \dfrac{8}{5}$, or $1\dfrac{3}{5}$

**EXAMPLE 3**   Ann bought 2 packages of ground beef.  One package was $2\frac{1}{2}$ lb, and the other package was $3\frac{1}{8}$ lb.  Ann divided the total amount of beef into 5 equal packages for the freezer.  How many pounds were in each package?

**Solution**

- The problem asks for the number of pounds in each of the 5 packages.

- Given facts: $2\frac{1}{2}$ lb and $3\frac{1}{8}$ lb of beef
  total divided into 5 equal packages

- To solve, first add to find the total amount of beef, and then divide to find the amount in each package.

$$2\frac{1}{2} + 3\frac{1}{8} = 2\frac{4}{8} + 3\frac{1}{8} = 5\frac{5}{8}$$

$$5\frac{5}{8} \div 5 = \frac{\overset{9}{\cancel{45}}}{8} \times \frac{1}{\underset{1}{\cancel{5}}} = \frac{9 \times 1}{8 \times 1} = \frac{9}{8}, \text{ or } 1\frac{1}{8}$$

Each package contained $1\frac{1}{8}$ lb of beef.

---

**Problem Solving Reminder**

When solving a problem, *review the problem solving strategies and tips* that you have learned.  As you work through the problems in the lesson, remember that you may need to supply additional information, eliminate extra information, or plan more than one step. Remember to reread the problem to be sure your answer is complete.

## Class Exercises

**State the reciprocal.**

**1.** 4       **2.** $-5$       **3.** $\frac{3}{4}$       **4.** $-\frac{5}{8}$       **5.** $\frac{2}{7}$       **6.** $\frac{11}{3}$

**7.** $-\frac{2}{3}$       **8.** $-\frac{6}{5}$       **9.** $\frac{11}{12}$       **10.** $\frac{3}{16}$       **11.** $-\frac{4}{7}$       **12.** $\frac{14}{5}$

**Complete.**

**13.** $\frac{2}{3} \div 5 = \frac{2}{3} \times \underline{\phantom{?}}$            **14.** $\frac{2}{3} \div \frac{1}{5} = \frac{2}{3} \times \underline{\phantom{?}}$

**15.** $\frac{6}{5} \div (-10) = \underline{\phantom{?}} \times \left(-\frac{1}{10}\right)$       **16.** $-\frac{3}{4} \div \frac{3}{10} = \underline{\phantom{?}} \times \underline{\phantom{?}}$

---

## Written Exercises

**Divide. Write the answer as a proper fraction in lowest terms or as a mixed number in simple form.**

**A**   **1.** $\frac{2}{5} \div \frac{3}{5}$      **2.** $\frac{7}{8} \div \frac{3}{8}$      **3.** $5 \div \frac{1}{5}$      **4.** $10 \div \frac{1}{2}$

     **5.** $\frac{5}{8} \div \frac{9}{5}$      **6.** $\frac{3}{4} \div \frac{7}{8}$      **7.** $\frac{\frac{5}{9}}{\frac{1}{7}}$      **8.** $\frac{\frac{2}{3}}{\frac{3}{8}}$

     **9.** $-\frac{21}{4} \div \left(-\frac{7}{8}\right)$    **10.** $-\frac{4}{5} \div \left(-\frac{36}{25}\right)$    **11.** $3\frac{1}{4} \div \frac{5}{8}$    **12.** $10\frac{1}{2} \div \frac{8}{9}$

     **13.** $-5\frac{5}{8} \div 10$    **14.** $11\frac{1}{9} \div 100$    **15.** $-6\frac{1}{3} \div \left(-\frac{19}{21}\right)$    **16.** $-10\frac{1}{3} \div \left(-\frac{31}{33}\right)$

**B**   **17.** $6\frac{1}{8} \div \left(\frac{8}{3} \times \frac{3}{7}\right)$      **18.** $-7 \div \left(\frac{21}{4} \times \frac{3}{7}\right)$      **19.** $\left(-1\frac{2}{3} \times \frac{18}{5}\right) \div 3$

     **20.** $-\left(\frac{35}{4} \times \frac{2}{7}\right) \div \left(-\frac{4}{3}\right)$    **21.** $-\left(3\frac{1}{5} \times \frac{5}{2}\right) \div \frac{8}{9}$    **22.** $\left(-2\frac{1}{6} \div \frac{1}{9}\right) \times \frac{1}{13}$

     **23.** $-\left(4\frac{1}{6} \div 5\right) \times \left(-\frac{2}{5}\right)$    **24.** $-3\frac{1}{8} \div 4 \div \left(-\frac{5}{4}\right)$    **25.** $5\frac{1}{3} \div 2\frac{2}{3} \div (-4)$

---

## Problems

**Solve.**

**A**   **1.** To the nearest million, the number of households in the United States having television sets was 4 million in 1950 and 76 million in 1980. Express the first number as a fraction of the second.

     **2.** A gasoline tank with a capacity of 15 gal is $\frac{3}{4}$ full. How many gallons will it take to fill the tank?

3. A town has raised $\frac{3}{8}$ of the $12,000 it needs to furnish its new library. How much more is it hoping to raise?

4. How many packages will $5\frac{1}{2}$ lb of raisins fill if each package holds 9 oz?

5. Karen Northrup worked $12\frac{1}{2}$ h last week and earned $50. What was her hourly rate of pay?

6. A television station released 300 balloons at an outdoor celebration. Of these, $\frac{3}{4}$ were orange. How many were orange?

7. Leo Delray earns $2 an hour for babysitting. If he works $3\frac{1}{4}$ h one evening, how much does he earn?

8. In a recent year there were 32,000 persons in the United States who had celebrated their 100th birthday. Of these, $\frac{3}{4}$ were women. How many were men?

B 9. One half of the class voted to have a picnic. One third of the class voted to hold a dinner instead. What fraction of the class wanted neither a picnic nor a dinner?

10. A picture measures $8\frac{3}{4}$ in. by 8 in. When framed it measures $10\frac{3}{4}$ in. by 10 in. How wide is each side of the frame?

C 11. Only $\frac{1}{5}$ of the downtown workers drive to work. Of those who do not drive, $\frac{3}{16}$ ride bicycles to work. What fraction of the workers ride bicycles to work?

12. Kevin's regular rate of pay is $4 per hour. When he works overtime, he earns $1\frac{1}{2}$ times as much per hour. How much will Kevin earn for $5\frac{1}{2}$ h of overtime work?

## Review Exercises

Use the inverse operation to write a related equation. Solve for the variable.

1. $n + 135 = 240$  2. $x - 45 = 52$  3. $4y = 156$  4. $9t = 504$

5. $12x = 432$  6. $r \div 17 = 27$  7. $38m = 418$  8. $a \div 22 = 33$

# 3-10 Fractions and Decimals

Any fraction can be represented as a decimal. You may recall that a fraction such as $\frac{3}{4}$ can be easily written as an equivalent fraction whose denominator is a power of 10, and then as a decimal. To represent $\frac{3}{4}$ as a decimal, we first write it as an equivalent fraction with denominator 100.

$$\frac{3}{4} = \frac{3 \times 25}{4 \times 25} = \frac{75}{100} = 0.75$$

For most fractions, however, we use the fact that $\frac{a}{b} = a \div b$ and divide numerator by denominator.

*EXAMPLE 1*  Write as a decimal: **a.** $-\frac{5}{16}$  **b.** $\frac{24}{55}$

*Solution*  **a.** First find $5 \div 16$.

$$\begin{array}{r} 0.3125 \\ 16\overline{)5.0000} \\ \underline{4\,8} \\ 20 \\ \underline{16} \\ 40 \\ \underline{32} \\ 80 \\ \underline{80} \\ 0 \end{array}$$

Therefore, $-\frac{5}{16} = -0.3125$.

The decimal $-0.3125$ is called a **terminating decimal** because the final remainder is 0 and the division ends.

**b.** Find $24 \div 55$.

$$\begin{array}{r} 0.43636 \\ 55\overline{)24.00000} \\ \underline{22\,0} \\ 2\,00 \\ \underline{1\,65} \\ 350 \\ \underline{330} \\ 200 \\ \underline{165} \\ 350 \\ \underline{330} \\ 20 \end{array}$$

Therefore, $\frac{24}{55} = 0.43636\ldots$.

The digits 36 continue to repeat without end. The decimal $0.43636\ldots$ is called a **repeating decimal.** We often write $0.43636\ldots$ as $0.4\overline{36}$, with a bar over the block of digits that repeats.

To say that $\frac{24}{55} = 0.43636\ldots$ means that the successive decimals 0.436, 0.4363, 0.43636, and so on, will come closer and closer to the value $\frac{24}{55}$.

We can predict when a fraction will result in a terminating decimal because the fraction in lowest terms has a denominator with no prime factors other than 2 and 5. Thus, the fraction $\frac{24}{55}$ does not result in a terminating decimal because its denominator has 11 as a prime factor.

When working with a mixed number, such as $-1\frac{5}{16}$ or $1\frac{24}{25}$, we may consider the mixed number as a sum of a whole number and a fraction, or we may rewrite the mixed number as an improper fraction and then divide.

If $a$ and $b$ are integers and $b \neq 0$, the quotient $a \div b$ is either a terminating decimal or a repeating decimal. The reason for this is that, for any divisor, the number of possible remainders at each step of the division is limited to the whole numbers less than the divisor. Sooner or later, either the remainder is 0 and the division ends, as in part (a) of Example 1, or one of the remainders reappears in the division as in part (b) of Example 1. Then the same block of digits will reappear in the quotient.

---

### Property

Every rational number can be represented by either a terminating decimal or a repeating decimal.

---

You already know how to write a terminating decimal as a fraction. Rewrite the decimal as a fraction whose denominator is a power of 10.

**EXAMPLE 2**   Write $-0.625$ as a fraction in lowest terms.

**Solution**   $-0.625 = -\dfrac{625}{1000} = -\dfrac{625 \div 125}{1000 \div 125} = -\dfrac{5}{8}$

The next example shows a method for writing a repeating decimal as a fraction.

**EXAMPLE 3**   Write $-1.\overline{21}$ as a fraction in lowest terms.

**Solution**   Let $n = 1.\overline{21}$.

Multiply both sides of the equation by a power of 10 determined by the number of digits in the block of repeating digits. Since there are

2 digits that repeat in the number $1.\overline{21}$, we multiply by $10^2$, or 100.

$$100n = 121.\overline{21}$$

Subtract: $\qquad \dfrac{n = \phantom{0}1.\overline{21}}{99n = 120}$

$$n = \frac{120}{99} = \frac{40}{33}$$

Thus, $-1.\overline{21} = -\dfrac{40}{33}$, or $-1\dfrac{7}{33}$.

---

## Property

Every terminating or repeating decimal represents a rational number.

---

Some decimals, such as those below, neither terminate nor repeat.

$$0.01001000100001\ldots \qquad\qquad 1.234567891011121314\ldots$$

The two decimals shown follow patterns, but they are not repeating patterns. The decimal on the right is made up of consecutive whole numbers beginning with 1.

Decimals that neither terminate nor repeat represent **irrational numbers.** Together, the rational numbers and the irrational numbers make up the set of **real numbers.** The number line that you have studied is sometimes called the **real number line.** For every point on the line, there is exactly one real number and for every real number there is exactly one point on the number line.

---

## Class Exercises

**Tell whether the decimal for the fraction is terminating or repeating. If the decimal is terminating, state the decimal.**

**1.** $\dfrac{1}{4}$      **2.** $\dfrac{5}{6}$      **3.** $2\dfrac{2}{5}$      **4.** $-\dfrac{9}{10}$      **5.** $-1\dfrac{1}{2}$      **6.** $\dfrac{13}{30}$

**State as a fraction in which the numerator is an integer and the denominator is a power of 10.**

**7.** $0.13$      **8.** $-0.9$      **9.** $1.4$      **10.** $-0.007$      **11.** $3.03$      **12.** $-5.001$

**State the repeating digit(s) for each decimal.**

**13.** $6.666\ldots$     **14.** $0.0444\ldots$     **15.** $6.050505\ldots$     **16.** $0.1666\ldots$

**17.** $5.1\overline{5}$     **18.** $0.\overline{422}$     **19.** $1.0\overline{6}$     **20.** $0.3\overline{64}$

---

## Written Exercises

**Write as a terminating or repeating decimal. Use a bar to show repeating digits.**

**A**  **1.** $\frac{1}{4}$     **2.** $\frac{1}{5}$     **3.** $\frac{2}{9}$     **4.** $\frac{3}{16}$     **5.** $\frac{9}{10}$     **6.** $-\frac{1}{18}$

    **7.** $-\frac{2}{3}$     **8.** $\frac{4}{9}$     **9.** $-\frac{3}{8}$     **10.** $\frac{3}{5}$     **11.** $-\frac{3}{25}$     **12.** $\frac{7}{15}$

    **13.** $1\frac{1}{10}$     **14.** $5\frac{2}{5}$     **15.** $\frac{7}{12}$     **16.** $-\frac{3}{11}$     **17.** $\frac{4}{15}$     **18.** $-4\frac{7}{8}$

    **19.** $-1\frac{7}{18}$     **20.** $2\frac{1}{9}$     **21.** $\frac{3}{20}$     **22.** $\frac{17}{36}$     **23.** $3\frac{2}{7}$     **24.** $\frac{5}{13}$

**Write as a proper fraction in lowest terms or as a mixed number in simple form.**

**25.** $0.05$     **26.** $0.005$     **27.** $-0.6$     **28.** $-2.1$     **29.** $2.07$

**30.** $-0.62$     **31.** $5.125$     **32.** $4.3$     **33.** $-1.375$     **34.** $-10.001$

**35.** $12.625$     **36.** $10.3$     **37.** $0.225$     **38.** $0.8375$     **39.** $-1.826$

**B**  **40.** $0.444\ldots$     **41.** $-0.555\ldots$     **42.** $0.0\overline{3}$     **43.** $-1.0\overline{1}$     **44.** $5.\overline{9}$

    **45.** $0.1515\ldots$     **46.** $-1.\overline{20}$     **47.** $0.3\overline{5}$     **48.** $0.7\overline{2}$     **49.** $-1.\overline{12}$

    **50.** $1.3\overline{62}$     **51.** $2.13\overline{4}$     **52.** $-8.0\overline{16}$     **53.** $0.\overline{123}$     **54.** $-5.\overline{862}$

**Tell whether the number is rational or irrational.**

**55.** $\frac{-13}{17}$      **56.** $1.515151\ldots$      **57.** $-3.72$      **58.** $2.121121112\ldots$

**Arrange the numbers in order from least to greatest.**

**59.** $3.0,\ 3.\overline{09},\ 3.00\overline{9},\ 3.1$        **60.** $0.182,\ 0.182\overline{5},\ 0.18\overline{2},\ 0.1\overline{8}$

**a. Express the first number as a fraction or mixed number.**
**b. Compare the first number with the second.**

**C**  **61.** $0.\overline{9};\ 1$     **62.** $0.4\overline{9};\ \frac{1}{2}$     **63.** $-1.2\overline{9};\ -\frac{5}{4}$     **64.** $2.3\overline{9};\ 2\frac{2}{5}$

## Self-Test B

**Perform the indicated operation. Write the answer as a proper fraction in lowest terms or as a mixed number in simple form.**

1. $\frac{1}{3} + \frac{1}{4}$

2. $\frac{2}{15} + \left(-\frac{5}{6}\right)$

3. $\frac{1}{5} - \frac{1}{3}$ 　　[3-6]

4. $17\frac{1}{3} + 5\frac{1}{9}$

5. $-6\frac{3}{8} - 3\frac{2}{3}$

6. $16\frac{5}{8} - \left(-\frac{3}{4}\right)$ 　　[3-7]

7. $\frac{3}{4} \times 5$

8. $\frac{1}{8} \times \left(-\frac{1}{3}\right)$

9. $-2\frac{4}{7} \times 3\frac{1}{6}$ 　　[3-8]

10. $\frac{5}{8} \div \frac{10}{24}$

11. $-\frac{11}{16} \div \frac{44}{8}$

12. $4\frac{1}{3} \div \left(-\frac{26}{27}\right)$ 　　[3-9]

**Write as a decimal. Use a bar to show repeating digits.**

13. $\frac{5}{8}$

14. $\frac{2}{11}$

15. $-\frac{1}{80}$

16. $\frac{7}{6}$ 　　[3-10]

**Write as a proper fraction in lowest terms or as a mixed number in simple form.**

17. $0.875$

18. $1.\overline{6}$

19. $-2.213$

20. $0.2\overline{3}$

---

*Self-Test answers and Extra Practice are at the back of the book.*

---

███ | **Computer Byte**

The following program will find the least common multiple of two numbers.

```
10   PRINT "TO FIND LCM:"
20   PRINT "INPUT A, B";
30   INPUT A,B
40   FOR X = 1 TO B
50   LET A1 = A * X
60   LET Q = A1 / B
70   IF Q = INT (Q) THEN 90
80   NEXT X
90   PRINT "LCM(";A;",";B;") = ";A1
100  END
```

**RUN the program to find the least common multiple of the following.**

1. 12, 25 　　2. 72, 84 　　3. 34, 60 　　4. 45, 80 　　5. 110, 240 　　6. 235, 180

# BASIC, A Computer Language

Computers are very powerful tools, but issuing an order such as "Do problem 12 on page 46 of my math book" will produce no results at all. There are many things computers can do more quickly and efficiently than people, but we need to communicate with computers in a special way to get them to work for us.

To tell a computer what to do, we write a set of instructions, called a *program,* using a *programming language.* Since most microcomputers use some version of BASIC (with slight differences), that is the language that we will use in this book. A BASIC program is made up of a set of *numbered lines* that provide step-by-step instructions for the computer. We can use any numbers from 1 to 99999 for line numbers, but we often use numbers in intervals of 10 so that we can insert other lines later if we need to.

A *statement* that tells the computer what to do follows each line number in a program. In the BASIC language, we use the symbols shown below to tell the computer to perform arithmetic operations.

| | | | |
|---|---|---|---|
| + | addition | — | subtraction |
| * | multiplication | / | division |

The symbol ↑ (or some similar symbol) is used to indicate exponentiation. Thus $3 \uparrow 6$ means $3^6$. When a statement contains more than one operation, the computer will perform all operations in parentheses first and will follow the order of operations that you learned in Chapters 1 and 2.

We use a **PRINT** statement to tell the computer to perform the operations listed in a statement and to print the result. We use an **END** statement to tell the computer that the program is over. The program shown below tells the computer to simplify the numerical expression and print the answer.

```
10   PRINT 5↑3 * (16 − 8 / 2)
20   END
```

After you have typed in this program (press RETURN or ENTER after each line), you type the *command* **RUN** to tell the computer to run (or *execute*) the program. The result, or *output,* is 1500.

A computer handles variables much as we do. We can ask it, for example, to give us the value of a variable expression when we give it a value of the variable in it. One way of doing this is to use an **INPUT** statement. This causes the computer to print a question mark and wait for the value to be typed in. Here is a simple program with a RUN shown at the right below.

```
10   INPUT X                          RUN
20   PRINT X↑2 + 2 * X + 4            ?10
30   END                                124
```

As you can see, we need some statement to tell the person using this program what is expected after the question mark. We do this by enclosing a descriptive expression in quotation marks in a PRINT statement, as in line 5 below. The semicolon at the end of line 5 will cause the question mark from line 10 to be printed right after the quoted expression. We have also inserted lines 12 and 15. After typing lines 5, 12, and 15, we can type the command **LIST** to see the revised program. A RUN is shown at the right below.

```
5    PRINT "WHAT IS YOUR VALUE OF X";
10   INPUT X                          RUN
12   PRINT "FOR X = ";X               WHAT IS YOUR VALUE OF X?10
15   PRINT "X↑2 + 2X + 4 = ";         FOR X = 10
20   PRINT X↑2 + 2 * X + 4            X↑2 + 2X + 4 = 124
30   END
```

1. Change lines 15 and 20 in the program above to evaluate another variable expression, with $x$ as the variable, that involves the operation or operations listed.
   a. subtraction
   b. multiplication
   c. division
   d. multiplication and addition
   e. division and subtraction

2. Change the program above to evaluate a variable expression using $m$ as the variable.

# Chapter Review

**Complete.**

**1.** 1, 2, 3, __?__, __?__, __?__, __?__, and __?__ are factors of 24.          [3–1]

**2.** 40 is divisible by __?__, __?__, __?__, __?__, __?__, __?__, __?__, and __?__.

**Write the letter of the correct answer.**

**3.** What is the next prime number after 47?          [3–2]

   **a.** 49          **b.** 51          **c.** 53          **d.** 57

**4.** What is the prime factorization of 72?

   **a.** $1 \cdot 72$          **b.** $6^2 \cdot 2$          **c.** $2^3 \cdot 3^2$          **d.** $3^3 \cdot 2^2$

**True or false?**

**5.** $8 \times \frac{1}{9} = \frac{9}{8}$          **6.** $4 \times \left(-\frac{1}{5}\right) = -\frac{4}{5}$          **7.** $\frac{-7}{11} = -\frac{7}{11}$          [3–3]

**8.** $-\frac{24}{148}$ in lowest terms is $-\frac{12}{74}$.   **9.** $7\frac{3}{8} = \frac{29}{8}$          [3–4]

**10.** The LCD of $\frac{2}{3}$ and $\frac{5}{11}$ is 33.   **11.** The LCD of $\frac{7}{30}$ and $-\frac{3}{35}$ is 150.          [3–5]

**Match.**

**12.** $\frac{5}{12} + \frac{11}{18}$          **13.** $\frac{1}{8} - \frac{3}{8}$          **A.** $-2\frac{11}{18}$          **B.** $-8\frac{17}{18}$          [3–6]

**14.** $-\frac{9}{16} - \left(-\frac{3}{4}\right)$          **15.** $\frac{2}{5} + \frac{3}{10}$          **C.** $\frac{37}{36}$          **D.** $\frac{3}{16}$

**16.** $-2\frac{1}{6} + \left(-\frac{4}{9}\right)$          **17.** $1\frac{3}{7} - \frac{5}{8}$          **E.** $\frac{27}{28}$          **F.** $-\frac{1}{4}$          [3–7]

**18.** $\frac{1}{4} \times \frac{1}{9}$          **19.** $\frac{4}{5} \times \frac{3}{8}$          **G.** $\frac{7}{10}$          **H.** $\frac{1}{36}$          [3–8]

**20.** $-7\frac{2}{3} \times 1\frac{1}{6}$          **21.** $\frac{2}{3} \times 5$          **I.** $3\frac{1}{3}$          **J.** $-1\frac{19}{30}$          [3–9]

**22.** $\frac{3}{14} \div \frac{2}{9}$          **23.** $1\frac{2}{5} \div \left(-\frac{6}{7}\right)$          **K.** $\frac{45}{56}$          **L.** $\frac{3}{10}$

**Complete. Use a bar to show repeating digits.**

**24.** $\frac{4}{3}$ written as a decimal is __?__.          [3–10]

# Chapter Test

**State which of the numbers 2, 3, 4, 5, 9, and 10 are factors of each number. Use the tests for divisibility.**

**1.** 822        **2.** 410        **3.** 315        **4.** 660      [3–1]

**Give the prime factorization of each number.**

**5.** 168        **6.** 96        **7.** 53        **8.** 111      [3–2]

**Complete.**

**9.** $5 \times \underline{\ ?\ } = \frac{5}{7}$      **10.** $6 \times \frac{1}{6} = \underline{\ ?\ }$      **11.** $\frac{-1}{4} = \frac{1}{-4} = \underline{\ ?\ }$      [3–3]

**Write as a fraction in lowest terms or as a mixed number in simple form.**

**12.** $\frac{14}{40}$      **13.** $-\frac{26}{52}$      **14.** $-\frac{136}{160}$      **15.** $\frac{15}{4}$      **16.** $\frac{-83}{9}$      [3–4]

**Write each pair of fractions as equivalent fractions with the least common denominator.**

**17.** $\frac{3}{7}, \frac{5}{14}$      **18.** $\frac{1}{9}, \frac{6}{11}$      **19.** $\frac{5}{8}, \frac{13}{36}$      **20.** $\frac{7}{12}, \frac{5}{24}$      [3–5]

**Perform the indicated operations. Write the answer as a proper fraction in lowest terms or as a mixed number in simple form.**

**21.** $\frac{3}{8} + \frac{1}{7}$      **22.** $-\frac{3}{10} + \left(-\frac{1}{4}\right)$      **23.** $\frac{1}{2} - \frac{5}{12}$      [3–6]

**24.** $1\frac{3}{4} + \left(-2\frac{1}{2}\right)$      **25.** $1\frac{1}{3} - \left(-2\frac{3}{4}\right)$      **26.** $-2\frac{7}{10} - \left(4\frac{9}{11}\right)$      [3–7]

**27.** $\frac{4}{7} \times \left(-\frac{4}{11}\right)$      **28.** $8\frac{1}{8} \times 2\frac{3}{5}$      **29.** $-\frac{1}{4} \times \left(-\frac{1}{6}\right)$      [3–8]

**30.** $\frac{2}{3} \div \frac{7}{9}$      **31.** $-\frac{6}{19} \div \frac{9}{38}$      **32.** $-1\frac{1}{6} \div \left(-4\frac{2}{3}\right)$      [3–9]

**Write as a decimal. Use a bar to show repeating digits.**

**33.** $\frac{6}{25}$      **34.** $\frac{5}{33}$      **35.** $\frac{9}{16}$      **36.** $\frac{4}{9}$      [3–10]

**Write as a fraction in lowest terms or as a mixed number in simple form.**

**37.** 0.04      **38.** −0.375      **39.** 6.125      **40.** $0.\overline{7}$

# Cumulative Review (Chapters 1–3)

## Exercises

**Give the solution of the equation for the given replacement set.**

**1.** $x - 12 = 46$; $\{58, 59, 60\}$      **2.** $8d = 2$; $\left\{\frac{1}{3}, \frac{1}{4}, \frac{1}{5}\right\}$

**3.** $3(6 + y) = 39$; $\{7, 14, 21\}$      **4.** $(5 - a) \div (a + 7) = \frac{1}{5}$; $\{1, 2, 3\}$

**5.** $16r - 11 = 69$; $\{0, 5, 10\}$      **6.** $m^2(7 + m) = 176$; $\{3, 4, 5\}$

**Evaluate the expression if $a = 3$, $b = 2$, and $c = 5$.**

**7.** $ab^2$      **8.** $a^2 + b^2$      **9.** $a^2c^2$      **10.** $bc^2$      **11.** $2b^2$

**12.** $a^2b^3$      **13.** $(a + b)^2$      **14.** $c(a^2 + b)$      **15.** $bc^3$      **16.** $\frac{10b^2}{c^2}$

**Use the symbol $>$ to order the numbers from greatest to least.**

**17.** 75.70, 75.40, 75.06      **18.** 19.05, 19.18, 19.50

**19.** 0.03, 0.30, 0.33      **20.** 105.07, 10.507, 1050.7

**List the integers that can replace $x$ to make the statement true.**

**21.** $|x| = 7$      **22.** $|x| = 18$      **23.** $|x| \geq 5$      **24.** $|x| < 7.3$

**25.** $4 < |x| < 6$      **26.** $6 > |x| > 0$      **27.** $31 < |x| < 40$      **28.** $17 \leq |x| \leq 25$

**What value of the variable makes the statement true?**

**29.** $-7 + y = 11$      **30.** $-3n = 51$      **31.** $-6(-b) = 72$

**32.** $17 - x = -5$      **33.** $8(-n) = 32$      **34.** $-a \div 15 = 3$

**Evaluate the expression when $x = 5$, $y = -7$, and $z = -5$.**

**35.** $\frac{y + 3}{x}$      **36.** $\frac{xy}{z}$      **37.** $\frac{x + y}{z}$      **38.** $\frac{yz}{x}$      **39.** $\frac{x - z}{y}$

**40.** $\frac{x^2 + y^2}{z^2}$      **41.** $\frac{x}{7} + \frac{y}{7}$      **42.** $\frac{2x}{15} + \frac{z}{3}$      **43.** $\frac{-y + z}{x}$      **44.** $\frac{-z - x}{-y}$

**Solve.** Write the answer as a proper fraction in lowest terms or as a mixed number in simple form.

**45.** $-5 \times \left(\frac{3}{8} \div \frac{1}{3}\right)$      **46.** $2\frac{1}{2} \div \left(\frac{5}{8} \times 1\frac{3}{4}\right)$      **47.** $7\frac{3}{8} + \left[-4\frac{1}{2} \div \left(-5\frac{2}{3}\right)\right]$

# Problems

**Solve.**

1. The Maxwell children have hired a caterer to provide food for an anniversary party for their parents. The caterer has quoted a price of $15.75 per person and is asking for an advance payment of $\frac{1}{4}$ of the total bill. If the estimated number of guests is 50, how much is the advance payment?

2. The Clean-as-a-Whistle Company provides a matching service for people looking for home cleaners and people wishing to clean homes. The fees include $6.50 per hour for the cleaner, plus $1.50 per hour for the agency. If you hire a cleaner from the company for 5 h, how much will you pay?

3. As a general rule for brick work, masons estimate 6.5 bricks per square foot. Based on this estimate, will 2500 bricks be enough for a patio that is 396 ft²?

4. The controller of a hospital found that laundry fees for a four-month period totaled $8755. Based on this total, what would be the estimated fee for an entire year?

5. Store owners at the Wagon Wheel Mall pay a monthly rental fee plus a maintenance fee. The maintenance fee is determined by the number of square feet occupied by the shop. The entire mall is 200,000 ft² and the annual fee for the entire mall is $63,000. What is the annual share of the maintenance fee for a store that occupies 2500 ft²?

6. An investor bought 12 acres of land for $70,000. She later subdivided the land into 22 lots that she sold for $4500 apiece. What was her profit on the sale?

7. Douglas bought 75 shares of Health Care Company (HCC) stock and 150 shares of Bowwow Brands (BWB) stock. Last year HCC paid a dividend of $1.85 per share and BWB paid a dividend of $2.04 per share. What was the total of the dividends that Douglas received?

# 4

# Solving Equations

NASA (National Aeronautics and Space Administration) is the government agency responsible for space exploration and experiments. The photograph shows a rocket about to be fired into space. It is carrying the space shuttle Columbia.

Aboard the shuttle is the Spacelab research station, where astronauts and other scientists will conduct in-flight scientific experiments. Much of the research will take months to analyze, but we know from earlier flights that the results will be of enormous importance in many fields, such as astronomy and medicine. For example, many of the experiments concern the effect of weightlessness on the human body.

## Career Note

Space scientists require a strong background in mathematics, physics, and related sciences. The work is demanding but exciting. The men and women who perform experiments in the Spacelab are selected because of their specialized knowledge in various fields, for example, in biology, chemistry, or medicine.

# 4-1 Equations: Addition and Subtraction

You have learned that an *equation* is a sentence that states that two expressions name the same number. A *replacement set,* consisting of values which may be substituted for the variable, is always stated or understood. For example, the equation

$$2x + 9 = 15$$

may have the replacement set $\{1, 3, 5\}$.

When a number from the replacement set makes an equation a true statement, it is called a *solution* of the equation. To determine whether a value in the replacement set $\{1, 3, 5\}$ is a solution of $2x + 9 = 15$, substitute it into the equation.

| | | |
|---|---|---|
| $2x + 9 = 15$ | $2x + 9 = 15$ | $2x + 9 = 15$ |
| $2(1) + 9 = 15$ | $2(3) + 9 = 15$ | $2(5) + 9 = 15$ |
| $2 + 9 = 15$ | $6 + 9 = 15$ | $10 + 9 = 15$ |
| $11 = 15$ | $15 = 15$ | $19 = 15$ |
| false | true | false |

Thus, 3 is a solution of the equation $2x + 9 = 15$.

If the replacement set for an equation is the set of whole numbers, it is not practical to use substitution to solve the equation. Instead, we **transform,** or change, the given equation into a simpler, **equivalent equation,** that is, one that has the same solution. When we transform the given equation, our goal is to arrive at an equivalent equation of the form

$$\text{variable} = \text{number}.$$

For example:

$$n = 5$$

The number, 5, is then the solution of the original equation. The following transformations can be used to solve equations.

---

Simplify numerical expressions and variable expressions.

*Transformation by addition:*   Add the same number to both sides.

*Transformation by subtraction:*   Subtract the same number from both sides.

---

**EXAMPLE 1**  Solve $x = 3 + 5$.

**Solution**  Simplify the numerical expression $3 + 5$.

$$x = 3 + 5$$
$$x = 8$$

The solution is 8.

**EXAMPLE 2**  Solve $x - 2 = 8$.

**Solution**  Our goal is to find an equivalent equation of the form

$$x = \text{a number.}$$

The left side of the given equation is $x - 2$. Recall that addition and subtraction are inverse operations. If we add 2 to both sides, the left side simplifies to $x$.

$$x - 2 = 8$$
$$x - 2 + 2 = 8 + 2$$
$$x = 10$$

The solution is 10.

**EXAMPLE 3**  Solve $x + 6 = -8$.

**Solution**  Subtract 6 from both sides of the equation to get an equivalent equation of the form $x = $ a number.

$$x + 6 = -8$$
$$x + 6 - 6 = -8 - 6$$
$$x = -14$$

The solution is $-14$.

In equations involving a number of steps, it is a good idea to check your answer. This can be done quite easily by substituting the answer in the original equation. Checking is illustrated in Example 4.

**EXAMPLE 4**  Solve $5 + x + 4 = 17$.

**Solution**

$$5 + x + 4 = 17$$
$$5 + 4 + x = 17$$
$$9 + x = 17$$
$$9 + x - 9 = 17 - 9$$
$$x = 8$$

Check: $5 + x + 4 = 17$
$$5 + 8 + 4 \stackrel{?}{=} 17$$
$$13 + 4 = 17 \ \checkmark$$

The solution is 8.

The following example shows how to solve an equation, such as $34 - x = 27$, in which the variable is being subtracted.

**EXAMPLE 5**   Solve $34 - x = 27$.

**Solution**   Add $x$ to both sides.
$$34 - x = 27$$
$$34 - x + x = 27 + x$$
$$34 = 27 + x$$

Subtract 27 from both sides.
$$34 - 27 = 27 + x - 27$$
$$7 = x$$

The solution is 7.

---

**Reading Mathematics:** *Study Skills*

Review the worked-out examples if you need help in solving any of the exercises. When doing so, be certain to read carefully and to make sure that you understand what is happening in each step.

You may be able to solve some of the equations in the exercises without pencil and paper. Nevertheless, it is important to show all the steps in your work and to make sure you can tell which transformation you are using in each step.

Throughout the rest of this chapter, if no replacement set is given for an equation, you should assume that the replacement set is the set of all numbers in our decimal system.

---

## Class Exercises

**State which transformation was used to transform the first equation into the second.**

1.     $x - 3 = 5$
       $x - 3 + 3 = 5 + 3$

2. $x = 8 + 7$
    $x = 15$

3. $-3 - 1 = 3 - x$
    $-4 = 3 - x$

4.     $x - 5 = 9$
       $x - 5 + 5 = 9 + 5$

**Complete each equation. State which transformation has been used.**

5. $x = 11 + 17$
    $x = \underline{\ ?\ }$

6.     $x - 11 = 12$
       $x - 11 + 11 = 12 + \underline{\ ?\ }$

7.     $x + 7 = 13$
       $x + 7 - 7 = 13 \underline{\ ?\ } 7$

8.     $4 - x = -2$
       $4 - x + x = -2 \underline{\ ?\ } x$

## Written Exercises

**Use transformations to solve each equation. Write down all the steps.**

**A**  **1.** $x + 15 = 27$

**3.** $x - 6 = -7$

**5.** $8 + (-12) = x$

**7.** $x = 38 - 15$

**9.** $x + 7 = 3(6 + 2)$

**11.** $23 = 30 - x$

**2.** $x - 8 = 21$

**4.** $19 + x = 35$

**6.** $3 + 16 = x$

**8.** $x = -24 + 15$

**10.** $34 = 4(3 - 1) + x$

**12.** $42 - x = -4$

**B**  **13.** $5(2 + 7) - x = 33$

**15.** $9 + 12 = (3 \times 5) + x$

**17.** $4(10 - 7) = x + 4$

**19.** $-\frac{3}{5} + n = -1$

**21.** $5 = 8\frac{2}{3} - a$

**23.** $n - 0.76 = 0.34$

**25.** $0.894 - y = 0.641$

**27.** $x - 0.323 = 0.873$

**29.** $3\frac{1}{2} + 5\frac{1}{4} = n - 1$

**31.** $n + 3\frac{1}{6} + 4\frac{1}{4} = 10$

**33.** $n + 0.813 - 0.529 = 0.642$

**14.** $6(8 - 3) = 48 - x$

**16.** $3(72 \div 12) - x = 5$

**18.** $22 + x = 4(39 \div 3)$

**20.** $n - \frac{3}{4} = 3$

**22.** $-4\frac{1}{5} + c = -2$

**24.** $n + 0.519 = 0.597$

**26.** $0.321 + r = 0.58$

**28.** $-0.187 + t = 0.67$

**30.** $4\frac{1}{3} - 1\frac{5}{6} = b + 1$

**32.** $n + 5\frac{7}{8} - 1\frac{1}{6} = 7$

**34.** $a + 0.952 - 0.751 = 0.7$

**C**  **35.** $8\left(4\frac{1}{10} - 3\frac{1}{4}\right) = y + 3$

**37.** $7(0.34 - 0.21) = b - 0.82$

**36.** $3\left(2\frac{1}{9} + 4\frac{5}{6}\right) = 25 - c$

**38.** $4(0.641 + 0.222) = n + 0.357$

## Review Exercises

**Complete.**

**1.** $5x \div \underline{\ ?\ } = x$

**4.** $\frac{x}{4} \times \underline{\ ?\ } = x$

**2.** $y \times 8 \div 8 = \underline{\ ?\ }$

**5.** $\frac{x}{9} \times \underline{\ ?\ } = x$

**3.** $3z \div \underline{\ ?\ } = z$

**6.** $\frac{z}{7} \times 7 = \underline{\ ?\ }$

# 4-2 Equations: Multiplication and Division

If an equation involves multiplication or division, the following transformations are used to solve the equation.

> *Transformation by multiplication:* Multiply both sides of the equation by the same nonzero number.
>
> *Transformation by division:* Divide both sides of the equation by the same nonzero number.

**EXAMPLE 1**  Solve $3n = 24$.

**Solution**    Our goal is to find an equivalent equation of the form
$$n = \text{a number.}$$
Use the fact that multiplication and division are inverse operations and that $3n \div 3 = n$.
$$3n = 24$$
$$\frac{3n}{3} = \frac{24}{3}$$
$$n = 8$$
The solution is 8.

**EXAMPLE 2**  Solve $-5x = 53$.

**Solution**    Divide both sides by 5.
$$-5x = 53$$
$$\frac{-5x}{-5} = \frac{53}{-5}$$
$$x = -10\frac{3}{5}$$
The solution is $-10\frac{3}{5}$.

**EXAMPLE 3**  Solve $\frac{n}{4} = 7$.

**Solution**    Multiply both sides by 4.
$$\frac{n}{4} = 7$$
$$\frac{n}{4} \times 4 = 7 \times 4$$
$$n = 28$$
The solution is 28.

## Class Exercises

a. State the transformation you would use to solve each equation.
b. Solve the equation.

**1.** $2n = 26$     **2.** $\frac{n}{2} = -5$     **3.** $\frac{n}{3} = 12$     **4.** $-3n = 12$

**5.** $5n = 35$     **6.** $33 = 11n$     **7.** $-7 = \frac{n}{4}$     **8.** $\frac{n}{5} = 20$

**9.** $4 + x = 7$     **10.** $4x = -24$     **11.** $28 = \frac{x}{7}$     **12.** $-11 = x - 2$

## Written Exercises

Use the transformations given in this chapter to solve each equation.
Show all steps. Check your solution.

**A**   **1.** $5n = 75$     **2.** $-87 = 3n$     **3.** $\frac{n}{3} = 6$     **4.** $-3n = 15$

    **5.** $7n = 42$     **6.** $\frac{n}{4} = -8$     **7.** $\frac{n}{6} = 9$     **8.** $-\frac{n}{5} = 6$

    **9.** $\frac{n}{4} = -21$     **10.** $10 = \frac{n}{6}$     **11.** $4 = \frac{n}{7}$     **12.** $5n = 55$

    **13.** $9n = 45$     **14.** $\frac{n}{8} = 9$     **15.** $11n = -110$     **16.** $8n = 96$

    **17.** $\frac{n}{13} = 7$     **18.** $\frac{n}{16} = -6$     **19.** $12n = 132$     **20.** $17n = -289$

**B**   **21.** $6x = 45$     **22.** $18x = 12$     **23.** $\frac{x}{17} = -13$     **24.** $\frac{x}{15} = 11$

    **25.** $-\frac{x}{21} = 12$     **26.** $\frac{x}{27} = 23$     **27.** $24x = -20$     **28.** $12x = 76$

    **29.** $9x = 80 + 7$     **30.** $-64x = 100 - 48$     **31.** $\frac{x}{26} = 2(3 + 4)$

    **32.** $42 + x = 179$     **33.** $x - 193 = 54$     **34.** $296 - x = -51$

**C**   **35.** $7x = \frac{14}{19}$     **36.** $11x = \frac{13}{20}$     **37.** $\frac{x}{16} = \frac{21}{32}$     **38.** $\frac{x}{9} = \frac{13}{84}$

## Review Exercises

Multiply or divide.

**1.** $\frac{5}{8} \times \frac{12}{35}$     **2.** $\frac{14}{30} \div \frac{2}{15}$     **3.** $\frac{17}{9} \times \frac{6}{85}$

**4.** $3\frac{2}{3} \times 5\frac{9}{11}$     **5.** $3\frac{5}{8} \div 1\frac{3}{16}$     **6.** $0.09 \times 3.74$

# 4-3 Equations: Decimals and Fractions

Sometimes the variable expression in an equation may involve a decimal. When this occurs, you can use the transformations that you have learned in the previous lessons.

**EXAMPLE 1**  Solve the equation $0.42x = 1.05$.

**Solution**  Divide both sides by 0.42.

$$0.42x = 1.05$$

$$\frac{0.42x}{0.42} = \frac{1.05}{0.42}$$

$$x = 2.5$$

The solution is 2.5.

**EXAMPLE 2**  Solve the equation $\frac{n}{0.15} = 92$.

**Solution**  Multiply both sides by 0.15.

$$\frac{n}{0.15} = 92$$

$$0.15 \times \frac{n}{0.15} = 0.15 \times 92$$

$$n = 13.80$$

The solution is 13.80.

The variable expression in an equation may also involve a fraction. To see how to solve an equation such as $\frac{2}{3}x = 6$, think how you would solve an equation such as $2x = 6$. (You would divide both sides by 2, getting $x = 3$.) Thus, to solve $\frac{2}{3}x = 6$, you would divide both sides by $\frac{2}{3}$. This is the same as multiplying by the reciprocal of $\frac{2}{3}$, or $\frac{3}{2}$. Therefore, we solve equations involving fractions in the following way.

> If an equation has the form
>
> $$\frac{a}{b}x = c,$$
>
> where both $a$ and $b$ are nonzero,
>
> multiply both sides by $\frac{b}{a}$, the reciprocal of $\frac{a}{b}$.

**EXAMPLE 3**   Solve the equation $\frac{1}{3}y = 18$.

**Solution**   Multiply both sides by 3, the reciprocal of $\frac{1}{3}$.

$$\frac{1}{3}y = 18$$

$$3 \times \frac{1}{3}y = 3 \times 18$$

$$y = 3 \times 18$$
$$y = 54$$

The solution is 54.

**EXAMPLE 4**   Solve the equation $\frac{6}{7}n = 8$.  Check.

**Solution**   Multiply both sides by $\frac{7}{6}$, the reciprocal of $\frac{6}{7}$.

$$\frac{6}{7}n = 8 \qquad\qquad \text{Check:} \qquad \frac{6}{7}n = 8$$

$$\frac{7}{6} \times \frac{6}{7}n = \frac{7}{6} \times 8 \qquad\qquad \frac{6}{7} \times \frac{28}{3} \overset{?}{=} 8$$

$$n = \frac{7}{\underset{3}{6}} \times \overset{4}{8} \qquad\qquad \frac{\overset{2}{6}}{\underset{1}{7}} \times \frac{\overset{4}{28}}{\underset{1}{3}} \overset{?}{=} 8$$

$$n = \frac{28}{3} \qquad\qquad \frac{2}{1} \times \frac{4}{1} = 8$$

The solution is $\frac{28}{3}$, or $9\frac{1}{3}$.

---

## Class Exercises

**a. State what number you would multiply or divide both sides of each equation by in order to solve it.**
**b. Solve the equation.**

**1.** $\frac{3}{4}x = 15$       **2.** $0.2x = 6$      **3.** $\frac{x}{0.4} = 1.7$      **4.** $\frac{x}{1.3} = 2.4$

**5.** $\frac{1}{8}x = 7$      **6.** $\frac{17}{12}x = 34$      **7.** $\frac{8}{3}x = 24$      **8.** $\frac{1}{9}x = 14$

**9.** $\frac{x}{1.8} = 2.9$      **10.** $\frac{x}{2.3} = 5$      **11.** $0.35x = 10.5$      **12.** $0.55x = 2.20$

**13.** $0.28x = 2.24$      **14.** $0.67x = 6.03$      **15.** $\frac{x}{2.6} = 3.7$      **16.** $\frac{x}{5.9} = 14.2$

## Written Exercises

Solve each equation.

**A**   **1.** $\frac{1}{8}x = 11$        **2.** $-\frac{1}{3}x = 13$        **3.** $0.4x = 8$        **4.** $0.6x = 24$

     **5.** $\frac{x}{0.3} = -5$        **6.** $\frac{x}{0.7} = 4$        **7.** $-\frac{2}{3}x = 16$        **8.** $\frac{3}{4}x = 21$

     **9.** $0.25x = 15$        **10.** $0.44x = -22$        **11.** $\frac{x}{1.5} = 13$        **12.** $-\frac{x}{2.2} = 22$

     **13.** $\frac{3}{2}x = 27$        **14.** $\frac{5}{9}x = -65$        **15.** $\frac{12}{5}x = 48$        **16.** $\frac{12}{7}x = 60$

     **17.** $-1.3x = 39$        **18.** $3.2x = 128$        **19.** $\frac{x}{4.5} = -11$        **20.** $\frac{x}{6.2} = 17$

**B**   **21.** $\frac{4}{3}n = 18$        **22.** $-\frac{6}{5}n = 20$        **23.** $\frac{8}{3}n = 28$        **24.** $\frac{6}{7}n = -21$

     **25.** $3.9 = 0.6n$        **26.** $3.6 = 1.6n$        **27.** $-1.5n = 1.2$        **28.** $1.25n = 3.5$

     **29.** $\left(2\frac{2}{5}\right)n = \frac{-4}{15}$        **30.** $\frac{3}{11} = \frac{9}{5}n$        **31.** $-\frac{7}{9} = \frac{14}{15}n$        **32.** $1\frac{5}{7}n = \frac{16}{35}$

     **33.** $\frac{n}{2.45} = 3.1$        **34.** $\frac{n}{6.31} = -2.12$        **35.** $\frac{n}{5.37} = 0.004$        **36.** $-\frac{n}{0.09} = 2.79$

**C**   **37.** $\frac{2}{3}x = 3.8$                **38.** $\frac{3}{4}x = -6.93$             **39.** $\frac{3}{5}x = 9.36$

     **40.** $\frac{x}{6.4} = \frac{5}{8}$               **41.** $\frac{x}{2.7} = \frac{11}{9}$              **42.** $\frac{x}{4.2} = \frac{17}{6}$

---

## Review Exercises

Solve.

  **1.** $x + 17 = 47$          **2.** $55 + x = 75$          **3.** $96 - x = 41$

  **4.** $82 - x = 37$          **5.** $x - 45 = 58$          **6.** $5x = 95$

  **7.** $3x = 51$          **8.** $7x = 91$          **9.** $4x = 76$

## ▮▮▮ Calculator Key-In

Using a calculator can greatly simplify the computations involved in solving an equation with decimals. Use a calculator to solve the following equations.

  **1.** $0.32x = 0.096$          **2.** $3.02x = 1.84$          **3.** $2.11x = 5.74$

  **4.** $\frac{x}{0.79} = 1.08$          **5.** $\frac{x}{1.91} = 1.77$          **6.** $\frac{x}{4.002} = 0.107$

# 4-4 Combined Operations

Many equations may be written in the form

$$ax + b = c,$$

where $a$, $b$, and $c$ are given numbers and $x$ is a variable. To solve such an equation, it is necessary to use more than one transformation.

**EXAMPLE 1**  Solve the equation $3n - 5 = 10 + 6$.

**Solution**  Simplify the numerical expression.

$$3n - 5 = 10 + 6$$
$$3n - 5 = 16$$

Add 5 to both sides.

$$3n - 5 + 5 = 16 + 5$$
$$3n = 21$$

Divide both sides by 3.

$$\frac{3n}{3} = \frac{21}{3}$$
$$n = 7$$

The solution is 7.

Example 1 suggests the following general procedure for solving equations.

> **1.** Simplify each side of the equation.
>
> **2.** If there are still indicated additions or subtractions, use the inverse operations to undo them.
>
> **3.** If there are indicated multiplications or divisions involving the variable, use the inverse operations to undo them.

It is important to remember that in using the procedure outlined above you must *always perform the same operation on both sides of the equation.* Also, you must use the steps in the procedure in the order indicated. That is, you first simplify each side of the equation, then undo additions and subtractions, and then undo multiplications and divisions.

**EXAMPLE 2**  Solve the equation $\frac{3}{2}n + 7 = -8$.

**Solution**  Subtract 7 from both sides.

$$\frac{3}{2}n + 7 = -8$$

$$\frac{3}{2}n + 7 - 7 = -8 - 7$$

$$\frac{3}{2}n = -15$$

Multiply both sides by $\frac{2}{3}$, the reciprocal of $\frac{3}{2}$.

$$\frac{2}{3} \times \frac{3}{2}n = \frac{2}{3} \times (-15)$$

$$n = \frac{2}{3} \times (-\overset{5}{15})$$

$$n = -10$$

The solution is $-10$.

**EXAMPLE 3**  Solve the equation $40 - \frac{5}{3}n = 15$.

**Solution**  Add $\frac{5}{3}n$ to both sides.

$$40 - \frac{5}{3}n = 15$$

$$40 - \frac{5}{3}n + \frac{5}{3}n = 15 + \frac{5}{3}n$$

$$40 = 15 + \frac{5}{3}n$$

Subtract 15 from both sides.

$$40 - 15 = 15 + \frac{5}{3}n - 15$$

$$25 = \frac{5}{3}n$$

Multiply both sides by $\frac{3}{5}$.

$$\frac{3}{5} \times 25 = \frac{3}{5} \times \frac{5}{3}n$$

$$\frac{3}{\overset{}{5}} \times \overset{5}{25} = n$$

$$15 = n$$

The solution is 15.

## Class Exercises

**State the two transformations you would use to find the solution of each equation. Be sure to specify which transformation you would use first.**

**1.** $3n + 2 = -10$

**2.** $4n - 1 = 19$

**3.** $\frac{1}{2}n - 6 = 1$

**4.** $\frac{1}{3}n + 5 = 7$

**5.** $\frac{2}{3}n - 6 = -12$

**6.** $\frac{5}{2}n + 2 = 13$

**7.** $3n - 6 = 15$

**8.** $7n + 21 = -63$

**9.** $\frac{3}{4}n - 8 = 12$

**10.** $\frac{1}{2}n + 2 = -5$

**11.** $2\frac{1}{3}n - 2 = 8$

**12.** $1\frac{2}{3}n + 15 = -21$

## Written Exercises

**Solve each equation.**

**A**

**1.** $2n - 5 = 17$

**2.** $3n + 6 = -24$

**3.** $5n + 6 = 41$

**4.** $4n - 15 = 9$

**5.** $6n + 11 = 77$

**6.** $8n - 13 = 51$

**7.** $50 - 3n = 20$

**8.** $42 - 5n = 7$

**9.** $29 - 6n = 11$

**10.** $-79 - 8n = -15$

**11.** $\frac{1}{4}n + 5 = 25$

**12.** $\frac{1}{8}n - 11 = 21$

**13.** $\frac{1}{2}n + 3 = 18$

**14.** $\frac{1}{3}n - 7 = -11$

**15.** $\frac{1}{5}n - 2 = 9$

**16.** $\frac{1}{4}n + 3 = 8$

**17.** $\frac{2}{3}n + 12 = 28$

**18.** $\frac{3}{5}n + 11 = -7$

**19.** $6n - 7 = 19$

**20.** $10n - 6 = -39$

**21.** $\frac{6}{5}n - 7 = 20$

**22.** $\frac{15}{4}n + 7 = -68$

**23.** $2\frac{2}{5}n + 5 = 23$

**24.** $1\frac{1}{7}n - 9 = 27$

**B**

**25.** $\frac{3}{5}n + \frac{2}{3} = \frac{8}{3}$

**26.** $\frac{2}{3}n - \frac{5}{6} = -\frac{1}{8}$

**27.** $\frac{3}{4}n - \frac{11}{15} = \frac{3}{5}$

**28.** $\frac{5}{6}n + \frac{1}{10} = \frac{29}{30}$

**29.** $\frac{7}{8}n - \frac{5}{6} = \frac{3}{4}$

**30.** $\frac{1}{3}n - \frac{11}{25} = \frac{3}{10}$

**31.** $\frac{2}{5}n + \frac{3}{7} = \frac{11}{5}$

**32.** $\frac{1}{6}n + \frac{3}{5} = \frac{7}{11}$

**33.** $1\frac{1}{3}n + \frac{5}{12} = \frac{3}{4}$

**34.** $2\frac{2}{3}n - \frac{4}{7} = \frac{8}{9}$

**35.** $\frac{11}{3}n - \frac{5}{9} = \frac{5}{6}$

**36.** $\frac{7}{2}n - \frac{11}{12} = \frac{5}{9}$

**37.** $1\frac{3}{8}n + \frac{1}{4} = \frac{7}{8}$

**38.** $\frac{3}{4}n - \frac{1}{12} = \frac{7}{3}$

**39.** $\frac{3}{7}n + \frac{4}{5} = \frac{6}{7}$

**C**  **40.** Solve $C = 2\pi r$ for $r$ if $C = 220$ and $\pi \approx \frac{22}{7}$.

**41.** Solve $P = 2l + 2w$ for $w$ if $P = 64$ and $l = 5$.

**42.** Solve $d = rt$ for $r$ if $d = 308$ and $t = 3.5$.

## Self-Test A

**Use transformations to solve each equation.**

**1.** $y + 7 = -17$

**2.** $x - 6 = 4$ [4-1]

**3.** $x + 6 = 27 - 12$

**4.** $5(7 + 2) = x - 11$

**5.** $7a = -91$

**6.** $-\frac{c}{4} = 17$ [4-2]

**7.** $-4t = -68$

**8.** $\frac{p}{3} = -11$

**9.** $\frac{1}{6}z = 25$

**10.** $-\frac{13}{3}y = 10$ [4-3]

**11.** $0.35a = -28$

**12.** $\frac{x}{0.03} = -58$

**13.** $8n - 40 = 180$

**14.** $\frac{2}{3}n + 18 = 98$ [4-4]

**15.** $\frac{1}{3}x + 4 = 6$

**16.** $5x - 2 = -17$

*Self-Test answers and Extra Practice are at the back of the book.*

---

■■■  **Calculator Key-In**

**Do the following on your calculator.**

**1.** Press any 3 digits.             852
**2.** Repeat the digits.           852,852
**3.** Divide by 7.                     ?
**4.** Divide by 11.                    ?
**5.** Divide by 13.                    ?

What is your answer? Now multiply your answer by 1001. Try again
using three different digits. Explain your results.

# 4-5 Writing Expressions for Word Phrases

In mathematics we often use symbols to translate word phrases into mathematical expressions. The same mathematical expression can be used to translate many different word expressions. Consider the phrases below.

Three more than a number $n$          The sum of three and a number $n$

Written as a variable expression, each of the phrases becomes
$$3 + n.$$

Notice that both the phrase *more than* and the phrase *the sum of* indicate addition.

The following are some of the word phrases that we associate with each of the four operations.

| + | − | × | ÷ |
|---|---|---|---|
| add | subtract | multiply | divide |
| sum | difference | product | quotient |
| plus | minus | times | |
| total | remainder | | |
| more than | less than | | |
| increased by | decreased by | | |

**EXAMPLE 1**   Write a variable expression for the word phrase.
a. A number $t$ increased by nine      b. Sixteen less than a number $q$
c. A number $x$ decreased by twelve, divided by forty
d. The product of sixteen and the sum of five and a number $r$

**Solution**   a. In this expression, the phrase *increased by* indicates that the operation is addition.     $t + 9$

b. In this expression, the phrase *less than* indicates that the operation is subtraction.     $q - 16$

c. In this expression, the phrases *decreased by* and *divided by* indicate that two operations, subtraction and division, are involved.     $(x - 12) \div 40$

d. In this expression, the words *product* and *sum* indicate that multiplication and addition are involved.
$16 \times (5 + r),$   or   $16(5 + r)$

**Reading Mathematics:** *Attention to Order*
Often a word expression contains more than one phrase that indicates an operation. Notice in Example 1 parts (c) and (d), on page 143 how parentheses were needed to represent the word phrase accurately. When translating from words to symbols, be sure to include grouping symbols if they are needed to make the meaning of an expression clear.

Many words that we use in everyday speech indicate operations or relationships between numbers. *Twice* and *doubled,* for example, indicate multiplication by 2. *Consecutive* whole numbers are whole numbers that differ by 1. The *preceding* whole number is the whole number *before* a particular number, and the *next* whole number is the whole number *after* a particular number.

**EXAMPLE 2**    If $2n$ is a whole number, represent (a) the preceding whole number and (b) the next four consecutive whole numbers.

**Solution**    **a.** The preceding whole number is 1 less than $2n$, or $2n - 1$.

**b.** Each of the next whole numbers is 1 more than the whole number before.
$$2n + 1, \; 2n + 2, \; 2n + 3, \; 2n + 4$$

## Class Exercises

**Match.**

1. A number $x$ multiplied by fourteen

2. The quotient of fourteen and a number $x$

3. Fourteen less than a number $x$

4. Seven increased by a number $x$

5. A number $x$ subtracted from fourteen

6. Fourteen more than a number $x$

7. Seven more than the product of fourteen and a number $x$

8. Twice the sum of a number $x$ and seven

9. The product of seven and the sum of fourteen and a number $x$

10. Fourteen divided by the difference between seven and $x$

**A.** $14 - x$

**B.** $2(x + 7)$

**C.** $14 \div (7 - x)$

**D.** $7 + 14x$

**E.** $14 \div x$

**F.** $7(14 + x)$

**G.** $x - 14$

**H.** $7 + x$

**I.** $14x$

**J.** $x + 14$

## Written Exercises

**Write a variable expression for the word phrase.**

A  1. The product of eight and a number $b$

2. A number $q$ divided by sixteen

3. A number $d$ subtracted from fifty-three

4. Four less than a number $f$

5. Thirty increased by a number $t$

6. Five times a number $c$

7. The sum of a number $g$ and nine

8. A number $k$ minus twenty-seven

9. Seventy-eight decreased by a number $m$

10. A number $y$ added to ninety

11. Nineteen more than a number $n$

12. Sixty-two plus a number $h$

13. The quotient when a number $d$ is divided by eleven

14. The difference when a number $a$ is subtracted from a number $b$

15. The remainder when a number $z$ is subtracted from twelve

16. The total of a number $x$, a number $y$, and thirteen

17. Fifteen more than the product of a number $t$ and eleven

18. The quotient when a number $b$ is divided by nine, decreased by seven

19. The sum of a number $m$ and a number $n$, multiplied by ninety-one

20. Forty-one times the difference when six is subtracted from a number $a$

21. A number $r$ divided by the difference between eighty-three and ten

22. The total of a number $p$ and twelve, divided by eighteen

23. The product of a number $c$ and three more than the sum of nine and twelve

24. The sum of a number $y$ and ten, divided by the difference when a number $x$ is decreased by five

**25.** The total of sixty, forty, and ten, divided by a number $d$

**26.** The product of eighteen less than a number $b$ and the sum of twenty-two and forty-five

**B** **27.** The greatest of four consecutive whole numbers, the smallest of which is $b$

**28.** The smallest of three consecutive whole numbers, the greatest of which is $q$

**29.** The greatest of three consecutive even numbers following the even number $x$

**30.** The greatest of three consecutive odd numbers following the odd number $y$

**31.** The value in cents of $q$ quarters

**32.** The number of inches in $f$ feet

**33.** The number of hours in $x$ minutes

**34.** The number of dollars in $y$ cents

**C** **35.** The difference between two numbers is ten. The greater number is $x$. Write a variable expression for the smaller number.

**36.** One number is six times another. The greater number is $a$. Write a variable expression for the smaller number.

## Review Exercises

**Use the inverse operation to solve for the variable.**

**1.** $x + 32 = 59$  **2.** $2y = 68$  **3.** $a \div 8 = 72$  **4.** $q - 14 = -23$

**5.** $-7c = 105$  **6.** $y - 11 = 21$  **7.** $n - 6 = 13$  **8.** $p + 3 = -39$

███ **Challenge**

In the set of whole numbers, there are two different values for $a$ for which this equation is true.

$$a + a = a \times a$$

What are they?

# 4-6 Word Sentences and Equations

Just as word phrases can be translated into mathematical expressions, word sentences can be translated into equations.

**EXAMPLE 1**  Write an equation for the word sentence.

    **a.** Twice a number $x$ is equal to 14.
    **b.** Thirty-five is sixteen more than a number $t$.

**Solution**
    **a.** First, write the phrase *twice a number* $x$ as the variable expression $2x$. Use the symbol $=$ to translate *is equal to*.

$$2x = 14$$

    **b.** Use the equals sign to translate *is*. Write *sixteen more than a number as* $t + 16$.

$$35 = t + 16$$

A word sentence may involve an unknown number without specifying a variable. When translating such a sentence into an equation, we may use any letter to represent the unknown number.

**EXAMPLE 2**  Write an equation for the word sentence.
    **a.** The sum of a number and seven is thirteen.
    **b.** A number increased by six is equal to three times the number.

**Solution**
    **a.** Let $n$ stand for the unknown number. $n + 7 = 13$
    **b.** Let $x$ stand for the unknown number. $x + 6 = 3x$

---

## Class Exercises

**Write a problem that each equation could represent. Use the words in parentheses as the subject of the problem.**

**1.** $x + 16 = 180$  (number of students in the eighth grade)

**2.** $0.59x = 2.36$  (buying groceries)

**3.** $48x = 630$  (traveling in a car)

**4.** $x - 25 = 175$  (number of cars in a parking lot)

**5.** $256 - x = 219$  (price reduction in a department store)

**6.** $10x = 26$  (running race)

**7.** $25x = 175$  (fuel economy in a car)

## Written Exercises

**Write an equation for the word sentence.**

**A**  1. Five times a number $d$ is equal to twenty.

2. A number $t$ increased by thirty-five is sixty.

3. Seven less than the product of a number $w$ and three equals eight.

4. The difference when a number $z$ is subtracted from sixteen is two.

5. Five divided by a number $r$ equals forty-two.

6. The sum of a number and seven is equal to nine.

7. A number decreased by one equals five.

8. Twelve equals a number divided by four.

9. Twice a number, divided by three, is fifteen.

10. The product of a number and eight, decreased by three, is equal to nine.

**B**  11. The quotient when the sum of four and $x$ is divided by two is thirty-four.

12. The sum of $n$ and twenty-two, multiplied by three, is seventy-eight.

13. Fifty-nine minus $x$ equals the sum of three and twice $x$.

14. Two increased by eight times $c$ is equal to $c$ divided by five.

**C**  15. The quotient when the difference between $x$ and 5 is divided by three is 2.

16. Twice a number is equal to the product when the sum of the number and four is multiplied by eight.

---

## Review Exercises

**Write a mathematical expression for each word phrase.**

1. Four less than a number

2. Five times a number

3. A number divided by seven

4. Ten more than a number

5. Forty minus a number

6. Twelve plus a number

7. A number times two

8. Ninety divided by a number

# 4-7 Writing Equations for Word Problems

In order to represent a word problem by an equation, we first read the problem carefully.

Next, we decide what numbers are being asked for. We then choose a variable and use it with the given conditions of the problem to represent the number or numbers asked for.

Now, we write an equation based on the given conditions of the problem. To do this, we write an expression involving the variable and set it equal to another variable expression or a number given in the problem that represents the same quantity.

The following example illustrates this procedure.

**EXAMPLE 1**   Write an equation for the following word problem.

Fran spent 3 times as long on her homework for English class as on her science homework. If she spent a total of 60 min on homework, how long did she spend on her science homework?

**Solution**
- The problem asks how long Fran spent on her science homework.

- Let $t$ = time spent on science. Since Fran spent 3 times as long on her English homework, she spent $3t$ on English. Therefore, the expression $t + 3t$ represents the amount of time spent on the two assignments. We are given that the total amount of time was 60 min.

- An equation that represents the conditions is

$$t + 3t = 60$$

## Class Exercises

a. **Name the quantity you would represent by a variable.**
b. **State an equation that expresses the conditions of the word problem.**

1. Jennifer bought 5 lb of apples for $3.45. What was the price per pound?

2. After Henry withdrew $350 from his account, he had $1150 left. How much money was in his account before this withdrawal?

3. A road that is 8.5 m wide is to be extended to 10.6 m wide. What is the width of the new paving?

4. In the seventh grade, 86 students made the honor roll. This is $\frac{2}{5}$ of the entire class. How many students are in the seventh grade?

**a. Name the quantity you would represent by a variable.**

**b. State an equation that expresses the conditions of the word problem.**

5. A 25-floor building is 105 m tall. What is the height of each floor if they are all of equal height?

6. A 150 L tank in a chemical factory can be filled by a pipe in 60 s. At how many liters per second does the liquid enter the tank?

## Problems

**Choose a variable and write an equation for each problem.**

**A** 1. On one portion of a trip across the country, the Oates family covered 1170 miles in 9 days. How many miles per day is this?

2. Carey's car went 224 km on 28 L of gas. How many kilometers per liter is this?

3. Marge has purchased 18 subway tokens at a total cost of $13.50. What is the cost per token?

4. After $525 was spent on the class trip, there was $325 left in the class treasury. How much was in the treasury before the trip?

5. By the end of one month a hardware store had 116 socket wrench sets left out of a shipment of 144. How many sets were sold during the month?

6. After depositing his tax refund of $350, Manuel Ruiz had $1580 in his bank account. How much money was in the account before the deposit?

7. If Mary Ling follows her usual route to work, she travels 12 km on one road and 18 km on another. If she takes a short cut her total distance to work is 19 km. How many kilometers does she save by taking the short cut?

8. A department store's total receipts from a sale on pillowcases were $251.64. If 36 pillowcases were sold, what was the price of each?

9. An investor has deposited $2000 into a special savings account. Two years later, the balance in the account is $2650. How much interest has been earned?

10. Deane's account in the company credit union had a balance of $3155. After she made a withdrawal to pay for car repairs, her balance was $2855. How much money did Deane withdraw?

**B** 11. In a school election $\frac{4}{5}$ of the students voted. There were 180 ballots. How many students are in the school?

12. In a heat-loss survey it was found that $\frac{3}{10}$ of the total wall area of the Gables' house consists of windows. The combined area of the windows is 240 ft². What is the total wall area of the house?

13. The difference between twice a number and thirty is 20. What is the number?

14. A bookstore received a shipment of books. Twenty were sold and $\frac{2}{5}$ of those remaining were returned to the publisher. If 48 books were returned, how many books were in the original shipment?

**C** 15. By mass, $\frac{1}{9}$ of any quantity of water consists of hydrogen. What quantity of water contains 5 g of hydrogen?

16. The balance in Peter Flynn's savings account is $6800. A withdrawal of $1000 is made, and the balance is to be withdrawn in 40 equal installments. What is the amount of each installment?

## Review Exercises

**Write an equation for each word sentence.**

1. Eight less than a number is forty-three.

2. Twelve times a number is one hundred eight.

3. Fourteen more than a number is seventy.

4. A number divided by nine is twenty-two.

5. A number minus seventeen is thirty-four.

6. Five times a number is sixty-five.

### ▮▮▮ | Challenge

A thoroughbred is 80 m ahead of a quarter horse, and is running at the rate of 27 m/s. The quarter horse is following at the rate of 31 m/s. In how many seconds will the quarter horse overtake the thoroughbred?

# 4-8 Solving Word Problems

The following five-step method will be helpful in solving word problems using an equation.

> ## Solving a Word Problem Using an Equation
>
> **Step 1** Read the problem carefully. Make sure that you understand what it says. You may need to read it more than once.
>
> **Step 2** Decide what numbers are asked for. Choose a variable and use it with the given conditions of the problem to represent the number(s) asked for.
>
> **Step 3** Write an equation based on the given conditions.
>
> **Step 4** Solve the equation and find the required numbers.
>
> **Step 5** Check your results with the words of the problem. Give the answer.

**EXAMPLE 1**  An evergreen in Sam's yard is now 78 in. tall. If it grows 6 in. each year, how many years will it take to grow to a height of 105 in.?

**Solution**

- The problem says
  present tree height, 78 in.
  tree growth per year, 6 in.
  future tree height, 105 in.

- The problem asks for
  number of years for tree to grow to 105 in.

  Let $n$ = number of years for tree to grow to 105 in.

  Since the tree grows 6 in. per year:
  height after 1 year, $78 + 6$
  height after 2 years, $78 + (6 \times 2)$
  height after 3 years, $78 + (6 \times 3)$
  height after $n$ years, $78 + 6n$

- We now have two expressions for the height of the tree after $n$ years. We set them equal to each other.

$$78 + 6n = 105$$

- Solve.
$$78 + 6n = 105$$
$$78 + 6n - 78 = 105 - 78$$
$$6n = 27$$
$$\frac{6n}{6} = \frac{27}{6}$$
$$n = 4\frac{1}{2}$$

- Check: If the tree grows 6 in. each year, in $4\frac{1}{2}$ years it will grow $4\frac{1}{2} \times 6$, or 27 in. The tree is now 78 in. tall. $78 + 27 = 105$. The result checks.

In $4\frac{1}{2}$ years the tree will grow to 105 in.

**EXAMPLE 2**  Two fifths of the members of the Riverview Sailing Club have signed up in advance for a club-wide race. On the day of the race, seven more members sign up, bringing the total of 41. How many members does the club have?

**Solution**
- The problem says
   two fifths of the members have signed up in advance
   the sum of this number and 7 is 41

- The problem asks for
   number of members in the club.

   Let $n$ = number of members in the club.
   Two fifths of the members, or $\frac{2}{5}n$, have signed up in advance.
   The sum of this number and 7, or $\frac{2}{5}n + 7$, is 41.

- We now have two expressions for the same number. We set them equal to each other.
$$\frac{2}{5}n + 7 = 41$$

- Solve.
$$\frac{2}{5}n + 7 = 41$$
$$\frac{2}{5}n + 7 - 7 = 41 - 7$$
$$\frac{2}{5}n = 34$$
$$\frac{5}{2} \times \frac{2}{5}n = \frac{5}{2} \times 34$$
$$n = 85$$

- Check: Two fifths of 85 is 34, and $34 + 7 = 41$. The result checks.

The club has 85 members.

## Problems

**Solve each problem using the five-step method.**

**A** 1. A mineralogist has learned that $\frac{2}{5}$ of a certain ore is pure copper. If a quantity of this ore yields 100 lb of pure copper, how large is the quantity?

2. Three tenths of the seats in a college's football stadium are reserved for alumni on Homecoming Weekend. This amounts to 4800 seats. What is the capacity of the stadium?

3. Three fourths of all the books in a school library are nonfiction. There are 360 nonfiction books. How many books are in the school library altogether?

4. Charles has $800 in a savings account. If he decides to deposit $40 into the account each week, how many weeks will it take for the account balance to reach $2000?

5. A department store received five cartons of shirts. One week later 25 shirts had been sold and 95 shirts were left in stock. How many shirts came in each carton?

6. The four walls in Fran's room have equal areas. The combined area of the doors and windows of the room is 12 m². If the total wall area (including doors and windows) is 68 m², what is the area of one wall?

**B** 7. Three fifths of those attending a club picnic decided to play touch football. After one more person decided to play, there were 16 players. How many people attended the picnic?

8. Four fifths of the athletic club treasury was to be spent on an awards banquet. After $450 was paid for food, there was $150 left out of the funds designated for the banquet. How much money had been in the treasury originally?

9. After using $\frac{2}{3}$ of a bag of fertilizer on his garden, Kent gave 8 lb to a neighbor. If Kent had 42 lb left, how much had been in the full bag?

**10.** Marcie biked to a point 5 km from her home. After a short rest, she then biked to a point 55 km from her home along the same road. If the second part of her trip took 4 hours, what was Marcie's speed, assuming her speed was constant?

**C 11.** When the gas gauge on her car was on the $\frac{3}{8}$ mark, Karen pumped 15 gal of gas into the tank in order to fill it. How many gallons of gas does the tank in Karen's car hold?

## Self-Test B

**Write a variable expression for the word phrase.**

**1.** The product of twelve and a number $x$                                [4-5]

**2.** A number $d$ subtracted from sixty

**Match each problem with one of the following equations.**

**a.** $x - 18 = 72$      **b.** $18x = 72$      **c.** $x + 18 = 72$      **d.** $72x = 18$

**3.** At 72 km/h how many hours would it take a car to travel 18 km?      [4-6]

**4.** Seth bought 18 more model cars for his collection. If he now has 72 cars, how many did he have before?

**5.** After the first day, 18 people were eliminated from the tournament. If 72 people were still playing, how many people started the tournament?

**Choose a variable and write an equation for each problem.**

**6.** Luann's car gets 21 miles per gallon. How many gallons of gasoline will Luann use to drive 189 miles?      [4-7]

**7.** John Silver made a withdrawal of $450 from his bank account. If there is $1845 left in the account, how much did he have in the bank before the withdrawal?

**Solve.**

**8.** At a recent tennis tournament, $\frac{3}{4}$ of the new balls were used. If 309 balls were used, how many new balls were there at the start of the tournament?      [4-8]

*Self-Test answers and Extra Practice are at the back of the book.*

# Balancing Equations in Chemistry

The basic chemical substances that make up the universe are called **elements.** Scientists often use standard symbols for the names of elements. Some of these symbols are given in the table below.

| Element | Symbol | Element | Symbol | Element | Symbol |
|---------|--------|---------|--------|---------|--------|
| Hydrogen | H | Helium | He | Carbon | C |
| Nitrogen | N | Oxygen | O | Sodium | Na |
| Aluminum | Al | Sulfur | S | Potassium | K |
| Chlorine | Cl | Copper | Cu | Iron | Fe |

The smallest particle of an element is called an **atom.** A pure substance made of atoms of two or more different elements is called a **compound.** In a compound the numbers of atoms of the elements always occur in a definite proportion. The formula for a compound shows this proportion. For example, the formula for water, $H_2O$, shows that in any sample of water there are twice as many hydrogen atoms as oxygen atoms.

In some elements and compounds, the atoms group together into **molecules.** The formula for a molecule is the same as the formula for the compound. The formula for oxygen is $O_2$ because a molecule of oxygen is made of two oxygen atoms. A molecule of water is made of two hydrogen atoms and one oxygen atom, so its formula is $H_2O$.

When compounds change chemically in a chemical reaction, we can describe this change by means of what chemists call an equation, although the equals sign is replaced by an arrow. An example of such a chemical equation is the following.

$$N_2 \quad + \quad 3\,H_2 \quad \longrightarrow \quad 2\,NH_3$$

nitrogen　　　　　hydrogen　　　　　ammonia

This equation indicates that a molecule of nitrogen (2 atoms) can combine with 3 molecules of hydrogen (2 atoms each) producing 2 molecules of ammonia.

Notice in the example that each side of the equation accounts for

$$\text{2 atoms of nitrogen: } N_2 \ldots \quad \longrightarrow \quad 2\,N \ldots$$

and　　　　　$\text{6 atoms of hydrogen: } \ldots 3\,H_2 \longrightarrow 2 \ldots H_3$

A chemical equation in which the same number of atoms of each element appears on both sides is said to be **balanced**. What number should replace the __?__ in order to balance the following equation?

$$S \quad + \quad 2\,H_2SO_4 \longrightarrow \underline{?}\,SO_2 \quad + \quad 2\,H_2O$$

sulfur　　　　sulfuric　　　　　sulfur　　　　water
　　　　　　　acid　　　　　　dioxide

To answer this question, let $n$ represent the unknown number. Then by equating the number of oxygen atoms on each side, we have

$$2 \times 4 = (n \times 2) + (2 \times 1)$$

or　　　　　　　　$8 = 2n + 2.$

Solving for $n$, we find that $n = 3$.

**Replace each __?__ with a whole number to produce a balanced equation.**

1. __?__ $NO_2$　　　　+　　　　$H_2O$ $\longrightarrow$ 2 $HNO_3$　　　　+　　　　NO
   nitric oxide　　　　　　　　water　　　　nitric acid　　　　　　nitrous acid

2. 4 $FeS_2$　　　　+　　　　__?__ $O_2$ $\longrightarrow$ 2 $Fe_2O_3$　　　　+　　　　8 $SO_2$
   iron sulfide　　　　　　　oxygen　　　　iron oxide　　　　　　sulfur dioxide

3. $Al(OH)_3$　　　　+　　　　3 HCl $\longrightarrow$ $AlCl_3$　　　　+　　　　__?__ $H_2O$
   aluminum　　　　　　　hydrochloric　　aluminum　　　　　　water
   hydroxide　　　　　　　acid　　　　　chloride

4. $C_2H_5OH$　　　　+　　　　__?__ $O_2$ $\longrightarrow$ 2 $CO_2$　　　　+　　　　3 $H_2O$
   ethanol　　　　　　　　oxygen　　　　carbon dioxide　　　　water

5. __?__ $C_2H_6$　　　　+　　　　__?__ $O_2$ $\longrightarrow$ 4 $CO_2$　　　　+　　　　6 $H_2O$
   ethane　　　　　　　　oxygen　　　　carbon dioxide　　　　water

# Chapter Review

**Complete.**

1. An equivalent equation for $2x - 4 = 16$ is $2x = $ __?__ .  [4-1]

2. The solution to $5(11 - 3 + 12) = 2x$ is __?__ .

3. To solve $4x = 82$, you would __?__ both sides by 4.  [4-2]

4. The solution to $\frac{n}{9} = 27$ is __?__ .

**Write the letter of the correct answer.**

5. Solve $\frac{4}{3}x = 60$.  6. Solve $\frac{n}{0.15} = 15$.  [4-3]

     **a.** 45    **b.** 80    **c.** $60\frac{3}{4}$      **a.** 1    **b.** 2.25    **c.** 22.5

7. Solve $\frac{3}{2}n - 5 = 70$.  8. Solve $\frac{n}{6} + 9 = 10.5$.  [4-4]

     **a.** 40    **b.** $112\frac{1}{2}$    **c.** 50      **a.** 9    **b.** 120    **c.** 96

9. Which expression represents the word phrase "the difference be-  [4-5]
tween seven and a number $n$"?
     **a.** $n - 7$      **b.** $7 - n$      **c.** $7n - 7$      **d.** $7 + n$

10. Which equation represents the problem?  [4-6]

     Luis bought 3 records on sale. The original cost of the records had
been $29.85, but Luis paid only $23.25 for all three. How much did
Luis save on each record?
     **a.** $29.85 + 3x = 23.25$    **b.** $23.25 - 3x = 29.85$    **c.** $29.85 - 3x = 23.25$

11. Write an equation for the following problem.  [4-7]

     A rectangle has a perimeter of 84 cm. Find the length if the width is
15 cm.
     **a.** $2l + 2w = 84$    **b.** $2l + 30 = 84$    **c.** $2l + 15 = 84$    **d.** $2l = 99$

12. Use the five-step method to solve the following problem.  [4-8]

     Tanya and Sara went biking. When they returned, they found that
they had gone 18 km in 0.75 h. What was their speed on the trip?
     **a.** 13.5 km/h    **b.** 32 km/h    **c.** 12 km/h    **d.** 24 km/h

# Chapter Test

**Use transformations to solve each equation.**

1. $x - 18 = 11$  2. $3(7 - 2) + x = 19$  [4-1]

3. $9n = 1$  4. $\frac{n}{11} = 13$  5. $15n = 12$  [4-2]

6. $\frac{2}{3}x = 24$  7. $0.55x = 11$  8. $\frac{x}{1.2} = 8.6$  [4-3]

9. $4n - 16 = 32$  10. $\frac{3}{4}n + 18 = 51$  11. $30 - \frac{1}{2}n = 11$  [4-4]

**Write an expression for each word phrase.**

12. The product of a number $n$ and twenty-one  [4-5]

13. Nine times the quotient of a number $n$ and 3

**Which equation represents the problem?**

14. Toby Baylor wrote a check for $12 to pay a bill. Two days later he deposited $20 in his checking account. If Toby had $88 in his account after these transactions, how much did he have originally?  [4-6]

   **a.** $20x - 12 = 88$  **b.** $x - 12 + 20 = 88$  **c.** $x - 88 = 12 + 20$

**Write an equation for the following problem.**

15. Annette Loo's car gets 23 miles per gallon. She is planning a trip to San Diego. If Annette lives 345 miles from San Diego, how many gallons of gasoline will she use driving to San Diego?  [4-7]

**Solve the following problem by the five-step method.**

16. Bill and Roberta are shipping boxes to their new house. It costs $35 to ship each box and there is also a charge of $50 for the entire shipment. If the cost of shipping the boxes, including the $50 charge, comes to $610, how many boxes are being shipped?  [4-8]

# Cumulative Review (Chapters 1–4)

## Exercises

Evaluate the expression when $x = 2$, $y = 4$, and $z = 1$.

**1.** $x + y - z$

**2.** $6(x + y) - 5z$

**3.** $10y \div (3x + 2z)$

**4.** $-x - y$

**5.** $-y - (-3z)$

**6.** $y + 5z - (-2x)$

**7.** $x^5$

**8.** $3y^3$

**9.** $y^x$

Replace __?__ with $=$, $>$, or $<$ to make a true statement.

**10.** $31 \underline{\phantom{?}?\phantom{?}} 24$

**11.** $206 \underline{\phantom{?}?\phantom{?}} 260$

**12.** $581 \underline{\phantom{?}?\phantom{?}} 519$

**13.** $-4.68 \underline{\phantom{?}?\phantom{?}} 3.2$

**14.** $0.6 \underline{\phantom{?}?\phantom{?}} -15.23$

**15.** $-8.99 \underline{\phantom{?}?\phantom{?}} -8.98$

**16.** $\frac{1}{2} \underline{\phantom{?}?\phantom{?}} \frac{4}{6}$

**17.** $-\frac{3}{7} \underline{\phantom{?}?\phantom{?}} -\frac{4}{9}$

**18.** $\frac{11}{35} \underline{\phantom{?}?\phantom{?}} -\frac{9}{40}$

What value of the variable makes the statement true?

**19.** $-6 + x = 2$

**20.** $x + 5 = -3$

**21.** $-11 + x = -31$

**22.** $-5.1 + x = -7.4$

**23.** $17.5 + x = -1$

**24.** $x + 28.3 = -4.7$

**25.** $3 - x = -2$

**26.** $x - 7 = -11$

**27.** $x - (-9) = -3$

Simplify.

**28.** $\frac{2}{3} + \left(-\frac{1}{6}\right)$

**29.** $-1\frac{1}{4} + \left(-2\frac{1}{3}\right)$

**30.** $\frac{5}{8} + (-6)$

**31.** $\frac{11}{8} - \frac{11}{4}$

**32.** $-2\frac{2}{3} - 1\frac{1}{5}$

**33.** $-11\frac{3}{4} - \left(-12\frac{9}{10}\right)$

**34.** $\frac{1}{7} \times \left(-\frac{1}{12}\right)$

**35.** $-12 \times 1\frac{3}{4}$

**36.** $-\frac{11}{16} \times \left(\frac{-9}{20}\right)$

**37.** $-11\frac{1}{9} \div 100$

**38.** $2\frac{1}{3} \div \left(-\frac{3}{7}\right)$

**39.** $-10\frac{1}{3} \div \left(-3\frac{1}{5}\right)$

Use transformations to solve each equation.

**40.** $y + 11 = 31$

**41.** $n - 8 = 12$

**42.** $29 = 5(6 - 3) + x$

**43.** $\frac{2}{3}x = 16$

**44.** $36 = 9y$

**45.** $-13n = 182$

**46.** $\frac{4}{3}t = -40$

**47.** $1.6c = 1.2$

**48.** $-1.5x = -22.5$

**49.** $4p - 7 = 37$

**50.** $\frac{1}{6}y - 7 = 14$

**51.** $\frac{-2}{3}n - 11 = -7$

# Problems

**Solve.**

1. Calculator batteries are being sold at 2 for 99¢. How much will 6 batteries cost?

2. "My new camera cost a fortune," boasted Frank. "Mine cost twice as much as yours," returned Eddie. If together the two cameras cost $545.25, how much did Frank's camera cost?

3. A newspaper with a circulation of 1.1 million readers estimates $\frac{1}{5}$ of its readers have subscriptions. About how many readers have subscriptions?

4. Amos Ellingsworth earns $455 per week. About $\frac{1}{4}$ of his pay is deducted for taxes, insurance, and Social Security. To the nearest dollar, how much does Amos take home each week?

5. Sally Gray takes home $378.50 per week. She will get a $50 per week raise in her next weekly check. If Sally takes home $\frac{4}{5}$ of her raise, what will be her new take-home pay?

6. The population of Elmwood was 26,547 in 1950, 31,068 in 1960, 30,327 in 1970, and 29,598 in 1980. What was the total increase in population between 1950 and 1980?

7. The Carpenters make annual mortgage payments of $6430.56 and property tax payments of $1446.00. What are the combined monthly payments for mortgage and taxes?

8. Cormo Corporation stock sells for $13\frac{5}{8}$ dollars a share. If Lorraine has $327, how many shares can she buy?

9. At a recent job fair, there were $\frac{2}{3}$ as many inquiries about jobs in health care as in electronics. A reported 250 inquires were made about both fields. How many inquiries were made about health care?

# 5

# Geometric Figures

The three fields shown in the center of the photograph are irrigated using the center-pivot irrigation system. Long arms, or booms, revolve around center pivots and distribute water over circular areas of land, such as those visible in the photograph. Although the booms miss the corners of the fields, traveling sprinklers would be much more expensive to use. The booms have the advantage of being able to control the amounts of water delivered with a minimum of labor.

Farmers have used irrigation for centuries, at least as far back as the Egyptians in 5000 B.C. Today we could not hope to feed the huge world population without the extension of water supplies by irrigation. According to a recent estimate, there are about 155,700,000 hectares of land under irrigation.

## Career Note

The demand for a greater variety of farm products, for improved farming methods and machinery, and for more careful environmental planning has led to an increase in the demand for agricultural engineers. Manufacturers of farm equipment look for engineers to design systems and machinery. Engineers also participate in research, production, sales, and management.

# 5-1 Points, Lines, Planes

All of the figures that we study in geometry are made up of **points.** We usually picture a single point by making a dot and labeling it with a capital letter.

$$P \qquad\qquad Q$$
$$\bullet \qquad\qquad \bullet$$
Point *P*      Point *Q*

    Among the most important geometric figures that we study are straight lines, or simply **lines.** You probably know this important fact about lines:

Two points determine exactly one line.

This means that through two points *P* and *Q* we can draw one line, and only one line, which we denote by $\overleftrightarrow{PQ}$ (or $\overleftrightarrow{QP}$).

$$P \qquad\qquad\qquad Q$$
Line *PQ*: $\overleftrightarrow{PQ}$, or $\overleftrightarrow{QP}$

Notice the use of arrowheads to show that a line extends without end in either direction.

    Three points may or may not lie on the same line. Three or more points that do lie on the same line are called **collinear.** Points not on the same line are called **noncollinear.**

    If we take a point *P* on a line and all the points on the line that lie on one side of *P*, we have a **ray** with **endpoint** *P*. We name a ray by naming first its endpoint and then any other point on it.

$$P \qquad\qquad Q \qquad\qquad\qquad A \qquad\qquad\qquad B$$
Ray *PQ*: $\overrightarrow{PQ}$        Ray *BA*: $\overrightarrow{BA}$

It is important to remember that the endpoint is always named first. $\overrightarrow{AB}$ is *not* the same ray as $\overrightarrow{BA}$.

    If we take two points *P* and *Q* on a line and all the points that lie between *P* and *Q*, we have a **segment** denoted by $\overline{PQ}$ (or $\overline{QP}$). The points *P* and *Q* are called the **endpoints** of $\overline{PQ}$.

$$P \qquad\qquad\qquad\qquad Q$$
Segment *PQ*: $\overline{PQ}$, or $\overline{QP}$

**EXAMPLE 1**  Name (a) one line, (b) two rays, (c) three segments, (d) three collinear points, and (e) three non-collinear points in the given diagram. (Various answers are possible.)

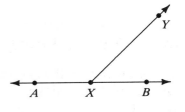

**Solution**  **a.** $\overleftrightarrow{AB}$  **b.** $\overrightarrow{XY}$, $\overrightarrow{AB}$  **c.** $\overline{AX}$, $\overline{XB}$, $\overline{XY}$  **d.** $A$, $X$, $B$  **e.** $A$, $Y$, $B$

Just as two points determine a line, three noncollinear points in space determine a flat surface called a **plane.** We can name a plane by naming any three noncollinear points on it. Because a plane extends without limit in all directions of the surface, we can show only part of it, as in the figure below.

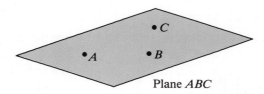

Plane $ABC$

Lines in the same plane that do not intersect are called **parallel lines.** Two segments or rays are parallel if they are parts of parallel lines. "$\overleftrightarrow{AB}$ is parallel to $\overleftrightarrow{CD}$" may be written as $\overleftrightarrow{AB} \parallel \overleftrightarrow{CD}$.

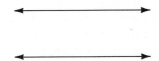

Intersecting lines
intersect in a point.

Parallel lines
do not intersect.

Planes that do not intersect are called **parallel planes.**

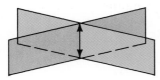

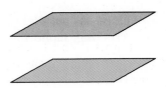

Intersecting planes
intersect in a line.

Parallel planes
do not intersect.

**EXAMPLE 2** Use the box to name (a) two parallel lines, (b) two parallel planes, (c) two intersecting lines, (d) two intersecting planes, and (e) two nonparallel lines that do not intersect. (Various answers are possible.)

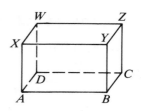

**Solution**

a. $\overleftrightarrow{AX}$ and $\overleftrightarrow{BY}$      b. plane $ABC$ and plane $XYZ$

c. $\overleftrightarrow{AB}$ and $\overleftrightarrow{AX}$      d. plane $ABC$ and plane $ABY$

e. $\overleftrightarrow{AD}$ and $\overleftrightarrow{BY}$

Two nonparallel lines that do not intersect, such as $\overleftrightarrow{AD}$ and $\overleftrightarrow{BY}$ in Example 2, are called **skew lines**.

---

## Class Exercises

**Tell how many endpoints each figure has.**

**1.** a segment      **2.** a line      **3.** a plane      **4.** a ray

**Exercises 5–9 refer to the diagram at the right in which $\overleftrightarrow{AB}$ and $\overleftrightarrow{CD}$ are parallel lines.**

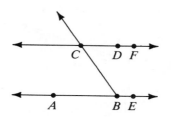

**5.** Name three collinear points.

**6.** Name two parallel rays.

**7.** Name two parallel segments.

**8.** Name two segments that are not parallel.

**9.** Name two rays that are not parallel.

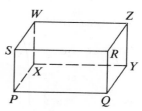

**Exercises 10–14 refer to the box at the right. Classify each pair of planes as parallel or intersecting. If the planes are intersecting, name the line of intersection.**

**10.** planes $QRY$ and $WSP$      **11.** planes $PQR$ and $PSW$

**12.** planes $SPW$ and $XYQ$      **13.** planes $PQR$ and $WXY$

**14.** In the box are $\overleftrightarrow{WX}$ and $\overleftrightarrow{RS}$ parallel, intersecting, or skew?

## Written Exercises

**In Exercises 1–4, give another name for the indicated figure.**

**A** 1. $\overline{XY}$

2. $\overleftrightarrow{BC}$

3. $\overrightarrow{PR}$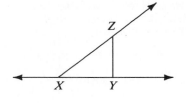

4. $\overrightarrow{VU}$

5. Name one line and three rays in the diagram below.

6. Name three rays and three segments in the diagram below.

7. Name three collinear points.

8. Name three noncollinear points.

9. Name three segments that intersect at $P$.

**Exercises 7–20 refer to the diagram at the right. $\overleftrightarrow{PQ}$ and $\overleftrightarrow{ST}$ are parallel. (There may be several correct answers to each exercise.)**

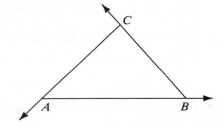

7. Name three collinear points.

8. Name three noncollinear points.

9. Name three segments that intersect at $P$.

10. Name two parallel rays.

11. Name two parallel segments.

12. Name two nonparallel segments that do not intersect.

13. Name two nonparallel rays that do not intersect.

14. Name two segments that intersect in exactly one point.

15. Name two rays that intersect in exactly one point.

16. Name a ray that is contained in $\overrightarrow{OT}$.

17. Name a ray that contains $\overline{SP}$.

18. Name the segment that is in $\overrightarrow{PQ}$ and $\overrightarrow{QP}$.

19. Name four segments that contain $O$.

20. Name two rays that intersect in more than one point.

**Exercises 21–24 refer to the box at the right.  (There may be several correct answers to each exercise.)**

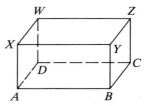

**21.** Name two intersecting lines and their point of intersection.

**22.** Name two intersecting planes and their line of intersection.

**23.** Name two parallel lines and the plane that contains them both.

**24.** Name two skew lines.

**In Exercises 25–32, tell whether the statement is true or false.**

**B** **25.** Two lines cannot intersect in more than one point.

**26.** Two rays cannot intersect in more than one point.

**27.** Two segments cannot intersect in more than one point.

**28.** If two rays are parallel, they do not intersect.

**29.** If two rays do not intersect, they are parallel.

**30.** If two points of a segment are contained in a line, then the whole segment is contained in the line.

**31.** If two points of a ray are contained in a line, then the whole ray is contained in the line.

**32.** If two points of a segment are contained in a ray, then the whole segment is contained in the ray.

**C** **33.** Draw $\overleftrightarrow{AB}$ and a point $P$ not on $\overleftrightarrow{AB}$.  Now draw a line through $P$ parallel to $\overleftrightarrow{AB}$.  How many such lines can be drawn?

**34.** Draw four points $A$, $B$, $C$, and $D$ so that $\overline{AB}$ is parallel to $\overline{CD}$ and $\overline{AD}$ is parallel to $\overline{BC}$.  Do you think that $\overline{AC}$ and $\overline{BD}$ must intersect?

**35.** We know that two points determine a line.  Explain what we mean by saying that two nonparallel lines in a plane determine a point.

---

## *Review Exercises*

**Round to the nearest tenth.**

| | | | |
|---|---|---|---|
| **1.** 2.87 | **2.** 6.32 | **3.** 4.56 | **4.** 7.08 |
| **5.** 2.98 | **6.** 11.753 | **7.** 9.347 | **8.** 6.482 |

## 5-2 Measuring Segments

In Washington, D.C., Jill estimated the length of a jet to be about 165 feet. When the plane landed in Paris, Jacques guessed the jet's length to be about 50 meters. Although the numbers 165 and 50 are quite different, the two estimates are about the same. This is so because of the difference in the units of measurement used.

The metric system of measurement uses the **meter (m)** as its basic unit of length. For smaller measurements we divide the meter into 100 equal parts called **centimeters (cm).**

We can measure a length to the nearest centimeter by using a ruler marked off in centimeters, as illustrated below.

We see that the length of $\overline{AB}$ is closer to 7 cm than to 8 cm. The length of $\overline{AB}$ is written $AB$. The symbol $\approx$ means *is approximately equal to.* Therefore, $AB \approx 7$ cm.

The drawing at the right shows that the length of $\overline{XY}$ is about 3 cm.

$$XY \approx 3 \text{ cm}$$

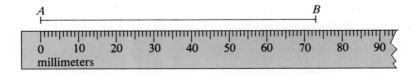

Measurements made with small units are more precise than those made with larger units. We can measure lengths more precisely by using a ruler on which each centimeter has been divided into ten equal parts called **millimeters (mm).** We see that to the nearest millimeter the length of $\overline{AB}$ is 73 mm.

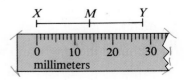

The drawing at the right shows that the length of $\overline{XY}$ is 28 mm and the lengths of $\overline{XM}$ and $\overline{MY}$ are 14 mm each. A point, such as $M$, that divides a segment into two other segments of equal length is called the **midpoint** of the segment. Thus, $M$ is the midpoint of $\overline{XY}$. Segments of equal length are called **congruent segments.** The symbol $\cong$ means *is congruent to.* Since $XM = MY$, $\overline{XM} \cong \overline{MY}$.

Usually centimeters and millimeters are marked on the same ruler, as shown below.

0 1 2 3 4 5 6 7 8 9 10 11 12
centimeters (with millimeters)

To measure longer lengths, such as distances between cities, we use **kilometers (km).** A kilometer is 1000 meters.

> 1 m = 100 cm = 1000 mm     1 cm = 10 mm
>
> 1 cm = 0.01 m     (*centi* means *hundredths*)
>
> 1 mm = 0.001 m     (*milli* means *thousandths*)
>
> 1 km = 1000 m     (*kilo* means *thousand*)

Never mix metric units of length. For example, do not write 3 m 18 cm; write 3.18 m or 318 cm instead.

## Class Exercises

**Copy and complete these tables.**

|  | 1. | 2. | 3. | 4. |
|---|---|---|---|---|
| Number of meters | 4 | ? | ? | ? |
| Number of centimeters | ? | 60 | 52 | ? |
| Number of millimeters | ? | ? | ? | 36 |

|  | 5. | 6. | 7. | 8. |
|---|---|---|---|---|
| Number of meters | 3500 | ? | 475 | ? |
| Number of kilometers | ? | 4.2 | ? | 0.034 |

**9.** Estimate the length and width of your desk top to the nearest centimeter.

**10.** Estimate the length and width of this book to the nearest centimeter.

**11.** Estimate the height of the classroom door to the nearest centimeter.

**12.** Estimate your height to the nearest centimeter.

**13.** Estimate the thickness of your pencil to the nearest millimeter.

**14.** Estimate the distance from your home to school in meters and in kilometers.

**In Exercises 15 and 16, $M$ is the midpoint of $\overline{AB}$. Draw a sketch to help you complete these sentences.**

**15.** If $AB = 18$ cm, then $AM = \underline{\ ?\ }$ cm and $MB = \underline{\ ?\ }$ cm.

**16.** If $AM = 4$ mm, then $MB = \underline{\ ?\ }$ mm and $AB = \underline{\ ?\ }$ mm.

**In Exercises 17 and 18, $P$ is a point of $\overline{XY}$. Draw a sketch to help you complete these sentences.**

**17.** If $XP = YP$, then $\overline{XP}$ is $\underline{\ ?\ }$ to $\overline{YP}$.

**18.** If $\overline{XP} \cong \overline{YP}$, then $P$ is the $\underline{\ ?\ }$ of $\overline{XY}$.

## Written Exercises

**Measure each segment (a) to the nearest centimeter and (b) to the nearest millimeter. (If you do not have a metric ruler, mark each segment on the edge of a piece of paper and use the ruler pictured earlier.)**

 **1.** $\overline{AE}$

**2.** $\overline{VZ}$

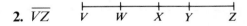 **3.** $\overline{MQ}$

**4.** $\overline{GJ}$

 **5.** $\overline{PT}$

 **6.** $\overline{NX}$

**7.** In Exercises 1–6, which pairs of the segments $\overline{AE}$, $\overline{VZ}$, $\overline{MQ}$, $\overline{GJ}$, $\overline{PT}$, and $\overline{NX}$ are approximately equal in length?

**8.** In Exercises 1–6, name the midpoints of $\overline{AE}$, $\overline{VZ}$, $\overline{MQ}$, $\overline{GJ}$, $\overline{PT}$, and $\overline{NX}$ given that the midpoint is named in each diagram.

**Copy and complete these tables.**

| | 9. | 10. | 11. | 12. | 13. | 14. |
|---|---|---|---|---|---|---|
| Number of meters | 4.5 | 1.63 | ? | ? | ? | ? |
| Number of centimeters | ? | ? | 250 | 82.6 | ? | ? |
| Number of millimeters | ? | ? | ? | ? | 60,000 | 368 |

| | 15. | 16. | 17. | 18. | 19. | 20. |
|---|---|---|---|---|---|---|
| Number of meters | 2000 | 20 | 625 | ? | ? | ? |
| Number of kilometers | ? | ? | ? | 3 | 4.5 | 0.25 |

Small units are often used to avoid decimals. However, it is sometimes easier to think about lengths given in meters. Change the following dimensions to meters.

**B**  **21.** 374 cm by 520 cm

**22.** 425 cm by 650 cm

**23.** 4675 mm by 7050 mm

**24.** 5925 mm by 8275 mm

Rewrite each measurement using a unit that will avoid decimals.

**25.** 2.7 cm          **26.** 4.32 km          **27.** 0.65 m          **28.** 10.6 cm

**For Exercises 29–32 use the following diagram and information.**

$C$ is the midpoint of $\overline{AB}$; $D$ is the midpoint of $\overline{AC}$;
$E$ is the midpoint of $\overline{AD}$; $F$ is the midpoint of $\overline{AE}$;
$G$ is the midpoint of $\overline{AF}$; $H$ is the midpoint of $\overline{AG}$.

**C**  **29.** If $AB = 140$ mm, $AH = \underline{\phantom{?}}$ mm.          **30.** If $BC = 70$ mm, $AF = \underline{\phantom{?}}$ mm.

**31.** If $AG = 4.375$ mm, $AD = \underline{\phantom{?}}$ mm.          **32.** If $FE = 8.75$ mm, $DC = \underline{\phantom{?}}$ mm.

## Review Exercises

Solve.

**1.** $x + 90 = 180$          **2.** $x + 20 = 90$          **3.** $x + 35 = 75$

**4.** $100 - x = 45$          **5.** $180 - x = 40$          **6.** $90 - x = 30$

**7.** $75 + x = 180$          **8.** $180 - x = 115$          **9.** $25 + x = 90$

# 5-3 Angles and Angle Measure

An **angle** is a figure formed by two rays with the same endpoint. The common endpoint is called the **vertex,** and the rays are called the **sides.**

We may name an angle by giving its vertex letter if this is the only angle with that vertex, or by listing letters for points on the two sides with the vertex letter in the middle. We use the symbol ∠ for *angle*. The diagram at the right shows several ways of naming an angle.

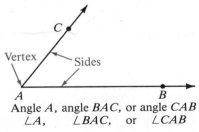

Angle *A*, angle *BAC,* or angle *CAB*
∠*A*, ∠*BAC,* or ∠*CAB*

To measure segments we used a ruler marked off in unit lengths. To measure angles, we use a **protractor** that is marked off in units of angle measure, called **degrees.** To use a protractor, place its center point at the vertex of the angle to be measured and one of its zero points on a side. In the drawing at the left below we use the outer scale and read the measure of ∠ *E* to be 60 degrees (60°). We write m ∠ *E* = 60°.

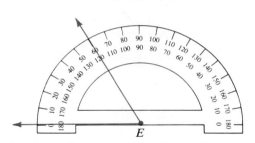

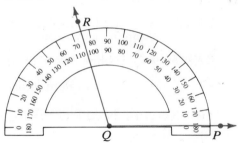

In the drawing on the right above the inner scale shows that m ∠ *PQR* = 105°.

We often label angles with their measures, as shown in the figures. Since ∠ *A* and ∠ *B* have equal measures we can write m∠ *A* = m∠ *B*. We say that ∠ *A* and ∠ *B* are **congruent angles** and we write ∠ *A* ≅ ∠ *B*.

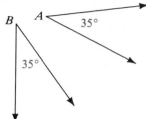

If two lines intersect so that the angles they form are all congruent, the lines are **perpendicular.** We use the symbol ⊥ to mean *is perpendicular to.* In the figure $\overline{WY} \perp \overline{XZ}$.

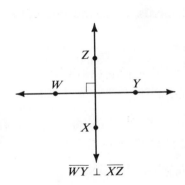

$\overline{WY} \perp \overline{XZ}$

Angles formed by perpendicular lines each have measure 90°. A 90° angle is called a **right angle.** A small square is often used to indicate a right angle in a diagram.

An **acute angle** is an angle with measure less than 90°. An **obtuse angle** has measure between 90° and 180°.

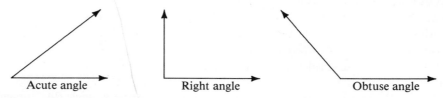

Acute angle          Right angle          Obtuse angle

Two angles are **complementary** if the sum of their measures is 90°. Two angles are **supplementary** if the sum of their measures is 180°.

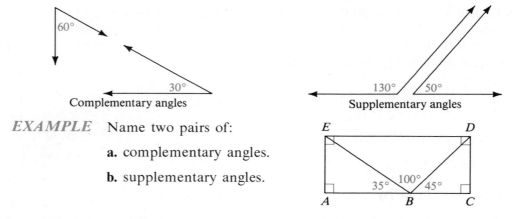

Complementary angles                    Supplementary angles

*EXAMPLE*   Name two pairs of:

      **a.** complementary angles.

      **b.** supplementary angles.

*Solution*   **a.** Since $\angle AEB$ and $\angle BED$ form a right angle, they are complementary. Similarly, $\angle CDB$ and $\angle BDE$ are complementary.

      **b.** Since the sum of m$\angle ABE$ and m$\angle EBC$ is 180°, they are supplementary. Similarly, $\angle ABD$ and $\angle DBC$ are supplementary.

Although the sides of angles are rays, we often show the sides as segments, as in the figure for the example.

---

**Reading Mathematics:** *Symbols*

When you read a mathematical sentence, be sure to give each symbol its complete meaning. For example:

$\overleftrightarrow{AB} \perp \overleftrightarrow{CD}$ is read as *line AB is perpendicular to line CD.*

$\overline{AB} \cong \overline{CD}$ is read as *segment AB is congruent to segment CD.*

$AB \approx 6$ cm is read as *the length of $\overline{AB}$ is approximately equal to six centimeters.*

m$\angle A = 10°$ is read as *the measure of angle A is equal to ten degrees.*

## Class Exercises

**1.** If an angle is named ∠ *EFG*, its vertex is ___?___.

**2.** If an angle is named ∠ *GEF*, its vertex is ___?___.

**Give three names for each angle.**

**3.**

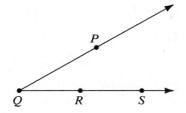

**4.**

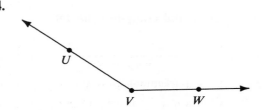

**5.** Use a protractor to find the measures of the angles in Exercises 3 and 4.

**Exercises 6–9 refer to the diagram at the right.**

**6.** Name five acute angles and one obtuse angle.

**7.** Name a pair of perpendicular segments.

**8.** Name a pair of complementary angles.

**9.** Name a pair of supplementary angles.

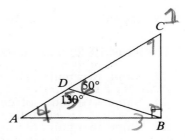

**State the measures of the complement and supplement of each angle.**

**10.** m∠ *F* = 70°   **11.** m∠ *G* = 15°   **12.** m∠ *H* = 45°   **13.** m∠ *J* = 60°

---

## Written Exercises

**Use a protractor to draw an angle having the given measure.**

**A**   **1.** 75°   **2.** 20°   **3.** 120°   **4.** 155°

**Use a protractor to measure the given angle. State whether the angle is acute or obtuse.**

**5.**

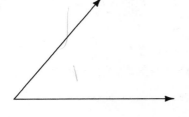

**6.**

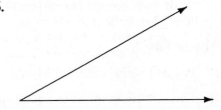

**Measure the given angle. Is the angle acute or obtuse?**

7.

8.

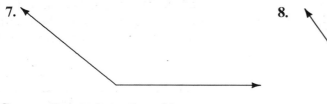

**Copy and complete the table.**

|  | 9. | 10. | 11. | 12. | 13. | 14. |
|---|---|---|---|---|---|---|
| ∠X | 43° | 9° | 135° | ? | ? | ? |
| Complement of ∠X | ? | ? |  | 12° | 71° | ? |
| Supplement of ∠X | ? | ? | ? | ? | ? | 150° |

**Use a protractor to draw an angle congruent to the angle in each given exercise.**

**15.** Exercise 5      **16.** Exercise 6      **17.** Exercise 7      **18.** Exercise 8

**Exercises 19–22 refer to the diagram at the right.**

**19.** Name two pairs of perpendicular lines.

**20.** Name two pairs of complementary angles.

**21.** Name two pairs of supplementary angles.

**22.** What is the sum of the measures of the four angles having vertex *E*?

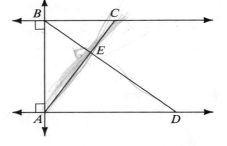

**Angles that share a common vertex and a common side, but with no common points in their interiors, are called *adjacent angles.***

**B**   **23.** Draw two adjacent complementary angles, one of which has measure 65°.

**24.** Draw two adjacent supplementary angles, one of which has measure 105°.

**25.** Draw two congruent adjacent supplementary angles. What is the measure of each?

**26.** Draw two congruent adjacent complementary angles. What is the measure of each?

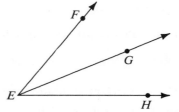

∠ *FEG* is adjacent to ∠ *GEH*
∠ *FEG* is *not* adjacent to ∠ *FEH*

**True or false?**

**27.** The supplement of an obtuse angle is acute.

**28.** The complement of an acute angle is obtuse.

**C** **29.** Measure the angles labeled 1, 2, 3, and 4. What general fact do your results suggest about angles formed by intersecting lines?

**30.** Measure the angles labeled 1, 2, 3, and 4. What general facts do your results suggest about two parallel lines intersected by a third line?

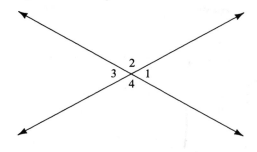

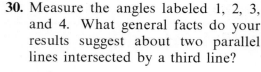
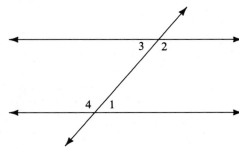

## Self-Test A

**Draw a sketch to illustrate each of the following.**

**1.** $\overline{CD}$        **2.** $\overleftrightarrow{AX}$        **3.** $\overrightarrow{RS}$            [5–1]

**Exercises 4–7 refer to the diagram below.**

**4.** Name two parallel lines.

**5.** Name two parallel planes.

**6.** Name two intersecting lines.

**7.** Name two intersecting planes.

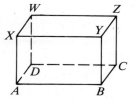

**Complete each statement.**

**8.** 4000 m = __?__ km          **9.** 87 cm = __?__ m       [5–2]

**10.** 785 mm = __?__ m        **11.** 109 mm = __?__ cm

**12.** If $M$ is the midpoint of $\overline{AB}$, then $AM =$ __?__ and $\overline{AM} \cong$ __?__.

**13.** A right angle has measure __?__.               [5–3]

**14.** Two angles with the same measures are __?__.

**15.** $\perp$ is the symbol for __?__.

**16.** A 37° angle is a(n) __?__ angle.

**17.** The complement of a 42° angle has measure __?__°.

**18.** The supplement of a 107° angle has measure __?__°.

*Self-Test answers and Extra Practice are at the back of the book.*

## 5-4 Triangles

A **triangle** is the figure formed when three points not on a line are joined by segments. The drawing at the right shows triangle *ABC*, written $\triangle ABC$, having the segments $\overline{AB}$, $\overline{BC}$, and $\overline{CA}$ as its **sides.** Each of the points *A*, *B*, and *C* is called a **vertex** (plural: *vertices*) of $\triangle ABC$. Each of the angles $\angle A$, $\angle B$, and $\angle C$ is called an **angle** of $\triangle ABC$.

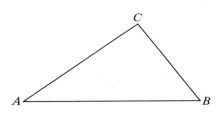

Suppose *A*, *B*, and *C* in the triangle above represent three points on a map. Do you think it is farther to travel from *A* to *B* and then to *C* or to travel directly from *A* to *C*? Measure to check. This illustrates the first fact about triangles stated below.

In any triangle:

**1.** The sum of the lengths of any two sides is greater than the length of the third side.

**2.** The sum of the measures of the angles is 180°.

You can verify the second fact by tearing off the corners of any paper triangle and fitting them together as shown at the right.

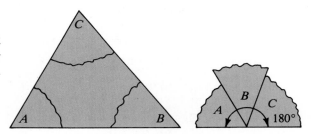

**EXAMPLE 1**   One angle of a triangle measures 40° and the other two angles have equal measures. Find the measures of the congruent angles.

***Solution***   The sum of the measures of the angles of a triangle is 180°. The sum of the measures of the congruent angles must be 180° − 40°, or 140°. Therefore each of the two congruent angles has measure 70°.

---

**Problem Solving Reminder**

Some problems do not give enough information. Sometimes you must *supply previously learned facts.* In the example above, you need to supply the additional information about the sum of the measures of the angles of a triangle.

There are several ways to name triangles. One way is by angles.

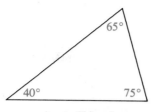

**Acute Triangle**
Three acute angles

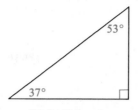

**Right Triangle**
One right angle

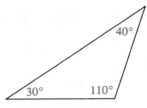

**Obtuse Triangle**
One obtuse angle

Triangles can also be classified by their sides.

**Scalene Triangle**
No two sides
congruent

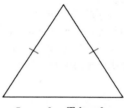

**Isosceles Triangle**
At least two sides
congruent

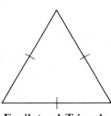

**Equilateral Triangle**
All three sides
congruent

As you might expect, the longest side of a triangle is opposite the largest angle, and the shortest side is opposite the smallest angle. Two angles are congruent if and only if the sides opposite them are congruent.

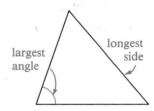

*EXAMPLE 2*  Classify each triangle by sides and by angles.

**a.**

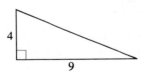

**b.**

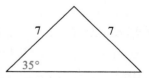

*Solution*

**a.** No two sides are congruent; the triangle is scalene.
There is one right angle; the triangle is a right triangle.
Scalene right triangle

**b.** Two sides are congruent; the triangle is isosceles.
The angles opposite the congruent sides are congruent; thus, the third angle has a measure of 110°; the triangle is obtuse.
Isosceles obtuse triangle

*Geometric Figures*  **179**

## Class Exercises

**How do you know, without measuring, that these triangles are labeled incorrectly?**

**1.**

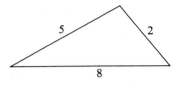

**2.**

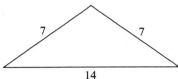

**3.**

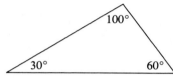

**4.**

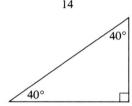

**Exercises 5 and 6 refer to the triangles below.**

**a.**

**b.**

**c.**

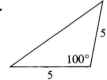

**d.**

**5.** Classify each triangle by sides.

**6.** Classify each triangle by angles.

**7.** Explain how you know that a triangle with two congruent angles is isosceles.

**8.** Explain how you know that a triangle with three congruent angles is equilateral.

**Exercises 9–13 refer to the diagram at the right.**

**9.** What segment is a common side of △ADC and △BCD?

**10.** What segment is a common side of △ABC and △BCD?

**11.** What angle is common to △ABC and △CAD?

**12.** Name two right triangles.

**13.** Name an obtuse triangle.

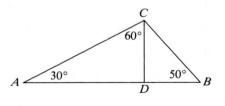

**180** *Chapter 5*

## Written Exercises

**The measures of two angles of a triangle are given. Find the measure of the third angle.**

**A**    **1.** 40°, 60°      **2.** 15°, 105°      **3.** 35°, 55°      **4.** 160°, 10°

**Classify each triangle by its sides.**

**5.** 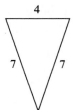     **6.** 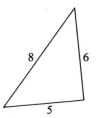     **7.** 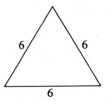     **8.**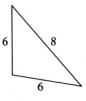

**Classify each triangle by its sides and by its angles.**

**9.** 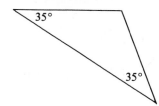         **10.**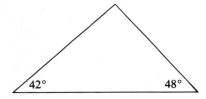

**11.** 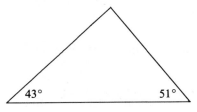         **12.**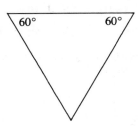

**Exercises 13–16 refer to the diagram at the right.**

**13.** Name three right triangles.

**14.** Name an isosceles triangle.

**15.** Name an acute scalene triangle.

**16.** Name an obtuse triangle.

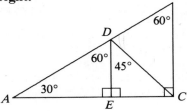

**17.** Use a ruler and a protractor to draw (a) a scalene acute triangle and (b) an isosceles obtuse triangle.

**18.** Use a ruler and a protractor to draw (a) a scalene obtuse triangle and (b) an isosceles acute triangle.

**B** **19.** What measures do the angles of an equilateral triangle have?

**20.** One of the congruent angles of an isosceles triangle has measure 40°. What measures do the other angles have?

**21.** One of the acute angles of a right triangle measures 75°. What measure does the other acute angle have?

**22.** What measures do the angles of an isosceles right triangle have?

**23.** An isosceles triangle has a 96° angle. What are the measures of its other angles?

**24.** An isosceles triangle has a 60° angle. What are the measures of its other angles?

**25.** One acute angle of a right triangle is 45°. What relationship, if any, is there between the two shorter sides?

**26.** Why is it not possible to have an equilateral right triangle?

**C** **27.** In $\triangle ABC$, $AB = 8$ and $BC = 5$. Then (a) $AC < \underline{\phantom{?}}$, and (b) $AC > \underline{\phantom{?}}$.

**28.** In $\triangle PQR$, $PR = 10$ and $RQ = 7$. Then (a) $PQ < \underline{\phantom{?}}$, and (b) $PQ > \underline{\phantom{?}}$.

**29.** Explain why in any triangle the difference of the lengths of any two sides cannot be greater than the length of the third side.

**30.** Draw a triangle. Draw rays that divide each of its angles into two congruent angles. Do this for several triangles of different shapes. What seems always to be true of the three rays?

**31.** Draw a triangle. Then draw segments joining each vertex to the midpoint of the opposite side. (These segments are called **medians** of the triangle.) Do this for several triangles of different shapes. What seems always to be true?

---

## Review Exercises

**Add.**

1. $3.75 + 4.92 + 6.41$    2. $7.83 + 6.91 + 5.29$    3. $8.36 + 4.95 + 2.21$

4. $5.3 + 6.21 + 7.3$    5. $8.02 + 5.1 + 7.21$    6. $3.07 + 4 + 5.93$

7. $11.27 + 6.513 + 4.09$    8. $10.03 + 5.7 + 4.93$    9. $12.004 + 4.9 + 7.864$

# 5-5 Polygons

A **polygon** is a closed figure formed by joining segments (**sides** of the polygon) at their endpoints (**vertices** of the polygon). We name polygons according to the number of sides they have.

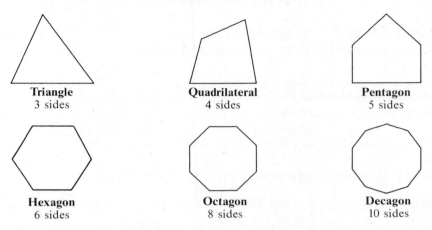

| Triangle | Quadrilateral | Pentagon |
| 3 sides | 4 sides | 5 sides |

| Hexagon | Octagon | Decagon |
| 6 sides | 8 sides | 10 sides |

A polygon is **regular** if all its sides are congruent and all its angles are congruent. As drawn above, the hexagon, the octagon, and the decagon are regular while the triangle, the quadrilateral, and the pentagon are not.

To name a polygon, we name its consecutive vertices in order. The quadrilateral shown at the right may be named quadrilateral $PQRS$.

A **diagonal** of a polygon is a segment joining two nonconsecutive vertices. Thus, $\overline{PR}$ and $\overline{QS}$ are the diagonals of quadrilateral $PQRS$.

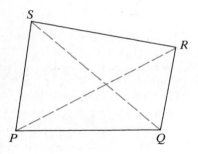

Certain quadrilaterals have special names.

A **parallelogram** has its opposite sides parallel and congruent.

A **trapezoid** has just one pair of parallel sides.

Certain parallelograms also have special names.

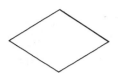

A **rhombus** has all its sides congruent.

A **square** has congruent sides and congruent angles.

A **rectangle** has all its angles congruent.

---

**Reading Mathematics:** *Vocabulary*
Many terms in mathematics have definitions with more than one condition. Be certain to read and learn the full definition. For example, a polygon is regular if (1) all its sides are congruent and (2) all its angles are congruent. Because the rhombus shown above does not meet condition 2, it is not a regular polygon. The square meets both conditions, so it is regular.

---

The **perimeter** of a figure is the distance around it. Thus, the perimeter of a polygon is the sum of the lengths of its sides.

*EXAMPLE*   Find the perimeter of each polygon.

a.
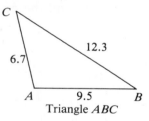
Triangle *ABC*

b.
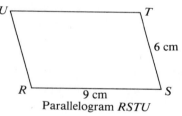
Parallelogram *RSTU*

*Solution*   **a.** Perimeter = 9.5 + 12.3 + 6.7 = 28.5

**b.** Because opposite sides of a parallelogram are congruent, the unlabeled sides have lengths 9 cm and 6 cm. Therefore:
$$\text{Perimeter} = (9 + 6 + 9 + 6) \text{ cm} = 30 \text{ cm}$$

---

## Class Exercises

**Name each polygon according to the number of sides.**

**1.**
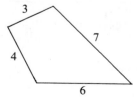

**2.**
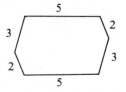

**Name each polygon according to the number of sides.**

3.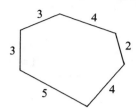

4.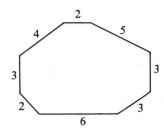

**5–8.** Find the perimeter of each polygon in Exercises 1–4.

**9–12.** State the number of diagonals that can be drawn from any one vertex of each figure in Exercises 1–4.

**Give the most special name for each quadrilateral.**

13.

Four congruent sides
Four congruent angles

14.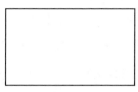

Four congruent angles

15.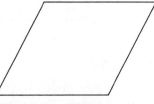

Opposite sides parallel
and congruent

16.

One pair of parallel sides

---

## Written Exercises

**Name the polygon having the given number of sides.**

**A**  **1.** 5  **2.** 4  **3.** 6  **4.** 10  **5.** 3  **6.** 8

**7.** What is another name for a regular quadrilateral?

**8.** What is another name for a regular triangle?

**Find the perimeter of each pentagon.**

**9.**

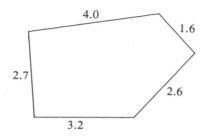

4.0
1.6
2.7
2.6
3.2

**10.**

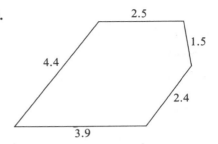

2.5
1.5
4.4
2.4
3.9

**Find the perimeter of a regular polygon whose sides have the given length.**

**11.** Hexagon, 52 cm

**12.** Pentagon, 43 mm

**13.** Triangle, 24.2 mm

**14.** Quadrilateral, 16.5 m

**15.** Decagon, 135.6 m

**16.** Octagon, 4.25 m

**17.** The sum of the measures of the angles of a pentagon is 540°. Find the measure of each angle of a regular pentagon.

**18.** The sum of the measures of the angles of a hexagon is 720°. Find the measure of each angle of a regular hexagon.

**19.** A STOP sign is a regular octagon 32 cm on a side. Express its perimeter in meters.

**20.** The Pentagon building in Washington, D.C., is in the form of a regular pentagon 276 m on a side. Express its perimeter in kilometers.

**21.** The perimeter of a regular pentagon is 60 m. How long is each side?

**Exercises 22–24 refer to the hexagon at the right. The shorter sides are half as long as the longer sides.**

**22.** Each shorter side is 3.2 cm long. What is the perimeter?

**23.** How many diagonals can be drawn from vertex $A$?

**B** **24.** How many diagonals are there in all?

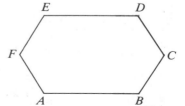

**Use a protractor and a ruler for Exercises 25 and 26. The sum of the measures of the angles of a hexagon is 720°.**

**25.** Draw a hexagon that is not regular, but has all its angles congruent.

**26.** Draw a hexagon that is not regular, but has all its sides congruent.

**27.** In the diagram below, $ABDE$ is a rhombus and $\angle DBC \cong \angle DCB$. Find the perimeter of trapezoid $ACDE$.

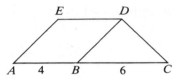

**For Exercises 28 and 29 draw several polygons with different numbers of sides. Pick a vertex and draw all the diagonals from this vertex.**

**28.** Count the number of triangles formed by the diagonals. How does the number of triangles compare to the number of sides of each polygon?

**29.** If the sum of the measures of the angles of the triangles formed equals the sum of the measures of the angles of the polygon, find the sum of the measures of the angles of the following.
  **a.** quadrilateral      **b.** decagon      **c.** trapezoid      **d.** octagon

**C 30.** Write a general formula for the sum of the measures of the angles of any polygon with $n$ sides. (*Hint:* See Exercises 28 and 29.)

**31.** Every pentagon has the same number of diagonals. How many? (*Hint:* First decide how many diagonals can be drawn from one vertex.)

**32.** Every octagon has the same number of diagonals. How many? (See the hint for Exercise 31.)

---

## Review Exercises

**Evaluate if $a = 7$, $b = 3.2$, and $c = 5.45$.**

| | | | |
|---|---|---|---|
| **1.** $ab$ | **2.** $ac$ | **3.** $a^2$ | **4.** $15b$ |
| **5.** $2c$ | **6.** $2bc$ | **7.** $2ab$ | **8.** $abc$ |

# 5-6 Circles

A **circle** is the set of all points in a plane at a given distance from a given point $O$ called the **center.** The drawing at the right shows how to use a **compass** to draw a circle with center $O$.

A segment, such as $\overline{OP}$, joining the center to a point on the circle is called a **radius** (plural: *radii*) of the circle. All radii of a given circle have the same length, and this length is called **the radius** of the circle.

A segment, such as $\overline{XY}$, joining two points on a circle is called a **chord,** and a chord passing through the center is a **diameter** of the circle. The ends of a diameter divide the circle into two **semicircles.** The length of a diameter is called **the diameter** of the circle.

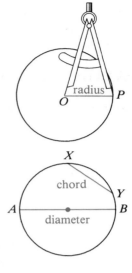

The perimeter of a circle is called the **circumference.** The quotient

$$\text{circumference} \div \text{diameter}$$

can be shown to be the same for all circles, regardless of their size. This quotient is denoted by the Greek letter $\pi$ (pronounced "pie"). No decimal gives $\pi$ exactly, but a fairly good approximation is 3.14.

If we denote the circumference by $C$ and the diameter by $d$, we can write

$$C \div d = \pi.$$

This formula can be put into several useful forms.

---

## *Formulas*

Let $C$ = circumference, $d$ = diameter, and $r$ = radius $(d = 2r)$. Then:

$$C = \pi d$$

$$C = 2\pi r$$

---

**EXAMPLE 1**   The diameter of a circle is 6 cm. Find the circumference.

**Solution**   We are given $d$ and asked to find $C$. We use the formula $C = \pi d$.

$$C = \pi d$$
$$C \approx 3.14 \times 6 = 18.84$$
$$C \approx 18.8 \text{ cm, or } 188 \text{ mm}$$

When using the approximation $\pi \approx 3.14$, give your answer to only three digits (as in Example 1) because the approximation is good only to three digits. That is, we round to the place occupied by the third digit from the left.

**EXAMPLE 2**   The circumference of a circle is 20. Find the radius.

**Solution**   To find the radius, use the formula $C = 2\pi r$.

$$C = 2\pi r$$
$$20 \approx (2 \times 3.14)r$$
$$20 \approx 6.28r$$
$$\frac{20}{6.28} \approx r$$
$$3.1847 \approx r$$

Since the third digit from the left is in the hundredths' place, round to the nearest hundredth. Thus, $r \approx 3.18$.

A polygon is **inscribed** in a circle if all of its vertices are on the circle. The diagram at the right shows a triangle inscribed in a circle.

It can be shown that three points *not on a line* determine a circle. This means that there is one circle, and only one circle, that passes through the three given points.

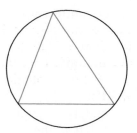

## Class Exercises

**Exercises 1–5 refer to the diagram below. $B$ is the center of the circle. Name each of the following.**

**1.** a diameter

**2.** three radii

**3.** five chords

**4.** two inscribed triangles

**5.** two isosceles triangles

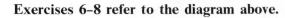

**Exercises 6–8 refer to the diagram above.**

**6.** If $BE = 8$, find $AC$.

**7.** If $AC = 10$, find $AB$.

**8.** If $BC = 20$, find the circumference of the circle.

**Draw a circle and an inscribed polygon of the specified kind.**

**9.** a pentagon

**10.** a hexagon

**11.** an octagon

## Written Exercises

**Use $\pi \approx 3.14$ and round to three digits unless otherwise specified.**

**Find the circumference of each circle with the given diameter or radius.**

**A**  1. diameter = 8 cm                    2. diameter = 20 km

     3. radius = 450 mm                  4. radius = 16 cm

     5. diameter = 42.6 m               6. radius = 278 mm

**Find the diameter of each circle.**

     7. circumference = 283 m           8. circumference = 175 cm

     9. circumference = 450 km         10. circumference = 468 mm

   11. circumference = 625 m          12. circumference = 180 km

**Find the radius of each circle.**

  13. circumference = 10 mm          14. circumference = 20 m

  15. circumference = 23.5 km        16. circumference = 33.3 km

  17. circumference = 17.5 cm        18. circumference = 27.2 m

19. The equator of Earth is approximately a circle of radius 6378 km. What is the circumference of Earth at the equator? Use the approximation $\pi \approx 3.1416$ and give your answer to five digits.

20. A park near Cristi's home contains a circular pool with a fountain at the center. Cristi paced off the distance around the pool and found it to be 220 m. What is the radius of the pool?

21. The diameter of a circular lake is measured and found to be 15 km. What is the circumference of the lake?

22. It is 45 m from the center of a circular field to the inside edge of the track surrounding it. The distance from the center of the field to the outside edge is 55 m. Find the circumference of each edge.

**B** 23. One circle has a radius of 15 m and a second has a radius of 30 m. How much larger is the circumference of the larger circle?

The curves in the diagrams below are parts of circles, and the angles are right angles. Find the perimeter of each figure.

**24.**

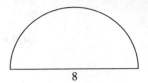

8

**25.**

10

**26.**

4

4

**27.**

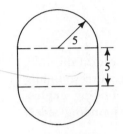

5

5

**28.**

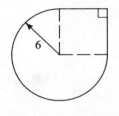

6

**29.**

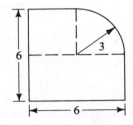

6

3

6

**30.** Find a formula that expresses the length, $S$, of a semicircle in terms of the radius, $r$.

**31.** Find a formula that expresses the length, $S$, of a semicircle in terms of the diameter, $d$.

**In Exercises 32 and 33 use the fact that three points not on a line determine a circle.**

**32.** Every triangle can be inscribed in some circle. Explain why this is so.

**33.** Explain how to draw a quadrilateral that cannot be inscribed in any circle.

**C** **34.** What is the radius of the semicircle that forms the curve of a 400 meter track if each straightaway is 116 m long?

**35.** Draw a circle and one of its diameters, $\overline{AB}$. Then draw and measure $\angle APB$, where $P$ is a point on the circle. Repeat this for several positions of $P$. What does this experiment suggest?

## Review Exercises

**Simplify.**

**1.** $6 + 4 \times 3$  **2.** $16 \div 2 + 2$  **3.** $3(4 + 5)$  **4.** $8(7 - 3)$

**5.** $64 \div (2 + 6)$  **6.** $(18 + 3)2$  **7.** $14 + 3 \times 2 - 6$  **8.** $52 - 18 \div 3 + 16$

## 5-7 Congruent Figures

Two figures are **congruent** if they have the same size and shape. Triangles *ABC* and *XYZ* shown below are congruent.

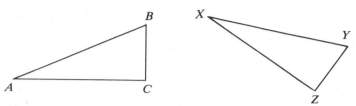

If we could lift △*ABC* and place it on △*XYZ*, *A* would fall on *X*, *B* on *Y*, and *C* on *Z*. These matching vertices are called **corresponding vertices.** Angles at corresponding vertices are **corresponding angles,** and the sides joining corresponding vertices are **corresponding sides.**

> Corresponding angles of congruent figures are congruent.
>
> Corresponding sides of congruent figures are congruent.

When we name two congruent figures, we list corresponding vertices in the same order. Thus, when we see

$$\triangle ABC \cong \triangle XYZ \quad \text{or} \quad \triangle CAB \cong \triangle ZXY,$$

we know that:

$$\angle A \cong \angle X, \qquad \angle B \cong \angle Y, \qquad \angle C \cong \angle Z$$
$$\overline{AB} \cong \overline{XY}, \qquad \overline{BC} \cong \overline{YZ}, \qquad \overline{CA} \cong \overline{ZX}$$

**EXAMPLE 1** pentagon *PQUVW* ≅ pentagon *LMHKT*

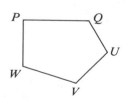
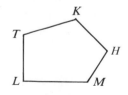

Complete these statements:

$$\angle W \cong \angle \underline{\ ?\ } \qquad \overline{QU} \cong \underline{\ ?\ } \qquad \angle H \cong \angle \underline{\ ?\ } \qquad \overline{TL} \cong \underline{\ ?\ }$$

**Solution** $\qquad \angle W \cong \angle T \qquad \overline{QU} \cong \overline{MH} \qquad \angle H \cong \angle U \qquad \overline{TL} \cong \overline{WP}$

If two figures are congruent, we can make them coincide (occupy the same place) by using one or more of these basic **rigid motions:**

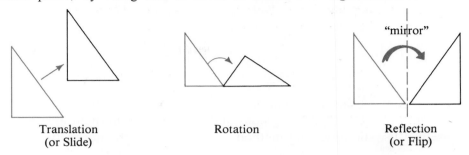

| Translation (or Slide) | Rotation | Reflection (or Flip) |

Consider the congruent trapezoids in panel (1) below. We can make $ABCD$ coincide with $PQRS$ by first reflecting $ABCD$ in the line $\overleftrightarrow{BC}$ as in panel (2), and then translating this reflection as shown in panel (3).

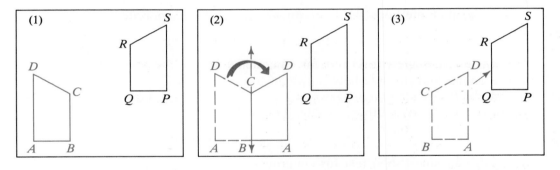

**EXAMPLE 2**  What type of rigid motion would make the red figure coincide with the black one?

a.

b.

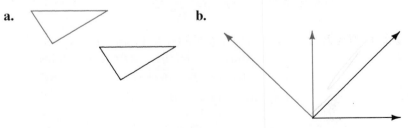

**Solution**  **a.** a translation: sliding the red triangle down and to the right would make it coincide with the black triangle

**b.** a rotation or a reflection: rotating the red angle around the vertex would make it coincide with the black angle; flipping the red angle over a line passing through the vertex would also make it coincide with the black angle

When working with triangles, we do not need to check all sides and all angles to establish congruence. Suppose that in the two triangles below, the sides and the angles marked alike are congruent.

If we were to match the congruent parts by using translation, we would find that all other corresponding sides and angles are congruent also. Thus we can use the following method to establish congruence in two triangles.

**The side-angle-side (SAS) test for congruence**
If two sides of one triangle and the angle they form (the *included angle*) are congruent to two sides and the included angle of another triangle, then the two triangles are congruent.

Two other methods that we can use to establish congruence in triangles are:

**The angle-side-angle (ASA) test for congruence**
If two angles and the side between them (the *included side*) are congruent to two angles and the included side of another triangle, then the two triangles are congruent.

**The side-side-side (SSS) test for congruence**
If three sides of one triangle are congruent to the three sides of another triangle, then the two triangles are congruent.

**EXAMPLE 3**   In the diagram, triangle $ABC$ is isosceles, with $\overline{AB} \cong \overline{CD}$.  $\overline{BD}$ bisects $\angle ABC$. Explain why $\triangle ABD \cong \triangle CBD$.

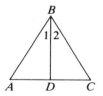

**Solution**   We know that $\overline{AB} \cong \overline{CB}$.
Since $\overline{BD}$ bisects $\angle ABC$, $\angle 1 \cong \angle 2$.
Also, $\overline{BD}$ is a side of both triangles.
Therefore, by the SAS test,
$\triangle ABD \cong \triangle CBD$.

## Class Exercises

Each figure in Exercises 1–8 is congruent to one of the figures $A - E$.
State which one.

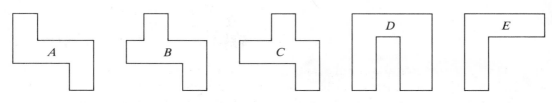

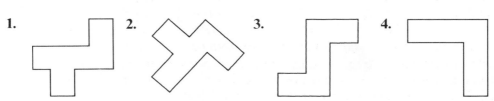

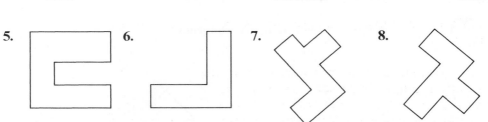

Complete the statements about each pair of congruent figures.

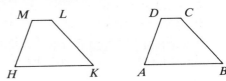

**9.**

a. $\triangle MNO \cong$ ___?___
b. $\angle N \cong$ _?_
c. $\overline{MO} \cong$ _?_

**10.**

a. Quadrilateral $HKLM \cong$ ___?___
b. $\angle B \cong$ _?_
c. $\overline{HM} \cong$ _?_

State which of the rigid motions is needed to match the vertices of the
triangles in each pair and give a reason why the triangles are congruent.

**11.**

**12.**

**13.**

## Written Exercises

**Which statement is correct?**

**A** **1.**

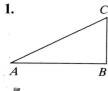 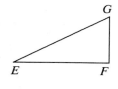

    **a.** $\triangle BCA \cong \triangle GEF$
    **b.** $\triangle ABC \cong \triangle EGF$
    **c.** $\triangle BCA \cong \triangle FGE$

**2.**

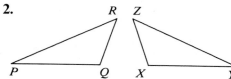

    **a.** $\triangle PQR \cong \triangle XYZ$
    **b.** $\triangle QRP \cong \triangle XZY$
    **c.** $\triangle PQR \cong \triangle XZY$

**Explain why the triangles in each pair are congruent.**

**3.**

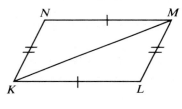

$$\triangle KLM \cong \triangle MNK$$

**4.**

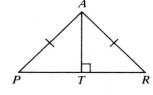

$$\triangle PAT \cong \triangle RAT$$

**What type of rigid motion would make the red figure coincide with the black one?**

**5.**

**6.**

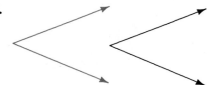

**7.**

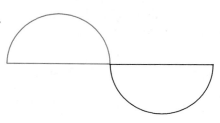

**8.**

**Name a pair of congruent triangles and explain why they are congruent.**

B  **9.**

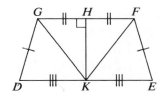

**10.**

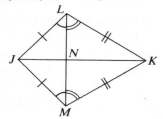

**Complete each statement.**

**11.** $\triangle XRL \cong \triangle NYS$

    **a.** $\angle X \cong$ ___?___
    **b.** $\angle R \cong$ ___?___
    **c.** $\overline{XL} \cong$ ___?___
    **d.** $\overline{YS} \cong$ ___?___

**12.** $PQTV \cong HJKM$

    **a.** $\angle Q \cong$ ___?___
    **b.** $\angle M \cong$ ___?___
    **c.** $\overline{VP} \cong$ ___?___
    **d.** $\overline{JK} \cong$ ___?___

**13.** $ABCD \cong EFGH$

    **a.** $\overline{BC} \cong$ ___?___
    **b.** $\overline{AD} \cong$ ___?___
    **c.** $\angle ABC \cong$ ___?___
    **d.** $\overline{GH} \cong$ ___?___

C  **14.** $ABCDEF$ is a regular hexagon. If all the diagonals from $F$ are drawn, name the following.

    **a.** all pairs of congruent triangles
    **b.** a pair of congruent quadrilaterals
    **c.** a pair of congruent pentagons

**15.** Let $\overline{AB}$ and $\overline{PQ}$ be corresponding sides of two congruent polygons. If one polygon is moved so that $\overline{AB}$ falls on $\overline{PQ}$, must the two polygons coincide?

---

## Review Exercises

**Solve.**

**1.** $6x = 42$
    **2.** $5x = 50$
    **3.** $y \times 4 = 44$
    **4.** $y \times 7 = 56$

**5.** $x \div 9 = 8$
    **6.** $x \div 11 = 6$
    **7.** $84 \div y = 21$
    **8.** $65 \div y = 13$

---

### Calculator Key-In

Many ancient civilizations used approximations for $\pi$. Use a calculator to determine the following approximations for $\pi$ as decimals. Which approximation is closest to the modern approximation of 3.14159265358?

**1.** Egyptian: $\frac{256}{81}$
    **2.** Greek: $\frac{223}{71}$
    **3.** Roman: $\frac{377}{120}$

**4.** Chinese: $\frac{355}{113}$
    **5.** Hindu: $\frac{3927}{1250}$
    **6.** Babylonian: $\frac{25}{8}$

# 5-8 Geometric Constructions

There is a difference between making a drawing and a **geometric construction.** For drawings, we may measure segments and angles; that is, we may use a ruler and a protractor to draw the figures. For geometric constructions, however, we may use only a compass and a straightedge. (We may use a ruler, but we must ignore the markings.)

Here are some important constructions. Construction I and Construction II involve dividing a segment or angle into two congruent parts. This process is called **bisecting** the segment or the angle.

**Construction I:** To bisect a segment $\overline{AB}$.

Use the compass to draw an arc (part of a circle) with center $A$ and radius greater than $\frac{1}{2}AB$. Using the same radius but with center $B$, draw another arc. Call the points of intersection $X$ and $Y$. $\overleftrightarrow{XY}$ is the **perpendicular bisector** of $\overline{AB}$ because it is perpendicular to $\overline{AB}$ and divides $\overline{AB}$ into two congruent segments.

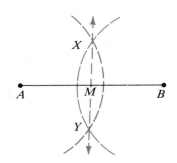

**Construction II:** To bisect an angle $BAC$.

Draw an arc with center $A$. Let $X$ and $Y$ be the points where the arc intersects the sides of the angle. Draw arcs of equal radii with centers $X$ and $Y$. Call the point of intersection $Z$. $\overrightarrow{AZ}$ is the **angle bisector** of $\angle BAC$, and $\angle CAZ \cong \angle ZAB$.

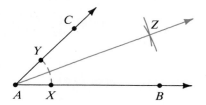

**Construction III:** To construct an angle congruent to a given angle $Y$.

Draw $\overrightarrow{MN}$. Draw an arc on $\angle Y$ with center $Y$. Let $X$ and $Z$ be the points where the arc intersects the sides of the angle. Draw an arc with center $M$ and the same radius as arc $XZ$. Let $S$ be the point where this arc intersects $\overrightarrow{MN}$. Call the other end of the arc $R$. With $S$ as center draw an arc with radius equal to $XZ$. Let $Q$ be the point where this arc intersects arc $RS$. Draw $\overrightarrow{MQ}$. $\angle NMQ \cong \angle Y$.

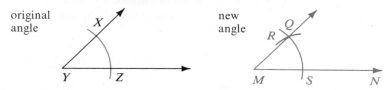

*EXAMPLE 1*  Construct a line that is perpendicular to $\overleftrightarrow{AB}$ and contains $A$.

*Solution*  1. Place the compass at point
$A$ and draw an arc intersect-
ing $\overleftrightarrow{AB}$ at two points. Call
these two points $X$ and $Y$,
respectively. $A$ is now the
midpoint of $\overline{XY}$.

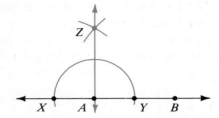

2. Place the compass at $X$ and, as in Construction I, draw an arc
with radius greater than $\overline{YA}$. Keeping the same radius, place the
compass at $Y$ and draw a second arc that intersects the first arc.
Call the point of intersection of the two arcs $Z$.

3. Draw $\overleftrightarrow{AZ}$. Since $\overleftrightarrow{AZ}$ is the perpendicular bisector of $\overline{XY}$, and thus
perpendicular to $\overleftrightarrow{AB}$, it is the required line.

*EXAMPLE 2*  Construct a 60° angle.

*Solution*  1. Draw a ray with endpoint $A$.

2. Draw an arc, with center $A$
and any radius, intersecting the ray at $B$.

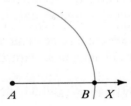

3. Draw an arc, with center $B$
and the same radius as in
step 2, intersecting the first
arc at $C$.

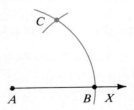

4. Draw $\overrightarrow{AC}$. Since $\triangle ABC$ is
equilateral,  $m \angle BAC =$
60°.

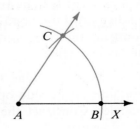

## Written Exercises

**In this exercise set use a compass and straightedge as your only construction tools.**

**A**  1. Construct a 45° angle. (Method: Construct a right angle as in Example 1 and then bisect it.)

2. Construct a 30° angle. (Method: Construct a 60° angle as in Example 2 and then bisect it.)

3. Construct a 22.5° angle. (Use Exercise 1.)

4. Construct a 15° angle. (Use Exercise 2.)

5. Use a protractor to draw an angle with measure 75°. Construct an angle congruent to this angle.

6. Use a protractor to draw an angle with measure 130°. Construct an angle congruent to this angle.

7. Draw a large isosceles triangle. Using this triangle, construct the perpendicular bisector of the base. Through what point does the perpendicular bisector appear to pass?

8. Draw a large scalene triangle. Bisect its three angles. Are the angle bisectors **concurrent;** that is, do all three have a point in common?

9. Draw a large scalene triangle. Construct the perpendicular bisectors of its sides. Are these bisectors concurrent?

**In Exercises 10 and 11, draw $\overleftrightarrow{ST}$ and a point $P$ not on $\overleftrightarrow{ST}$.**

**B**  10. Construct a line through $P$ perpendicular to $\overleftrightarrow{ST}$. (Method: Draw an arc with center $P$ to intersect $\overleftrightarrow{ST}$ in two points, $A$ and $B$. Construct the perpendicular bisector of $\overline{AB}$.)

11. Construct a line through $P$ parallel to $\overleftrightarrow{ST}$. (Method: 1. Construct $\overleftrightarrow{PQ}$ perpendicular to $\overleftrightarrow{ST}$ as in Exercise 10. 2. Construct $\overleftrightarrow{PR}$ perpendicular to $\overleftrightarrow{PQ}$ as in Example 1.)

12. Draw a large scalene triangle. A line through a vertex that is perpendicular to the opposite side is called an **altitude** of the triangle. Construct the three altitudes of the triangle (see Exercise 10). Are they concurrent?

13. Draw a large scalene triangle. A line through a vertex and the midpoint of the side opposite the vertex is called a **median** of the triangle. Construct the three medians (see Construction I). Are they concurrent?

**C**  **14.** Draw three noncollinear points, *A*, *B*, and *C*. Construct the circle that passes through these points. (*Hint:* The perpendicular bisectors of $\overline{AB}$ and $\overline{BC}$ both pass through the center of the circle.)

**15.** Use a compass to construct a regular hexagon. (*Hint:* The length of each side of a regular hexagon inscribed in a circle equals the radius of the circle.)

## Self-Test B

**Complete each statement.**

**1.** A triangle with three congruent sides is ___?___.                    [5-4]

**2.** The sum of the measures of the angles of a triangle is _?_°.

**3.** An acute triangle has ___?___ acute angle(s).

**4.** A(n) ___?___ has eight sides.                                        [5-5]

**5.** A ___?___ has its opposite sides parallel and congruent.

**6.** A trapezoid has sides of 7 cm, 5 cm, 7 cm, and 14 cm. Find its perimeter.

**7.** The radius of a circle is 16 cm. Find its circumference. Use         [5-6]
$\pi \approx 3.14$ and round to three digits.

**True or false?**

**8.** A diameter cuts a circle into two semicircles.

**9.** A radius is a chord.

**10.** Pentagon *ABCDE* ≅ Pentagon *FGHIJ*. Complete each statement.       [5-7]
   **a.** $\overline{AB} \cong$ _?_     **b.** $\angle E \cong \angle$ _?_     **c.** $\angle DEA \cong \angle$ _?_

**11.** Construct an isosceles right triangle.                              [5-8]

**12.** Construct an equilateral triangle.

*Self-Test answers and Extra Practice are at the back of the book.*

![] █ █ █    **Challenge**

You have your choice of your height in nickels that are stacked or in quarters that are laid side by side. Which would you choose?

# More Programming in BASIC

In Chapter 3 we learned that to enter different values of a variable in BASIC we can use the INPUT statement.

To assign a value to a variable that will be repeated over and over again, we use the **LET** statement. For example, the statement

```
20   LET K = 4037
```

assigns the value 4037 to the variable $K$. This statement tells the computer to store 4037 in its memory at location $K$. The value of a variable can be changed by assigning a new value. When we write

```
20   LET K = 0.025
```

the original value, 4037, is replaced by the new value, 0.025.

The program below converts miles to kilometers by using the fact that 1 mi = 1.61 km.

```
10   PRINT "FROM MILES TO KILOMETERS"
20   PRINT "DISTANCE IN MILES";
30   INPUT X
40   LET A = 1.61
50   PRINT X;" MI = ";A*X;" KM"
60   END
```

Let us use the program to convert the approximate distance in miles from the planet Saturn to the Sun. That is, convert 887,000,000 mi to kilometers.

```
RUN
FROM MILES TO KILOMETERS
DISTANCE IN MILES? 887000000      ←──  do not use commas
887000000 MI = 1428070000 KM            to enter the distance
```

**Use the program on the previous page to complete the table.**

| | Planet | Distance (in mi) from the Sun | Distance (in km) from the Sun |
|---|---|---|---|
| **1.** | Mercury | 36,000,000 | ? |
| **2.** | Venus | 67,000,000 | ? |
| **3.** | Earth | 93,000,000 | ? |

Instead of running the program three times, we can modify it to repeat lines 20 through 50 so that all three distances are converted in one RUN. To do this, we use the **FOR** and **NEXT** statements to create a *loop*. The loop starts with the FOR statement, and ends with the NEXT statement. These two statements tell the computer how many times to repeat a group of statements located between them. The program below is now modified to repeat the loop three times. The output for Exercises 1–3 is shown at the right.

```
10   PRINT "FROM MILES TO KILOMETERS"        RUN
15   FOR I = 1 TO 3                          FROM MILES TO KILOMETERS
20   PRINT "DISTANCE IN MILES";              DISTANCE IN MILES? 36000000
30   INPUT X                                 36000000 MI = 57960000 KM
40   LET A = 1.61                            DISTANCE IN MILES? 67000000
50   PRINT X;" MI = ";A*X;" KM"              67000000 MI = 107870000 KM
55   NEXT I                                  DISTANCE IN MILES? 93000000
60   END                                     93000000 MI = 149730000 KM
```

Depending on the computer you are using, the output displayed for the conversions above may be expressed in *scientific notation*. That is, a number such as 376770000 may be expressed as

$$3.7677E+08.$$

The code $E+08$ means "times 10 raised to the power of 8." Therefore

$$3.7677E+08 \text{ means } 3.7677 \times 10^8, \text{ or } 376770000.$$

**Complete.**

**4.** $37,492,000,000 = 3.7492E+\underline{\ ?\ }$

**5.** $5,491,000,000,000 = 5.491E+\underline{\ ?\ }$

**6.** $9.4678E+07 = \underline{\ ?\ }$

**7.** $3.2186E+10 = \underline{\ ?\ }$

**8.** Write a program to print out the multiples of 2 from one to ten.

**9.** Write a program to print out the distance traveled at a constant rate of 760 mi/h for 15, 27, 31, 40, and 55 hours. Use the formula $d = rt$.

# Chapter Review

**Complete.**

1. Points on the same line are called ___?___.                           [5-1]

2. A ___?___ has one endpoint.

3. $Z$ divides $\overline{XY}$ into two congruent segments. $Z$ is called the ___?___   [5-2]
   of $\overline{XY}$.

4. 927 mm = ___?___ cm = ___?___ m

**True or false?**

5. A right angle is obtuse.                                              [5-3]

6. In $\angle ABC$, $A$ is the vertex.

7. The supplement of a 40° angle has measure 140°.

8. A triangle that has three sides of different lengths is called scalene.   [5-4]

9. In a triangle, the longest side is opposite the smallest angle.

10. All quadrilaterals are parallelograms.                               [5-5]

**Write the letter of the correct answer.**

11. A hexagon is regular. One side has length 8 cm. What is the perim-
    eter?
    **a.** 40 cm        **b.** 64 cm        **c.** 80 cm        **d.** 48 cm

12. Name the segment joining the center of a circle to a point on the     [5-6]
    circle.
    **a.** diameter        **b.** chord        **c.** radius        **d.** circumference

13. A circle has diameter 16 cm. Use $\pi \approx 3.14$ to find the circumfer-
    ence and round to three digits.
    **a.** 50.24 cm        **b.** 50.3 cm        **c.** 50.2 cm        **d.** 50 cm

14. Which is the symbol for congruence?                                  [5-7]
    **a.** $\cong$        **b.** $\perp$        **c.** $\angle$        **d.** $\triangle$

15. Construct a 120° angle.                                             [5-8]

16. Construct a right triangle.

# Chapter Test

**Exercises 1–3 refer to the diagram at the right.**

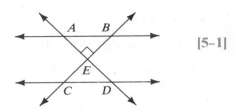

1. Name a pair of perpendicular lines.

2. Name two rays that are not parallel, but do not intersect.

3. Name three collinear points.

[5-1]

**Complete.**

4. 27 m = _?_ cm

5. 3.6 km = _?_ m

[5-2]

6. 5000 cm = _?_ km

7. 4 mm = _?_ m

8. Give the measures of the complement and the supplement of ∠ A if m ∠ A = 27°.

[5-3]

9. If ∠ X ≅ ∠ Y, then m ∠ X = _?_.

10. True or false? The sides of a right angle are perpendicular.

11. One angle of an isosceles triangle has measure 98°. Find the measures of the two congruent angles.

[5-4]

12. An obtuse triangle has how many obtuse angles?

13. True or false? A square is a rhombus.

[5-5]

14. A quadrilateral has sides of length 8 cm, 13 cm, 9 cm, and 16 cm. Find the perimeter.

15. A regular hexagon has perimeter 84 mm. Find the length of each side.

16. The diameter of a circle is 18 cm. Find the radius.

[5-6]

17. A circle has radius 50 mm. Find the circumference. Use $\pi \approx 3.14$.

18. True or false? Every triangle can be inscribed in a circle.

19. If quadrilateral $ABCD$ ≅ quadrilateral $WXYZ$, then $\overline{AD}$ ≅ _?_.

[5-7]

20. Draw a segment. Construct the perpendicular bisector. Then construct the bisector of one of the right angles.

[5-8]

# Cumulative Review (Chapters 1–5)

## Exercises

**Evaluate the expression if $x = 3$ and $y = 5$.**

**1.** $x + y$      **2.** $2x + y$      **3.** $2y - x$      **4.** $x + 6y$

**5.** $y + x + 4$      **6.** $-x - y$      **7.** $-3x - 2y$      **8.** $5x + (-y) - 7$

**9.** $x^2$      **10.** $xy^2$      **11.** $(-x)^2 y$      **12.** $-(xy)^2$

**Solve using transformations.**

**13.** $x + 36 = 50$      **14.** $x - 11 = -4$      **15.** $-4x = 75$

**16.** $-7x = -105$      **17.** $\frac{x}{8} = 9$      **18.** $\frac{x}{7} = -8$

**19.** $\frac{2}{3}x = 16$      **20.** $-\frac{5}{6}x = 25$      **21.** $0.45x = 13.5$

**22.** $\frac{x}{1.8} = 2.7$      **23.** $\frac{4}{5}x + 8 = 20$      **24.** $\frac{15}{4}x + 6 = -4$

**Write in lowest terms.**

**25.** $\frac{32}{40}$      **26.** $\frac{56}{80}$      **27.** $-\frac{12}{108}$

**28.** $-\frac{13}{182}$      **29.** $\frac{98}{147}$      **30.** $\frac{72}{96}$

**Write as a terminating or repeating decimal. Use a bar to indicate repeating digits.**

**31.** $\frac{2}{5}$      **32.** $\frac{3}{4}$      **33.** $\frac{7}{16}$

**34.** $\frac{3}{10}$      **35.** $-\frac{5}{18}$      **36.** $-\frac{2}{3}$

**Write as a proper fraction in lowest terms or as a mixed number in simple form.**

**37.** $0.07$      **38.** $0.007$      **39.** $1.\overline{9}$

**40.** $4.\overline{20}$      **41.** $-1.\overline{24}$      **42.** $-4.\overline{862}$

**True or false?**

**43.** An equilateral triangle is acute.      **44.** A diameter is a chord.

# Problems

---

**Problem Solving Reminders**

Here are some reminders that may help you solve some of the problems on this page.
- Determine which facts are necessary to solve the problem.
- Determine whether more than one operation is needed.
- Estimate to check your answers.

---

**Solve.**

1. Susan purchased the following items for the school dance: streamers, $2.89; tape, $4.59; bunting, $8.88; paper decorations, $14.75. How much did Susan spend?

2. Fred's Fish Farm started the week with 2078 fish. On Monday Fred sold 473 fish, on Tuesday he sold 509 fish, and on Wednesday 617 fish were sold. Fred bought 675 fish on Thursday and sold 349 on Friday. How many fish did Fred have at the end of the week?

3. Mr. Chou was putting a certain amount into his savings account each month. Last month he increased the amount by $38. If Mr. Chou deposited $162 into his savings account last month, how much was he putting in before the increase?

4. Yonora bought 7 gallons of paint at $16.95 a gallon, 3 brushes at $6.99 each, 4 rollers at $2.95 each, and a dropcloth for $7.88. How much did Yonora spend on painting supplies?

5. A side of a square is 13 m long. Find the perimeter.

6. If a number is multiplied by 3, the result is 51. Find the number.

7. The seventh grade sold greeting cards to raise money for a trip. There were 12 cards and 12 envelopes in each box. If the class sold 1524 cards with envelopes, how many boxes did they sell?

8. Elgin had $150 in his checking account. He wrote checks for $17.95, $23.98, $45.17, and $31.26. How much does Elgin have left in his account?

9. Becky bought a round pool that is 7 m in diameter. What is the circumference of the pool to the nearest meter?

10. Juanita is 7 years older than her brother Carlos, who is 3 years older than their sister Maria. If Carlos is 6 years old, how old are Juanita and Maria?

# 6

# Ratio, Proportion, and Percent

Before the invention of the microscope, objects appeared to consist only of those materials seen with the unaided eye. Today the ability to magnify objects, such as the salt crystals shown at the right, has enabled scientists to understand the structures of various compounds in detail. The most advanced microscopes in use at the present time are electron microscopes. Electron microscopes can magnify objects hundreds of thousands of times by using beams of focused electrons.

The visibility of fine detail in a magnification depends on several factors, including the light and the magnifying power of the microscope. For a simple microscope, the magnifying power can be expressed as this ratio:

$$\frac{\text{size of the image on the viewer's eye}}{\text{size of the object seen without a microscope}}.$$

In this chapter, you will learn how scales and ratios are used.

## Career Note

When you think of photography, you probably think of it as a means of portraying people and places. When used in conjunction with a microscope (photomicrography), or with infrared or ultraviolet light, photography can become an important research tool. Scientific photographers must have a knowledge of film, filters, lenses, illuminators, and all other types of camera equipment. They must also have a thorough understanding of scale and proportion in order to find the best composition for a particular photograph.

# 6-1 Ratios

At Fair Oaks Junior High School there are 35 teachers and 525 students. We can compare the number of teachers to the number of students by writing a quotient.

$$\frac{\text{number of teachers}}{\text{number of students}} = \frac{35}{525}, \text{ or } \frac{1}{15}$$

The indicated quotient of one number divided by a second number is called the **ratio** of the first number to the second number. We can write the ratio above in the following ways.

$$\frac{1}{15} \qquad 1:15 \qquad 1 \text{ to } 15$$

All of these expressions are read *one to fifteen*. If the colon notation is used, the first number is divided by the second. A ratio is said to be in **lowest terms** if the two numbers are relatively prime. You do not change an improper fraction to a mixed number if the improper fraction represents a ratio.

**EXAMPLE 1**   There are 9 players on a baseball team. Four of these are infielders and 3 are outfielders. Find each ratio in lowest terms.

**a.** infielders to outfielders

**b.** outfielders to total players

**Solution**   **a.** $\dfrac{\text{infielders}}{\text{outfielders}} = \dfrac{4}{3}$,   or   $4:3$,   or   4 to 3

**b.** $\dfrac{\text{outfielders}}{\text{total players}} = \dfrac{3}{9} = \dfrac{1}{3}$,   or   $1:3$,   or   1 to 3

Some ratios compare measurements. In these cases, we must be sure that the measurements are expressed in the same unit.

**EXAMPLE 2**   It takes Herb 4 min to mix some paint. Herb can paint a room in 3 h. What is the ratio of the time it takes Herb to mix the paint to the time it takes Herb to paint the room?

**Solution**   Use minutes as a common unit for measuring time.

$$3 \text{ h} = 3 \times 60 \text{ min} = 180 \text{ min}$$

The ratio is $\dfrac{\text{min to mix}}{\text{min to paint}} = \dfrac{4}{180} = \dfrac{1}{45}$, or $1:45$.

## Class Exercises

**Express each ratio as a fraction in lowest terms.**

**1.** 5 to 7

**2.** 11 to 6

**3.** 10:30

**4.** 12:24

**5.** 8 to 2

**6.** 32 to 4

**7.** 68:17

**8.** 45:18

**Rewrite each ratio so that the numerator and denominator are expressed in the same unit of measure.**

**9.** $\dfrac{2 \text{ dollars}}{50 \text{ cents}}$

**10.** $\dfrac{5 \text{ months}}{2 \text{ years}}$

**11.** $\dfrac{35 \text{ cm}}{1 \text{ m}}$

**12.** $\dfrac{12 \text{ min}}{2 \text{ h}}$

## Written Exercises

**For each diagram below, name each ratio as a fraction in lowest terms.**
**a. The number of shaded squares to the number of unshaded squares**
**b. The number of shaded squares to the total number of squares**
**c. The total number of squares to the number of unshaded squares**

**A  1.**

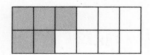

**2.**

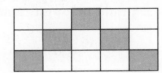

**3.**

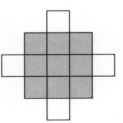

**4.**

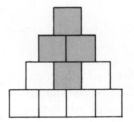

**Express each ratio as a fraction in lowest terms.**

**5.** 18 hours to 2 days

**6.** 25 cm to 3 m

**7.** 4 days:2 weeks

**8.** 48 s:5 min

**9.** 3 kg:800 g

**10.** 6 lb:24 oz

**Find each ratio as a fraction in lowest terms.**

**B  11. a.** The number of vowels to the number of consonants in the alphabet (consider y a consonant)
  **b.** The number of consonants to the total number of letters
  **c.** The total number of letters to the number of vowels

**12. a.** The number of weekdays to the number of weekend days (Saturdays and Sundays) in the month of February (not in a leap year)
  **b.** The number of weekdays in the month of February (not a leap year) to the number of days in February
  **c.** The number of days to the number of Sundays in the month of February (not a leap year)

**13. a.** The number of diagonals drawn in the figure at the right to the total number of segments
  **b.** The number of sides of the figure to the number of diagonals
  **c.** The total number of segments to the number of sides

**14. a.** The number of prime numbers between 10 and 25 to the number of whole numbers between 10 and 25
  **b.** The number of prime numbers between 10 and 25 to the number of composite numbers between 10 and 25
  **c.** The number of whole numbers between 10 and 25 to the number of composite numbers between 10 and 25

**In Exercises 15–18, $AB = 7\frac{1}{5}$, $CD = 10\frac{1}{2}$, $EF = 12$, and $GH = 6\frac{3}{4}$. Express each ratio in lowest terms.**

**C  15.** $\dfrac{AB}{EF}$      **16.** $\dfrac{EF}{GH}$      **17.** $\dfrac{CD}{GH}$      **18.** $\dfrac{GH}{AB}$

---

## Problems

**Solve.**

**A  1.** The *mechanical advantage* of a simple machine is the ratio of the weight lifted by the machine to the force necessary to lift it. What is the mechanical advantage of a jack that lifts a 3200-lb car with a force of 120 lb?

**2.** The *C*-string of a cello vibrates 654 times in 5 seconds. How many vibrations per second is this?

**3.** At sea level, 4 ft³ of water weighs 250 lb. What is the density of water in pounds per cubic foot?

**4.** The *index of refraction* of a transparent substance is the ratio of the speed of light in space to the speed of light in the substance. Using the table, find the index of refraction of
**a.** glass.　　**b.** water.

| Substance | Speed of Light (in km/s) |
|-----------|--------------------------|
| space     | 300,000                  |
| glass     | 200,000                  |
| water     | 225,000                  |

**5.** A share of stock that cost $88 earned $16 last year. What was the price-to-earnings ratio of this stock?

**In Exercises 6 and 7, find the ratio in lowest terms.**

**B**　**6. a.** $\dfrac{AB}{DE}$

　　**b.** $\dfrac{\text{Perimeter of } \triangle ABC}{\text{Perimeter of } \triangle DEF}$

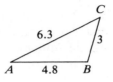

**7. a.** $PQ:TU$
　**b.** $QR:UV$
　**c.** $\dfrac{\text{Perimeter of } PQRS}{\text{Perimeter of } TUVW}$

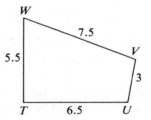

**For Exercises 8–11, refer to the table to find the ratios in lowest terms.**

**8.** The population of Centerville in 1980 to its population in 1970

**9.** The growth in the population of Easton to its 1980 population

| Population (in thousands) | | |
|------|------|------|
| Town | 1970 | 1980 |
| Centerville | 36 | 44 |
| Easton | 16 | 28 |

**10.** The total population of both towns in 1970 to their total population in 1980

**11.** The total growth in the population of both towns to their total 1980 population

---

## Review Exercises

**Solve.**

**1.** $6x = 54$　　　　**2.** $11x = 99$　　　　**3.** $5x = 45$　　　　**4.** $3x = 39$

**5.** $20x = 100$　　　**6.** $10x = 80$　　　　**7.** $12x = 144$　　　**8.** $9x = 72$

# 6-2 Rates

Some ratios are of the form

$$40 \text{ miles per hour} \quad \text{or} \quad 5 \text{ for a dollar.}$$

These ratios involve quantities of different kinds and are called **rates.**
Rates may be expressed as decimals or mixed numbers. Rates should
be simplified to a *per unit* form.

**EXAMPLE 1**  Alice's car went 258 mi on 12 gal of gasoline. Express the rate of
fuel consumption in miles per gallon.

**Solution**  The rate of fuel consumption is

$$\frac{258}{12} = \frac{43}{2} = 21\frac{1}{2} \text{ miles per gallon.}$$

Alice's car consumes 1 gal of gasoline every $21\frac{1}{2}$ mi.

Some of the units in which rates are given are:

| | | | |
|---|---|---|---|
| mi/gal (or mpg) | miles per gallon | km/L | kilometers per liter |
| mi/h (or mph) | miles per hour | km/h | kilometers per hour |

If a rate is the price of one item, it is called the **unit price.** For
example, if 2 peaches sell for 78¢, the unit price is $\frac{78}{2}$, or 39, cents
per peach.

**EXAMPLE 2**  If 5 oranges sell for 95¢, what is the cost of 12 oranges?

**Solution**  You can plan to find the price of one orange and then multiply by
12 to answer the question in the problem.

$$\text{unit price} = \frac{95}{5} = 19 \text{ (cents per orange)}$$

$$\text{cost of 12 oranges} = 12 \times 19 = 228 \text{ (cents)}$$

The cost of 12 oranges is 228¢, or $2.28.

---

## Class Exercises

**Express each rate in per unit form.**

**1.** 120 km in 3 h

**2.** 70 mi on 5 gal of gasoline

**3.** $1000 in 4 months

**4.** 30 km on 3 L of gasoline

**5.** Three melons for $1.59

**6.** Two cans of tennis balls for $5.88

**7.** Six cans of water per 2 cans of juice

**8.** 28 bicycles sold in 7 days

**9.** 192.5 km in 3.5 h

**10.** A dozen eggs for $1.08

**11.** Ten oranges for $1.65

**12.** 225 m in 25 s

**13.** 28 teachers per 56 students

**14.** 18°F in 2 hours

**15.** 11 tickets for $30.80

**16.** Seven days for $399

---

## Problems

**Give the unit price of each item.**

**A**  **1.** 7 oz of crackers for $1.19

**2.** 14 oz of cottage cheese for $1.19

**3.** 5 yd of upholstery fabric for $80

**4.** 16 boxes of raisins for $5.60

**Solve.**

**5.** A certain kind of lumber costs $3.00 for 8 ft. At this rate, how much does a piece that is 14 ft long cost?

**6.** A package of 3 lb of ground beef costs $5.10. How much ground beef could you buy for $3.40 at this rate?

**7.** A car travels for 3 h at an average speed of 65 km/h. How far does the car travel?

**8.** A boat covers 48 km in 2 h. What is the boat's average speed?

**9.** A homing pigeon flies 180 km in 3 h. At this average speed, how far could the pigeon travel in 4 h?

**10.** A bicyclist travels for 2 h at an average speed of 12 km/h. How far does the bicyclist travel? At this speed, how long will it take the bicyclist to travel 54 km?

**Determine the better buy based on unit price alone.**

**B**  **11.** A can of 35 oz of Best Brand Pear Tomatoes is on sale for 69¢. A can of 4 lb of Sun Ripe Pear Tomatoes costs $1.88. Which brand is the better buy?

**12.** A can of Favorite Beef Dog Food holds $14\frac{1}{2}$ oz. Four cans cost $1.00. Three cans of Delight Beef Dog Food, each containing 12 oz, cost $.58. Which is the better buy?

**13.** Three bottles of Bright Shine Window Cleaner, each containing 15 oz, cost $2.75. Two bottles of Sparkle Window Cleaner, each containing 18.75 oz, can be purchased for $1.98. Which is the better buy?

**14.** A bottle of Harvest Time Apple Juice contains 64 oz and costs 99¢. Farm Fresh Juice is available in bottles that contain 1 gal for $1.88 each. Which is the better buy?

**Solve.**

**15.** A car uses 3 gal of gasoline every 117 mi. During the first part of a trip, the car traveled 130 mi in 4 h. If the car continues to travel at the same average speed and the trip takes a total of 6 h, how many gallons of gasoline will be consumed?

**16.** Emily Depietro purchased 3 trays of strawberry plants for $16.47. Emily also purchased 10 trays of ivy plants. If the ratio of the price per tray of strawberry plants to the price per tray of ivy plants is 3 to 2, what is the total cost of the plants?

**C** **17.** In the time that it takes one car to travel 93 km, a second car travels 111 km. If the average speed of the second car is 12 km/h faster than the speed of the first car, what is the speed of each car?

## Review Exercises

**Find the perimeter of a regular polygon whose sides have the given length.**

**1.** pentagon, 4 cm

**2.** square, 6.2 m

**3.** octagon, 10.5 cm

**4.** triangle, 8.9 m

**5.** rhombus, 35.6 m

**6.** hexagon, 21.75 cm

**7.** quadrilateral, 14.35 m

**8.** decagon, 64.87 cm

**9.** rhombus, 12.36 cm

# 6-3 Proportions

The seventh grade at Madison Junior High School has 160 students and 10 teachers. The seventh grade at Jefferson Junior High School has 144 students and 9 teachers. Let us compare the two teacher-student ratios.

$$\frac{10}{160} = \frac{1}{16} \qquad \frac{9}{144} = \frac{1}{16}$$

Thus, the two ratios are equal.

$$\frac{10}{160} = \frac{9}{144}$$

An equation that states that two ratios are equal is called a **proportion.** The proportion above may be read as

$$10 \text{ is to } 160 \text{ as } 9 \text{ is to } 144.$$

The numbers 10, 160, 9, and 144 are called the **terms** of the proportion.

Sometimes one of the terms of a proportion is a variable. If, for example, 192 students will be in the seventh grade at Madison Junior High next year, how many teachers will be needed if the teacher-student ratio is to remain the same?

Let $n$ be the number of teachers needed next year. Then, if the teacher-student ratio is to be the same, we must have

$$\frac{n}{192} = \frac{10}{160}.$$

To **solve** this proportion, we find the value of the variable that makes the equation true. This can be done by finding equivalent fractions with a common denominator. For example:

$$\frac{160 \times n}{160 \times 192} = \frac{10 \times 192}{160 \times 192}$$

Since the denominators are equal, the numerators also must be equal.

$$160 \times n = 10 \times 192$$

Notice that this result could also be obtained by **cross-multiplying** in the original proportion.

$$\frac{n}{192} \diagdown\!\!\!\!\diagup \frac{10}{160}$$

$$160 \times n = 10 \times 192$$
$$160n = 1920$$
$$n = 1920 \div 160 = 12$$

Therefore, the seventh grade will need 12 teachers next year.

The example on the previous page illustrates the following property of proportions.

> ## Property
>
> If $\frac{a}{b} = \frac{c}{d}$, with $b \neq 0$ and $d \neq 0$, then $ad = bc$.

In the proportion $\qquad \frac{a}{b} = \frac{c}{d}$

the terms $a$ and $d$ are called the **extremes,** and the terms $b$ and $c$ are called the **means.** The property above can therefore be stated:

The product of the means equals the product of the extremes.

*EXAMPLE*   Solve $\frac{3}{8} = \frac{12}{n}$.

*Solution* $\qquad\qquad \frac{3}{8} = \frac{12}{n}$

$$3 \times n = 8 \times 12$$
$$3n = 96$$
$$n = 96 \div 3 = 32$$

It is a simple matter to check your answer when solving a proportion. You merely substitute your answer for the variable and cross-multiply. For instance, in the example above:

$$\frac{3}{8} \overset{?}{=} \frac{12}{32}$$
$$3 \times 32 \overset{?}{=} 8 \times 12$$
$$96 = 96$$

## Class Exercises

**Cross-multiply and state an equation that does not involve fractions.**

1. $\frac{n}{9} = \frac{2}{3}$      2. $\frac{3}{5} = \frac{n}{20}$      3. $\frac{2}{7} = \frac{6}{n}$      4. $\frac{3}{n} = \frac{6}{10}$

5. $\frac{12}{15} = \frac{n}{5}$      6. $\frac{2}{n} = \frac{3}{9}$      7. $\frac{n}{16} = \frac{3}{4}$      8. $\frac{8}{3} = \frac{24}{n}$

## Written Exercises

**Solve and check.**

**A** 1. $\dfrac{n}{3} = \dfrac{12}{9}$

2. $\dfrac{7}{2} = \dfrac{x}{10}$

3. $\dfrac{21}{r} = \dfrac{3}{8}$

4. $\dfrac{3}{75} = \dfrac{2}{m}$

5. $\dfrac{8}{5} = \dfrac{56}{u}$

6. $\dfrac{14}{n} = \dfrac{7}{9}$

7. $\dfrac{80}{c} = \dfrac{4}{3}$

8. $\dfrac{b}{24} = \dfrac{15}{9}$

9. $\dfrac{8}{7} = \dfrac{x}{63}$

10. $\dfrac{d}{20} = \dfrac{14}{8}$

11. $\dfrac{15}{11} = \dfrac{n}{33}$

12. $\dfrac{13}{11} = \dfrac{26}{m}$

13. $\dfrac{5}{r} = \dfrac{2}{3}$

14. $\dfrac{4}{3} = \dfrac{n}{7}$

15. $\dfrac{17}{20} = \dfrac{v}{10}$

16. $\dfrac{c}{2} = \dfrac{6}{18}$

**B** 17. $\dfrac{x}{5} = \dfrac{20}{10}$

18. $\dfrac{3}{m} = \dfrac{9}{27}$

19. $\dfrac{4}{n} = \dfrac{12}{36}$

20. $\dfrac{a}{16} = \dfrac{4}{8}$

21. $\dfrac{20}{25} = \dfrac{16}{y}$

22. $\dfrac{v}{50} = \dfrac{18}{30}$

23. $\dfrac{25}{x} = \dfrac{15}{9}$

24. $\dfrac{49}{16} = \dfrac{n}{4}$

25. $\dfrac{n}{8} = \dfrac{7}{10}$

26. $\dfrac{15}{4} = \dfrac{9}{r}$

27. $\dfrac{9}{10} = \dfrac{b}{5}$

28. $\dfrac{9}{n} = \dfrac{15}{7}$

29. If $\dfrac{x}{7} = \dfrac{3}{21}$, what is the ratio of $x$ to 3?

30. If $\dfrac{27}{m} = \dfrac{9}{2}$, what is the ratio of $m$ to 27?

31. If $\dfrac{3}{5} = \dfrac{12}{n}$, what is the ratio of $n$ to 5?

**C** 32. Choose nonzero whole numbers $a$, $b$, $c$, $d$, $x$, and $y$ such that $\dfrac{a}{b} = \dfrac{x}{y}$ and $\dfrac{c}{d} = \dfrac{x}{y}$. Use these numbers to check whether $\dfrac{a+c}{b+d} = \dfrac{x}{y}$.

33. Find nonzero whole numbers $a$, $b$, $c$, and $d$ to show that if $\dfrac{a+c}{b+d} = \dfrac{x}{y}$, it may not be true that $\dfrac{a}{b} = \dfrac{x}{y}$ and $\dfrac{c}{d} = \dfrac{x}{y}$.

## Review Exercises

**Multiply.**

1. $2\dfrac{2}{3} \times 5$

2. $3\dfrac{5}{8} \times 6$

3. $5 \times 4\dfrac{5}{9}$

4. $7 \times 6\dfrac{3}{4}$

5. $3\dfrac{4}{9} \times 2\dfrac{1}{2}$

6. $4\dfrac{1}{3} \times 5\dfrac{1}{4}$

7. $6 \times 7\dfrac{1}{4}$

8. $5 \times 8\dfrac{3}{5}$

9. $9 \times 6\dfrac{7}{10}$

# 6-4 Solving Problems with Proportions

Proportions can be used to solve problems. The following steps are helpful in solving problems using proportions.

1. Decide which quantity is to be found and represent it by a variable.
2. Determine whether the quantities involved can be compared using ratios (rates).
3. Equate the ratios in a proportion.
4. Solve the proportion.

**EXAMPLE** Linda Chu bought 4 tires for her car at a total cost of $264. How much would 5 tires cost at the same rate?

**Solution** Let $c$ = the cost of 5 tires. Set up a proportion.

$$\frac{4}{264} = \frac{5}{c} \longleftarrow \text{ number of tires} \atop \longleftarrow \text{ cost}$$

Solve the proportion.

$$\frac{4}{264} = \frac{5}{c}$$
$$4c = 5 \times 264$$
$$4c = 1320$$
$$c = 330$$

Therefore, 5 tires would cost $330.

Notice that the proportion in the Example could also be written as:

$$\frac{264}{4} = \frac{c}{5} \longleftarrow \text{ cost} \atop \longleftarrow \text{ number of tires}$$

---

## Class Exercises

**State a proportion you could use to solve each problem.**

1. If 4 bars of soap cost $1.50, how much would 8 bars cost?

2. If you can buy 4 containers of cottage cheese for $4.20, how many could you buy for $9.45?

3. If a satellite travels 19,500 km in 3 h, how far does it travel in 7 h?

4. If a car uses 5 gal of gasoline to travel 160 mi, how many gallons would the car use in traveling 96 mi?

**5.** A recipe for 20 rolls calls for 5 tablespoons of butter. How many tablespoons are needed for 30 rolls?

**6.** If 9 kg of fertilizer will feed 300 m² of grass, how much fertilizer would be required to feed 500 m²?

**7.** If 2 cans of paint will cover a wall measuring 900 ft², what area will 3 cans cover?

## Problems

**Solve.**

**A**  **1.** A train traveled 720 km in 9 h.
   **a.** How far would it travel in 11 h?
   **b.** How long would it take to go 1120 km?

**2.** Five pounds of apples cost $3.70.
   **a.** How many pounds could you buy for $5.92?
   **b.** How much would 9 lb cost?

**3.** Eight oranges cost $1.50.
   **a.** How much would 20 oranges cost?
   **b.** How many oranges could you buy for $5.25?

**4.** Due to Earth's rotation, a point on the equator travels about 40,000 km every 24 h.
   **a.** How far does a point on the equator travel in 33 h?
   **b.** How long does it take a point on the equator to travel 95,000 km?

**5.** Seventy-five cubic centimeters of maple sap can be boiled down to make 2 cm³ of maple syrup.
   **a.** How much maple syrup would 200 cm³ of sap make?
   **b.** How much sap would be needed to make 9 cm³ of syrup?

**6.** A long-playing record revolves 100 times every 3 min.
   **a.** How many revolutions does it make in 2.25 min?
   **b.** How long does it take for 275 revolutions?

**7.** Three and a half pounds of peaches cost $1.68. How much would $2\frac{1}{2}$ lb of peaches cost?

**Solve.**

8. A type of steel used for bicycle frames contains 5 g of manganese in every 400 g of steel. How much manganese would a 2200 g bicycle frame contain?

9. Five cans of paint will cover 130 m² of wall space. How many cans will be needed to cover 208 m²?

10. To obtain the correct strength of a medicine, 5 cm³ of distilled water is added to 12 cm³ of an antibiotic. How much water should be added to 30 cm³ of the antibiotic?

11. A receipe that serves 8 calls for 15 oz of cooked tomatoes. How many ounces of tomatoes will be needed if the recipe is reduced to serve 6 people? How many servings can be made with 20 oz of cooked tomatoes?

12. A geologist found that silt was deposited on a river bed at the rate of 4 cm every 170 years. How long would it take for 5 cm of silt to be deposited? How much silt would be deposited in 225 years?

13. A printing press can print 350 sheets in 4 min. How long would it take to print 525 sheets?

14. A pharmacist combines 5 g of a powder with 45 cm³ of water to make a prescription medicine. How much powder should she mix with 81 cm³ of water to make a larger amount of the same medicine?

**B** 15. A baseball team has won 8 games and lost 6. If the team continues to have the same ratio of wins to losses, how many wins will the team have after playing 21 games?

16. The ratio of cars to trucks passing a certain intersection is found to be 7:2. If 63 vehicles (cars and trucks) pass the intersection, how many might be trucks?

17. Five vests can be made from $2\frac{1}{2}$ yd of fabric. How many vests can be made from 6 yd of fabric?

18. A fruit punch recipe calls for 3 parts of apple juice to 4 parts of cranberry juice. How many liters of cranberry juice should be added to 4.5 L of apple juice?

19. A 3 lb bag of Fairlawn's Number 25 grass seed covers a 4000 ft² area. How great an area will 16 oz of the Number 25 grass seed cover?

**20.** A wall hanging requires 54 cm of braided trim. How many wall hangings can be completed if 3 m of braided trim is available?

**C 21.** In a recent election, the ratio of votes *for* a particular proposal to votes *against* the proposal was 5 to 2. There were 4173 more votes for the proposal than against the proposal. How many votes were for and how many votes were against the proposal?

**22.** A certain soil mixture calls for 8 parts of potting soil to 3 parts of sand. To make the correct mixture, Vern used 0.672 kg of sand and 2 bags of potting soil. How much potting soil was in each bag?

**23.** In the third century B.C., the Greek mathematician Eratosthenes calculated that an angle of $7\frac{1}{2}°$ at Earth's center cuts off an arc of about 1600 km on Earth's surface. From this information compute the circumference of Earth.

**24.** In the diagram $\overline{PQ}$ is perpendicular to $\overline{MN}$. It can be shown that $\frac{a}{x} = \frac{x}{b}$. If $a = 50$ and $b = 2$, find $x$.

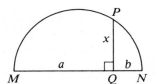

## Review Exercises

**Rewrite each ratio so that the numerator and denominator are expressed in the same unit of measure.**

**1.** $\dfrac{2\text{ m}}{15\text{ cm}}$

**2.** $\dfrac{20\text{ mm}}{7\text{ cm}}$

**3.** $\dfrac{10\text{ yd}}{5\text{ ft}}$

**4.** $\dfrac{8\text{ in.}}{3\text{ ft}}$

**5.** $\dfrac{2\text{ km}}{450\text{ m}}$

**6.** $\dfrac{85\text{ cm}}{4\text{ m}}$

**7.** $\dfrac{1\text{ yd}}{20\text{ in.}}$

**8.** $\dfrac{310\text{ m}}{3\text{ km}}$

▌▌▌ **Calculator Key-In**

It is fairly simple to find the reciprocal of a fraction or a whole number. It is harder to find the reciprocal of a decimal. However, many calculators have a reciprocal key to carry out this procedure.

**Use a calculator with a reciprocal key to find each reciprocal.**

**1.** 0.67     **2.** 0.579     **3.** 0.2539     **4.** 1.564     **5.** 4.7851

# 6-5 Scale Drawing

In the drawing of the house the actual height of 9 m is represented by a length of 3 cm, and the actual length of 21 m is represented by a length of 7 cm. This means that 1 cm in the drawing represents 3 m in the actual building. Such a drawing in which all lengths are in the same ratio to actual

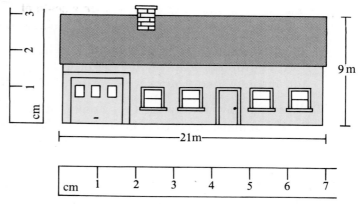

lengths is called a **scale drawing.** The relationship of length in the drawing to actual length is called the **scale.** In the drawing of the house the scale is 1 cm:3 m.

We can express the scale as a ratio, called the scale ratio, if a common unit of measure is used. Since 3 m equals 300 cm, the scale ratio above is $\frac{1}{300}$.

*EXAMPLE*   Find the length and width of the room shown if the scale of the drawing is 1 cm:1.5 m.

*Solution*   Measuring the drawing, we find that it has length 4 cm and width 3 cm.

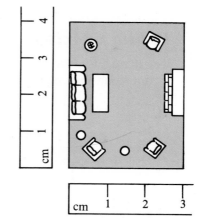

*Method 1*  Write a proportion for the length.

$\frac{1}{1.5} = \frac{4}{l}$ ← unit lengths in the drawing
          ← actual length

$l = 1.5 \times 4 = 6$
The room is 6 m long.

Write a proportion for the width.  $\frac{1}{1.5} = \frac{3}{w}$

$w = 1.5 \times 3 = 4.5$   The room is 4.5 m wide.

*Method 2*   Use the scale ratio: $\frac{1 \text{ cm}}{1.5 \text{ m}} = \frac{1 \text{ cm}}{150 \text{ cm}} = \frac{1}{150}$

The actual length is 150 times the length in the drawing.

$l = 150 \times 4 = 600 \text{ cm} = 6 \text{ m}$        $w = 150 \times 3 = 450 \text{ cm} = 4.5 \text{ m}$

## Class Exercises

**A drawing of a bureau is to be made with a scale of 1 cm to 10 cm. Find the dimension on the drawing if the actual dimension is given.**

1. Height of bureau (70 cm)
2. Width of bureau (80 cm)
3. Height of legs (17.5 cm)
4. Width of top (75 cm)
5. Height of top drawer (10 cm)
6. Height of second drawer (12.5 cm)
7. Height of third drawer (15 cm)
8. Width of drawer (72.5 cm)

## Written Exercises

**An O-gauge model railroad has a scale of 1 in. : 48 in. Find the actual length of each railroad car, given the scale dimension.**

**A**
1. Flat car: 23 in.
2. Freight car: 11 in.
3. Tank car: 12 in.
4. Caboose: 9 in.
5. Passenger car: 20 in.
6. Refrigerator car: 15 in.

**Exercises 7–14, on the next page, refer to the map below.**
**a. Measure each distance in the map shown to the nearest 0.5 cm.**
**b. Compute the actual distance to the nearest 100 km.**

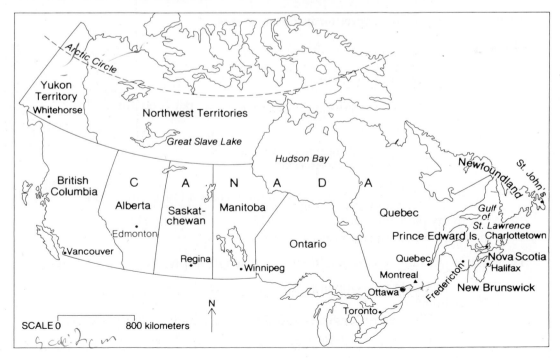

**7.** Vancouver to Edmonton  *2 cm*  **8.** Toronto to Winnipeg  *4 cm*

**9.** Whitehorse to Montreal  *40.5 cm*  **10.** Ottawa to Charlottetown  *2.5 cm*

**11.** Toronto to Halifax  *3 cm*  **12.** Fredericton to St. John's  *2.5 cm*

**B 13.** By how many kilometers would an airplane route from Winnipeg to Montreal be extended if the plane went by way of Toronto?

**14.** How much farther is it by air from Vancouver to Edmonton to Whitehorse than it is from Vancouver directly to Whitehorse?

**If Earth had the diameter of a peppercorn (5 mm), the sun would have the diameter of a large beach ball (54.5 cm) and it would be about the length of two basketball courts (58.75 m) away. Assuming that the diameter of Earth is about 12,700 km, compute each measurement.**

**15.** The actual diameter of the sun

**16.** The distance from Earth to the sun

**17.** In the scale described above, the diameter of the planet Jupiter would be 55 mm. What is the actual diameter of Jupiter?

---

## Self-Test A

**Express each ratio as a fraction in lowest terms.**

**1.** $12:8$  **2.** 51 to 27  **3.** $\frac{18}{99}$  [6–1]

**Give the unit price of each item.**

**4.** 8 gal of gasoline for \$9.20  **5.** 11 cans of pet food for \$6.72  [6–2]

**Solve.**

**6.** $\frac{x}{3} = \frac{8}{12}$  **7.** $\frac{12}{9} = \frac{n}{3}$  **8.** $\frac{180}{n} = \frac{4}{3}$  [6–3]

**9.** Find the price per gram of a metal that costs \$154.10 for 230 g.  [6–4]

**10.** A company paid a dividend of \$30 on 12 shares of stock. How much will it pay on 44 shares?

**11.** On a map, 3 in. represents 16 ft. What length represents $5\frac{1}{3}$ ft?  [6–5]

**12.** What is the scale in a drawing in which a vase 28 cm tall is drawn 1.75 cm high?

---

*Self-Test answers and Extra Practice are at the back of the book.*

## 6-6 Percents and Fractions

During basketball season, Alice made 17 out of 25 free throws, while Nina made 7 out of 10. To see who did better, we compare the fractions representing each girl's successful free throws:

$$\frac{17}{25} \quad \text{and} \quad \frac{7}{10}$$

In comparing fractions it is often convenient to use the common denominator 100, even if 100 is not the LCD of the fractions.

$$\frac{17}{25} = \frac{17 \times 4}{25 \times 4} = \frac{68}{100} \qquad \frac{7}{10} = \frac{7 \times 10}{10 \times 10} = \frac{70}{100}$$

Since Alice makes 68 free throws per hundred and Nina makes 70 per hundred, Nina is the better free-throw shooter.

The ratio of a number to 100 is called a **percent.** We write percents by using the symbol %. For example,

$$\frac{17}{25} = \frac{68}{100} = 68\% \qquad \text{and} \qquad \frac{7}{10} = \frac{70}{100} = 70\%.$$

---

### *Rule*

To express the fraction $\frac{a}{b}$ as a percent, solve the equation

$\frac{n}{100} = \frac{a}{b}$ for the variable $n$ and write $n\%$.

---

***EXAMPLE 1*** Express $\frac{17}{40}$ as a percent.

***Solution***

$$\frac{n}{100} = \frac{17}{40}$$

Cross-multiply.

$$40 \times n = 17 \times 100$$

$$n = \frac{17}{40} \times 100 = \frac{17}{40} \times \overset{5}{\cancel{100}} = \frac{85}{2} = 42\frac{1}{2}$$

Therefore, $\frac{17}{40} = 42\frac{1}{2}\%$, or 42.5%.

**EXAMPLE 2**   Express $7\frac{1}{2}\%$ as a fraction in lowest terms.

**Solution**      $7\frac{1}{2}\% = 7.5\% = \frac{7.5}{100} = \frac{7.5 \times 10}{100 \times 10} = \frac{75}{1000} = \frac{3}{40}$

Since a percent is the ratio of a number to 100, we can have percents that are greater than or equal to 100%. For example,

$$\frac{100}{100} = 100\% \qquad \text{and} \qquad \frac{165}{100} = 165\%.$$

**EXAMPLE 3**   Write 250% as a mixed number in simple form.

**Solution**      $250\% = \frac{250}{100}$

$$= 2\frac{50}{100} = 2\frac{1}{2}$$

**EXAMPLE 4**   A certain town spends 42% of its budget on education. What percent is used for other purposes?

**Solution**      The whole budget is represented by 100%. Therefore, the part used for other purposes is

$$100 - 42, \text{ or } 58\%.$$

## Class Exercises

**Express as a fraction in lowest terms or as a mixed number in simple form.**

| | | | |
|---|---|---|---|
| **1.** 17% | **2.** 90% | **3.** 50% | **4.** 25% |
| **5.** 20% | **6.** 100% | **7.** 4% | **8.** 150% |
| **9.** 300% | **10.** 30% | **11.** 35% | **12.** 210% |

Express as a percent.

**13.** $\frac{1}{50}$      **14.** $\frac{1}{10}$      **15.** $\frac{7}{10}$      **16.** 1

**17.** 2      **18.** $\frac{1}{20}$      **19.** $3\frac{1}{2}$      **20.** $\frac{9}{10}$

**21.** $\frac{3}{4}$      **22.** $4\frac{1}{2}$      **23.** $\frac{2}{25}$      **24.** $\frac{3}{20}$

---

## Written Exercises

**Express as a fraction in lowest terms or as a mixed number in simple form.**

**A**
**1.** 75%      **2.** 60%      **3.** 45%      **4.** 95%

**5.** 12%      **6.** 76%      **7.** 125%      **8.** 220%

**9.** $15\frac{1}{2}\%$      **10.** $8\frac{4}{5}\%$      **11.** $10\frac{3}{4}\%$      **12.** $5\frac{3}{8}\%$

**Express as a percent.**

**13.** $\frac{4}{5}$      **14.** $\frac{1}{4}$      **15.** $\frac{3}{10}$      **16.** $\frac{1}{25}$

**17.** $\frac{12}{25}$      **18.** $\frac{17}{20}$      **19.** $\frac{31}{50}$      **20.** $1\frac{3}{4}$

**21.** $2\frac{1}{5}$      **22.** $3\frac{11}{25}$      **23.** $\frac{51}{50}$      **24.** $\frac{31}{25}$

**B**
**25.** $\frac{7}{8}$      **26.** $\frac{7}{40}$      **27.** $\frac{1}{200}$      **28.** $\frac{12}{125}$

**29.** $\frac{3}{400}$      **30.** $\frac{9}{250}$      **31.** $\frac{121}{40}$      **32.** $\frac{25}{8}$

*EXAMPLE*    Express $33\frac{1}{3}\%$ as a fraction in lowest terms.

*Solution*      $33\frac{1}{3}\% = \dfrac{33\frac{1}{3}}{100} = 33\frac{1}{3} \div 100$

$$= \frac{100}{3} \times \frac{1}{100} = \frac{1}{3}$$

**Express each percent as a fraction in lowest terms.**

**C**
**33.** $16\frac{2}{3}\%$      **34.** $66\frac{2}{3}\%$      **35.** $41\frac{2}{3}\%$      **36.** $83\frac{1}{3}\%$

## Problems

A  **1.** In a public opinion poll 62% of the questionnaires sent out were returned. What percent were not returned?

**2.** The efficiency of a machine is the percent of energy going into the machine that does useful work. A turbine in a hydroelectric plant is 92% efficient. What percent of the energy is wasted?

**3.** Of the 300 acres on Swanson's farm, 180 acres are used to grow wheat. What percent of the land is used to grow wheat?

**4.** Of the selling price of a pair of gloves, 42% pays the wholesale cost of the gloves, 33% pays store expenses, and the rest is profit. What percent is profit?

B  **5.** In a 500 kg metal bar, 475 kg is iron and the remainder is impurities. What percent of the bar is impurities?

**6.** A baseball team won 42 games out of its first 80. What percent of the games did the team win?

**7.** One year the Caterpillars lost 5 games out of 16. If the Caterpillars also tied 1 game, what percent of their games did they win?

C  **8.** The Bears won 40 games and lost 24, while the Bulls won 32 games and lost 18. Which team had the higher percent of wins?

**9.** At a company sales conference there were 4 executives, 28 salespeople, and 8 marketing consultants. What percent of the people were executives? Salespeople? Not marketing consultants?

---

## Review Exercises

**Change each fraction to a decimal.**

**1.** $\frac{9}{20}$     **2.** $\frac{19}{25}$     **3.** $\frac{11}{40}$     **4.** $\frac{17}{80}$     **5.** $\frac{1}{25}$

**6.** $\frac{13}{30}$     **7.** $\frac{8}{11}$     **8.** $\frac{19}{22}$     **9.** $\frac{43}{75}$     **10.** $\frac{83}{90}$

# 6-7 Percents and Decimals

By looking at the following examples, you may be able to see a general relationship between decimals and percents.

$$57\% = \frac{57}{100} = 0.57 \qquad\qquad 0.79 = \frac{79}{100} = 79\%$$

$$113\% = \frac{113}{100} = 1\frac{13}{100} = 1.13 \qquad 0.06 = \frac{6}{100} = 6\%$$

These examples suggest the following rules.

---

### Rules

1. To express a percent as a decimal, move the decimal point two places to the left and remove the percent sign.

$$57\% = 0.57 \qquad 113\% = 1.13$$

2. To express a decimal as a percent, move the decimal point two places to the right and add a percent sign.

$$0.79 = 79\% \qquad 0.06 = 6\%$$

---

**EXAMPLE 1**  Express each percent as a decimal.
   **a.** 83.5%     **b.** 450%     **c.** 0.25%

**Solution**     **a.** 83.5% = 0.835

   **b.** 450% = 4.50 = 4.5

   **c.** 0.25% = 0.0025

**EXAMPLE 2**  Express each decimal as a percent.

   **a.** 10.5     **b.** 0.0062     **c.** 0.574

**Solution**     **a.** 10.5 = 1050%

   **b.** 0.0062 = 00.62%

   **c.** 0.574 = 57.4%

In the previous lesson you learned one method of changing a fraction to a percent. The ease of changing a decimal to a percent suggests the alternative method shown on the following page.

## Rule

To express a fraction as a percent, first express the fraction as a decimal and then as a percent.

**EXAMPLE 3**  Express $\frac{7}{8}$ as a percent.

**Solution**  Divide 7 by 8.

$$\begin{array}{r} 0.875 \\ 8\overline{)7.000} \\ \underline{6\,4} \\ 60 \\ \underline{56} \\ 40 \\ \underline{40} \\ 0 \end{array}$$

$\frac{7}{8} = 0.875 = 87.5\%$

**EXAMPLE 4**  Express $\frac{1}{3}$ as a percent. Round to the nearest tenth of a percent.

**Solution**  Divide 1 by 3 to the ten-thousandths' place.

Round the quotient to the nearest thousandth.

$0.3333 \approx 0.333$

Express the decimal as a percent.

$0.333 = 33.3\%$

To the nearest tenth of a percent, $\frac{1}{3} = 33.3\%$.

$$\begin{array}{r} 0.3333 \\ 3\overline{)1.0000} \\ \underline{9} \\ 10 \\ \underline{9} \\ 10 \\ \underline{9} \\ 10 \\ \underline{9} \\ 1 \end{array}$$

---

## Class Exercises

**Express each percent as a decimal.**

**1.** 39%    **2.** 4%    **3.** 150%    **4.** 0.8%    **5.** 1080%    **6.** 1%

**Express each decimal as a percent.**

**7.** 0.56    **8.** 0.005    **9.** 0.07    **10.** 1.6    **11.** 5.3    **12.** 0.0001

**Express each of the following first as a decimal and then as a percent.**

**13.** $\frac{1}{2}$    **14.** $\frac{1}{4}$    **15.** $\frac{3}{5}$    **16.** $2\frac{1}{2}$    **17.** $\frac{1}{1000}$    **18.** $\frac{23}{1000}$

## Written Exercises

**Express each percent as a decimal.**

**A**
1. 93%
2. 46%
3. 114%
4. 175%

5. 260%
6. 1150%
7. 49.5%
8. 78.2%

9. 0.6%
10. 99.44%
11. 0.05%
12. 0.032%

**Express each decimal as a percent.**

13. 0.59
14. 0.87
15. 0.09
16. 0.075

17. 2.6
18. 10.6
19. 12.83
20. 5.01

21. 0.007
22. 0.033
23. 0.0867
24. 0.0026

**Express each fraction as a decimal and then as a percent. Round to the nearest tenth of a percent if necessary.**

**B**
25. $\frac{3}{8}$
26. $\frac{9}{125}$
27. $\frac{3}{500}$
28. $\frac{27}{40}$

29. $1\frac{5}{8}$
30. $2\frac{37}{40}$
31. $\frac{47}{80}$
32. $\frac{7}{11}$

33. $\frac{17}{24}$
34. $\frac{19}{12}$
35. $\frac{2}{7}$
36. $\frac{16}{9}$

**Express each fraction as an exact percent.**

*EXAMPLE*  $\frac{1}{3}$

*Solution*  Divide 1 by 3 to the hundredths' place: $\qquad$ $1 \div 3 = 0.33$ R 1, or $0.33\frac{1}{3}$

Express the quotient as a percent: $\qquad$ $0.33\frac{1}{3} = 33\frac{1}{3}\%$

Thus, as a percent, $\frac{1}{3} = 33\frac{1}{3}\%$.

**C**
37. $\frac{1}{6}$
38. $\frac{2}{3}$
39. $\frac{8}{9}$
40. $\frac{1}{15}$
41. $\frac{5}{6}$

---

## Review Exercises

**Solve.**

1. $x + 2.47 = 5.42$
2. $x - 3.43 = 1.91$
3. $x \div 3.72 = 4.15$

4. $2.48x = 8.1096$
5. $x \div 4.03 = 5.92$
6. $1.37x = 11.4806$

# 6-8 Computing with Percents

The statement 20% of 300 is 60 can be translated into the equations

$$\frac{20}{100} \times 300 = 60 \qquad \text{and} \qquad 0.20 \times 300 = 60.$$

Notice the following relationship between the words and the symbols.

$$
\begin{array}{ccccc}
\underline{20\%} & \underline{\text{of}} & 300 & \underline{\text{is}} & 60 \\
\downarrow & \downarrow & & \downarrow & \\
\left.\begin{array}{c} \frac{20}{100} \\[4pt] 0.20 \end{array}\right\} & \times & 300 & = & 60
\end{array}
$$

A similar relationship occurs whenever a statement or a question involves a number that is a percent of another number.

**EXAMPLE 1**   What number is 8% of 75?

**Solution**   Let $n$ represent the number asked for.

$$
\begin{array}{ccc}
\underline{\text{What number}} & \underline{\text{is}}\ 8\% & \underline{\text{of}}\ 75? \\
\downarrow & \downarrow & \downarrow \\
n & = 0.08 & \times\ 75
\end{array}
$$

6 is 8% of 75.

**EXAMPLE 2**   What percent of 40 is 6?

**Solution**   Let $n\%$ represent the percent asked for. We can translate the question into an equation as follows.

$$
\begin{array}{cccc}
\underline{\text{What percent}} & \underline{\text{of}}\ 40 & \underline{\text{is}} & 6? \\
\downarrow & \downarrow & \downarrow & \\
n\% & \times\ 40 & = & 6
\end{array}
$$

$$n\% \times 40 = 6$$

$$n\% = \frac{6}{40}$$

$$\frac{n}{100} = \frac{3}{20}$$

$$n = \frac{3}{20} \times 100 = 15$$

15% of 40 is 6.

***EXAMPLE 3***   140 is 35% of what number?

***Solution***         Let $n$ represent the number asked for.

140          is   35%   of   what number?

140          $= 0.35$   $\times$        $n$

$$140 = 0.35n$$
$$140 \div 0.35 = 0.35n \div 0.35$$
$$140 \div 0.35 = n$$
$$400 = n$$

140 is 35% of 400

## Class Exercises

**State an equation involving a variable that expresses the conditions of the question.**

1. What number is 10% of 920?

2. What percent of 650 is 130?

3. 28 is 80% of what number?

4. 120 is what percent of 150?

5. 40% of 25 is what number?

6. What percent of 80 is 100?

7. 60 is 75% of what number?

8. What number is 125% of 160?

## Written Exercises

**Answer each question by writing an equation and solving it. Round your answer to the nearest tenth of a percent if necessary.**

**A**  1. What percent of 225 is 90?

2. What number is 76% of 350?

3. 45% of 600 is what number?

4. What percent of 150 is 48?

5. 52 is 4% of what number?

6. 56 is 4% of what number?

7. What number is 36% of 15?

8. 96% of 85 is what number?

9. What percent of 36 is 30?

10. 48 is what percent of 72?

**B** 11. What is 110% of 95?

12. 0.5% of what number is 15?

13. 116% of 75 is what number?

14. What percent of 40 is 86?

15. What is 0.35% of 256?

16. 12 is 150% of what number?

**Answer each question by writing an equation and solving it. Round your answer to the nearest tenth of a percent if necessary.**

**17.** What percent of 21 is 24?

**18.** What is 81% of 60?

**19.** 12.5% of what number is 28?

**20.** What percent of 45 is 600?

C **21.** 18 is $33\frac{1}{3}$% of what number?

(*Hint:* Write the percent as a fraction in lowest terms.)

**22.** What is $41\frac{2}{3}$% of 300?

**23.** 231 is $91\frac{2}{3}$% of what number?

## Problems

**Solve.**

A **1.** Lisa earns $250 a week and lives in a state that taxes income at 5%. How much does Lisa pay in state tax each week?

**2.** A baseball park has 40,000 seats of which 13,040 are box seats and 18,200 are reserved seats. What percent of the seats are box seats? Reserved seats?

**3.** A basketball player made 62 out of 80 free-throw shots. What percent of the free throws did she make?

**4.** A sweater is 65% wool by weight. If the sweater weighs 12.4 ounces, how much wool is in the sweater?

B **5.** In a class of 40 students, 36 received passing grades on a geography test. What percent of the students in the class did not receive passing grades?

**6.** A service contract offered by a washing machine manufacturer can be purchased for 9% of the price of a washing machine. If a certain washer costs $580, what is the price of the service contract?

**7.** Of the 427 people responding to a public opinion poll, 224 answered *Yes* to a certain question and 154 answered *No*. What percent of those corresponding were undecided? Give your answer to the nearest tenth of a percent.

**8.** In 1979 the United States produced 3112 million barrels of oil. This was 13.7% of the world oil output. What was the world oil output to the nearest million barrels?

**9.** It is estimated that 60% of the people of the world live in Asia. If there are 2.4 billion people living in Asia, what is the population of the world?

**C  10.** In a recent election, 45% of the eligible voters actually voted. Of these, 55% voted for the winner.
  **a.** What percent of eligible voters voted for the winning candidate?
  **b.** Suppose 495 people voted for the winner. How many eligible voters were there?

**11.** Of a city's 180,000 workers, 25% use the subway system. Of these, 18,000 use the subway between 7 A.M. and 9:30 A.M.
  **a.** What percent of the city's total number of workers ride the subway between 7 A.M. and 9:30 A.M.?
  **b.** What percent of the total number of subway riders use the subway between 7 A.M. and 9:30 A.M.?

## Self-Test B

**Express as a fraction in lowest terms or as a mixed number in simple form.**

**1.** 27%        **2.** 83%        **3.** 164%        **4.** 290%    [6–6]

**Express as a percent.**

**5.** $\frac{1}{20}$        **6.** $\frac{3}{8}$        **7.** 4        **8.** $3\frac{1}{4}$

**Express as a decimal.**

**9.** 45%        **10.** 78%        **11.** 348%        **12.** 0.8%    [6–7]

**Express as a percent.**

**13.** 0.64        **14.** 0.81        **15.** 7.85        **16.** 0.068

**17.** What percent of 56 is 14?        **18.** 82 is what percent of 40?    [6–8]

**19.** What is 44% of 25?        **20.** 70% of what number is 84?

*Self-Test answers and Extra Practice are at the back of the book.*

# Fibonacci Numbers

In the thirteenth century, an Italian mathematician named Leonardo of Pisa, nicknamed Fibonacci, discovered a sequence of numbers that has many interesting mathematical properties, as well as applications to biology, art, and architecture. Fibonacci defined the sequence as the number of pairs of rabbits you would have, starting with one pair, if each pair produced a new pair after two months and another new pair every month thereafter.

The diagram below illustrates the process that Fibonacci described for the first six months:

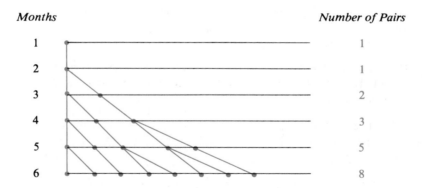

The sequence formed by the numbers of pairs is called the **Fibonacci sequence.** Note that any number in the sequence is the sum of the two numbers that precede it (for example, $2 + 3 = 5$, $3 + 5 = 8$). Thus the sequence would continue:

$$1, \ 1, \ 2, \ 3, \ 5, \ 8, \ 13, \ 21, \ 34, \ 55, \ 89, \ 144, \ 233, \ 377, \ 610, \ \ldots$$

To see one of the many mathematical properties of this sequence, study the sums of the squares of some pairs of consecutive Fibonacci numbers:

$$2^2 + 3^2 = 4 + 9 = 13$$
$$5^2 + 8^2 = 25 + 64 = 89$$
$$13^2 + 21^2 = 169 + 441 = 610$$

Note that in each case the sum is another number in the sequence.

The Fibonacci sequence also relates to a historically important number called the **Golden Ratio,** or the **Golden Mean.** The Golden Ratio is the ratio of the length to the width of a "perfect" rectangle. A rectangle was considered to be "perfect" if the ratio of its length to its width was the same as the ratio of their sum to its length, as written in the proportion at the right. If we let the width of the rectangle be 1, the length is the nonterminating, nonrepeating decimal 1.61828.... Thus, the Golden Ratio is the ratio 1.61828 ... to 1.

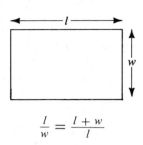

$$\frac{l}{w} = \frac{l + w}{l}$$

It is remarkable that the ratios of successive terms of the Fibonacci sequence get closer and closer to this number. For example,

$$\frac{8}{5} = 1.6 \qquad\qquad \frac{13}{8} = 1.625 \qquad\qquad \frac{21}{13} = 1.615384\ldots$$

$$\frac{34}{21} = 1.61904\ldots \qquad\qquad \frac{55}{34} = 1.61764\ldots \qquad\qquad \frac{89}{55} = 1.61818\ldots$$

1. Examine the fourth, eighth, twelfth, and sixteenth terms of the Fibonacci sequence. What do these numbers have in common? Try the fifth, tenth, and fifteenth terms.

**Find some other numerical patterns by performing these calculations with the numbers in the Fibonacci sequence listed on the previous page. Tell what you notice about each new pattern.**

2. Find the difference of two numbers that are two places apart in the sequence (the first and the third terms, the second and the fourth terms, the third and the fifth terms, and so on).

3. Subtract the squares of two numbers that are two places apart (as in Exercise 2).

4. Multiply two consecutive numbers (the first and second terms), multiply the next two consecutive numbers (the second and third terms), and then add the products.

5. Multiply two numbers that are two places apart (begin with the second and fourth terms), multiply two other numbers that "straddle" one of these (the first and third terms), and then add the products.

**Career Activity** Botanists have discovered that Fibonacci numbers occur naturally in many plant forms. For example, a pine cone is made up of 8 spirals swirling upward in one direction and 13 spirals swirling upward in the opposite direction. Find some other instances of Fibonacci numbers that botanists have found in nature.

# Chapter Review

**Match.**

1. 2 min to 90 sec    **2.** 1 lb : 8 oz    **a.** 4 : 3    **b.** 22 mi/gal    [6–1]

3. 682 mi on 31 gal   **4.** 40 km in 30 min   **c.** 2 : 1    **d.** $1\frac{1}{3}$ km/min    [6–2]

**Write the letter of the correct answer.**

5. Solve $\frac{6}{x} = \frac{9}{75}$.                                        [6–3]

  **a.** 450            **b.** 54              **c.** 50           **d.** 37.5

6. A bicyclist went 7 mi in 30 min. How far would the cyclist go in    [6–4]
   45 min?

  **a.** 4.2 mi         **b.** 14 mi          **c.** $10\frac{1}{2}$ mi   **d.** $\frac{14}{3}$ mi

7. Three cans of paint will cover 80 m² of wall. What area will 5 cans
   cover?

  **a.** $133\frac{1}{3}$ m²      **b.** 48 m²            **c.** 1200 m²   **d.** 400 m²

8. A model truck has a scale of 1 cm : 25 cm. The model is 14 cm high.    [6–5]
   How high is the actual truck?
     **a.** 60 cm         **b.** 160 cm        **c.** 24 cm      **d.** 350 cm

9. On an engineering diagram the scale is 5 mm : 1 mm. A circuit
   measures 4 mm across. What is the length of the circuit in the
   drawing?
     **a.** 0.8 mm       **b.** 20 mm        **c.** 1.25 mm   **d.** 4 cm

**Match.**

10. 75%                **11.** $\frac{13}{20}$           **a.** 600%      **b.** 7.5%      [6–6]

12. 6                  **13.** 0.6%        **c.** 65%       **d.** 60%

14. $\frac{3}{5}$               **15.** 0.075      **e.** 0.006     **f.** $\frac{3}{4}$        [6–7]

**Write the letter of the correct answer.**

16. What is 65% of 140?                                   [6–8]
     **a.** 215           **b.** 9100         **c.** 0.46      **d.** 91

17. 18 is what percent of 40?
     **a.** 222%        **b.** 0.45         **c.** 45%      **d.** 7.2%

# Chapter Test

**Express each ratio as a fraction in lowest terms.**

**1.** 35 min : 3 h

**2.** 2 m to 85 cm

[6–1]

**Give the unit price of each item.**

**3.** 5 basketballs for $59.75

**4.** 12 oz of cereal for $1.32

[6–2]

**Solve.**

**5.** $\frac{n}{40} = \frac{3}{8}$

**6.** $\frac{24}{5} = \frac{x}{10}$

**7.** $\frac{5}{11} = \frac{20}{a}$

[6–3]

**8.** A car traveled 162 mi in 3 h. How many hours would it take to travel 351 mi?

[6–4]

**9.** On a map, 2 in. represents 300 mi. If two points are separated by 5 in. on the map, what is the actual distance between them?

[6–5]

**Express as a fraction in lowest terms or as a mixed number in simple form.**

**10.** 48%

**11.** 6%

**12.** 215%

**13.** 190%

[6–6]

**Express as a percent.**

**14.** $\frac{11}{20}$

**15.** $\frac{5}{8}$

**16.** 3

**17.** $2\frac{1}{8}$

**Express as a decimal.**

**18.** 93%

**19.** 42%

**20.** 259%

**21.** 0.86%

[6–7]

**Express as a percent.**

**22.** 0.81

**23.** 0.07

**24.** 2.91

**25.** 1.01

**Solve.**

**26.** What percent of 85 is 51?

[6–8]

**27.** 27 is what percent of 60?

**28.** What is 30% of 80?

**29.** 20% of what number is 25?

# Cumulative Review (Chapters 1–6)
## Exercises

Evaluate the expression when $x = 2$, $y = 3$, and $z = -2$.

**1.** $5y - 10$

**2.** $6x + 3y$

**3.** $2z + 6$

**4.** $8 - z$

**5.** $4xy \div (-2z)$

**6.** $3x + 3y \div z$

**7.** $5y \div (2 + y)$

**8.** $10x - (3y + 1)$

**9.** $x^x$

**10.** $2y^2$

**11.** $(2y)^2$

**12.** $(3x)^{-2}$

True or false?

**13.** $2.1(3 \times 4.7) = (2.1 \times 3)(2.1 \times 4.7)$

**14.** $9.51 + (-6.21) = (-6.21) + 9.51$

**15.** $1(-10.3) = 10.3$

**16.** $7.4 \times 0 = 0$

Write as equivalent fractions using the LCD.

**17.** $\frac{2}{3}, \frac{1}{5}$

**18.** $\frac{3}{4}, \frac{6}{7}$

**19.** $-\frac{1}{8}, \frac{1}{3}$

**20.** $\frac{7}{10}, -\frac{4}{9}$

**21.** $\frac{2}{5}, \frac{1}{6}, \frac{1}{4}$

**22.** $\frac{3}{11}, \frac{1}{2}, \frac{3}{4}$

Use transformations to solve each equation.

**23.** $x + 7 = 12$

**24.** $x - 3 = -6$

**25.** $6x = 18$

**26.** $-9x = 27$

**27.** $\frac{3}{4}x = 12$

**28.** $2x + 9 = 27$

Solve.

**29.** $\frac{x}{4} = \frac{2}{8}$

**30.** $\frac{x}{7} = \frac{9}{21}$

**31.** $\frac{3}{8} = \frac{24}{x}$

**32.** $\frac{48}{176} = \frac{8}{x}$

**33.** $\frac{13}{11} = \frac{3}{x}$

**34.** $\frac{148}{4} = \frac{x}{5}$

Complete.

**35.** A __?__ is a parallelogram with four equal sides.

**36.** If two lines are __?__ , they form four right angles.

**37.** A __?__ is a chord of a circle which is twice as long as the radius.

**38.** To the nearest hundredth, the ratio of the circumference of a circle to its diameter is __?__ .

# Problems

**Solve.**

1. Donald bought a pair of hiking boots for $35.83, a sweater for $24.65, and a backpack for $18. The tax on his purchase was $.90. How much did Donald spend?

2. This week, Marisa worked $1\frac{1}{2}$ h on Monday, $2\frac{1}{4}$ h on Tuesday, $1\frac{1}{3}$ h on Wednesday, and $7\frac{1}{2}$ h on Saturday. How many hours did she work this week?

3. An airplane flying at an altitude of 25,000 ft dropped 4000 ft in the first 25 s and rose 2500 ft in the next 15 s. What was the altitude of the airplane after 40 s?

4. Carolyn Cramer spent 3 h 15 min mowing and raking her lawn. She spent twice as long raking as she did mowing. How long did she spend on each task?

5. At a milk processing plant 100 lb of farm milk are needed to make 8.13 lb of nonfat dry milk. To the nearest pound, how many pounds of farm milk are needed to produce 100 lb of nonfat dry milk?

6. A highway noise barrier that is 120 m long is constructed in two pieces. One piece is 45 m longer than the other. Find the length of each piece.

7. The perimeter of a rectangle is 40 m. The length of the rectangle is 10 m greater than the width. Find the length and the width.

8. Oak Hill School, Longview School, and Peabody School participated in a clean-up campaign to collect scrap aluminum. Oak Hill School collected 40% more scrap aluminum than Longview School. Longview School collected 25% more than Peabody School. If the total collected by the 3 schools was 560 kg, how much did Oak Hill School collect?

9. Eladio invested $200 at 6% annual interest and $350 at 5.75% annual interest, both compounded annually. If he makes no deposits or withdrawals, how much will he have after two years?

# 7

# Percents and Problem Solving

The photograph shows a strand of DNA, or deoxyribonucleic acid, as seen through the center of the strand. In 1953 James Watson and Francis Crick developed a model for the DNA. They called it the double helix. In 1962 Watson and Crick were awarded the Nobel Prize in Medicine for their work with the double helix model.

According to the model, a strand of DNA is built much like a spiral staircase with phosphates and sugars forming the frame of the staircase. The four bases, adenine, guanine, thymine, and cytosine, form the steps. In analysis of DNA obtained from different organisms, it can be shown that the percentages of the four bases vary considerably. Thus, the sequence of the four bases are thought to determine individual heredity.

In this chapter, you will learn some interesting applications of percents to business, consumer, and financial situations.

## Career Note

Chemists analyze the structure, composition, and nature of matter. They can specialize in a variety of fields from developing new products to organic analysis of moon rocks. Their work involves quantitative and qualitative analyses, as well as practical applications of basic research. A strong background in science and mathematics and an inquisitive mind are essential for this career.

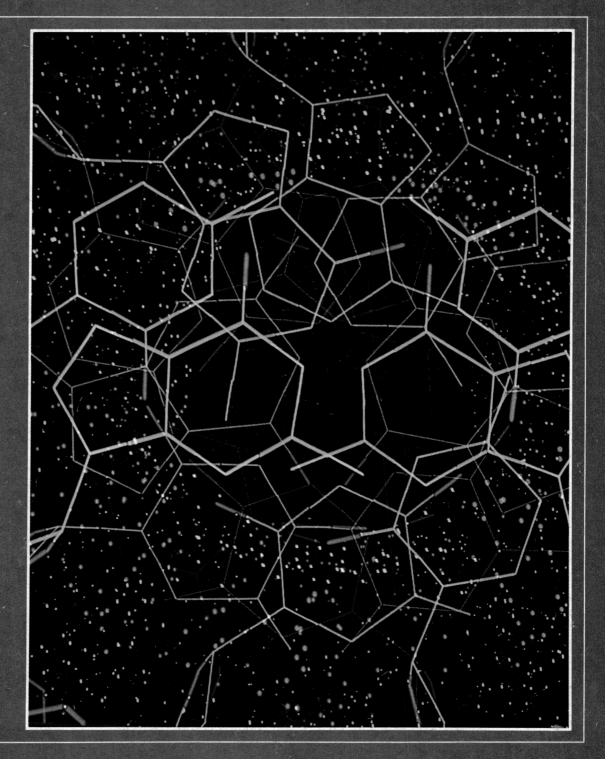

# 7-1 Percent of Increase or Decrease

A department store has a sale on audio equipment. An amplifier that originally sold for $260 is selling for $208. To find the **amount of change** in the price, we subtract the sale price from the original price.

$$\$260 - \$208 = \$52$$

To find the **percent of change** in the price we divide the amount of change by the original price and express the result as a percent.

$$\frac{52}{260} = 0.20 = 20\%$$

---

### *Formula*

$$\text{percent of change} = \frac{\text{amount of change}}{\text{original amount}}$$

---

The denominator in the formula above is always the *original* amount, whether smaller or larger than the new amount.

**EXAMPLE 1**  Find the percent of increase from 20 to 24.

**Solution**  amount of change $= 24 - 20 = 4$

percent of change $= \frac{\text{amount of change}}{\text{original amount}} = \frac{4}{20} = 0.2 = 20\%$

The formula above can be rewritten to find the amount of change when the original amount and the percent of change are known.

---

### *Formula*

amount of change $=$ percent of change $\times$ original amount

---

**EXAMPLE 2**  Find the new number when 75 is decreased by 26%.

**Solution**  First find the amount of change.

amount of change $=$ percent of change $\times$ original amount
$$= \qquad 26\% \qquad \times \qquad 75$$
$$= 0.26 \times 75 = 19.5$$

Since the original number is being decreased, we subtract to find the new number.  $75 - 19.5 = 55.5$

**EXAMPLE 3**   The population of Eastown grew from 25,000 to 28,000 in 3 years. What was the percent of increase for this period?

**Solution**   The problem asks for the percent of increase.

amount of change $= 28,000 - 25,000 = 3000$

percent of change $= \dfrac{\text{amount of change}}{\text{original amount}}$

$= \dfrac{3000}{25,000} = \dfrac{3}{25} = 0.12 = 12\%$

The percent of increase was 12%.

---

**Problem Solving Reminder**

Sometimes *extra information* is given in a problem. The time period, 3 years, is not needed for the solution of the problem in Example 3.

---

## Class Exercises

a. State the amount of change from the first number to the second.
b. State the percent of increase or decrease from the first number to the second.

| | | | |
|---|---|---|---|
| **1.** 10 to 12 | **2.** 4 to 3 | **3.** 12 to 6 | **4.** 6 to 12 |
| **5.** 2 to 5 | **6.** 5 to 7 | **7.** 25 to 4 | **8.** 1 to 4 |
| **9.** 4 to 7 | **10.** 100 to 55 | **11.** 100 to 160 | **12.** 125 to 100 |

Find the new number produced when the given number is increased or decreased by the given percent.

**13.** 120; 20% decrease          **14.** 30; 10% decrease

**15.** 48; 50% increase          **16.** 24; 25% increase

---

## Written Exercises

Find the percent of increase or decrease from the first number to the second. Round to the nearest tenth of a percent if necessary.

A
| | | | |
|---|---|---|---|
| **1.** 20 to 17 | **2.** 25 to 12 | **3.** 70 to 98 | **4.** 16 to 10 |
| **5.** 40 to 73 | **6.** 63 to 79 | **7.** 32 to 17 | **8.** 8 to 19 |
| **9.** 125 to 124 | **10.** 160 to 380 | **11.** 12 to 8.7 | **12.** 240 to 245.5 |

**Find the new number produced when the given number is increased or decreased by the given percent.**

**13.** 165; 20% decrease

**14.** 76; 25% increase

**15.** 65; 12% decrease

**16.** 250; 63% decrease

**17.** 84; 145% increase

**18.** 260; 105% increase

**19.** 125; 0.4% decrease

**20.** 1950; 0.8% increase

**Find the new number produced when the given number is changed by the first percent, and then the resulting number is changed by the second percent.**

**B**  **21.** 80; increase by 50%; decrease by 50%

**22.** 128; decrease by 25% increase by 25%

**23.** 150; increase by 40%; increase by 60%

**24.** 480; decrease by 35%; decrease by 65%

**25.** 350; increase by 76%; decrease by 45%

**26.** 136; decrease by 85%; increase by 175%

**Find the original number if the given number is the result of increasing or decreasing the original number by the given percent.**

**C**  **27.** 80; original number increased by 25%

**28.** 63; original number increased by 75%

**29.** 78; original number decreased by 35%

**30.** 30; original number decreased by 85%

## Problems

**Solve. Round to the nearest tenth of a percent if necessary.**

**A**  **1.** The number of employees at a factory was increased by 5% from its original total of 1080 workers. What was the new number of employees?

**2.** The cost of a basket of groceries at the Shopfast Supermarket was $62.50 in April. In May the cost of the same groceries had risen by 0.8%. What was the cost in May?

3. The attendance at a baseball stadium went from 1,440,000 one year to 1,800,000 the next year. What was the percent of increase?

4. The number of registered motor vehicles in Smalltown dropped from 350 in 1978 to 329 in 1979. What percent of decrease is this?

5. The new Maple City Library budget will enable the library to increase its collection of books by 3.6%. If the library now has 7250 books, how many will it have after the increase?

**B**  6. The number of students at Center State University is 22,540. Ten years ago there were only 7000 students. What is the percent of increase?

7. The annual budget of Brictown was $9,000,000 last year. Currently, the budget is only $7,500,000. Find the percent of decrease.

8. Last year the population of Spoon Forks grew from 1250 to 1300. If the population of the town grows by the same percent this year, what will the population be?

9. This year, Village Realty sold 289 homes. Last year the realty sold 340 homes. If sales decrease by the same percent next year, how many homes can Village Realty expect to sell?

**C**  10. Contributions to the annual Grayson School fund raising campaign were 10% greater in 1984 than they were in 1983. In 1983, contributions were 15% greater than they were in 1982. If contributions for 1984 total $8855, what was the total in 1982?

11. In July the price of a gallon of gasoline at Quick Sale Service Station rose 12%. In August it fell 15% of its final July price, ending the month at $1.19. What was its price at the beginning of July?

## Review Exercises

**Solve.**

1. $p = 0.36 \times 27$

2. $1.89 = r \times 7$

3. $144.5 = 8.5n$

4. $21.65 \times 0.7 = m$

5. $156 = 0.3q$

6. $a \times 0.15 = 4.125$

7. $\frac{x}{21} = 10.5$

8. $91.53 \times 0.4 = t$

9. $\frac{324}{y} = 8100$

# 7-2 Discount and Markup

A **discount** is a decrease in the price of an item. A **markup** is an increase in the price of an item. Both of these changes can be expressed as an amount of money or as a percent of the original price of the item. For example, a store may announce a discount of $3 off the original price of a $30 basketball, or a discount of 10%.

**EXAMPLE 1**  A warm-up suit that sold for $42.50 is on sale at a 12% discount. What is the sale price?

**Solution**  *Method 1* Use the formula:

amount of change = percent of change × original amount
= 12% × 42.50

Therefore, the discount is 0.12 × 42.50, or $5.10.

The amount of the discount is $5.10.
The sale price is 42.50 − 5.10, or $37.40.

*Method 2* Since the discount is 12%, the sale price is 100% − 12%, or 88%, of the original price. The sale price is 0.88 × 42.50, or $37.40.

As shown in the solutions to Example 1, when you know the amount of discount you subtract to find the new price. When dealing with a markup, you add to find the new price.

**EXAMPLE 2**  The price of a new car model was marked up 6% over the previous year's model. If the previous year's model sold for $7800, what is the cost of the new car?

**Solution**  *Method 1* Use the formula:

amount of change = percent of change × original amount
= 6% × 7800

Therefore, the markup is 0.06 × 7800, or $468.

The amount of the markup is $468.
The new price is 7800 + 468, or $8268.

*Method 2* Since the markup is 6%, the new price is 100% + 6%, or 106%, of the original price. The new price is 1.06 × 7800, or $8268.

A method similar to the second method of the previous examples can be used to solve problems like the one in the next example.

**EXAMPLE 3**    This year a pair of ice skates sells for $46 after a 15% markup over last year's price. What was last year's price?

**Solution**    This year's price is $100 + 15$, or 115%, of last year's price. Let $n$ represent last year's price.

$$46 = \frac{115}{100} \times n$$

$$46 = 1.15 \times n$$

$$\frac{46}{1.15} = n$$

$$40 = n$$

The price of the skates last year was $40.

Example 4 illustrates how to find the original price if you know the discounted price.

**EXAMPLE 4**    A department store advertised electric shavers at a sale price of $36. If this is a 20% discount, what was the original price?

**Solution**    The sale price is $100 - 20$, or 80%, of the original price. Let $n$ represent the original price.

$$36 = \frac{80}{100} \times n$$

$$36 = 0.8 \times n$$

$$36 \div 0.8 = n$$

$$45 = n$$

The original price was $45.

---

**Problem Solving Reminder**

Be sure that your *answers are reasonable*. In Example 3, last year's price should be less than this year's marked-up price. In Example 4, the original price should be greater than the sale price.

---

We can use the following formula to find the percent of discount or the percent of markup.

$$\text{percent of change} = \frac{\text{amount of change}}{\text{original amount}}$$

For example, if the original price of an item was $25 and the new price is $20, the amount of discount is $5 and the percent of discount is $5 \div 25 = 0.2$, or 20%.

## Class Exercises

**Copy and complete the following table.**

|     | Old price | Percent of change | Amount of change | New price |
|-----|-----------|-------------------|------------------|-----------|
| 1.  | $12       | 25% discount      | ?                | ?         |
| 2.  | $60       | 10% markup        | ?                | ?         |
| 3.  | $50       | ?                 | ?                | $60       |
| 4.  | $120      | ?                 | ?                | $240      |
| 5.  | $200      | ?                 | $30 markup       | ?         |
| 6.  | $250      | ?                 | $100 discount    | ?         |
| 7.  | ?         | 12% discount      | $48 discount     | ?         |
| 8.  | ?         | 5% markup         | $2.50 markup     | ?         |
| 9.  | ?         | 20% discount      | ?                | $160      |
| 10. | ?         | 150% markup       | ?                | $50       |

## Problems

**Solve.**

**A**  1. A basketball backboard set that sold for $79 is discounted 15%. What is the new price?

2. A parka that sold for $65 is marked up to $70.20. What is the percent of markup?

3. A stereo tape deck that sold for $235 was on sale for $202.10. What was the percent of discount?

4. At the end-of-summer sale, an air conditioner that sold for $310 was discounted 21%. What was the sale price?

5. Because of an increase of 8% in wholesale prices, a shoe store had to mark up its new stock by the same percent. What was the new price of a pair of shoes that had sold for $24.50?

6. A department store has a sale on gloves. The sale price is 18% less than the original price, resulting in a saving of $2.97. What was the original price of the gloves? What is the sale price?

7. A coat that originally cost $40 was marked up 50%. During a sale the coat was discounted 50%. What was the sale price?

8. A 7% sales tax added $3.15 to the selling price of a pair of ski boots. What was the selling price of the boots? What was the total price including the tax?

**B** 9. A tape recorder that cost $50 was discounted 20% for a sale. It was then returned to its original price. What percent of markup was the original price over the sale price?

10. At an end-of-season sale a power lawnmower was on sale for $168. A sign advertised that this was 20% off the original price. What was the original price?

11. At a paint sale a gallon can of latex was discounted 24%. If the sale price of a gallon is $9.50, what was the original price?

12. The cost of a record album was $10.53, including an 8% sales tax. What was the price of the album without the tax?

**C** 13. An item is discounted 20%. Then another 10% discount is given on the new price. What percent of the original price is the final price?

14. An item is marked up 20% and then discounted 15% based on the new price. What percent of the original price is the final price?

15. Which of the following situations will produce a lower final price on a given item?
    a. The item is marked up 30% and then discounted 30% of the new price.
    b. The item is discounted 30% and then marked up 30% of the new price.

---

## Review Exercises

**Complete.**

1. circumference $= 2\pi \times$ __?__

2. amount of change $=$ __?__ $\times$ original amount

3. circumference $= \pi \times$ __?__

4. percent of change $= \dfrac{?}{\text{original amount}}$

# 7-3 Commission and Profit

In addition to a salary, many salespeople are paid a percent of the price of the products they sell. This payment is called a **commission.** Like a discount, a commission can be expressed as a percent or as an amount of money. The following formula applies to commissions.

### Formula

amount of commission = percent of commission $\times$ total sales

**EXAMPLE 1**  Maria Bertram sold $42,000 worth of insurance in January. If her commission is 3% of the total sales, what was the amount of her commission in January?

**Solution**
$$\text{amount of commission} = \text{percent} \times \text{total sales}$$
$$= 0.03 \times 42,000 = 1260$$

Her commission was $1260.

**Profit** is the difference between total income and total operating costs.

### Formula

profit = total income $-$ total costs

The **percent of profit** is the percent of total income that is profit.

### Formula

percent of profit $= \dfrac{\text{profit}}{\text{total income}}$

**EXAMPLE 2**  In April, a shoe store had an income of $8600 and operating costs of $7310. What percent of the store's income was profit?

**Solution**
$$\text{profit} = \text{total income} - \text{total costs} = 8600 - 7310 = 1290$$
$$\text{percent of profit} = \frac{\text{profit}}{\text{total income}} = \frac{1290}{8600} = 0.15$$

The percent of profit was 15%.

## Class Exercises

**Copy and complete the following table.**

|     | Total sales | Percent of commission | Amount of commission |
|-----|-------------|-----------------------|----------------------|
| 1.  | $3250       | 10%                   | ?                    |
| 2.  | $680        | 20%                   | ?                    |
| 3.  | $900        | ?                     | $225                 |
| 4.  | $2500       | ?                     | $125                 |
| 5.  | ?           | 15%                   | $300                 |
| 6.  | ?           | 10%                   | $14.50               |

**a. State the amount of profit.**
**b. Give an equation that could be used to find the percent of profit.**
**c. Find the percent of profit.**

7. Income $12,000; costs $9000

8. Income $10,000; costs $8700

9. Income $14,000; costs $12,600

10. Income $3000; costs $2850

11. Income $5000; costs $4950

12. Income $24,000; costs $16,800

13. Income $12,300; costs $9840

14. Income $105,360; costs $63,216

## Problems

**Solve.**

**A** 1. Esther Simpson receives a 15% commission on magazine subscriptions. One week her sales totaled $860. What was her commission for the week?

2. The Greenwood Lumber Company had an income of $7680 for one week in May. If the company's profit for this period was 15% of its income, what were its costs for the week?

3. Margaret DeRosa's day-care service makes a profit of $222 per week. What are her costs if this profit is 18.5% of her total income?

**4.** Harvey Williams sold new insurance policies worth $5120 in August. If he receives a 4.5% commission on new policies, how much did he earn in commissions in August?

**B**  **5.** Sole Mates Shoes has expenses of $9592 per month. What must the store's total income be if it is to make a 12% profit?

**6.** The Top Drawer Furniture Company made a profit of $4360 in one month. What were its operating costs if this profit was 16% of its total income?

**Nina Perez is a real estate agent who receives a commission of 6% of the selling price of each house she sells. The seller of the house pays Nina's commission out of the selling price and keeps the remainder. What should be the selling price of each house if the seller wants to keep the amount indicated below?**

**7.** $61,000          **8.** $66,000          **9.** $55,000          **10.** $70,500

**C**  **11.** In May Sal's Bakery had operating costs of $6630 and made a profit of $1170. In June the operating costs are expected to be $6273. What must the bakery's income be if its profit is to remain the same percent of its income?

**12.** Each month Fran Parks receives a 6% commission on all her sales of barber supplies up to $15,000. She receives 8% commission on the portion of her sales that are above $15,000. Her commission for March was $1260. What were her sales?

**13.** Mildred Hofstadter receives a 5% commission on her sales of exercise equipment and a 6% commission on her sales of weight-training equipment. One month she sold $7900 worth of exercise equipment and made a total of $650 in commissions. How much were her sales of weight-training equipment that month?

**14.** Norman's Natural Foods has weekly expenses of $1075 and makes a profit of 14% of sales. If his weekly expenses increase to $1225 and he wants to make the same dollar profit as before, what percent of sales will this profit represent?

## Review Exercises

**Evaluate if $w = 8$, $x = 0.25$, $y = 0.8$, and $z = 5$.**

**1.** $wxz$                    **2.** $xyz$                    **3.** $wx + yz$          **4.** $wy - xz$

**5.** $(wyz) \div x$          **6.** $(wy) \div (xz)$          **7.** $0.2wx$            **8.** $1.3wyz$

# 7-4 Percents and Proportions

In Chapter 6, you learned how to use a proportion to write a fraction as a percent. You can also use proportions to solve problems involving percents. The following discussion shows how we may write a proportion that relates percentage, rate, and base.

The statement *9 is 15% of 60* can be written as $9 = 15\% \times 60$. Because a percent is an amount *per hundred*, we can think of 15% as 15 per hundred, or the ratio of 15 to 100, $\frac{15}{100}$. Substituting, we can write the original statement as

$$9 = \frac{15}{100} \times 60.$$

Dividing both sides of the equation by 60, we obtain the proportion

$$\frac{9}{60} = \frac{15}{100},$$

or 9 is to 60 as 15 is to 100.

Note that $\frac{9}{60}$ is the ratio of the percentage ($p$) to the base ($b$).

---

Percentage, base, and rate are related as shown in the following proportion:

$$\frac{p}{b} = \frac{n}{100},$$

where $\frac{n}{100}$ is the rate expressed as an amount per hundred.

---

**EXAMPLE 1**  143 is 65% of what number?

**Solution**  The rate, 65%, expressed in the form $\frac{n}{100}$, is $\frac{65}{100}$.

The percentage, $p$, is 143. Write a proportion to find the base, $b$.

$$\frac{p}{b} = \frac{n}{100}$$

$$\frac{143}{b} = \frac{65}{100}$$

$$65b = 100 \times 143$$

$$b = \frac{14{,}300}{65} = 220$$

**EXAMPLE 2**  In a recent survey, 38 of 120 people preferred the Bright Light disposable flashlight to other flashlights. What percent of the people surveyed preferred Bright Light?

**Solution**  The question asks what percent, or how many people per hundred, prefer Bright Light.

Let $n$ equal the number per hundred.
You know that $p$ is 38 and $b$ is 120.
Set up a proportion and then solve for $n$.

$$\frac{p}{b} = \frac{n}{100}$$

$$\frac{38}{120} = \frac{n}{100}$$

$$38 \times 100 = 120n$$

$$\frac{3800}{120} = n$$

$$n = 31\frac{2}{3}$$

In the survey, $31\frac{2}{3}\%$ of the people preferred Bright Light.

**EXAMPLE 3**  The price of a pocket cassette player has been discounted 25%. The original price was $59. What is the sale price?

**Solution**  The question asks you to find the sale price.

First find the amount of the discount.
$p$ is to be found, $b$ is 59, and $n$ is 25.
Set up a proportion and then solve for $p$.

$$\frac{p}{b} = \frac{n}{100}$$

$$\frac{p}{59} = \frac{25}{100}$$

$$100p = 59 \times 25$$

$$p = \frac{1475}{100} = 14.75$$

The amount of the discount is $14.75.

Next find the sale price.

$$59 - 14.75 = 44.25$$

The sale price is $44.25.

## Class Exercises

**For each exercise, set up a proportion to find the number or percent. Do not solve the proportion.**

1. What percent of 32 is 20?

2. 14 is 25% of what number?

3. What is 58% of 24?

4. A digital clock is selling at a discount of 15%. The original price was $8.98. How much money will you save by buying the clock at the sale price?

5. Catherine answered 95% of the test questions correctly. If she answered 38 questions correctly, how many questions were on the test?

6. This year, 360 people ran in the town marathon. Only 315 people finished the race. What percent of the people finished the race?

## Written Exercises

**Use a proportion to solve.**

**A**  1. What is 16% of 32?

3. What percent of 20 is 16?

5. 15 is 37.5% of what number?

2. 306 is 51% of what number?

4. What is 104% of 85?

6. What percent of 72 is 18?

## Problems

**Use a proportion to solve.**

**A**  1. The Plantery received a shipment of 40 plants on Wednesday. By Friday, 33 of the plants had been sold. What percent of the plants were sold?

2. The Eagles have won 6 of the 8 games they have played this season. What percent of this season's games have the Eagles won?

3. The library ordered 56 new books. 87.5% of the books are nonfiction. How many books are nonfiction?

4. This year 18.75% more students have joined the Long Hill High School Drama Club. The records show that 6 new students have joined. How many students were members last year?

5. A water purifier is discounted 20% of the original price for a saving of $6.96. What is the original price of the water purifier? What is the sale price?

6. The original price of a Sportsmaster 300 fishing pole was $37.80. The price has now been discounted 15%. What is the amount of the discount? What is the sale price?

**B** 7. Steve recently received a raise of 6% of his salary at his part-time job. He now earns $38.69 per week. How much did he earn per week before his raise?

8. Since the beginning of the year, the number of subscriptions to *New Tech Magazine* has increased by 180%. The number of subscriptions is now 140,000. What was the number of subscriptions at the beginning of the year?

*EXAMPLE* This **pie chart,** or **circle graph,** shows the distribution of the Valley Springs annual budget. Use a proportion to find the measure of the angle for the wedge for education.

*Solution* The number of degrees in a circle is 360°. The circle represents the total budget. 45% of the total budget is for education. Therefore,

$$\frac{45}{100} = \frac{n}{360}$$

$$100n = 45 \times 360$$

$$n = \frac{16,200}{100}$$

$$n = 162$$

There are 162° in the wedge for education.

**VALLEY SPRINGS BUDGET**
**$12 MILLION**

POLICE AND FIRE 30%

EDUCA-TION 45%

OTHER 25%

**Use proportions to answer these questions about the pie chart above.**

9. **a.** What is the measure of the angle for the wedge for police and fire expenses?
   **b.** How much money is budgeted for police and fire expenses?

10. **a.** What is the measure of the angle for the wedge for other expenses?
    **b.** How much money is budgeted for other expenses?

**For Exercises 11–13, (a) use a compass and protractor to draw a pie chart to represent the budget, and (b) use proportions to find the dollar value of each expense.**

C  **11.** The monthly budget of the Miller family allows for spending 25% on food, 15% on clothing, 30% on housing expenses, 10% on medical expenses, and 20% on other expenses. Their total budget is $1800 per month.

**12.** The total weekly expenses of Daisy's Diner average $8400. Of this total, 20% is spent for employee's salaries, 60% is spent on food, 5% is spent on advertising, 10% is spent on rent, and 5% is used for other expenses.

**13.** The Chimney Hill school budget allows 72% for salaries, 8% for maintenance and repair, and 5% for books and supplies. The remainder is divided equally among recreation, after-school programs, and teacher training. The budget totals $680,000.

## Self-Test A

**Find the percent of increase or decrease.**

**1.** 12 to 15             **2.** 20 to 13             **3.** 200 to 246             [7–1]

**4.** A sleeping bag that cost $85 is marked up 22%. What is the new price?             [7–2]

**5.** A toy that had sold for $16.50 was on sale for $11.55. What was the percent of discount?

**6.** A real estate agency charges a commission of 6% on sales. How much is the commission on a house that sells for $106,000?             [7–3]

**7.** In August, Middle Mountain Mines had an income of $82,600. If profit for August was 12% of income, what were the company's costs that month?

**Use a proportion to solve.**

**8.** What is 81% of 540?             **9.** 64 is what percent of 80?             [7–4]

**10.** Of the traffic violations cited by Officer Huang, 64% were for speeding. If Officer Huang cited 48 motorists for speeding, how many violations were there altogether?

*Self-Test answers and Extra Practice are at the end of the book.*

# 7-5 Simple Interest

When you lease a car or an apartment, you pay the owner rent for the use of the car or the apartment. When you borrow money, you pay the lender **interest** for the use of the money. The amount of interest you pay is usually a percent of the amount borrowed figured on a yearly basis. This percent is called the **annual rate.** For example, if you borrow $150 at an annual rate of 12%, you pay:

$$\$150 \times 0.12 = \$18 \text{ interest for one year}$$
$$\$150 \times 0.12 \times 2 = \$36 \text{ interest for two years}$$
$$\$150 \times 0.12 \times 3 = \$54 \text{ interest for three years}$$

When interest is computed year by year in this manner we call it **simple interest.** The example above illustrates the following formula.

---

### Formula

Let $I$ = simple interest charged
$\quad P$ = amount borrowed, or **principal**
$\quad r$ = annual rate
$\quad t$ = time in years for which the amount is borrowed

Then, interest = principal × rate × time, or $I = Prt$.

---

**EXAMPLE 1**  How much simple interest do you pay if you borrow $640 for 3 years at an annual rate of 15%?

**Solution**  Use the formula: $I = Prt$
$$I = 640 \times 0.15 \times 3 = 288$$

The interest is $288.

**EXAMPLE 2**  Sarah Sachs borrowed $3650 for 4 years at an annual rate of 16%. How much money must she repay in all?

**Solution**  Use the formula: $I = Prt$
$$I = 3650 \times 0.16 \times 4 = 2336$$

The interest is $2336.
The total to be repaid is the principal plus the interest: 3650 + 2336, or $5986.

**Problem Solving Reminder**

When solving a problem, be certain to *answer the question asked.* In Example 2, you are asked to find the total amount to be repaid, not just the interest. To get the total amount to be repaid, you must add the interest to the amount borrowed.

*EXAMPLE 3*    Renny Soloman paid $375 simple interest on a loan of $1500 at 12.5%. What was the length of time for the loan?

*Solution*        Let $t$ = time.

Use the formula: $I = Prt$
$$375 = 1500 \times 0.125 \times t$$
$$375 = 187.5t$$

$$\frac{375}{187.5} = t$$

$$t = 2$$

The loan was for 2 years.

*EXAMPLE 4*    George Landon paid $585 simple interest on a loan of $6500 for 6 months. What was the annual rate?

*Solution*        Let $r$ = annual rate.

Use the formula: $I = Prt$

$$585 = 6500 \times r \times \tfrac{1}{2}$$

$$585 = 3250r$$

$$r = \frac{585}{3250} = 0.18$$

The annual rate was 18%.

In Example 4 notice that the time, 6 months, is expressed as $\frac{1}{2}$ year, since the rate of interest in the formula is the *annual or yearly rate.*

## Class Exercises

**Give the simple interest on each loan at the given annual rate for 1 year, 3 years, and 6 months.**

**1.** $100 at 8%          **2.** $200 at 12%          **3.** $5000 at 10%

**4.** $400 at 5%          **5.** $1500 at 6%          **6.** $100 at 8.4%

**Find the annual rate of interest for each loan.**

7. principal: $1100, time: 2 years, simple interest: $220

8. principal: $1600, time: $1\frac{1}{2}$ years, simple interest: $120

**Find the length of time for each loan.**

9. principal: $1200, interest rate: 6%, simple interest: $144

10. principal: $500, interest rate: 8%, simple interest: $30

**Find the total amount that must be repaid on each loan.**

11. principal: $200, interest rate: 15%, time: 2 years

12. principal: $600, interest rate: 9%, time: 3 years

## Written Exercises

**Find the simple interest on each loan and the total amount to be repaid.**

A  1. $1280 at 15% for 2 years

2. $4250 at 12% for 3 years

3. $2760 at 18% for 1 year, 6 months

4. $3500 at 16% for 9 months

5. $5640 at 7.5% for 4 years

6. $7250 at 12.8% for $2\frac{1}{2}$ years

7. $6380 at 14.5% for 6 years

8. $14,650 at 16.4% for $3\frac{1}{2}$ years

**Find the annual rate of interest for each loan.**

9. $4360 for 2 years, 6 months; simple interest: $1526

10. $1240 for 3 years, 6 months; simple interest: $651

11. $2600 for 4 years; total to be repaid: $3484

12. $5760 for 2 years, 9 months; total to be repaid: $8532

13. $6520 for 3 years, 3 months; simple interest: $3390.40

14. $3980 for $4\frac{1}{2}$ years; total to be repaid: $7024.70

**Find the length of time for each loan.**

15. $3775 at 12%; simple interest: $226.50

16. $7850 at 6.5%; simple interest: $510.25

**Find the original amount (principal) of the given loan.**

*EXAMPLE*   9% for 4 years; total to be repaid: $7140

*Solution*        Let $P$ = the original amount of the loan.

$$\text{amount to be repaid} = P + Prt$$
$$= P + P \times 0.09 \times 4$$
$$= P + 0.36P$$
$$= P(1 + 0.36)$$
$$7140 = 1.36P$$
$$P = \frac{7140}{1.36} = 5250$$

The original amount of the loan was $5250.

**B**  **17.** 8% for $4\frac{1}{2}$ years; total to be repaid: $6052

**18.** 13.5% for 5 years; total to be repaid: $3417

**19.** 16% for 2 years, 3 months; total to be repaid: $9139.20

**20.** 12.4% for 3 years, 6 months; total to be repaid: $2222.70

**21.** 9.6% for 4 years; total to be repaid: $2560.40

**22.** 10.8% for 2 years, 9 months; total to be repaid: $4734.05

**Solve.**

**C**  **23.** If the simple interest on $250 for 1 year, 8 months is $30, how much is the interest on $425.50 for 3 years, 4 months?

**24.** If $150 earns $28.75 simple interest in 1 year and 8 months, what principal is required to earn $747.50 interest in 2 years, 6 months?

## Problems

**Solve.**

**A**  **1.** Lois Pocket owns bonds worth $10,500 that pay 11% annual interest.  The interest is paid semiannually in two equal amounts.  How much is each payment?

**2.** A car loan of $4650 at an annual rate of 16% for 2 years is to be repaid in 24 equal monthly payments, including principal and interest.  How much is each of these payments?

**3.** Gilbert White wants to borrow $2250 for 3 years to remodel his garage. The annual rate is 18%. If the principal and interest are repaid in equal monthly installments, how much will each installment be?

**4.** Fernando Lopez can borrow $5640 at 12.5% for 4 years, or he can borrow the same amount for 3 years at 15%. Find the total amount to be repaid on each loan. Which amount is smaller?

**B**  **5.** An education loan of $8400 for ten years is to be repaid in monthly installments of $122.50. What is the annual rate of this loan, computed as simple interest?

**6.** Will Darcy's three-year home improvement loan is to be repaid in monthly installments of $375.90 each. If the annual rate (as simple interest) is 14.4%, what is the principal of the loan?

**C**  **7.** In how many years would the amount to be repaid on a loan at 12.5% simple interest be double the principal of the loan?

**8.** At what rate of simple interest would the amount to be repaid on a loan be triple the principal of the loan after 25 years?

---

## Review Exercises

**Multiply. Round to the nearest thousandth if necessary.**

**1.** $\frac{1}{2} \times 0.75$     **2.** $\frac{1}{4} \times 0.3$     **3.** $\frac{1}{5} \times 0.25$

**4.** $\frac{3}{4} \times 0.61$     **5.** $\frac{4}{5} \times 1.87$     **6.** $\frac{3}{10} \times 2.03$

**7.** $\frac{2}{3} \times 1.25$     **8.** $\frac{7}{9} \times 2.375$     **9.** $\frac{5}{6} \times 2.875$

---

### ▮▮▮ Calculator Key-In

**Some calculators have a percent key. Use a calculator with a percent key to do the following exercises.**

**1.** What is 8% of 200?

**2.** 35 is 20% of what number?

**3.** What percent of 75 is 36?

**4.** Express $\frac{17}{20}$ as a percent.

**5.** Express $\frac{13}{15}$ as a percent.

**6.** Express $\frac{7}{11}$ as a percent.

# 7-6 Compound Interest

If you deposited $100 in an account that paid 10% interest, you would have $10 interest after one year for a total of $110. You could then withdraw the $10 interest or leave it in the account. By withdrawing the interest, your principal would remain $100, and you would again receive simple interest. If, however, you left the $10 interest in the account, your principal would become $110, and you would accumulate **compound interest,** that is, interest on principal plus interest. The following chart illustrates these alternatives.

| | Simple Interest | | Compound Interest | |
|---|---|---|---|---|
| After | Interest | Principal | Interest | Principal |
| 0 years | 0 | $100 | 0 | $100 |
| 1 year | 10% of $100 = $10 | $100 | 10% of $100 = $10 | $100 + $10 = $110 |
| 2 years | 10% of $100 = $10 | $100 | 10% of $110 = $11 | $110 + $11 = $121 |
| 3 years | 10% of $100 = $10 | $100 | 10% of $121 = $12.10 | $121 + $12.10 = $133.10 |
| | Total interest: $30 | | Total interest: $33.10 | |

Notice that because the principal increases as interest is compounded, the total amount of interest paid on $100 after 3 years is greater than what is paid at the same rate of simple interest.

Interest is often compounded in one of the following manners.

monthly: 12 times a year
quarterly: 4 times a year
semiannually: 2 times a year

**EXAMPLE**  If $500 is deposited in an account paying 8% interest compounded quarterly, how much will the principal amount to after 1 year?

**Solution**  Use the formula $I = Prt$.

After 1 quarter:
$I = 500 \times 0.08 \times \frac{1}{4} = \$10$        $P = 500 + 10 = \$510$

After 2 quarters:
$I = 510 \times 0.08 \times \frac{1}{4} = \$10.20$        $P = 510 + 10.20 = \$520.20$

After 3 quarters:
$I = 520.20 \times 0.08 \times \frac{1}{4} \approx \$10.40$        $P = 520.20 + 10.40 = \$530.60$

After 4 quarters:
$I = 530.60 \times 0.08 \times \frac{1}{4} \approx \$10.61$        $P = 530.60 + 10.61 = \$541.21$

## Class Exercises

**$8000 is deposited in a bank that compounds interest annually at 5%. Find the following.**

**1.** The interest paid at the end of the first year

**2.** The new principal at the beginning of the second year

**3.** The interest paid at the end of the second year

**4.** The new principal at the beginning of the third year

**5.** The interest paid at the end of the third year

**$1000 is invested at 12%, compounded quarterly. Find the following.**

**6.** The interest paid after 3 months    **7.** The new principal after 3 months

**8.** The interest paid after 6 months    **9.** The new principal after 6 months

## Written Exercises

**How much will each principal amount to if it is deposited for the given time at the given rate? (If you do not have a calculator, round to the nearest penny at each step.)**

**A**   **1.** $6400 for 2 years at 5%, compounded annually

**2.** $500 for 1 year at 8%, compounded semiannually

**3.** $1500 for 6 months at 8%, compounded quarterly

**4.** $8000 for 2 months at 6%, compounded monthly

**5.** $2500 for 18 months at 16%, compounded semiannually

**6.** $1280 for 9 months at 10%, compounded quarterly

**7.** $1800 for 1 year, 3 months at 14%, compounded quarterly

**8.** $3475 for $2\frac{1}{2}$ years at 6%, compounded semiannually

**How much will each principal amount to if it is deposited for the given time at the given rate? (If you do not have a calculator, round to the nearest penny at each step.)**

**B** **9.** $3200 for 3 years at 7.5%, compounded annually

**10.** $3125 for 3 months at 9.6%, compounded monthly

**11.** $8000 for 2 years at 10%, compounded semiannually

**12.** $6250 for 1 year at 16%, compounded quarterly

**13.** What is the difference between the simple and compound (compounded semiannually) interest on $250 in 2 years at 14% per year?

**14.** What is the difference between the interest on $800 in 1 year at 8% per year compounded semiannually and compounded quarterly?

**C** **15.** How long would it take $2560 to grow to $2756.84 at 10% compounded quarterly?

**16.** What principal will grow to $1367.10 after 1 year at 10% compounded semiannually?

---

## Problems

**Solve. (If you do not have a calculator, round to the nearest penny at each step.)**

**A** **1.** The River Bank and Trust Company pays 10%, compounded semiannually, on their 18-month certificates. How much would a certificate for $1600 be worth at maturity?

**2.** Mike Estrada invested $1250 at 9.6%, compounded monthly. How much was his investment worth at the end of 3 months?

**B** **3.** Jennifer Thornton can invest $3200 at 12% simple interest or at 10% interest compounded quarterly. Which investment will earn more in 9 months?

**4.** Which is a better way to invest $6400 for $1\frac{1}{2}$ years: 16% simple interest or 15% compounded semiannually?

**C** **5.** A money market fund pays 12.8% per year, compounded semiannually. For a deposit of $5000, this compound interest rate is equivalent to what simple interest rate for 1 year? Give your answer to the nearest tenth of a percent.

# 7-7 Percents and Problem Solving

Percents are used frequently to express facts about situations we encounter. The skills acquired earlier for working with percents can be used to solve problems that are involved in these situations.

**EXAMPLE 1**   Eve Malik has a yearly salary of $27,600. She is paid in equal amounts twice a month. She and her employer each pay 6.7% of her salary to her social security account. What is the combined amount paid to social security for Eve each pay period?

**Solution**   The problem asks for the combined amount paid to social security.
Given facts: yearly salary $27,600
              2 paychecks a month
              6.7% each for social security

2 payments a month for 12 months: $2 \times 12 = 24$ payments
$27,600 \div 24 = \$1150$, salary per payment
social security paid by each is 6.7% of $1150

$$0.067 \times 1150 = \$77.05$$

Combined amount paid to social security is $2 \times \$77.05$, or $154.10.

**EXAMPLE 2**   Property taxes in Jackson County are $1.42 per $100 of assessed value of the property. Assessed value is 60% of the actual market value. If taxes on a home are $681.60, what is the market value of the property?

**Solution**   The problem asks for the market value.
Given facts: Taxes are $681.60
              Tax rate is $1.42 per $100 of assessed value.
              Assessed value is 60% of market value.

Divide $681.60 by $1.42 to find how many $100's are in the assessed value: $681.60 \div 1.42 = 480$
Assessed value is $480 \times 100 = 48,000$
Assessed value is 60% ($= 0.6$) of market value.

$$48,000 = 0.6 \times m$$

$$\frac{48,000}{0.6} = m$$

$$80,000 = m$$

The market value is $80,000.

**EXAMPLE 3**    Central Electric Co. charges $.055 per kW·h for the first 500 kW·h and $.05 per kW·h for the next 2000 kW·h. There is a discount of 2% for paying bills promptly. In March the Gronskis used 1450 kW·h. How much did they pay if they paid in time to receive the discount for prompt payment?

**Solution**    The problem asks how much the Gronskis paid.

Given facts: Cost is $.55 per kW·h for the first 500 kW·h,
$.05 per kW·h for the next 2000 kW·h.
1450 kW·h were used.
Discount is 2%.

First find the total charge.
500 kW·h at $.055 each: $500 \times 0.055 = \$27.50$
$1450 - 500 = 950$
950 kW·h at $.05 each: $950 \times 0.05 = \$47.50$
Total charge: $\$27.50 + \$47.50 = \$75.00$

Since the discount is 2%, they only paid 100% − 2%, or 98%, of the actual charge.

$$75 \times 0.98 = 73.50$$

The Gronskis paid $73.50.

---

**Problem Solving Reminder**

In some problems, it is necessary to use several operations. Notice that Example 3 above required that the charge for the first 500 kW·h and the charge for the next 950 kW·h be found separately and then added to find the total charge. Finally, the discounted price was calculated.

---

## Problems

**Solve.**

**A**   **1.** Jim Damato charged $86.50 on his FasterCharge card and got a cash advance of $120. There is a finance charge of 1.5% on purchases and 1% on cash advances. How much was his total finance charge?

**2.** Esther Upperman invested $1250 in 50 shares of BU + U stock, which pays an annual dividend of $1.90 per share. Which investment gives a greater yield, the BU + U shares or a certificate of deposit that pays 7% interest?

3. A guitarist is to be paid a royalty of 6% of the selling price of a new album. If 46,000 copies of the album were sold at $7.95 each, how much should the guitarist be paid?

4. The publisher of *Lost in the Jungle* receives $12.60 from each copy sold. The remaining portion of the $15 selling price goes to the author. What royalty rate does the author earn?

5. At a clearance sale, a gas-powered lawn mower was discounted 32%. The lawn mower was sold for $117.81, including a sales tax of 5% of the sale price. What was the original price of the mower?

6. The Ransoms intend to reduce their electricity use by 15% in September. In August they used 2200 kW·h. Their electric company charges $0.062 per kW·h for the first 1000 kW·h, $0.055 per kW·h for the next 1000 kW·h, and $0.05 per kW·h for any additional use. How much can the Ransoms expect to save?

**B** 7. Anita Ramirez owns two bonds, one paying 8.5% interest and the other paying 9% interest. Every six months Anita receives a total of $485 in interest from both bonds. If the 8.5% bond is worth $4000, how much is the 9% bond worth?

8. Mike Robarts owns two bonds, one worth $3000, the other worth $5000. The $3000 bond pays 12% interest. Every 6 months, Mike receives a total of $530 interest from both bonds. What is the annual rate of the $5000 bond?

**C** 9. The property tax rate in Glendale is $2.15 per $100 of assessed value. Assessed value is 60% of market value. This year the Prestons' tax bill was $129 more than last year, when the market value of their house was $84,000. What is the present assessed value of their house?

**10.** In the primary election, 56% of eligible voters voted. Smith received 65% of the votes, Jones got 30%, and 5% of the voters chose other candidates. Smith won by a margin of 5488 votes over Jones. How many eligible voters are there?

**11.** David Cho deposited $10,000 in a six-month savings certificate that paid simple interest at an annual rate of 12%. After 6 months, David deposited the total value of his investment in another six-month savings certificate. David made no other deposits or withdrawals. After 6 months, David's investment was worth $11,130. What was the annual simple interest rate of the second certificate?

**12.** A savings account pays 5% interest, compounded annually. Show that if $1000 is deposited in the account, the account will contain

$1000 \times 1.05$ dollars after 1 year.
$1000 \times (1.05)^2$ dollars after 2 years.
$1000 \times (1.05)^3$ dollars after 3 years, and so on.

## Self-Test B

**Solve.**

**1.** A loan of $6400 at 12.5% simple interest is to be repaid in $4\frac{1}{2}$ years. What is the total amount to be repaid? [7–5]

**2.** A $6500 automobile loan is to be repaid in 3 years. The total amount to be repaid is $9230. If the interest were simple interest, what would be the annual rate?

**3.** You borrow $2500 at 16% interest, compounded semiannually. No interest is due until you repay the loan. If you repay the loan at the end of 18 months, how much will you pay? [7–6]

**4.** Julian Dolby invested $5000 in a special savings account that pays 8% interest, compounded quarterly. How long must Julian keep his money in the account to earn at least $250 in interest?

**5.** Mary Barnes earns $9.00 an hour. In 1984 she worked an average of 38 h a week. She pays 5% of her wages to a pension plan, and her employer pays an additional 7.2% of her wages to the same plan. How much was paid to the plan for Mary in 1984? [7–7]

*Self-Test answers and Extra Practice are at the back of the book.*

# Credit Balances

Before stores or banks issue credit cards, they do a complete credit check of the prospective charge customer. Why? Because issuing credit is actually lending money. When customers use credit cards they are getting instant loans. For credit loans, interest, in the form of a *finance charge,* is paid for the favor of the loan.

To keep track of charge account activity, monthly statements are prepared. The sample statement below is a typical summary of activity.

| Customer's Statement<br>Account Number 35–0119–4G | | | | Billing Cycle Closing Date 9/22<br>Payment Due Date 10/17 | |
|---|---|---|---|---|---|
| Date | Dept. No. | Description | Purchases & Charges | Payments & Credits | |
| 9/3 | 753 | TOYS | 30.00 | | |
| 9/14 | 212 | HOUSEWARES | 12.50 | | |
| 9/17 | | PAYMENT | | 20.00 | |

| Previous Balance | Payments & Credits | Unpaid Balance | Finance Charge | Purchases & Charges | New Balance | Minimum Payment |
|---|---|---|---|---|---|---|
| 40.99 | 20.00 | 20.99 | .61 | 42.50 | 64.10 | 20.00 |

Statements usually show any finance charges and a complete list of transactions completed during the billing cycle. Finance charges are determined in a variety of ways. One common method is to compute the amount of the unpaid balance and make a charge based on that amount. Then new purchases are added to compute the new balance on which a minimum payment is due.

To manage the masses of data generated by a credit system, many merchants use computerized cash registers. The cash registers work in the following way. When a customer asks to have a purchase charged, the salesperson enters the amount of the purchase and the customer's credit card number into the computer. The computer then retrieves the account balance from memory, adds in the new purchase, and compares the total to the limit allowed for the account. If the account limit has not been reached, the transaction is completed and the amount of the new purchase is stored in memory. At the end of the billing cycle, the computer totals the costs of the new purchases, deducts payments or credits, adds any applicable finance charges, and prints out a detailed statement.

**Copy and complete. Use the chart below to find the minimum payment.**

|    | Previous Balance | Payments and Credits | Unpaid Balance | Finance Charge | Purchases and Charges | New Balance | Minimum Payment |
|----|------------------|----------------------|----------------|----------------|-----------------------|-------------|-----------------|
| 1. | 175.86 | 50.00 | ? | 1.89 | 35.00 | ? | ? |
| 2. | 20.00 | 20.00 | ? | 0 | 30.00 | ? | ? |
| 3. | 289.75 | 40.00 | ? | 5.25 | 68.80 | ? | ? |
| 4. | 580.00 | 52.00 | ? | 8.00 | 189.65 | ? | ? |
| 5. | 775.61 | 110.00 | ? | 10.48 | 0 | ? | ? |

**Solve.**

6. Debra Dinardo is comparing her sales receipts to her charge account statement for the month. The billing cycle closing date for the statement is 4/18. Debra has sales receipts for the following dates and amounts.
3/30  $17.60          4/8  $21.54
4/16  $33.12          4/26  $9.75
On April 14, Debra paid the balance on her last statement with a check for $139.80. Her statement shows the information below. Is the statement correct? Explain.

| New Balance | Minimum Payment |
|-------------|-----------------|
| Up to $20.00 | New Balance |
| $ 20.01 to $200.00 | $20.00 |
| $200.01 to $250.00 | $25.00 |
| $250.01 to $300.00 | $30.00 |
| $300.01 to $350.00 | $35.00 |
| $350.01 to $400.00 | $40.00 |
| $400.01 to $450.00 | $45.00 |
| $450.01 to $500.00 | $50.00 |
| Over $500.00 | $50.00 plus $10.00 for each $50.00 (or fraction thereof) of New Balance over $500 |

| Previous Balance | Payments & Credits | Unpaid Balance | Finance Charge | Purchases & Charges | New Balance | Minimum Payment |
|------------------|--------------------|----------------|----------------|---------------------|-------------|-----------------|
| 139.80 | 139.80 | 0 | 0 | 72.26 | 72.26 | 20.00 |

**Research Activity**   Find out why stores and banks are willing to extend credit. How are the expenses of running the credit department paid for? Why do some stores offer discounts to customers who pay cash?

# Chapter Review

**Complete each statement.**

[7-1]

1. The percent of increase from 24 to 30 is __?__.

2. The percent of decrease from 50 to 37 is __?__.

3. The percent of increase from 76 to __?__ is 25%.

**Write the letter of the correct answer.**

4. A pair of shoes that regularly sells for $32 is on sale for $24. What is the percent of discount?

[7-2]

    **a.** $33\frac{1}{3}\%$      **b.** 25%      **c.** 8%      **d.** 125%

5. The Vico Manufacturing Company makes a 12% profit. If the company sold $15,250 worth of goods in April, what was its profit?

[7-3]

    **a.** $13,420      **b.** $15,250      **c.** $1830      **d.** $18,300

6. The Ski Club has 50 members. If 36 members attend the ski trip, what percent attends the ski trip? Which of the following proportions could be used to solve this problem?

[7-4]

    **a.** $\dfrac{36}{100} = \dfrac{n}{50}$          **b.** $\dfrac{n}{100} = \dfrac{36}{50}$

**True or false?**

7. If George's aunt gives him a loan of $650 for one year at 12% simple interest, George will owe $728 at the end of the year.

[7-5]

8. An investment of $500 for 2 years at 12% simple interest will pay more than one of $500 at $8\frac{1}{2}\%$ for 30 months.

**Write the letter of the correct answer.**

9. Lorraine Eldar invested $6000 at 8% interest compounded semiannually. How long will it take her investment to exceed $7000?

[7-6]

    **a.** 1 year      **b.** $1\frac{1}{2}$ years      **c.** 2 years      **d.** $2\frac{1}{2}$ years

10. Anita Ramirez owns two bonds, one paying 8.5% interest and the other paying 9% interest. Every six months Anita receives a total of $485 in interest from both bonds. If the 8.5% bond is worth $4000, how much is the 9% bond worth?

[7-7]

    **a.** $7000      **b.** $4000      **c.** $1611.11      **d.** $10,777.78

# Chapter Test

**Solve.**

1. Sam Golden's salary increased $30 a week. If his salary was $400 a week before the raise, by what percent did his salary increase? [7–1]

2. A clock ratio that sells for $30 is on sale at a 20% discount. What is the sale price? [7–2]

3. Edna's Autos sold $24,000 worth of cars last week. If Edna's costs were $16,800, how much profit did Edna make and what was the percent of profit? [7–3]

4. Computer Village had sales of $16,000 for December. The store's profits were $2880. What was the percent of profit?

**Use proportions to solve Exercises 5 and 6.**

5. The Pro Shop sold 6 of 20 tennis rackets in May. What percent of the tennis rackets were sold? [7–4]

6. This term, 36% of the students in one class are on the honor roll. If 9 students are on the honor roll, how many students are in the class?

**Solve.**

7. John Anthony borrowed $1200 for $1\frac{1}{2}$ years at 13% simple interest. How much must he repay when the loan is due? [7–5]

8. Joan Wu invested $5000 in an account that pays 13% simple interest. If the interest is paid 4 times a year, how much is each payment?

9. Billtown Bank pays 8% interest compounded quarterly. To the nearest penny, how much will $2500 earn in one year? [7–6]

10. You deposit money in a savings account and make no other deposits or withdrawals. How much is in the account at the end of three months if you deposit $350 at 6% interest, compounded monthly?

11. A major credit card charges a 1.5% interest rate each month on unpaid balances. If you were charged $8.50 in interest in June, what was your balance? [7–7]

# Cumulative Review (Chapters 1–7)

## Exercises

**Evaluate the expression if $x = 3$ and $y = 5$.**

**1.** $x + y$  **2.** $2x + y$  **3.** $2y - x$  **4.** $x + 6y$

**5.** $y + x + 4$  **6.** $-x - y$  **7.** $-3x - 2y$  **8.** $5x + (-y) - 7$

**9.** $x^2$  **10.** $xy^2$  **11.** $(-x)^2 y$  **12.** $-(xy)^2$

**Write the numbers in order from least to greatest.**

**13.** $3, -2.5, 0, -7, 6.4, -2.2$  **14.** $4, -6.2, -7.3, 0, 3.5, 2.5$

**15.** $-1.6, -8.4, -3, -1.0, 0.5$  **16.** $-4.7, -11.9, 1, 3.8, 0.7, -0.6$

**Tell whether the statement is true or false for the given value of the variable.**

**17.** $t - 3 \le -6;\ -4$  **18.** $-m - 3 > 0;\ -3$  **19.** $2b < 3;\ 1$

**20.** $3 - x = 2;\ 5$  **21.** $n + 1 \ge -7;\ -6$  **22.** $r < -2r - 4;\ -2$

**Write the fractions as equivalent fractions having the least common denominator (LCD).**

**23.** $\frac{2}{3}, \frac{5}{9}$  **24.** $-\frac{3}{5}, \frac{7}{20}$  **25.** $-\frac{8}{3}, \frac{9}{11}$  **26.** $-\frac{5}{7}, -\frac{12}{17}$

**27.** $\frac{7}{3}, \frac{3}{4}$  **28.** $\frac{8}{45}, \frac{11}{75}$  **29.** $\frac{17}{32}, -\frac{3}{128}$  **30.** $\frac{3}{34}, \frac{5}{39}$

**Solve.**

**31.** $\frac{n}{4} = \frac{36}{12}$  **32.** $\frac{5}{n} = \frac{60}{12}$  **33.** $\frac{90}{n} = \frac{4}{8}$

**34.** $\frac{4}{6} = \frac{n}{30}$  **35.** $\frac{21}{28} = \frac{n}{84}$  **36.** $\frac{20}{10} = \frac{10}{n}$

**State which of the rigid motions are needed to match the vertices of the triangles and explain why the triangles are congruent.**

**37.** 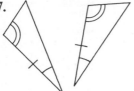  **38.** 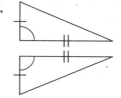  **39.**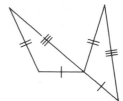

# Problems

**Solve.**

1. The perimeter of an equilateral triangle is 45 cm. What is the length of each side?

2. A photographer works 35 h a week and earns $295. What is the hourly rate to the nearest cent?

3. Elena commutes to work. She travels 2.6 km by subway and 1.8 km by bus. How far is that in all?

4. The temperature at 11:30 P.M. was 7° below zero. By the next morning, the temperature had fallen 6°. What was the temperature then?

5. During one week, the stock of DataTech Corporation had the following daily changes in price: Monday, up $1\frac{1}{2}$ points; Tuesday, down 2 points; Wednesday, down $\frac{3}{4}$ of a point; Thursday, up $3\frac{1}{8}$ points; Friday, up $2\frac{3}{4}$ points. What was the change in the price of the stock for the week?

6. A rope 25 m long is cut into 2 pieces so that one piece is 9 m shorter than the other. Find the length of each piece.

7. A water purification device can purify 15.5 L of water in one hour. How many liters can it purify in $3\frac{3}{4}$ hours?

8. The Onagas have a wall at the back of their property. They want to fence off a rectangular garden using 7 m of the existing wall for the back part of the fence. If they have 15 m of fencing, what will be the length of each of the shorter sides?

9. Maria Sanchez gave 8% of her mathematics students a grade of A. If she had 125 students in her 4 classes, how many of these students received an A?

# 8

# Equations and Inequalities

The trains pictured at the right were introduced in France. They can travel at speeds of up to 160 km/h on conventional tracks and nearly triple that on their own continuously welded tracks. The speed of a train on an actual trip depends, of course, on the number of curves and the type of track. A trip of 425 km from Paris to Lyon that takes about 4 h in a conventional train, for instance, can be completed in about 2.5 h in one of these trains.

The time it takes a train to complete a trip depends on the speed of the train and the distance traveled. Some relationships, such as the relationship between time, rate, and distance, can be expressed as equations. Other relationships can be expressed as inequalities. You will learn more about equations and inequalities and methods to solve both in this chapter. You will also learn how to use equations to help solve problems.

## Career Note

Reporters representing newspapers, magazines, and radio and television stations often attend newsworthy events such as the introduction of new trains. Reporters need a wide educational background. They must report facts, not their own opinions. They must be willing to meet deadlines and often to work at irregular hours.

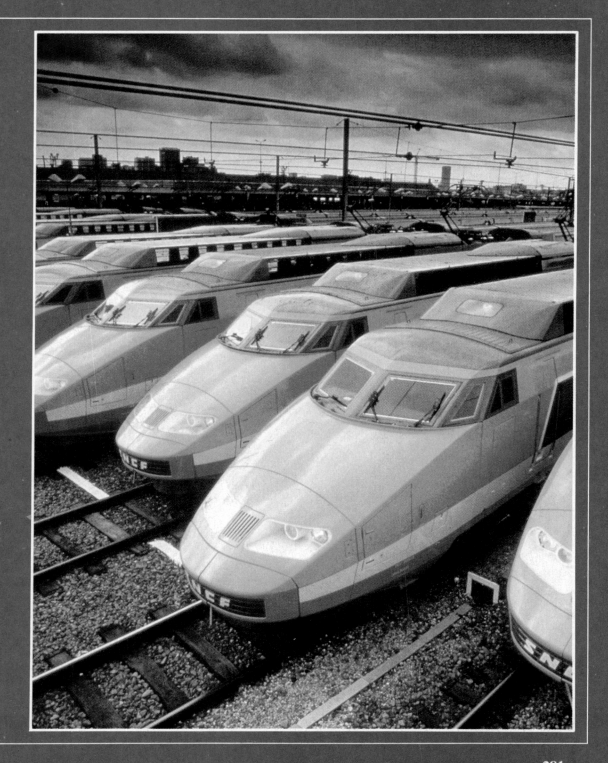

# 8-1 Equations: Variable on One Side

In Chapter 4, you learned the transformations that are used in solving equations. In this chapter, you will learn how these transformations can be used to solve more difficult equations.

Two terms are called **like terms** if their variable parts are the same. For example, $x$ and $11x$ are like terms, as are $-4y$ and $7y$. Because their variable parts are different, $4x$, $3x^2$, and $6xy$ are not like terms.

We can use the properties of addition and multiplication to simplify expressions with like terms. To simplify $3a + 11 + 4a$, we use the commutative and distributive properties.

$$\begin{aligned} 3a + 11 + 4a &= (3a + 4a) + 11 \\ &= (3 + 4)a + 11 \\ &= 7a + 11 \end{aligned}$$

The terms $7a$ and $11$ cannot be combined because they are not like terms. When an expression is in *simplest form,* it contains no like terms.

We can use the distributive property $a(b + c) = ab + ac$ to simplify certain variable expressions that involve parentheses.

*EXAMPLE 1*   Simplify $-2(x - 6) + 8x$.

*Solution*
$$\begin{aligned} -2(x - 6) + 8x &= -2x + 12 + 8x \\ &= -2x + 8x + 12 \\ &= 6x + 12 \end{aligned}$$

These procedures may be used when solving an equation.

*EXAMPLE 2*   Solve $2x + 15 + 8x = 20$.

*Solution*
$$\begin{aligned} 2x + 15 + 8x &= 20 \\ 2x + 8x + 15 &= 20 \\ 10x + 15 &= 20 \\ 10x + 15 - 15 &= 20 - 15 \\ 10x &= 5 \\ \frac{10x}{10} &= \frac{5}{10} \\ x &= \frac{1}{2} \end{aligned}$$

The solution is $\frac{1}{2}$.

**EXAMPLE 3**   Solve $3(x + 3) + 6x = 12$.

**Solution**

$$3(x + 3) + 6x = 12$$
$$3x + 9 + 6x = 12$$
$$3x + 6x + 9 = 12$$
$$9x + 9 = 12$$
$$9x + 9 - 9 = 12 - 9$$
$$9x = 3$$
$$\frac{9x}{9} = \frac{3}{9}$$
$$x = \frac{1}{3}$$

The solution is $\frac{1}{3}$.

## Class Exercises

**Which pairs of terms are like terms?**

**1.** $7a, 12a$         **2.** $4x, -6x$         **3.** $3r, 4rs$         **4.** $6a, 6b$

**5.** $-8y, y$         **6.** $-11, -11x$         **7.** $a^2, 7a$         **8.** $3a^2, 3a$

**Simplify.**

**9.** $4c + 7c$         **10.** $13y - 8y$         **11.** $z - 7z$

**12.** $2t + t + 4t$         **13.** $-5x + 6x - x$         **14.** $4p - p + 5p$

**15.** $2(b + 5)$         **16.** $-3(x + 7)$         **17.** $9(c - 4)$

**18.** $4(n + 4) + 3$         **19.** $-5(x - 1) - 6$         **20.** $3(y + 4) + 2y$

**21.** $12(d - 4) - 10d$         **22.** $-10(p - 3) + 6p$         **23.** $15(y + 1) + 10(y - 2)$

## Written Exercises

**Simplify.**

**A**   **1.** $10m - 8m$         **2.** $-3a + 11a$         **3.** $-8c + c$

**4.** $-3n + 8 + 8n$         **5.** $-5 + 6c - 5c$         **6.** $12 + 10z - 12z$

**7.** $-7(y - 11)$         **8.** $6(n + 3) + 2n$         **9.** $-10(a + 3) + 6a$

**10.** $8(p + 4) + 6(p - 3)$         **11.** $4(x - 3) + 2(x - 1)$         **12.** $-5(n - 5) + 7(n + 1)$

Solve.

**13.** $4x + 8x = 6$        **14.** $5z - 7z = -4$        **15.** $-2y + 6y = 36$

**16.** $p - 10p = -27$        **17.** $-7c - 9c = 20$        **18.** $-z - 10z = -3$

**19.** $3n + 2n - 8 = 17$        **20.** $-6a + 7a - 3 = 8$        **21.** $-x + 9 - 5x = 12$

**22.** $4t - 8 + 8t = 32$        **23.** $5 - 4y - 7y = 16$        **24.** $-7 + 3n + 9n = 29$

**25.** $12 = 8y + 16y$        **26.** $-9 = 3y - 9y$        **27.** $15 = x + 6 - 10x$

**28.** $-4 = 3z - 10 + 4z$        **29.** $-14 = 6 + z - 9z$        **30.** $3 = t - 4t + 15$

**B**   **31.** $2(x + 1) = 4$        **32.** $-3(y + 4) = -6$        **33.** $8(t - 1) = -16$

     **34.** $-7(c - 3) = 35$        **35.** $3(7 - p) = 3$        **36.** $-6(4 - z) = 20$

     **37.** $2(n + 4) + 6n = 2$        **38.** $6(c - 2) + 4c = 8$        **39.** $-5(d + 3) - 7d = 5$

     **40.** $-4(x - 5) - 16x = 10$      **41.** $2(3 - y) + 6y = 12$      **42.** $-3(2 - z) - 8z = 15$

**C**   **43.** $3(a + 2) - (a - 1) = 17$           **44.** $5(c - 3) - (5 - c) = 0$

     **45.** $4(p + 2) - 2(1 - p) - 4p = 0$        **46.** $5(v - 2) - 2(v + 4) = 6$

     **47.** $2(y - 5) + 7 = 4(y + 7) - 3$        **48.** $-3(x + 2) + 3 = 5(x - 1) + 10$

## Review Exercises

Solve.

**1.** $\dfrac{n}{4} = \dfrac{8}{32}$        **2.** $\dfrac{n}{48} = \dfrac{5}{6}$        **3.** $\dfrac{6}{n} = \dfrac{24}{60}$

**4.** $\dfrac{50}{n} = \dfrac{5}{15}$        **5.** $\dfrac{2}{11} = \dfrac{n}{66}$        **6.** $\dfrac{1}{6} = \dfrac{n}{18}$

**7.** $\dfrac{2}{3} = \dfrac{7}{n}$        **8.** $\dfrac{12}{84} = \dfrac{1}{n}$        **9.** $\dfrac{36}{n} = \dfrac{72}{8}$

## Challenge

Fill in the blanks.

**1.** 6, 10, 15, 21, __?__, 36, 45        **2.** 103, __?__, 305, 406, 507

**3.** 5, 3, 4, 2, __?__, 1, 2        **4.** 132, 243, 354, 465, __?__

**5.** 17, 16, 18, __?__, 19, 14, 20, 13        **6.** 6, 11, 21, 41, __?__

# 8-2 Equations: Variable on Both Sides

Some equations have variables on each side. To solve such an equation, add a variable expression to each side, or subtract a variable expression from each side.

**EXAMPLE 1**  Solve $8c = c + 14$.

**Solution**  Subtract $c$ from both sides.

$$8c = c + 14$$
$$8c - c = c + 14 - c$$
$$7c = 14$$
$$\frac{7c}{7} = \frac{14}{7}$$
$$c = 2$$

The solution is 2.

**EXAMPLE 2**  Solve $7c = 3 - 2c$.

**Solution**  Add $2c$ to both sides.

$$7c = 3 - 2c$$
$$7c + 2c = 3 - 2c + 2c$$
$$9c = 3$$
$$\frac{9c}{9} = \frac{3}{9}$$
$$c = \frac{1}{3}$$

The solution is $\frac{1}{3}$.

**EXAMPLE 3**  Solve $2(a - 3) = 5(a + 3)$.

**Solution**
$$2(a - 3) = 5(a + 3)$$
$$2a - 6 = 5a + 15$$
$$2a - 6 + 6 = 5a + 15 + 6$$
$$2a = 5a + 21$$
$$2a - 5a = 5a + 21 - 5a$$
$$-3a = 21$$
$$\frac{-3a}{-3} = \frac{21}{-3}$$
$$a = -7$$

The solution is $-7$.

## Class Exercises

**Solve.**

**1.** $2y = y + 1$     **2.** $t + 3 = 2t$     **3.** $4a + 2 = 5a$     **4.** $3n + 5 = 4n$

**5.** $a - 1 = 2a$     **6.** $6p = 5p - 4$     **7.** $2a - 7 = 3a$     **8.** $6t + 1 = 7t$

---

## Written Exercises

**Solve.**

**A**

**1.** $6c = 4c + 10$     **2.** $13z = 15 + 8z$     **3.** $49 + a = 8a$

**4.** $8x + 16 = 4x$     **5.** $7y = 5y - 12$     **6.** $17t = 8t - 36$

**7.** $32y = 24y - 4$     **8.** $17z = 15z - 3$     **9.** $3s = -6s - 36$

**10.** $-14p = 6p + 10$     **11.** $11n + 9 = -4n$     **12.** $-4x + 6 = -2x$

**13.** $2c + 6 = 5c - 6$     **14.** $6y + 3 = 8y + 2$     **15.** $-3d - 8 = 7d + 17$

**16.** $10a - 1 = 2a - (-5)$     **17.** $15 - 7x = 8x - 30$     **18.** $y - 11 = 24 - 9y$

**B**    **19.** $2(x + 1) = 4x$     **20.** $-3(y + 4) = -6y$     **21.** $8(t - 1) = -16t$

**22.** $-5(d + 3) = 7d + 5$     **23.** $6(c - 2) + 4c = 8c$     **24.** $-4(x - 5) = 16x + 10$

**25.** $6(d + 1) = 4(d + 2)$     **26.** $8(p + 4) = 6(p - 2)$

**27.** $-2(c - 3) = 4(c + 6)$     **28.** $-2(x - 4) = 5(x - 2)$

**29.** $10(y + 2) + 4 = 6(y - 3) + 8$     **30.** $-12(x + 3) + 2 = 6(x - 1) + 8$

**C**    **31.** $\frac{1}{2}(x + 4) = \frac{1}{4}(x - 8)$     **32.** $\frac{1}{3}(y - 6) = \frac{2}{9}(y + 18)$

**33.** $\frac{2}{5}(a + 10) = \frac{1}{4}(a + 20)$     **34.** $\frac{5}{8}(t - 24) = \frac{1}{4}(t + 12)$

**35.** $\frac{9}{10}(n - 30) = \frac{1}{2}(n + 10)$     **36.** $\frac{1}{3}(p + 21) = \frac{1}{6}(p - 9)$

---

## Review Exercises

**Use inverse operations to solve.**

**1.** $x + 8 = 12$     **2.** $x + (-5) = -3$     **3.** $x - 5 = 1$     **4.** $x - (-3) = 4$

**5.** $6x = 72$     **6.** $-9x = 81$     **7.** $\frac{x}{4} = 6$     **8.** $-\frac{x}{7} = 11$

# 8-3 Equations in Problem Solving

The five-step method that was shown in Chapter 4 may be applied with the procedures shown in this chapter.

## Solving a Word Problem Using an Equation

**Step 1** Read the problem carefully. Make sure that you understand what it says. You may need to read it more than once.

**Step 2** Decide what numbers are asked for. Choose a variable and use it to represent the number(s) asked for.

**Step 3** Write an equation based on the given conditions.

**Step 4** Solve the equation and find the required number(s).

**Step 5** Check your results with the words of the problem. Give the answer.

**EXAMPLE 1** An electrician has a length of wire that is 120 m long. The wire is cut into two pieces, one 30 m longer than the other. Find the lengths of the two pieces of wire.

**Solution**

- The problem says: length of wire is 120 m, one piece is 30 m longer than the other

- The problem asks for: the lengths of the two pieces of wire
  Let $l$ = the length of shorter piece of wire.
  The length of the other piece of wire $= l + 30$.

- We set the sum of the lengths ($l + l + 30$) equal to 120.

$$l + l + 30 = 120$$

- Solve.

$$l + l + 30 = 120$$
$$2l + 30 = 120$$
$$2l + 30 - 30 = 120 - 30$$
$$2l = 90$$
$$\frac{2l}{2} = \frac{90}{2}$$
$$l = 45, \; l + 30 = 75$$

- Check: $45 + 75 = 120$, and $75 - 45 = 30$. The result checks.

  The lengths are 45 m and 75 m.

**EXAMPLE 2** The sum of two consecutive integers is 137. Name the larger integer.

**Solution**
- The problem says: the sum of two consecutive integers is 137
- The problem asks for: the larger of the two integers

  Let $n$ = the smaller integer.
  Then the larger integer = $n + 1$.
  The sum of the two integers is $n + n + 1$.
  The sum of the two integers is 137.

- We set the expression for the sum of the two integers ($n + n + 1$) equal to 137.

$$n + n + 1 = 137$$

- Solve.
$$n + n + 1 = 137$$
$$2n + 1 = 137$$
$$2n + 1 - 1 = 137 - 1$$
$$2n = 136$$
$$\frac{2n}{2} = \frac{136}{2}$$
$$n = 68$$
$$n + 1 = 69$$

- Check: $68 + 69 = 137$.

  The larger integer is 69.

---

**Reading Mathematics: Study Skills**
When you find a reference in the text to material you learned earlier, reread the material in the earlier lesson to help you understand the new lesson. For example, page 287 includes a reference to the five-step method learned in Chapter 4. Turn back to Chapter 4 to review how the five-step method was used.

---

## Problems

**Solve.**

**A**  **1.** A 40 ft board is cut into two pieces so that one piece is 8 ft longer than the other piece. Find the lengths of the two pieces.

**2.** The sum of two numbers is 60. One number is 16 less than the other number. What is the larger number?

3. Marisa Gonzalez is training for a cross-country race. Yesterday, it took her 80 minutes to jog to the reservoir and back. The return trip took 10 minutes longer than the trip to the reservoir. How long did it take her each way?

4. Mark Johnson's weight is 18 kg greater than his sister's. If they weigh a total of 110 kg, what are their weights?

5. The perimeter of an isosceles triangle is 50 cm. If the congruent sides of the triangle are each twice as long as the remaining side, what is the length of each side of the triangle?

6. Find two consecutive integers whose sum is 115.

7. West High School defeated Central High by eleven points in the city championship basketball game. If a total of 159 points were scored in the game, how many points did Central High score?

8. Together, a table and a set of four chairs cost $499. If the table costs $35 more than the set of chairs, what is the cost of the table?

**B** 9. A 50 ft rope is cut into two pieces. One piece is 2 ft longer than twice the length of the other. Find the length of the longer piece.

10. Twice the sum of two consecutive integers is 42. What are the two integers?

11. A collection of jewelry includes two rings, one 18 years old and the other 46 years old. In how many years will the older ring be twice as old as the newer ring?

12. Neil is 4 years older than Andrea. If Neil was twice as old as Andrea four years ago, how old was Andrea?

**C** 13. A sailing race takes place on a 3000 m course. The second leg of the course is 100 m longer than the first, and the third leg is 100 m longer than the second. How long is the second leg?

14. A teller's supply of dimes and nickels totals $3.60. If the teller's supply of dimes numbers one more than three times the number of nickels, how many dimes are there?

## Self-Test A

**Solve.**

**1.** $3x + 9x = 48$          **2.** $2a - 4a = 1$      [8–1]

**3.** $-y + 6 - 5y = 18$      **4.** $6 - z - 4z = -4$

**5.** $5(y - 6) = -60$      **6.** $-4(x - 3) - 6x = 10$

**7.** $2(z - 4) - z = 8$      **8.** $-4(n + 3) + 2n = 6$

**9.** $3c = c + 10$      **10.** $9t = 4t - 20$      [8–2]

**11.** $3p - 1 = 8p + 9$      **12.** $3(a + 1) = 19 - 5a$

**13.** $2(p - 1) = 5(p + 2)$      **14.** $-4(t + 2) = 5(t - 3)$

**15.** $3(c + 4) + c = 2(c - 5)$      **16.** $5(x - 3) = 3(x + 5) - 2x$

**17.** A 60 ft long piece of chain link fence is cut into two pieces, one      [8–3]
twice as long as the other. What is the length of the shorter piece?

**18.** The sum of two consecutive integers is 211. What is the larger
integer?

*Self-Test answers and Extra Practice are at the back of the book.*

 **Challenge**

Flanders, Fulton, and Farnsworth teach two subjects each in a small
junior high school. The courses are mathematics, science, carpentry,
music, social studies, and English.

**1.** The carpentry and music teachers are next-door neighbors.

**2.** The science teacher is older than the mathematics teacher.

**3.** The teachers ride together going to school and coming home.
Farnsworth, the science teacher, and the music teacher each drive
one week out of three.

**4.** Flanders is the youngest.

**5.** When they can find another player, the English teacher, the mathematics teacher, and Flanders spend their lunch period playing
bridge.

What subjects does each teach?

## 8-4 Writing Inequalities

In previous chapters, you have learned how to use the *inequality symbols* $<$ (less than) and $>$ (greater than). Sentences written using these symbols are called **inequalities.**

**EXAMPLE**   Write an inequality for each word sentence.

   **a.** Two is less than ten.
   **b.** A number $2 + x$ is greater than a number $t$.
   **c.** A number $n$ is between 6 and 12.

**Solution**   **a.** $2 < 10$
   **b.** $2 + x > t$
   **c.** $6 < n < 12$, or $12 > n > 6$

---

## Class Exercises

**Suppose the numbers given in Exercises 1–6 have been graphed on a number line. Replace the first __?__ with "left" or "right." Replace the second __?__ with $<$ or $>$.**

**1.** The graph of 9 is to the __?__ of the graph of 2.     9 __?__ 2

**2.** The graph of 4 is to the __?__ of the graph of 7.     4 __?__ 7

**3.** The graph of 3 is to the __?__ of the graph of 8.     3 __?__ 8

**4.** The graph of 10 is to the __?__ of the graph of 0.   10 __?__ 0

**5.** The graph of 0 is to the __?__ of the graph of 5.     0 __?__ 5

**6.** The graph of 6 is to the __?__ of the graph of 1.     6 __?__ 1

---

## Written Exercises

**Write an inequality for each word sentence.**

**A**   **1.** Twelve is less than twenty-two.     **2.** Nineteen is greater than nine.

   **3.** Six is greater than zero.     **4.** Twenty-two is less than thirty-three.

   **5.** On a number line eight is between zero and ten.

**B**   **6.** On a number line twenty-five is between fifty-two and five.

Pictured below is a portion of a number line showing the graph of a whole number *m*. Copy this number line and graph each of the following numbers.

**7.** $m + 1$  **8.** $m - 2$  **9.** $m + 2 - 3$  **10.** $m - 2 + 3$

**Write an inequality for each word sentence.**

**11.** Six is greater than a number *t*.

**12.** Twenty-seven is less than a number $3m$.

**13.** A number *p* is greater than a number *q*.

**14.** A number *a* is less than a number *b*.

**15.** The value in cents of *d* dimes is less than the value in cents of *n* nickels.

**16.** The value in cents of *m* pennies is greater than the value in cents of *y* quarters.

**17.** On a number line a number $2n$ is between 6 and 8.

**18.** On a number line a number $r + 1$ is between 10 and 4.

**19.** On a number line a number *a* is between a number *x* and a number *y*, where $x < y$.

**20.** On a number line a number $4x$ is between a number *m* and a number *n*, where $m > n$.

C **21.** On a number line 6 is between 2 and 10, and 20 is between 10 and 50.

**22.** On a number line a number *x* is between 0 and 10, and a number *y* is between 10 and 14.

**23.** On a number line 5 is between a number *a* and a number *b*, where $a < b$, and 8 is also between *a* and *b*.

---

## *Review Exercises*

**Solve using transformations.**

**1.** $x + 4 = 9$  **2.** $x + 11 = 14$  **3.** $x - 3 = 2$  **4.** $x - 6 = -4$

**5.** $4x = 36$  **6.** $-5x = 70$  **7.** $\frac{x}{7} = 11$  **8.** $-\frac{x}{5} = 6$

# 8-5 Equivalent Inequalities

To solve an inequality, we transform the inequality into an **equivalent inequality.** The transformations that we use are similar to those we use to solve equations.

> Simplify numerical expressions and variable expressions.
>
> Add the same number to, or subtract the same number from, both sides of the inequality.
>
> Multiply or divide both sides of the inequality by the same *positive* number.
>
> Multiply or divide both sides of the inequality by the same *negative number and reverse the inequality sign.*

Notice that we do not multiply the sides of an inequality by 0.

It is easy to see that when the inequality $-6 < 5$ is multiplied by 2, the inequality sign does not change. The following shows why we must reverse the inequality sign when multiplying or dividing by a negative number.

$$-6 < 5$$
$$-2(-6) \ ? \ -2(5)$$
$$12 > -10$$

**EXAMPLE 1** Solve each inequality.

    **a.** $x + 7 \leq -18$          **b.** $y - 4.5 > 32$

**Solution**     **a.** Subtract 7 from both sides of the inequality.

$$x + 7 \leq -18$$
$$x + 7 - 7 \leq -18 - 7$$
$$x \leq -25$$

The solutions are all the numbers less than or equal to $-25$.

**b.** Add 4.5 to both sides of the inequality.

$$y - 4.5 > 32$$
$$y - 4.5 + 4.5 > 32 + 4.5$$
$$y > 36.5$$

The solutions are all the numbers greater than 36.5.

**EXAMPLE 2**  Solve $2\frac{1}{4} - \frac{1}{2} \geq n$.

**Solution**  We may exchange the sides of the inequality and *reverse the inequality sign* before we simplify the numerical expression.

$$2\frac{1}{4} - \frac{1}{2} \geq n$$

$$n \leq 2\frac{1}{4} - \frac{1}{2}$$

$$n \leq \frac{7}{4}, \text{ or } 1\frac{3}{4}$$

The solutions are all the numbers less than or equal to $\frac{7}{4}$.

**EXAMPLE 3**  Solve each inequality.

**a.** $7a < 91$  **b.** $-3x \geq 18$  **c.** $26 \leq \frac{y}{4}$  **d.** $\frac{d}{-9} > -108$

**Solution**  **a.** Divide both sides by 7.

$$7a < 91$$

$$\frac{7a}{7} < \frac{91}{7}$$

$$a < 13$$

The solutions are all the numbers less than 13.

**b.** Divide both sides by $-3$ and *reverse the inequality sign.*

$$-3x \geq 18$$

$$\frac{-3x}{-3} \leq \frac{18}{-3}$$

$$x \leq -6$$

The solutions are all the numbers less than or equal to $-6$.

**c.** Multiply both sides by 4.

$$26 \leq \frac{y}{4}$$

$$4 \times 26 \leq 4 \times \frac{y}{4}$$

$$104 \leq y$$

The solutions are all the numbers greater than or equal to 104.

**d.** Multiply both sides by $-9$ and *reverse the inequality sign.*

$$\frac{d}{-9} > -108$$

$$-9 \times \frac{d}{-9} < -9 \times (-108)$$

$$d < 972$$

The solutions are all the numbers less than 972.

## Class Exercises

**Identify the transformation used to transform the first inequality into the second.**

**1.** $e + 5 > 8$

$\quad e + 5 - 5 > 8 - 5$

**2.** $q - 3 < 7$

$\quad q - 3 + 3 < 7 + 3$

**3.** $\frac{1}{2}k < 4$

$\quad 2 \times \frac{1}{2}k < 2 \times 4$

**4.** $6a > 12$

$\quad \frac{6a}{6} > \frac{12}{6}$

**5.** $-\frac{1}{4}u > 3$

$\quad -4 \times -\frac{1}{4}u < -4 \times 3$

**6.** $-9m < 27$

$\quad \frac{-9m}{-9} > \frac{27}{-9}$

**Identify each transformation and complete the equivalent inequality.**

**7.** $n - 6 > 9$

$\quad n - 6 + 6 \underline{\ \ ?\ \ } 9 + 6$

**8.** $d + 4 < -3$

$\quad d + 4 - 4 \underline{\ \ ?\ \ } -3 - 4$

**9.** $3r < 15$

$\quad \frac{3r}{3} \underline{\ \ ?\ \ } \frac{15}{3}$

**10.** $\frac{1}{5}v > 6$

$\quad 5 \times \frac{1}{5}v \underline{\ \ ?\ \ } 5 \times 6$

**11.** $-4z > 32.8$

$\quad \frac{-4z}{-4} \underline{\ \ ?\ \ } \frac{32.8}{-4}$

**12.** $-\frac{1}{3}s < 2\frac{1}{3}$

$\quad -3 \times -\frac{1}{3}s \underline{\ \ ?\ \ } -3 \times \frac{7}{3}$

## Written Exercises

**Use transformations to solve the inequality. Write down all the steps.**

**A**

**1.** $-4 + 16 > k$

**2.** $g < -18 - 9$

**3.** $(17 + 4)2 < j$

**4.** $f \geq 7(23 - 9)$

**5.** $a + 7 < 10$

**6.** $c + 8 > 13$

**7.** $e - 11 \geq 9$

**8.** $m - 5 \leq 15$

**9.** $-16 > n + 2$

**10.** $-13 < q - 6$

**11.** $8w < 56$

**12.** $7y > 42$

**13.** $-5t > 35$

**14.** $-6a < 18$

**15.** $-48 \leq -4b$

**16.** $-49 \geq -7m$

**17.** $\frac{u}{3} \leq 5$

**18.** $\frac{w}{4} \geq 9$

**Use transformations to solve the inequality. Write down all the steps.**

**19.** $\dfrac{f}{-2} > 13$

**20.** $\dfrac{n}{-5} < 12$

**21.** $-10 \le \dfrac{y}{-6}$

**22.** $-9 \ge \dfrac{d}{4}$

**23.** $\dfrac{4h - 2h}{2} \le -6 - 5$

**24.** $\dfrac{16}{4} > \dfrac{7n - 2n}{5}$

**B**  **25.** $-8(13.4 + 7.6) > a$

**26.** $12 \le p - 4.27$

**27.** $3r > 63.9$

**28.** $10\left(3\frac{1}{2} + 4\frac{2}{5}\right) \ge g$

**29.** $j + 3\frac{4}{7} < 9$

**30.** $d + 5\frac{1}{4} > -8$

**31.** $-10 \ge n - 6.13$

**32.** $c < (18.7 - 7.6)4$

**33.** $e \le \left(5\frac{1}{3} + 2\frac{1}{2}\right)6$

**34.** $42.4 < 4t$

**35.** $\dfrac{d}{0.6} \le -18$

**36.** $\frac{1}{5}w \ge \frac{2}{5}$

**37.** $\frac{3}{4} \le \frac{1}{4}n$

**38.** $\dfrac{m}{3.1} \ge 12$

**39.** $-7 > -\frac{1}{3}d$

**40.** $-\frac{1}{2}f \le -13$

**41.** $-6y < 24.6$

**42.** $42.5 > -5b$

**43.** $\dfrac{a}{-13.2} \ge 7$

**44.** $\dfrac{x}{-0.3} < -9$

**45.** $h - 10\frac{2}{3} > 12\frac{1}{5}$

**46.** $-8.27 \le k + 17.41$

**47.** $22.5 \ge -2.5m$

**48.** $3.2u \ge 25.6$

**C**  **49.** $\frac{3}{4}n + \frac{1}{2}n > \frac{2}{3} \times \frac{3}{8}$

**50.** $\dfrac{-5}{8}p - \frac{1}{8}p > \frac{1}{4} \div 7$

**51.** $3.4(21.2 - 18.9) < 4.6(-3b + 2b)$

**52.** $\dfrac{5(-1.6t)}{2} < -14.3 - 9.7$

---

## Review Exercises

**Write the numbers in order from least to greatest.**

**1.** $-1.4,\ 1.2,\ -2.6,\ 3.2$

**2.** $3.7,\ 4.02,\ -3.07,\ -4.1,\ 3$

**3.** $12,\ -12.2,\ -12.09,\ 112,\ -11.2$

**4.** $9.8,\ -8.09,\ 0.89,\ -0.98,\ -9$

**5.** $-5.03,\ 3.05,\ -30.05,\ 0.35,\ -5.3$

**6.** $-0.2,\ -0.02,\ -2,\ -20,\ 200$

**7.** $2.89,\ 2.089,\ -2.8,\ 28.9,\ -2.89$

**8.** $-7.6,\ -0.76,\ -7.06,\ -0.076,\ -76.0$

## 8-6 Solving Inequalities by Several Transformations

We may need to use more than one transformation to solve some inequalities. It is helpful to follow the same steps that we use for solving equations.

> **1.** Simplify each side of the inequality.
>
> **2.** Use the inverse operations to undo any indicated additions or subtractions.
>
> **3.** Use the inverse operations to undo any indicated multiplications or divisions.

**EXAMPLE 1**   Solve $-4r + 6 + 2r - 18 < -8$.

**Solution**

Simplify the left side.
$$-4r + 6 + 2r - 18 < -8$$
$$-2r - 12 < -8$$

Add 12 to both sides.
$$-2r - 12 + 12 < -8 + 12$$
$$-2r < 4$$

Divide both sides by $-2$ and reverse the inequality sign.
$$\frac{-2r}{-2} > \frac{4}{-2}$$
$$r > -2$$

The solutions are all the numbers greater than $-2$.

The solutions to the inequality in Example 1 include all numbers greater than $-2$. We show the graph of the solutions on the number line in the following way.

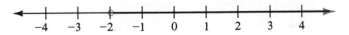

The open dot at $-2$ indicates that $-2$ is not on the graph of the solutions of the inequality.

We use a solid dot to show that a number is on the graph of the solutions. We graph the solutions of $r \geq -2$ in the following way.

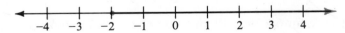

**EXAMPLE 2**   Solve $3(5 + x) \leq -3$ and graph the solutions.

**Solution**   Simplify the left side.

$$3(5 + x) \leq -3$$
$$3(5) + 3x \leq -3$$
$$15 + 3x \leq -3$$

Subtract 15 from both sides.

$$15 + 3x - 15 \leq -3 - 15$$
$$3x \leq -18$$

Divide both sides by 3.

$$\frac{3x}{3} \leq \frac{-18}{3}$$
$$x \leq -6$$

The solutions are all the numbers less than or equal to $-6$.

Because $-6$ is on the graph, we used a solid dot to graph the solutions of the inequality.

Like equations, inequalities may have variables on both sides. We may add a variable expression to both sides or subtract a variable expression from both sides to obtain an equivalent inequality.

**EXAMPLE 3**   Solve $-z + 7 \leq 3z - 18$.

**Solution**   Add $z$ to both sides.

$$-z + 7 \leq 3z - 18$$
$$z + (-z) + 7 \leq z + 3z - 18$$
$$7 \leq 4z - 18$$

Add 18 to both sides.

$$7 + 18 \leq 4z - 18 + 18$$
$$25 \leq 4z$$

Divide both sides by 4.

$$\frac{25}{4} \leq \frac{4z}{4}$$
$$6.25 \leq z$$

The solutions are all the numbers greater than or equal to 6.25.

## Class Exercises

**State which transformations you would use to solve the inequality. State them in the order in which you would use them.**

**1.** $8x + 7 - 5x + 2 \leq 30$

**2.** $-29 > 5 - 2y - 10 - 4y$

**3.** $7 \geq \frac{-4}{5}m + 6 + \frac{3}{5}m - 4$

**4.** $\frac{6}{7}k - 8 - \frac{5}{7}k + 3 < -3$

**5.** $4(h - 3) > 16$

**6.** $-4 \geq \frac{1}{2}(w + 12)$

**7.** $2 < (k + 45)\frac{-1}{5}$

**8.** $(3 - s)7 \leq 42$

**9.** $-u + 8 \geq 4u - 10$

**10.** $a - 6 < -5a - 18$

---

## Written Exercises

**Solve and graph the solutions.**

**A**

**1.** $3q + 5 < -13$

**2.** $-7 + 5m > 28$

**3.** $30 \geq -4b - 6$

**4.** $-57 \leq -8z - 9$

**5.** $-\frac{w}{2} + 8 > 23$

**6.** $-\frac{c}{3} + 7 < -18$

**7.** $-12 \leq -5 + \frac{r}{6}$

**8.** $13 \geq 6 + \frac{e}{4}$

**Solve.**

**9.** $3v + 11 - 8v - 5 < -21$

**10.** $9 - 12u + 4 + 6u > -5$

**11.** $12 \geq \frac{3}{5}f + 6 - \frac{4}{5}f + 4$

**12.** $19 \leq 7 + \frac{1}{3}k + 2 - \frac{2}{3}k$

**13.** $h > -5h + 18 - 6$

**14.** $t < 17 - 4t - 2$

**15.** $38 - 8 - 7n \leq 3n$

**16.** $-9q + 18 - 3 \geq 6q$

**17.** $\frac{1}{4}(20 - x) < 6$

**18.** $\frac{1}{6}(42 - p) > 12$

**19.** $21 \geq 7(m - 2)$

**20.** $-25 \leq 5(w + 3)$

**21.** $j + 8 > 4j - 16$

**22.** $-6f + 7 < 2f - 9$

**23.** $10 + \frac{3}{4}a \leq \frac{-7}{8}a + 5$

**24.** $\frac{5}{6}g - 4 \geq \frac{2}{3}g + 6$

**B 25.** $-12.8c + 8 + 7.8c - 6 \geq 17$

**26.** $4 + 7.5y - 9 + 2.5y > 25$

**Solve.**

**27.** $1\frac{4}{5} < 8m + 4\frac{1}{5} - 3m - 7\frac{2}{5}$

**28.** $-2\frac{2}{3} \le 5\frac{2}{3} - 9u + 4\frac{2}{3} + 3u$

**29.** $0.2(40k + 62.5) \le -3.5$

**30.** $-0.5(12v - 18.6) < 21.3$

**31.** $-2\frac{7}{8} > \left(\frac{5}{8} - t\right)\frac{1}{5}$

**32.** $5\frac{3}{4} \le \frac{1}{3}\left(x - \frac{3}{4}\right)$

**33.** $2.7e + 8.2 < -9.3e + 32.2$

**34.** $5.4 - 10.9w \ge 9.1w - 14.6$

**35.** $\frac{5}{6}f - 2\frac{4}{5} > 1\frac{1}{5} + \frac{2}{3}f$

**36.** $-2\frac{1}{9}d + \frac{3}{5} \le -3\frac{2}{9}d - 4\frac{2}{5}$

**C 37.** $0.11(360 + 25n) < (1.86n \div 6) - 3.06n$

**38.** $\frac{1}{5}\left(w - 5\frac{5}{9}\right) + \frac{2}{9} + \frac{3}{4}w \ge \frac{1}{8}w - 1 + \frac{5}{8}w$

**39.** $[(250.25x - 10.5) \div -5] - 14.6 \le -0.05x + 502.5$

**40.** $9\frac{4}{5} + \frac{2}{7}k - 7\frac{3}{5} > -\frac{11}{12} + \left[\left(\frac{4}{7}k + 2\frac{1}{3}\right) \div \frac{4}{5}\right]$

## Review Exercises

**Use $n$ to write a variable expression for the word phrase.**

**1.** A number increased by nine

**2.** Twice a number

**3.** Sixteen less than a number

**4.** Half of a number

**5.** The sum of a number and three

**6.** Six more than a number

**7.** A number divided by five

**8.** Eight times a number

## Challenge

Each letter in each exercise stands for a digit from 0 through 9. Find the value of each letter to make the computation correct. (There may be more than one correct answer.)

**1.**
```
   H O W
 + A R E
 -------
   Y O U
```

**2.**
```
   T I M E
 + S U R E
 ---------
   F L I E S
```

**3.**
```
   M A T H
 -      I S
 ---------
   F U N
```

# 8-7 Inequalities in Problem Solving

The five-step method is also helpful in solving problems that require inequalities.

**EXAMPLE** The sum of two consecutive integers is less than 40. What pair of integers with this property has the greatest sum?

**Solution**
- The problem says:
  the sum of two consecutive integers is less than 40

- The problem asks for:
  the greatest pair of integers whose sum is less than 40
  Let $n$ = the smaller of the two integers.
  Then the other integer $= n + 1$.
  The sum of the two integers is $n + n + 1$.
  The sum of the two integers is less than 40.

- We now have enough information to write an inequality.

$$n + n + 1 < 40$$

- Solve.

$$n + n + 1 < 40$$
$$2n + 1 < 40$$
$$2n + 1 - 1 < 40 - 1$$
$$2n < 39$$
$$\frac{2n}{2} < \frac{39}{2}$$
$$n < 19.5$$
$$n = 19$$
$$n + 1 = 20$$

- Check: $19 + 20 = 39$, $39 < 40$. The result checks.

  The two integers are 19 and 20.

---

## Class Exercises

**Using $x$ as the variable, write an inequality based on the given information. Do not solve.**

1. A used-car dealer has sold 30 of his compact cars and now has fewer than 70 left.

2. Alejandro Mendoza, who had more than 60 base hits during his school team's baseball season, had 25 more hits than his teammate Peter Evans.

**3.** Deborah's bowling score was 8 less than half of Lydia's score. Deborah's score was less than 95.

**4.** A house and lot together cost more than $120,000. The cost of the house was $2000 more than six times the cost of the lot.

**5.** The number of students at Ivytown High School who study computer science is twice the number who study home economics. The total number of students enrolled in these courses exceeds 600.

**6.** The sum of two consecutive integers is greater than 30.

## Problems

**Solve.**

A  **1.** The sum of two consecutive integers is less than 75. Find the pair of integers with the greatest sum.

**2.** Of all pairs of consecutive integers whose sum is greater than 100, find the pair whose sum is the least.

**3.** Two trucks start from the same point traveling in different directions. One truck travels at a speed of 54 mi/h, the other at 48 mi/h. How long must they travel to be at least 408 mi apart?

**4.** A purse contains 30 coins, all either quarters or dimes. The total value of the coins is greater than $5.20. At least how many of the coins are quarters? At most how many are dimes?

**5.** A home and adjoining lot cost more than $160,000 together. The cost of the house was $1000 more than six times the cost of the lot. What is the smallest possible cost of the lot?

**6.** Paul is two fifths as old as Janet. Five years from now, he will be at least half as old as Janet. At most how old is Paul now?

B  **7.** The sum of three consecutive integers, decreased by five, is greater than twice the smallest of the integers. What are the three least positive integers with this property?

**8.** A pair of consecutive integers has the property that five times the smaller is less than four times the greater. Find the greatest pair of integers with this property.

**9.** The number of Software Services employees who use public transportation to commute to work is 200 more than twice the number who drive their own cars. If there are 1400 employees, how many drive their own cars?

**C** **10.** Bonanza Rent-A-Car rents cars for $40 per day and 10¢ for every mile driven. Autos Unlimited rents cars for $50 per day with no extra charge for mileage. How many miles per day can you drive a Bonanza car if it is to cost you less than an Auto Unlimited car?

**11.** A bank offers two types of checking accounts. Account A has a $4 maintenance fee each month and charges 10¢ for each check cashed. Account B has a $6 maintenance fee and charges 6¢ for each check. What is the least number of checks that can be written each month for the "A" account to be more expensive?

## Self-Test B

**Write an inequality for each word sentence.**

**1.** A number $a$ is less than a number $b$. [8-4]

**2.** Thirty-five is greater than a number $4z$.

**Use transformations to solve the inequality.**

**3.** $m - 8 \leq 20$      **4.** $t + 3 > 6$      **5.** $c + 7 \geq 14$      [8-5]

**6.** $9p > 42$      **7.** $-4b < 56$      **8.** $-7x < -35$

**Solve.**

**9.** $2x + 4 + 6x - 2 < 18$      **10.** $12 - 6u + 4 + 8u > 20$      [8-6]

**11.** $\frac{1}{2}(30 - 6y) \geq 15$      **12.** $-30 \leq 6(t + 1)$

**13.** The sum of two consecutive integers is less than 150. Find the greatest pair of integers with this property. [8-7]

███ **Challenge**

Using each of the digits 1, 3, 5, 7, and 9 exactly once, create a three-digit integer and a two-digit integer. The integers may be either positive or negative.

**1.** What is the greatest possible sum of the two integers? the least possible sum?

**2.** What is the greatest possible difference between the two integers? the least possible difference?

**3.** What is the greatest possible product of the two integers? the least possible product?

# People in Mathematics

## Archimedes (287–212 B.C.)

Archimedes is considered to be one of the greatest mathematicians of all time. He was a native of the Greek city of Syracuse, although he did spend some time at the University of Alexandria in Egypt.

Archimedes is the subject of many stories and legends. The most famous story about Archimedes is that of King Hieron's crown. The crown was supposedly all gold, but the king suspected that it contained silver and he asked Archimedes to determine whether it was pure gold. Archimedes hit upon the solution, while bathing, by discovering the first law of hydrostatics. The story relates that he jumped from the bath and ran through the streets shouting "Eureka."

Archimedes discovered many important mathematical facts. He found formulas for the volumes and surface areas of many geometric solids. He invented a method for approximating $\pi$ and studied spirals, one of which bears his name. His work, as shown by a paper not found until 1906, even contained the beginning of calculus, a branch of mathematics not developed until the seventeenth century.

## Sonya Kovaleski (1850–1891)

Sonya Kovaleski was one of the great mathematicians of the nineteenth century. Her work includes papers on such diverse topics as partial differential equations (a theorem is named in her honor), Abelian integrals, and the rings of Saturn. Her other work included research in the topics of analysis and physics. In 1888 her research paper entitled *On the Rotation of a Solid Body about a Fixed Point* was awarded the Prix Bordin of the French Academy of Sciences. It was considered so outstanding that the prize of 3000 francs was doubled.

Kovaleski's rise to prominence was far from easy. In order to leave Russia to study at a foreign university, she had to arrange a marriage, at age eighteen, to Vladimir Kovaleski. In 1868 they went to Heidelberg where she studied with Kirchhoff and Helmholtz, two famous physicists. In 1871 she went to Berlin to study with Karl Weierstrass. Since women were not admitted to university lectures, all her studying was done privately. Finally, in 1874, the University of Gottingen awarded her a doctorate *in absentia*. However, she was unable to find an academic position for ten years despite strong letters of recommendation from Weierstrass. Finally, in 1884, she was appointed as a lecturer at the University of Stockholm where she was made a full professor five years later.

## Emmy Noether (1882–1935)

Emmy Noether is considered one of the brilliant mathematicians of the twentieth century. Her most important work was done in the field of advanced algebra, a branch of mathematics that deals with structures called "groups" and "rings." An important theorem in advanced algebra is called the Noether-Lasker Decomposition Theorem. Noether is also known for work she did on Einstein's theory of relativity.

Although her father was a mathematician, Noether faced many of the same obstacles as Sonya Kovaleski. She sat in on courses at the University of Erlangen (shown in the photograph) and the University of Gottingen from 1900 to 1903, but was not allowed to officially enroll until 1904 when Erlangen changed its policy toward women. She received her doctorate in 1907.

In 1915 Noether was invited to Gottingen by David Hilbert. There, she worked with Hilbert and Felix Klein, two prominent mathematicians, although she was not appointed to the faculty until 1922. She left Gottingen in 1933 and took a professorship at Bryn Mawr College in Pennsylvania.

## Research Activities

1. Look up the statement of the law of physics known as Archimedes' principle. Using a measuring cup, devise a simple experiment to verify this law.

2. Archimedes discovered a way to calculate the number $\pi$ very accurately. Look up the history of $\pi$ in an encyclopedia. Find what values ancient civilizations thought it had.

3. Two other people considered important in the history of mathematics are Hypatia and Maria Agnesi. Find out about the lives and work of these mathematicians.

# Chapter Review

**True or false?**

1. If $2y + 8y = -20$, $y = 2$.

2. If $p + 6 - 6p = -8$, $p = -3$.    [8-1]

3. If $4(n - 3) = 6$, $n = 4\frac{1}{2}$.

4. If $-4(a + 3) = 15$, $a = \frac{3}{4}$.

5. If $5z + 4 = 3z + 6$, $z = 1$.

6. If $6x - 11 = 12x - 7$, $x = 4$.    [8-2]

7. If $2(x - 4) = 5(x + 3)$, $x = 20$.

8. If $4(y + 7) = 3(y - 6)$, $y = -2$.

**Write the letter of the correct answer.**

9. The perimeter of an isosceles triangle is 63 cm. If the congruent    [8-3]
   sides of the triangle are each three times as long as the remaining
   side, how long are the congruent sides?
   **a.** 7 cm        **b.** 27 cm        **c.** 9 cm        **d.** 18 cm

10. A 60 ft board with a thickness of 1 in. is to be cut into two pieces,
    one three times as long as the other. Find the lengths of the two
    pieces.
    **a.** 30 ft, 30 ft    **b.** 40 ft, 20 ft    **c.** 45 ft, 15 ft    **d.** 48 ft, 12 ft

11. Which inequality represents the word sentence? Thirty is greater    [8-4]
    than a number $x$.
    **a.** $30 < x$        **b.** $30 \leq x$        **c.** $30 > x$        **d.** $30 \geq x$

**Match the equivalent inequalities.**

12. $-8 + 12 < x$    13. $6x > 54$        **A.** $x \leq 8$    **B.** $x > 8$    [8-5]

14. $-64 \leq -8x$    15. $2x < 8$        **C.** $x > 4$    **D.** $x > 9$

16. $\frac{1}{2}x + 6 < 26$    17. $\frac{x}{-5} + 5 \geq 10$    **E.** $x \leq -25$    **F.** $x < 40$    [8-6]

18. $-17 \leq -5 + \frac{x}{3}$    19. $-\frac{x}{6} + 5 > -3$    **G.** $x < 48$    **H.** $x \geq -36$

**Write the letter of the correct answer.**

20. The sum of two consecutive integers is greater than 60. Find the    [8-7]
    least pair of integers having this property.
    **a.** 29, 30        **b.** 20, 21        **c.** 30, 31        **d.** 31, 32

21. NewBank offers two checking accounts. Account A has a monthly
    fee of $2 and charges 15¢ for each check cashed. Account B has a
    monthly fee of $5 and charges 10¢ for each check cashed. What is
    the greatest number of checks that can be cashed for Account A to
    be less expensive than Account B?
    **a.** 60        **b.** 59        **c.** 61        **d.** 20

# Chapter Test

**Solve.**

1. $9a + 6a = 120$                 2. $-6z - (-4z) = -5$     [8-1]

3. $4 + (-4a) + 20 = 16$        4. $-7n - 6 + 8n = -6$

5. $4(x - 5) = 10$               6. $8(y + 5) = -20$

7. $5y + 4 = 6y - 3$          8. $6(x + 4) = -2x$         [8-2]

9. $-3(a + 1) = 6a + 12$     10. $6(p + 1) = 4(p + 2)$

11. $-3(n - 3) = 2n + 5$      12. $4(y + 2) = -6y + 8$

**Use an equation to solve.**

13. The sum of two numbers is 100. One number is 24 less than the     [8-3]
other number. What is the larger number?

**Write an inequality for each word sentence.**

14. Sixty-eight is less than eighty.                               [8-4]

15. Nineteen is between fourteen and thirty.

16. Zero is less than a number $y$.

17. A number $z$ is greater than a number $r$.

**Solve.**

18. $a + 12 < 19$       19. $\frac{m}{4} > -8$         20. $-3y \leq 42$       [8-5]

21. $\frac{w}{-6} \geq 9$           22. $75 < 5p$            23. $27 \geq n - 16$

**Solve and graph the solutions.**

24. $\frac{1}{4}d - 2 \leq 18$                 25. $-35 \leq -5(r + 8)$       [8-6]

26. $\frac{a}{9} + 17 \geq 20$                27. $8x - 4 < 5x + 23$

**Solve.**

28. Two trucks start from the same point traveling in opposite direc-     [8-7]
tions. One truck travels at a speed of 50 mi/h, the other travels at a
speed of 45 mi/h. How long must they travel to be 380 mi apart?

# Cumulative Review (Chapters 1–8)

## Exercises

Evaluate the expression when $a = 2$, $b = -4$, and $c = 3$.

**1.** $a + b - c$

**2.** $c - b + a$

**3.** $3ac$

**4.** $-2b$

**5.** $4c - 5a$

**6.** $10a - 3b$

**7.** $8c \div (-3b)$

**8.** $50 \div (a + c)$

**9.** $a^2$

**10.** $b^2$

**11.** $(-b)^3$

**12.** $(ab)^2$

Solve.

**13.** $x + 9 = 12$

**14.** $x - 4 = -11$

**15.** $2x = 14$

**16.** $-5x = 35$

**17.** $3x + 7 = 22$

**18.** $-7x + 3 = -32$

**19.** $2x + 3x = 20$

**20.** $10x - 18x = 4$

**21.** $-7x - 9x + 7 = 11$

**22.** $5x = 30 - 10x$

**23.** $7x - 3 = 3x + 9$

**24.** $x - 14 = 26 - 9x$

Write as equivalent fractions using the LCD.

**25.** $\frac{1}{2}, \frac{7}{8}$

**26.** $\frac{3}{5}, \frac{9}{20}$

**27.** $\frac{1}{5}, \frac{3}{4}$

**28.** $\frac{2}{7}, \frac{5}{9}$

**29.** $\frac{2}{3}, \frac{1}{4}, \frac{5}{12}$

**30.** $\frac{21}{24}, \frac{5}{8}, \frac{7}{9}$

Complete.

**31.** Two nonparallel lines that do not intersect are called ___?___ lines.

**32.** In a ___?___ triangle, no two sides are congruent.

**33.** The number $\pi$ is the ratio of the circumference of a circle to its ___?___.

**34.** The ___?___ of a figure is the distance around it.

**35.** The common endpoint of the two rays of an angle is called the ___?___.

**36.** A ___?___ of a polygon is a segment joining two nonconsecutive vertices.

Solve.

**37.** What percent of 60 is 55?

**38.** What is 81% of 120?

**39.** 30 is 15% of what number?

**40.** What percent of 100 is 49?

# Problems

**Solve.**

1. Donald bought a pair of hiking boots for $35.83, a sweater for $24.65, and a backpack for $18. The tax on his purchase was $.90. How much did Donald spend?

2. This week, Marisa worked $1\frac{1}{2}$ h on Monday, $2\frac{1}{4}$ h on Tuesday, $1\frac{1}{3}$ h on Wednesday, and $7\frac{1}{2}$ h on Saturday. How many hours did she work this week?

3. An airplane flying at an altitude of 25,000 ft dropped 4000 ft in the first 25 s and rose 2500 ft in the next 15 s. What was the altitude of the airplane after 40 s?

4. The difference between twice a number and 50 is 10. What is the number?

5. At a milk processing plant 100 lb of farm milk are needed to make 8.13 lb of nonfat dry milk. To the nearest pound, how many pounds of farm milk are needed to produce 100 lb of nonfat dry milk?

6. A highway noise barrier that is 120 m long is constructed in two pieces. One piece is 45 m longer than the other. Find the length of each piece.

7. A side of a square is 13 cm long. What is the perimeter of the square?

8. Oak Hill School, Longview School, and Peabody School participated in a clean-up campaign to collect scrap aluminum. Oak Hill School collected 40% more scrap aluminum than Longview School. Longview School collected 25% more than Peabody School. If the total collected by the 3 schools was 560 kg, how much did Oak Hill School collect?

9. Eladio invested $200 at 6% annual interest and $350 at 5.75% annual interest, both compounded annually. If he makes no deposits or withdrawals, how much will he have after two years?

# 9

# The Coordinate Plane

One of the most fascinating aspects of computer science is the field of computer-aided design (C.A.D.). Computers are used to design a wide variety of goods ranging from televisions and cars to skyscrapers, airplanes, and spacecraft. Computers are also used to generate designs for decorative purposes, such as the rug pattern shown at the right. A particular advantage of using the computer is the ease with which the design can be prepared and modified. The designer may use a light pen to create the design on the computer screen and may type in commands to tell the computer to enlarge, reduce, or rotate the design. Although the actual programs for computer graphics are often complicated, the underlying idea is based on establishing a grid with labeled reference points. The grid is actually an application of the coordinate plane that is presented in this chapter.

## Career Note

If you enjoy working with fabrics and drawing your own patterns, you might consider a career as a textile designer. Textile designers create the graphic designs printed or woven on all kinds of fabrics. In addition to having a good understanding of graphics, textile designers must possess a knowledge of textile production. Another important part of textile design is the ability to appeal to current tastes.

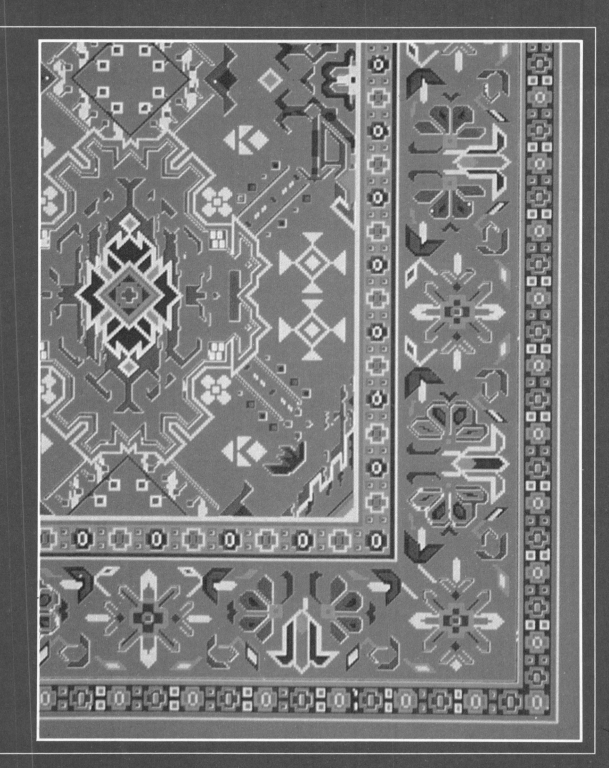

# 9-1 The Coordinate Plane

We describe the position of a point on a number line by stating its coordinate. Similarly, we can describe the position of a point in a plane by stating a pair of coordinates that locate it in a **rectangular coordinate system.** This system consists of two number lines perpendicular to each other at point $O$, called the **origin.** The horizontal line is called the **x-axis,** and the vertical line is called the **y-axis.** The positive direction is to the right of the origin on the $x$-axis and upward on the $y$-axis. The negative direction is to the left of the origin and downward.

> **Reading Mathematics:** *Diagrams*
> As you read text that is next to a diagram, stop after each sentence and relate what you have read to what you see in the diagram. For example, as you read the first paragraph next to the diagram below, locate point $M$ on the graph, and find the vertical line from $M$ to the $x$-axis.

To assign a pair of coordinates to point $M$ located in the **coordinate plane,** first draw a vertical line from $M$ to the $x$-axis. The point of intersection on the $x$-axis is called the **x-coordinate,** or **abscissa.** The abscissa of $M$ is 5.

Next draw a horizontal line from $M$ to the $y$-axis. The point of intersection on the $y$-axis is called the **y-coordinate,** or **ordinate.** The ordinate of $M$ is 4.

Together, the abscissa and ordinate form an **ordered pair of numbers** that are the coordinates of a point. The coordinates of $M$ are (5, 4).

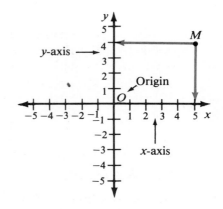

**EXAMPLE 1** Give the coordinates of each point.
      **a.** $F$   **b.** $S$   **c.** $P$   **d.** $T$

**Solution**     **a.** Point $F$ has $x$-coordinate 4 and $y$-coordinate $-3$. $F$ has coordinates $(4, -3)$.

        Similarly,

        **b.** $S$ has coordinates (0, 2).

        **c.** $P$ has coordinates $(-3, 4)$.

        **d.** $T$ has coordinates $(-5, -5)$.

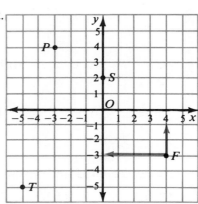

Notice that $(4, -3)$ and $(-3, 4)$ are coordinates of different points. In an ordered pair of numbers the $x$-coordinate is listed first, followed by the $y$-coordinate, in the form **(x, y)**.

Just as each point in the plane is associated with exactly one ordered pair, each ordered pair of numbers determines exactly one point in the plane.

**EXAMPLE 2**  Graph these ordered pairs:
$(-6, 5)$, $(5, -4)$, $(0, 3)$, $(-3, -2)$, and $\left(2\frac{1}{2}, 1\right)$.

*Solution*  First locate the $x$-coordinate on the $x$-axis. Then move up or down to locate the $y$-coordinate.

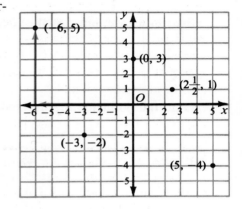

The $x$- and $y$-axes divide the coordinate plane into **Quadrants I, II, III,** and **IV.** The ranges of values for the $x$-coordinate and $y$-coordinate of any point in each quadrant are shown at the right.

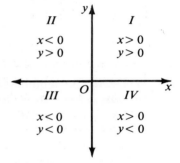

## Class Exercises

**Give the coordinates of the point.**

| | | |
|---|---|---|
| **1.** $A$ | **2.** $K$ | **3.** $E$ |
| **4.** $J$ | **5.** $L$ | **6.** $F$ |

**Name the point for the ordered pair.**

| | |
|---|---|
| **7.** $(5, 2)$ | **8.** $(-6, 0)$ |
| **9.** $(4, -5)$ | **10.** $(-4, -2)$ |
| **11.** $(-3, 5)$ | **12.** $(-6, 3)$ |

**Name the quadrant containing the point.**

| | | | |
|---|---|---|---|
| **13.** $F$ | **14.** $D$ | **15.** $H$ | **16.** $J$ | **17.** $K$ |

## Written Exercises

**For Exercises 1–18, use the graph at the right.**

**Give the coordinates of the point.**

**A**  **1.** $N$    **2.** $P$    **3.** $T$    **4.** $R$

**5.** $Q$    **6.** $M$    **7.** $S$    **8.** $V$

**Name the point for the ordered pair.**

**9.** $(4, -5)$               **10.** $(-3, -2)$

**11.** $(-5, 1)$              **12.** $\left(\frac{5}{2}, -3\right)$

**13.** $\left(-\frac{7}{2}, 3\right)$        **14.** $(1, -2)$        **15.** $\left(\frac{3}{2}, 1\right)$

**16.** $(-1, 4)$             **17.** $(0, -6)$        **18.** $(0, 3)$

**a. Graph the given ordered pairs on a coordinate plane.**
**b. Draw line segments to connect the points in the order listed and to connect the first and last points.**
**c. Name the closed figure as specifically as you can.**

*EXAMPLE*    $(-3, -2), (-1, 2), (4, 2), (2, -2)$

*Solution*    **a.** 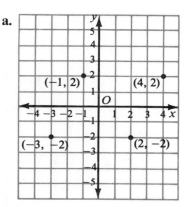    **b.**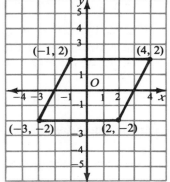

**c.** parallelogram

**B**  **19.** $(1, -2), (3, -2), (3, 4), (1, 4)$        **20.** $(0, 1), (3, -2), (6, 1), (3, 4)$

**21.** $(0, -5), (0, 2), (-3, 6), (-3, -1)$        **22.** $(-5, 0), (-1, 2), (1, 6), (-3, 4)$

We can translate a figure on a coordinate plane by changing the ordered pairs. We can reflect a figure on a coordinate plane by changing the signs of the ordered pairs.

**a. Graph the given ordered pairs on a coordinate plane, connect the points in the order listed, and connect the first and last points.**
**b. Change the values of the coordinates as directed.**
**c. Graph and connect all the new points to form a second figure.**
**d. Identify the change as translation or reflection.**

C **23.** $(-5, 4)$, $(-2, 1)$, $(2, 1)$, $(2, 4)$
Decrease all $y$-coordinates by 3.
(*Hint:* $(-5, 4)$ becomes $(-5, 1)$)

**24.** $(1, 5)$, $(1, 1)$, $(5, 3)$
Decrease all $x$-coordinates by 2.

**25.** $(-2, -2)$, $(2, -2)$, $(4, -4)$, $(4, -6)$, $(1, -4)$, $(-1, -4)$, $(-4, -6)$, $(-4, -4)$
Increase all $y$-coordinates by 5.

**26.** $(-8, 4)$, $(-4, 4)$, $(-4, 5)$, $(-1, 3)$, $(-4, 1)$, $(-4, 2)$, $(-8, 2)$
Increase all $x$-coordinates by 9.

**27.** $(6, 5)$, $(2, 1)$, $(9, 3)$
Multiply all $x$-coordinates by $-1$.

**28.** $(1, 7)$, $(4, 2)$, $(7, 3)$, $(4, 8)$
Multiply all $y$-coordinates by $-1$.

## Review Exercises

**Complete.**

**1.** If $x + 5 = 13$, then $x = $ _?_ .

**2.** If $8 + x = -1$, then $x = $ _?_ .

**3.** If $9x = 243$, then $x = $ _?_ .

**4.** If $-4x = 52$, then $x = $ _?_ .

**5.** If $2x + 11 = 1$, then $x = $ _?_ .

**6.** If $99 - 9x = 18$, then $x = $ _?_ .

**7.** If $-\frac{4}{5}x = 12$, then $x = $ _?_ .

**8.** If $9 + \frac{2}{3}x = 15$, then $x = $ _?_ .

### ▮▮▮▮ Challenge

Set up a pair of coordinate axes on graph paper. Connect the following points in the order given:

$$(-9, 3), (-4, 3), (-1.5, 1), (2, -1),$$
$$(2, -4), (8.5, -5), (9.5, -1), (2, -1)$$

Do you recognize the figure? (*Hint:* It is a well-known group of stars that is part of the constellation *Ursa Major*.)

# 9-2 Equations in Two Variables

The equation

$$x + y = 5$$

has two variables, $x$ and $y$. A solution to this equation consists of two numbers, one for each variable. The solution can be expressed as an ordered pair of numbers, $(x, y)$. There are many ordered pairs that satisfy this equation. Some solutions are

$$(-3, 8), (5, 0), (4, 1), (0, 5), (11, -6).$$

In fact, there are *infinitely* many ordered pairs that satisfy this equation.

**EXAMPLE 1**   Tell whether each ordered pair is a solution of the equation $2x + y = 7$.

    **a.** $(4, -1)$                       **b.** $(-4, 1)$

**Solution**   Substitute the given values of $x$ and $y$ in the equation.

    **a.**           $2x + y = 7$         **b.**           $2x + y = 7$

$$2(4) + (-1) \overset{?}{=} 7 \qquad\qquad 2(-4) + 1 \overset{?}{=} 7$$

$$8 - 1 \overset{?}{=} 7 \qquad\qquad\qquad -8 + 1 \overset{?}{=} 7$$

$$7 = 7 \qquad\qquad\qquad\qquad -7 \neq 7$$

                 $(4, -1)$ is a solution.               $(-4, 1)$ is not a solution.

An equation in the two variables $x$ and $y$ establishes a correspondence between values of $x$ and values of $y$. To find a solution of a given equation in $x$ and $y$, we can choose any value for $x$, substitute it in the equation, and solve for the corresponding value for $y$.

**EXAMPLE 2**   Give one solution of the equation $x - 3y = 2$.

**Solution**   Choose any value for $x$. For example, if $x = 14$:

$$x - 3y = 2$$
$$14 - 3y = 2$$
$$-3y = 2 - 14$$
$$y = \frac{-12}{-3} = 4$$

The values $x = 14$ and $y = 4$ correspond.

$(14, 4)$ is one solution of the equation.

To find the value of $y$ corresponding to any given value of $x$, we could substitute the value of $x$ into a given equation and solve for $y$, as in Example 2. An easier method is to solve for $y$ in terms of $x$ first, and then substitute, as in Example 3.

**EXAMPLE 3**  Find the solutions for $2x + 3y = 6$ for the following values of $x$: $-9$, $-3$, $0$, $3$, $6$.

**Solution**  First solve for $y$ in terms of $x$ by writing an equation with $y$ on one side and $x$ on the other.

$$2x + 3y = 6$$
$$3y = 6 - 2x$$
$$y = 2 - \frac{2}{3}x$$

Then substitute the values of $x$ in the new equation and solve for the corresponding values of $y$.

| $x$ | $y = 2 - \frac{2}{3}x$ | $(x, y)$ |
|---|---|---|
| $-9$ | $2 - \frac{2}{3}(-9) = 8$ | $(-9, 8)$ |
| $-3$ | $2 - \frac{2}{3}(-3) = 4$ | $(-3, 4)$ |
| $0$ | $2 - \frac{2}{3}(0) = 2$ | $(0, 2)$ |
| $3$ | $2 - \frac{2}{3}(3) = 0$ | $(3, 0)$ |
| $6$ | $2 - \frac{2}{3}(6) = -2$ | $(6, -2)$ |

Thus $(-9, 8)$, $(-3, 4)$, $(0, 2)$, $(3, 0)$, and $(6, -2)$ are solutions of the equation for the given values of $x$.

Notice in the table above that each given value of $x$ corresponds to exactly one value of $y$. In general, any set of ordered pairs such that no two different ordered pairs have the same $x$-coordinate is called a **function.** An equation, as the one above, that produces such a set of ordered pairs defines a function.

## Class Exercises

**Tell whether the ordered pair is a solution of the given equation.**

$2x + y = 7$
  **1.** (2, 3)              **2.** (1, 5)              **3.** (7, 0)              **4.** (0,7)

$x - 3y = 1$
  **5.** (2, 1)              **6.** (4, 1)              **7.** (7, 2)              **8.** (−2, 1)

**Solve the equation for $y$ in terms of $x$.**
  **9.** $x - y = 5$              **10.** $x - y = 9$              **11.** $-3x + y = 7$

**If $y = x - 12$, give the value of $y$ such that the ordered pair is a solution of the equation.**

**12.** (9, ?)              **13.** (−4, ?)              **14.** (0, ?)              **15.** (−6, ?)

---

## Written Exercises

**Tell whether the ordered pair is a solution of the given equation.**

**A**   $-x + 3y = 15$
  **1.** (0, 5)              **2.** (6, −7)              **3.** (6, 7)              **4.** (−3, 6)

$x - 4y = 12$
  **5.** (4, 2)              **6.** (−6, −12)              **7.** (0, 3)              **8.** $\left(13, \frac{1}{4}\right)$

$3x - 2y = 8$
  **9.** (1, 6)              **10.** (0, 4)              **11.** (4, 0)              **12.** $\left(3, \frac{1}{2}\right)$

$-5x - 2y = 18$
  **13.** (−2, −4)              **14.** (−4, 1)              **15.** $\left(1, -\frac{23}{2}\right)$              **16.** (0, 0)

**Solve the equation for $y$ in terms of $x$.**

**17.** $2x + y = 7$              **18.** $-8x + 5y = 10$              **19.** $-12x + 3y = 48$

**20.** $4x - 9y = 36$              **21.** $x - \frac{3}{2}y = 3$              **22.** $5x - 4y = 20$

**a. Solve the equation for $y$ in terms of $x$.**
**b. Find the solutions of the equation for the given values of $x$.**

**23.** $y - x = 7$
    values of $x$: 2, −5, 7

**24.** $x + y = -1$
    values of $x$: 3, 1, −2

**25.** $-x + 2y = 10$
values of $x$: 4, 0, 6

**26.** $x - 3y = 12$
values of $x$: $-3$, 6, 12

**27.** $x + 4y = 20$
values of $x$: 4, $-8$, 0

**28.** $-x - 2y = 8$
values of $x$: $-6$, $-2$, 10

**B 29.** $3x - 2y = 6$
values of $x$: 4, $-2$, $-8$

**30.** $-2x + 3y = 12$
values of $x$: 3, $-3$, $-12$

**31.** $2y - x = 5$
values of $x$: 3, 5, $-1$

**32.** $-3y + x = 7$
values of $x$: 1, 7, $-5$

**33.** $2y - 5x = 6$
values of $x$: 2, 1, $-3$

**34.** $4y - 3x = 1$
values of $x$: 9, $-1$, $-5$

**Give any three ordered pairs that are solutions of the equation.**

**35.** $y + x = -1$

**36.** $x - y = 6$

**37.** $x + y = 0$

**38.** $4y - x = 7$

**39.** $5x - y = -5$

**40.** $-3x - 2y = 0$

**41.** $2x - 3y = 12$

**42.** $-\frac{1}{4}y - x = 0$

**43.** $-x + \frac{1}{2}y = 3$

**Write an equation that expresses a relationship between the coordinates of each ordered pair.**

*EXAMPLE*    (1, 4), (2, 3), (5, 0), (−1, 6)

*Solution*    Notice that all the coordinates share a common property: the sum of the coordinates in each pair is 5. You can express this relationship by the equation $x + y = 5$.

**C 44.** (7, 6), (11, 10), (4, 3)

**45.** (3, −3), (6, −6), (−4, 4)

**46.** (−2, −4), (−3, −5), (4, 2)

**47.** (3, −6), (−3, 0), (6, −9)

---

## Review Exercises

**Solve.**

**1.** $8a = 24$

**2.** $6 - 5m = 11$

**3.** $6t - 31 = 137$

**4.** $-7x + 4 = -10$

**5.** $9c = 18 + 3c$

**6.** $\frac{4}{5}y = 16$

**7.** $2 + \frac{7}{8}d = -47$

**8.** $-10 - \frac{2}{3}f = 12$

**9.** $5(8 + 2h) = 0$

# 9-3 Graphing Equations in the Coordinate Plane

The equation in two variables

$$x + 2y = 6$$

has infinitely many solutions. The **table of values** below lists some of the solutions. When we graph the solutions on a coordinate plane, we find that they all lie on a straight line.

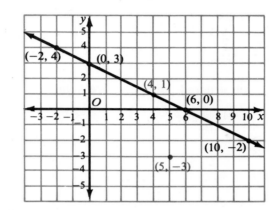

| $x + 2y = 6$ | |
|:---:|:---:|
| $x$ | $y$ |
| $-2$ | $4$ |
| $0$ | $3$ |
| $6$ | $0$ |
| $10$ | $-2$ |

If we choose any other point on this line, we will find that the coordinates also satisfy the equation. For example, the ordered pair (4, 1) is a solution:

$$x + 2y = 6$$
$$4 + 2(1) = 6$$

If we choose a point *not* on this line, such as the graph of (5, −3), we find that its coordinates *do not* satisfy the equation:

$$x + 2y = 6$$
$$5 + 2(-3) \neq 6$$

The graph of an ordered pair is on the line if and only if it is a solution of the equation. The set of all points that are the graphs of solutions of a given equation is called the **graph of the equation.**

Note that the graph of $x + 2y = 6$ crosses the $y$-axis at (0, 3). The $y$-coordinate of a point where a graph crosses the $y$-axis is called the **$y$-intercept** of the graph. In this case, the $y$-intercept is 3. Since the graph also crosses the $x$-axis at (6, 0), the **$x$-intercept** of the graph is 6.

In general, any equation that can be written in the form

$$ax + by = c$$

where $x$ and $y$ are variables and $a$, $b$, and $c$ are numbers (with $a$ and $b$

not both zero), is called a **linear equation in two variables** because its graph is always a straight line in the plane.

In order to graph a linear equation, we need to graph only two points whose coordinates satisfy the equation and then join them by means of a line. It is wise, however, to graph a third point as a check.

*EXAMPLE*   Graph the equation $x + 3y = 6$.

*Solution*   First find three points whose coordinates satisfy the equation. It is usually easier to start with the $y$-intercept and the $x$-intercept, that is, $(0, y)$ and $(x, 0)$.

If $x = 0$:   $0 + 3y = 6$          If $y = 0$:   $x + 3(0) = 6$
$3y = 6$                                        $x = 6$
$y = 2$

As a check, if $x = 3$:   $3 + 3y = 6$
$3y = 3$
$y = 1$

$x + 3y = 6$

| $x$ | $y$ |
|-----|-----|
| 0   | 2   |
| 6   | 0   |
| 3   | 1   |

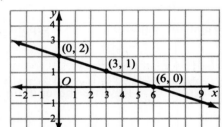

---

## Class Exercises

**Find three solutions of the linear equation. Include those that are in the form $(0, y)$ and $(x, 0)$.**

**1.** $x - y = 7$          **2.** $y + x = -4$          **3.** $-x - 2y = 8$

**4.** $3y - x = 9$          **5.** $2x + y = 10$          **6.** $3x - 2y = 3$

**7.** $y = x$          **8.** $y = 3x$          **9.** $y = \frac{1}{2}x$

---

## Written Exercises

**A  1–9.** Graph each equation in Class Exercises 1–9. Use a separate set of coordinate axes for each equation.

**Graph the equation on a coordinate plane. Use a separate set of axes for each equation.**

10. $x + y = 5$

11. $x - 6 = y$

12. $y - x = 4$

13. $y + x = -3$

14. $2x - y = 10$

15. $8x + 2y = 8$

16. $3y + x = 0$

17. $3x - y = -6$

18. $-2x + 3y = 12$

19. $2y - \frac{1}{2}x = 4$

20. $\frac{1}{3}y + 2x = 3$

21. $4y - 3x = 12$

22. $4x - 3y = 6$

23. $-2x + 5y = 5$

24. $\frac{1}{2}y + 2x = 3$

**B** 25. $\frac{x + y}{3} = 2$

26. $\frac{x - y}{4} = -1$

27. $\frac{3x - y}{2} = 3$

28. $\frac{2x + y}{3} = 1$

29. $\frac{x - 2y}{3} = -4$

30. $\frac{x + 3}{4} = y$

31. $\frac{3x - 1}{2} = y$

32. $\frac{-3y + 2x}{4} = -2$

33. $\frac{-4x + 3y}{6} = -4$

**Graph the equation. Use a separate set of axes for each equation.**

*EXAMPLE*   $y = 6$

*Solution*   Rewrite the equation in the form

$$y = 0x + 6.$$

Find three solutions.

If $x = -1$: $y = 0(-1) + 6 = 6$

If $x = 2$: $y = 0(2) + 6 = 6$

If $x = 3$: $y = 0(3) + 6 = 6$

Three solutions are $(-1, 6)$, $(2, 6)$, and $(3, 6)$. Graph these points and draw the line.

The graph of $y = 6$ is a horizontal line.

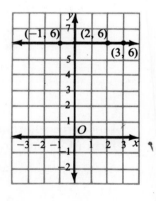

34. $y = 2$

35. $x = -4$

36. $y = 0$

37. $y = -3$

38. $x = 6$

39. $y = \frac{3}{2}$

**C** 40. Graph the equation $y = x^2$ by graphing the points with $x$-coordinates $-3$, $-2$, $-1$, 0, 1, 2, and 3. Join the points by means of a curved line. (This curve is called a **parabola**.)

**41.** Graph the equation $xy = 12$ by graphing the points with $x$-coordinates 1, 3, 6, and 12. Join the points by means of a curved line. On the same set of axes, graph the points with $x$-coordinates $-1$, $-3$, $-6$, and $-12$. Join the points by means of a curved line. (This two-branched curve is called a **hyperbola.**)

Graph the equation using the following values of $x$: $-5$, $-2$, $0$, $2$, $5$. Join the points by means of a straight line.

**42.** $y = |x|$       **43.** $y = -|x|$       **44.** $y = |x - 2|$       **45.** $y - 4 = |x|$

For what value of $k$ is the graph of the given ordered pair in the graph of the given equation?

**46.** $(3, -2)$; $2y + kx = 14$          **47.** $(-5, 4)$; $3x - ky = -12$

---

## Self-Test A

Exercises 1–6 refer to the diagram.
Give the coordinates of each point.

**1.** $M$       **2.** $Z$       **3.** $L$

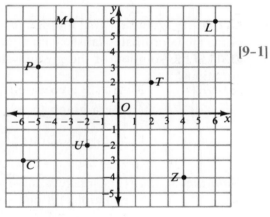

[9–1]

Name the point for the given ordered pair.

**4.** $(-6, -3)$    **5.** $(2, 2)$    **6.** $(-5, 3)$

Graph the given ordered pairs on one set of axes.

**7.** $(-3, 5)$     **8.** $(9, 7)$     **9.** $(-4, -8)$

**10.** $(11, -10)$    **11.** $(0, -2)$    **12.** $(-4, 2)$

Find the solutions of the equation for the given values of $x$.

**13.** $4y - x = 2$                       **14.** $y + 2x = 6$                [9–2]

     values of $x$: 0, $-4$, 9                  values of $x$: $-3$, $\frac{1}{2}$, 11

Graph each equation on a coordinate plane. Use a separate set of axes for each equation.

**15.** $y = x - 1$      **16.** $y = 2x$      **17.** $4x - y = 9$      **18.** $x = 7$      [9–3]

---

*Self-Test answers and Extra Practice are at the back of the book.*

## 9-4 Graphing a System of Equations

Two equations in the same variables are called a **system of equations.** If the graphs of two linear equations in a system have a point in common, the coordinates of that point must be a solution of both equations. We can find a solution of a system of equations such as

$$x - y = 2$$
$$x + 2y = 5$$

by finding the *point of intersection* of the graphs of the two equations.

The graphs of the two equations above are shown on the same set of axes. The point with coordinates (3, 1) appears to be the point of intersection of the graphs. To check whether (3, 1) satisfies the given system, we substitute its coordinates in both equations:

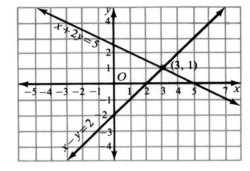

| $x - y = 2$ | $x + 2y = 5$ |
|---|---|
| $3 - 1 = 2$ | $3 + 2(1) = 5$ |

Since the coordinates satisfy both equations, (3, 1) is the solution of the given system of equations.

**EXAMPLE 1**   Use a graph to solve the system of equations:

$$y - x = 4$$
$$3y + x = 8$$

**Solution**   First make a table of values for each equation, then graph both equations on one set of axes.

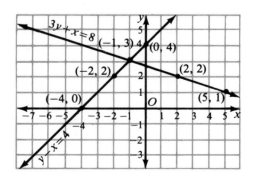

| $y - x = 4$ | |
|---|---|
| $x$ | $y$ |
| 0 | 4 |
| −4 | 0 |
| −2 | 2 |

| $3y + x = 8$ | |
|---|---|
| $x$ | $y$ |
| −1 | 3 |
| 5 | 1 |
| 2 | 2 |

The point of intersection appears to be $(-1, 3)$.

To check, substitute the coordinates $(-1, 3)$ in both equations:

$$y - x = 4 \qquad\qquad 3y + x = 8$$
$$3 - (-1) \stackrel{?}{=} 4 \qquad\qquad 3(3) + (-1) \stackrel{?}{=} 8$$
$$3 + 1 = 4 \ \checkmark \qquad\qquad 9 - 1 = 8 \ \checkmark$$

The solution for the given system is $(-1, 3)$.

**EXAMPLE 2** Use a graph to solve the system of equations: $\quad 2y - x = 2$
$$2y - x = -4$$

**Solution** Make a table of values for each equation and graph both equations on one set of axes.

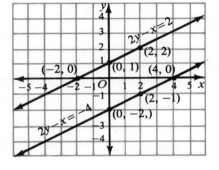

$2y - x = 2$

| $x$ | $y$ |
|-----|-----|
| 0 | 1 |
| -2 | 0 |
| 2 | 2 |

$2y - x = -4$

| $x$ | $y$ |
|-----|-----|
| 0 | -2 |
| 4 | 0 |
| 2 | -1 |

The graphs do not intersect; they are **parallel lines.** Thus, the system has *no solution.*

A system may have infinitely many solutions, as the following example illustrates.

**EXAMPLE 3** Use a graph to solve the system of equations: $\quad 6x + 3y = 18$
$$2x + y = 6$$

**Solution** Make a table of values for each equation and graph both equations on one set of axes.

$6x + 3y = 18$

| $x$ | $y$ |
|-----|-----|
| 0 | 6 |
| 3 | 0 |
| 1 | 4 |

$2x + y = 6$

| $x$ | $y$ |
|-----|-----|
| 0 | 6 |
| 3 | 0 |
| 1 | 4 |

The graphs *coincide.* The coordinates of all points on the line satisfy both equations. This system has *infinitely many solutions.*

To solve a system of equations, first graph the equations on one set of axes. Then consider:

> If the graphs intersect, the system has one solution. The coordinates of the point of intersection form the solution.

> If the graphs are parallel, the system has no solution.

> If the graphs coincide, the system has infinitely many solutions.

## Class Exercises

**Is the ordered pair a solution of the system of equations?**

1. $(4, -1)$   $x + y = 3$
              $x - y = 5$

2. $(5, -1)$   $x + y = -6$
              $x - y = 4$

3. $(3, 1)$   $x + 2y = 5$
             $x - y = 2$

4. $(-4, -4)$   $2x + y = -12$
               $x - y = 0$

5. A system of two linear equations has no solution. What does the graph of this system look like?

6. The coordinates $(-3, -1)$, $(0, 0)$, and $(3, 1)$ are solutions of a system of two linear equations. What does the graph of this system look like?

7. The graphs of two equations appear to intersect at $(-4, 5)$. How can you check to see whether $(-4, 5)$ satisfies the system?

## Written Exercises

**Use a graph to solve the system. Do the lines intersect or coincide, or are they parallel?**

**A**

1. $x + y = 6$
    $x - y = 0$

2. $x + y = 5$
    $3y + 3x = 15$

3. $x - y = -3$
    $x - y = 2$

4. $x - y = 4$
    $x + y = 0$

5. $-2y - 2x = -6$
    $x + y = 3$

6. $2x - y = 4$
    $y + 2x = 2$

7. $x - 2y = 8$
   $2y - x = 4$

8. $6y + 4x = 24$
   $2x + 3y = 12$

9. $3x - y = -6$
   $y - x = 6$

10. $x - 2y = -8$
    $2y - 3x = 12$

11. $7x - 14y = 70$
    $x - 2y = 10$

12. $3x + 2y = 6$
    $3x + 2y = 12$

**B** 13. $2x + \frac{3}{2}y = 2$
    $x + 2y = 6$

14. $x + 3y = 2$
    $2x + 5y = 3$

15. $\frac{2}{3}x - y = 2$
    $6y - 4x = 18$

16. $3x - 5y = 9$
    $\frac{1}{2}x - 2y = 5$

17. $\frac{3}{2}x + 2y = 4$
    $2x + 5y = 3$

18. $-7x - 2y = 7$
    $-\frac{7}{2}x - y = 1$

**C** 19. Find the value of $k$ in the equations

$$6x - 4y = 12$$
$$3x - ky = 6$$

such that the system has infinitely many solutions.

20. Find the value of $k$ in the equations

$$5x - 3y = 15$$
$$kx - 9y = 30$$

such that the system has no solution.

## Review Exercises

**Write in lowest terms.**

1. $\frac{18}{16}$

2. $\frac{12}{27}$

3. $\frac{49}{56}$

4. $\frac{18}{81}$

5. $\frac{24}{30}$

6. $\frac{48}{144}$

7. $\frac{13}{169}$

8. $\frac{17}{101}$

▮▮▮▮ **Calculator Key-In**

On Roger's first birthday he received a dime from his parents. For each birthday after that, they doubled the money. How much will Roger receive for his eighteenth birthday?

# 9-5 Using Graphs to Solve Problems

In the drawing at the left below, the slope of the hill changes. It becomes steeper partway up, and then flattens out near the top.

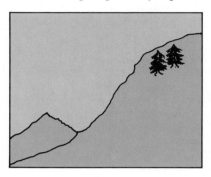
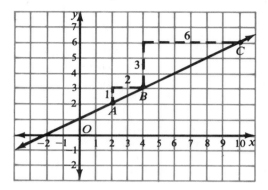

On the other hand, the slope of the straight line shown in the graph above does not change. It remains constant. The **slope of a line** is the ratio of the change in the $y$-coordinate to the change in the $x$-coordinate when moving from one point on the line to another.

Moving from $A$ to $B$: slope $= \dfrac{\text{change in } y}{\text{change in } x} = \dfrac{1}{2}$

Moving from $B$ to $C$: slope $= \dfrac{\text{change in } y}{\text{change in } x} = \dfrac{3}{6} = \dfrac{1}{2}$

A basic property of a straight line is that its slope is constant.

A straight-line graph sometimes expresses the relationship between two physical quantities. For example, such a graph can represent the conditions of temperature falling at a constant rate or of a hiker walking at a steady pace. If we can locate two points of a graph that is known to be a straight line, then we can extend the graph and get more information about the relationship.

**EXAMPLE 1** The temperature at 8 A.M. was 3°C. At 10 A.M. it was 7°C. If the temperature climbed at a constant rate from 6 A.M. to 12 noon, what was it at 6 A.M.? What was it at 12 noon?

**Solution** Set up a pair of axes.

Let the coordinates on the horizontal axis represent the hours after 6 A.M.

Let the vertical axis represent the temperature.

Use the information given to plot the points.

At 8 A.M., or 2 h after 6 A.M., the temperature was 3°C. This gives us the point (2, 3).

At 10 A.M., or 4 h after 6 A.M., the temperature was 7°C. This gives us the point (4, 7).

Since we know that the temperature climbed at a constant rate, we can first graph the two points and then draw a straight line through them.

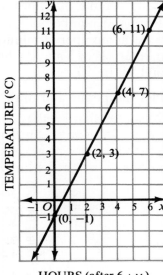

HOURS (after 6 A.M.)

We can see that the line crosses the vertical axis at −1. Therefore, at 6 A.M., or 0 h after 6 A.M., the temperature was −1°C. At 12 noon, or 6 h after 6 A.M., the temperature was 11°C.

Sometimes we are given information about the relationship between two quantities. We can use the information to write and graph an equation. We can then read additional information from the graph.

**EXAMPLE 2**  Maryanne uses 15 Cal of energy in stretching before she runs and 10 Cal for every minute of running time. Write an equation that relates the total number of Calories ($y$) that she uses when she stretches and goes for a run to the number of minutes ($x$) that she spends running. Graph this equation. From your graph, determine how long Maryanne must run after stretching to use a total of 55 Cal.

(*The solution is on the next page.*)

**Solution**        Consider the facts given:

         15 Cal used in stretching
         10 Cal/min used in running

Use the variables suggested.

    If $x$ = running time in minutes, then $10x$ = Cal used in running for $x$ min.

    If $y$ = Cal used in stretching *and* running for $x$ min, then $y = 15 + 10x$.

Make a table of values to locate two points. Because the number of Calories is large compared with the number of minutes, mark the vertical axis in intervals of 5 and the horizontal axis in intervals of 1.

$$y = 15 + 10x$$

| $x$ | $y$ |
|---|---|
| 0 | 15 |
| 2 | 35 |

Plot the points and draw the graph.

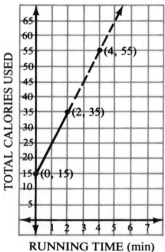

From the graph, you can see that Maryanne must run 4 min after stretching to use a total of 55 Cal.

---

**Problem Solving Reminder**

To help identify the conditions of a problem, it may be useful strategy to *rewrite the facts in simpler form.* In Example 2, we list the given facts before writing an equation based on the conditions of the problem.

## Class Exercises

**Complete for each graph.**

**1.** change in $y$ = _?_
change in $x$ = _?_
slope = _?_

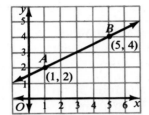

**2.** change in $y$ = _?_
change in $x$ = _?_
slope = _?_

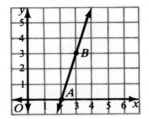

**3.** change in $y$ = _?_
change in $x$ = _?_
slope = _?_

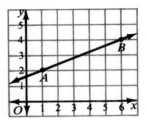

**Use a straight-line graph to complete the ordered pairs, then find the slope.**

**4.** $A(1, 3)$, $B(5, 5)$, $C(9, ?)$, $D(?, -1)$

**5.** $A(0, -3)$, $B(2, 2)$, $C(4, ?)$, $D(?, -8)$

**6.** $A(8, 4)$, $B(-1, -8)$, $C(5, ?)$, $D(?, -4)$

**7.** $A(0, 1)$, $B(2, 3)$, $C(-2, ?)$, $D(?, 0)$

**8.** $A(-1, 6)$, $B(0, 4)$, $C(2, ?)$, $D(?, -4)$

## Problems

**Solve.  Use the same intervals for both axes.**

**A**

**1.** A pack of greeting cards costs $1.50.  Five packs will cost $7.50. Draw a straight-line graph to show the relationship between the number of packs and the cost.  Let the $x$-axis represent the number of packs and the $y$-axis represent the cost in dollars.  What is the slope of the line?

**2.** An object of 2 g suspended from a spring stretches the spring to a length of 6 cm.  An object of 5 g stretches the spring to a length of 11 cm.  Draw a straight-line graph to show the relationship between the number of grams of the object and the length of the stretched spring.  Let the $x$-axis represent grams, and the $y$-axis represent centimeters.  What is the slope of the line?

**Use a graph to solve the problem.**

**B**  3. At 1 A.M. the temperature was −6°C. At 5 A.M. it was −1°C. If the temperature continues to rise steadily, what will the temperature be at 9 A.M.?

4. Hernando is saving money for a camping trip. Each week he deposits $5 in the bank. He had $25 to start the account. In 3 weeks he had $40 in the account.
   **a.** How much money will Hernando have by the seventh week?
   **b.** If the trip costs $85, how long will it take him to save enough money?

5. At 9 A.M. a test car driving at a constant speed passes a marker 50 mi from its starting point. At noon the car is about 130 mi from the marker. If the test drive ends at 1:30 P.M., how far will the car be from its starting point?

6. Between 1970 and 1980, the Rockfort Corporation used a 10-year expansion plan to increase its annual earnings steadily. The corporation had earnings of $3.6 million in 1975 and $6 million in 1977.
   **a.** About how much were the corporation's annual earnings in 1972?
   **b.** About how much did it earn by the end of the 10-year plan?

**a. Write an equation relating the quantity labeled y to the quantity labeled x.**
**b. Graph the equation.**
**c. Use the graph to answer the question.**

**C**  7. A butcher charges $4.50 a pound for the best cut of beef. For an additional $2.00, any order will be delivered. Relate the total cost (y) to the amount of beef (x) ordered and delivered. How many pounds of beef can be ordered and delivered for $33.50?

8. It takes a work crew 15 min to set up its equipment, and 3 min to paint each square meter of a wall. Relate the number of minutes (y) it takes to complete a job to the number of square meters (x) to be painted. If it took the crew 33 hours to complete the job, about how many square meters was the wall?

9. A word processor can store documents in its memory. It takes about 10 s to get the information, and about 35 s to print each page of it. Therefore, in about 360 s, or 6 min, the processor can complete a 10-page document. Relate the time ($y$) in minutes it takes to complete a document to the number of pages ($x$) to be typed. How many minutes will it take to complete an 18-page document?

10. Shaoli Hyatt earns a salary of $215 a week, plus a $15 commission for each encyclopedia she sells. Relate her total pay for one week ($y$) to the number of encyclopedias she sells during the week ($x$). How many encyclopedias must she sell to receive $350 for one week?

**Use a graph to solve.**

11. The velocity of a model rocket is 3 km/min at 1 s after takeoff. The velocity decreases to 2 km/min at 2.5 s after takeoff. When the rocket reaches its maximum height, the velocity will be 0. Assume that the decrease in velocity is constant. How long will it take the rocket to reach its maximum height?

## *Review Exercises*

**Solve.**

1. $3x + 4 > 8$

2. $5y - 10 < 0$

3. $7 + 3y \geq -4y$

4. $2(8x + 3) < -2$

5. $6(2y + 7) \leq 42$

6. $-3\left(9x - \frac{2}{3}\right) > -7$

7. $4(9x + 10) \geq 4$

8. $5(4y + 9) \leq 15$

9. $12\left(\frac{x}{6} - \frac{5}{6}\right) > 72$

▮▮▮ **Calculator Key-In**

Use a calculator to find the product.

$$1^2 = \underline{\quad?\quad}$$
$$11^2 = \underline{\quad?\quad}$$
$$111^2 = \underline{\quad?\quad}$$
$$1111^2 = \underline{\quad?\quad}$$

Use the pattern to predict the products $11,111^2$, $111,111^2$, $1,111,111^2$, $11,111,111^2$, and $111,111,111^2$.

# 9-6 Graphing Inequalities

When we graph a linear equation such as

$$y = x + 1$$

we see that the graph separates the coordinate plane into three sets of points:

(1) those above the line, such as $(-4, 5)$,

(2) those below the line, such as $(3, 1)$, and

(3) those on the line, such as $(2, 3)$.

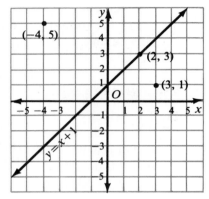

The region *above* the line is the graph of the set of solutions of the inequality

$$y > x + 1.$$

The region *below* the line is the graph of the set of solutions of the inequality

$$y < x + 1.$$

The line $y = x + 1$ forms the **boundary line** of the graphs of the inequalities $y > x + 1$ and $y < x + 1$. For any inequality we can get the boundary line by replacing the inequality symbol with the "equals" symbol.

   Since the graph of an inequality consists of all the points above or below a boundary line, we use shading to indicate the region. If the boundary line is part of the graph, it is drawn with a solid line. If the boundary line is not part of the graph, use a dashed line.

| $y > x + 1$ | $y \geq x + 1$ | $y < x + 1$ | $y \leq x + 1$ |
|---|---|---|---|
| 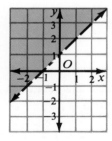 | 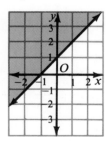 | 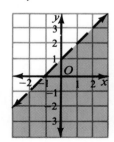 | 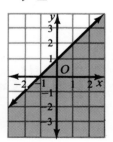 |

   To check whether the shading is correct, choose a point in the shaded region. Then substitute its coordinates for $x$ and $y$ in the inequality. If the coordinates satisfy the inequality, then the shading is correct.

   Any set of ordered pairs is a **relation.** Each of the open sentences

$y = x + 1$, $y > x + 1$, $y \geq x + 1$, $y < x + 1$, and $y \leq x + 1$ defines a relation. The equation $y = x + 1$ is a special kind of relation because it defines a function. Not all relations are functions. Recall that for a function, every value of $x$ has only one corresponding value of $y$. Notice in the shaded region of each graph above, more than one value of $y$ may be associated with each value of $x$. For example, the following ordered pairs all satisfy the inequalities $y > x + 1$ and $y \geq x + 1$:

$$(-2, 0), \ (-2, 1), \ (-2, 2), \ (-2, 3)$$

Therefore, the relation defined by the inequalities $y > x + 1$ and $y \geq x + 1$ is not a function.

**EXAMPLE 1**  Graph $y - x < 4$.

**Solution**  First transform $y - x < 4$ into an equivalent inequality with $y$ alone on one side.

$$y < 4 + x$$

Then locate the boundary line by graphing

$$y = 4 + x.$$

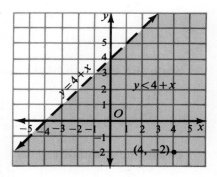

Use a dashed line to draw the boundary line, and shade the region below.

Check:  Use $(4, -2)$.
$$y - x \overset{<}{} 4$$
$$-2 - 4 \overset{?}{<} 4$$
$$-6 < 4 \quad \checkmark$$

**EXAMPLE 2**  Graph $x \geq -7$.

**Solution**  Locate the boundary line by graphing

$$x = -7.$$

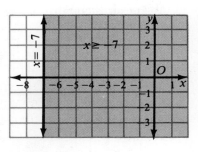

Use a solid line to draw the boundary. Shade the region to the right since all values of $x$ greater than $-7$ lie to the right of the boundary line.

## Class Exercises

**Transform the inequality into an equivalent inequality with $y$ alone on one side. State the equation of the boundary line.**

**1.** $4x + y < 8$

**2.** $y - 9x < 2$

**3.** $4y + 2x > 10$

**4.** $6x - y > 0$

**5.** $3 \leq 5x + y$

**6.** $9y \geq 18$

**State whether each point belongs to the graph of the given inequality.**

**7.** $y \leq x + 2$  (0, 1), (3, 6)

**8.** $-x + 2y \geq 0$  (4, 4), (−2, 1)

**9.** $x \leq -7$  (−4, −10), (−8, 0)

**10.** $5 \leq -x + y$  (−6, −5), (2, 7)

**11.** $0 \geq x + y - 5$  (2, 1), (4, 1)

**12.** $y \geq 3$  (9, 4), (0, 8)

---

## Written Exercises

**State the equation of the boundary line.**

**A**   **1.** $3x + y > -6$

**2.** $x + y \geq 2$

**3.** $x - y \leq 2$

**4.** $2y - 4 > 0$

**5.** $-x \geq y - 3$

**6.** $2y - 6x < 0$

**Graph the inequality.**

**7.** $x + y \geq 9$

**8.** $y > -3x + 3$

**9.** $-2x + y \geq 2$

**10.** $x + 6y \leq -5$

**11.** $4x + 2y \geq 8$

**12.** $3y - x < -3$

**13.** $y \geq 2x + 5$

**14.** $x + 2y \leq 6$

**15.** $3y - 4 > 2x - 5$

**B**   **16.** $3(x - y) > 6$

**17.** $y \leq 8$

**18.** $6x + 2y + 3 < 2x - 1$

**19.** $4x + 3y \leq x - 3$

**20.** $3y - 6 > 0$

**21.** $x \leq -3$

**22.** $y > 0$

**23.** $2(x + y) < 6x + 10$

**24.** $3y - 6 \geq 3(x + 2y)$

**Write an inequality for the graph shown.**

**25.**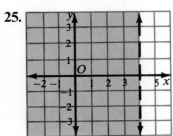

**26.**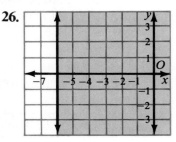

**27.**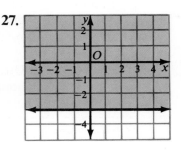

**336**   *Chapter 9*

**Graph the solutions of each system of inequalities.**

*EXAMPLE*   $y \geq -3 + x$
$y \leq 2 - x$

*Solution*   First graph $y \geq -3 + x$. Use blue to shade the region. Then graph $y \leq 2 - x$. Use gray to shade the region.

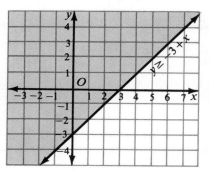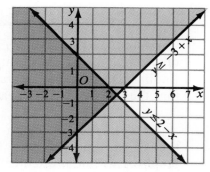

The graph of the solutions of the system is the region where the blue and gray shading overlap.

**C**   **28.** $y \geq -4$         **29.** $y > 5 - x$         **30.** $x \leq 9 - y$         **31.** $x \geq -6$
    $x \leq 2$                 $y \geq x + 5$               $y < x + 3$                 $x < 3$

---

## Self-Test B

**Use a graph to solve the system. Do the lines intersect or coincide, or are they parallel?**

**1.** $x + y = 6$         **2.** $x + 2y = 6$         **3.** $2x + y = 3$         [9–4]
    $y - x = -2$             $6y + 3x = 18$           $-2x - y = 2$

**Use a graph to solve.**

**4.** When Fritz enrolled in a speed-reading class, he read about 250       [9–5]
words per minute. Two weeks later, his reading speed was about
750 words per minute. If his reading speed increases steadily, how
many weeks after enrollment will it take Fritz to read about 1250
words per minute?

**Graph each inequality.**

**5.** $x + y \leq 10$         **6.** $3y - x > 9$         **7.** $x - 5y \geq 5$         [9–6]

---

*Self-Test answers and Extra Practice are at the back of the book.*

# The Computer and Linear Equations

For a science project, Nina watered 6 identical lima bean gardens by different amounts. The crop yields were as recorded below.

| Water applied daily in cm ($x$) | 0.5 | 1.0 | 1.5 | 2.0 | 2.5 | 3.0 |
|---|---|---|---|---|---|---|
| Yield of lima beans in kg ($y$) | 1.6 | 1.8 | 2.0 | 2.4 | 2.6 | 2.8 |

Can Nina use these data to predict crop yields for other amounts of daily watering? If we plot her findings as shown on the graph at the right, we see that the points are very nearly on a line.

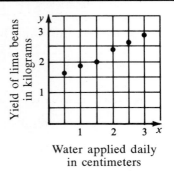

Water applied daily in centimeters

The computer program below gives the equation of the line that best fits the points on the graph. This line is often called the **line of best fit.** The instruction in line 40 provides the program with all the coordinates of the points. Compare the numbers in line 40 to the numbers in the chart.

```
10   DIM L(50)
20   PRINT "NUMBER OF POINTS ON GRAPH IS";
30   INPUT N
40   DATA .5,1.6,1,1.8,1.5,2,2,2.4,2.5,2.6,3,2.8
50   FOR I = 1 TO N
60   READ X,Y
70   LET A = A + X
80   LET B = B + Y
90   LET C = C + X * X
100  LET D = D + X * Y
110  NEXT I
120  LET Q = N * C - A * A
130  LET R =  INT (100 * (N * D - A * B) / Q + .5) / 100
140  LET S =  INT (100 * (B * C - A * D) / Q + .5) / 100
145  PRINT
150  PRINT "EQUATION OF BEST FITTING LINE IS"
155  PRINT " Y = ";R;"X";
160  IF S >  = 0 THEN 190
170  PRINT S
180  STOP
190  PRINT " ";"+ ";S
200  END
```

1. RUN the program to find the equation of the best fitting line for Nina's data.

**2.** Draw the graph of the equation.

**3.** Use the graph to predict the yield ($y$) if Nina applied 3.3 cm of water daily ($x$).

**Use the computer program to find the equation of the line of best fit for the data given in each chart. Graph each equation and use the equation to answer each question.**

**4.** The chart below shows the distance a spring stretches when different masses are hung from it.

| Mass in kg | 0.3 | 0.6 | 0.9 | 1.2 | 1.5 | 1.8 |
|---|---|---|---|---|---|---|
| Stretch in cm | 2.1 | 4.9 | 6.0 | 7.1 | 8.9 | 10.8 |

About how much stretch would a 1 kg mass produce?

**5.** The chart below shows temperature changes as a cold front approached.

| Time in hours from 1st reading | 0 | 1 | 2 | 3 | 4 |
|---|---|---|---|---|---|
| Temperature in °C | 21 | 17 | 14 | 12 | 7 |

**a.** Estimate the temperature 1.5 hours from the first reading.

**b.** If the temperature continues to decrease steadily, about how many hours will it take to reach 0°C?

**6.** The chart below shows the profit earned by a bookstore on the sale of a bestseller.

| No. sold | 3 | 4 | 7 | 9 | 11 |
|---|---|---|---|---|---|
| Profit | $13.50 | $18.00 | $31.50 | $40.50 | $49.50 |

How many sales will it take to earn a profit of at least $80?

**7.** The chart below shows the cost of college education for the past 5 years.

| Year | 1 | 2 | 3 | 4 | 5 |
|---|---|---|---|---|---|
| Cost (in thousands) | $7.8 | $8.6 | $9.4 | $9.9 | $10.1 |

Predict the cost of college education for the next two years.

# Chapter Review

**True or false?**

1. The coordinates of point $G$ are (3, 2).

2. The abscissa of point $W$ is 2.

3. The ordinate of point $T$ is 1.

4. Point $B$ is associated with the ordered pair (2, −3).

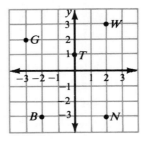

[9–1]

**Match each equation with an ordered pair that satisfies the equation.**

5. $2y - 7x = 4$      **A.** $(-18, 1)$

6. $10x + \frac{5}{9}y = 0$      **B.** $(9, 2)$

7. $-2x + 9y = 0$      **C.** $(2, 9)$

8. $y - \frac{1}{3}x = 7$      **D.** $(1, -18)$

[9–2]

**Graph the equation on a coordinate plane. Use a separate set of axes for each equation.**

9. $x + y = 0$      10. $2y - 3x = -6$      11. $y - 2x = 2$

[9–3]

**Use a graph to solve the system of equations. Match the system with the word that describes its graphs.**

12. $y + x = -3$
    $2y - 3x = 4$

13. $y - 3x = -6$
    $y - 3x = 3$

14. $5y - 2x = 4$
    $15y - 6x = 12$

[9–4]

**A.** intersect      **B.** coincide      **C.** parallel

**Complete.**

15. A basic property of a straight line is that its slope remains __?__.

[9–5]

16. The slope of a line is the ratio of the change in the __?__-coordinate to the change in the __?__-coordinate when moving from one point on the line to another.

17. The boundary line for the graph of $y > x + 6$ is a __?__ line.

[9–6]

18. If the boundary line is part of the graph of an inequality, it is drawn with a __?__ line.

# Chapter Test

**Give the coordinates of the point.**

**1.** $I$      **2.** $M$      **3.** $E$      **4.** $R$

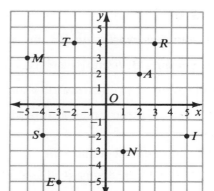

[9-1]

**Name the point for the ordered pair.**

**5.** $(-2, 4)$            **6.** $(1, -3)$

**7.** $(-4, -2)$        **8.** $(5, -2)$

**a.** Solve the equation for $y$ in terms of $x$.
**b.** Find solutions of the equation for the given values of $x$: **−3, 0, 4.**

**9.** $y + 7x = 23$      **10.** $2y - 4x = 1$      **11.** $3y - 6x = 0$      [9-2]

**Graph the equation on a coordinate plane.  Use a separate set of axes for each equation.**

**12.** $y - 3x = 7$      **13.** $4y - x = 8$      **14.** $2y + x = -10$      [9-3]

**Use a graph to solve the system.  Do the lines intersect or coincide, or are they parallel?**

**15.** $x - 2y = 0$      **16.** $3x - y = -4$      **17.** $5x + y = 2$      [9-4]
      $2x + y = 5$            $3x - y = 3$           $10x + 2y = 4$

**Use a graph to solve the problem.**

**18.** One hour after the start of an experiment, the temperature of a solution was $-15°C$.  Three hours later it was $-6°C$.  If the temperature continues to rise steadily, about how many hours will it take for the temperature to reach $0°C$?

    [9-5]

**19.** It takes the window washers 24 min to get ready and 4 min to wash a 6 m by 6 m window.
    **a.** Write an equation relating the total time ($y$) required on the job to the number of windows ($x$) to be washed.
    **b.** Graph the equation.
    **c.** About how many 6 m by 6 m windows can they wash in 2 h?

**a.** State the equation of the boundary line.
**b.** Graph the inequality.

**20.** $y \geq 2x + 1$      **21.** $x + y > -4$      **22.** $y - 3x < 0$      [9-6]

# Cumulative Review (Chapters 1–9)

## Exercises

**Write a variable expression for the word phrase.**

**1.** The product of $b$ and five        **2.** Eight times a number $x$

**3.** Twelve less than the sum of $y$ and $z$     **4.** A number $q$ divided by seven

**5.** The sum of three times a number $p$ and seven

**Round to the nearest hundredth.**

**6.** 46.871        **7.** 288.005        **8.** 0.7826        **9.** 100.758        **10.** 33.663

**Perform the indicated operation.**

**11.** $-8.709 + 13.6001$      **12.** $7615.7 - 333.61$      **13.** $606.08 + (-51.99)$

**14.** $272.65 - (-0.88)$      **15.** $37.61 \times 0.08$      **16.** $1.5798 \div 0.03$

**17.** $-11.56 \times 36.77$      **18.** $72.5 \div 5$      **19.** $6.7 \times (-1.22)$

**20.** $-\frac{1}{3} + \frac{4}{5}$      **21.** $\frac{4}{7} \times \left(-\frac{3}{8}\right)$      **22.** $6\frac{1}{2} \div \left(-2\frac{1}{2}\right)$

**23.** $\frac{7}{8} - \left(-\frac{1}{4}\right)$      **24.** $3\frac{1}{2} - 1\frac{7}{8}$      **25.** $-3\frac{4}{5} \times 1\frac{3}{10}$

**Write an equation or inequality for the word sentence and solve.**

**26.** The sum of $x$ and three is seven.

**27.** A number $t$ is the product of negative three fourths and one fifth.

**28.** The quotient of negative eight divided by $w$ is seven.

**29.** The sum of $x$ and twelve is greater than the product of $x$ and 2.

**Find the circumference of the circle described. Use $\pi \approx 3.14$.**

**30.** diameter $= 28$ cm      **31.** radius $= 3.3$ m      **32.** diameter $= 88.5$ km

**Solve the proportion.**

**33.** $\frac{n}{4.8} = \frac{0.4}{1.2}$      **34.** $\frac{1.5}{n} = \frac{1}{4}$      **35.** $\frac{13}{14} = \frac{39}{n}$      **36.** $\frac{1.2}{2.0} = \frac{n}{3.0}$

**Tell whether the ordered pair is a solution of the given equation.**

$2x + y = 15$      **37.** $(0, 12)$      **38.** $(-5, 25)$      **39.** $(5, 5)$      **40.** $(4, 13)$

# Problems

> **Problem Solving Reminders**
> Here are some reminders that may help you solve some of the problems on
> this page.
> * Sometimes more than one method can be used to solve.
> * Consider whether making a sketch will help.
> * When rounding an answer to division, consider whether it is reasonable to
>   round up or round down.

**Solve.**

1. Bill Murphy earns $4.85 an hour working in a day care center. Last
   week he worked from 1:00 P.M. to 4:30 P.M. on Monday through
   Friday. How much did he earn?

2. "I feel like I ran a mile," gasped Marge. If Marge ran 5 times
   around a circular track with a diameter of 210 ft, did she really run a
   mile? (*Hint:* 5280 ft = 1 mi)

3. Light travels at a speed of 297,600 km/s. If the circumference of
   Earth is about 39,800 km, about how many times could light travel
   around Earth in 1 s?

4. Michael Woolsey bought $5671 worth of stock. The commission
   rate was $28 plus 0.6% of the dollar amount. How much was the
   commission?

5. A car dealer is expecting a price increase of between $2\frac{1}{2}\%$ and 5%
   over the price of last year's models. If a certain car sold for $9560
   last year, what is the least it can be expected to sell for this year? the
   most?

6. Myra Daley was given an advance of $18,000 on royalties expected
   from a book she wrote. If the selling price of the book is $15.95 and
   her royalty rate is 5%, how many books must be sold before her
   royalties exceed her advance?

7. A baby that weighed 7 lb 6 oz at birth weighed 10 lb 10 oz at 6
   weeks of age. To the nearest tenth of a percent, what was the percent of increase?

8. It took Thomas 25 min longer to do his math homework than to do
   his French homework. He spent a total of 2.25 h on both subjects.
   How much time did he spend on math?

9. If a microwave oven uses 1.5 kilowatt hours (kW·h) of electricity in
   15 min, how much electricity does it use in 5 min?

# 10

# Areas and Volumes

Pyramids are one type of space figure in geometry. The bottom, or base, of a pyramid is in the shape of a polygon and the sides are always triangular. When your study of measurement is extended to include area, volume, and capacity, you can then describe the pyramid shown in the photograph in more mathematical terms.

The pyramid in the photograph is one of the pyramids built by the Egyptian kings, or Pharaohs, as a monumental tomb. The largest is the Great Pyramid, built by Pharaoh Khufu around 2600 B.C. The Great Pyramid has a square base measuring about 230 m on a side and originally rose to an approximate height of 150 m. When built, it contained about 2,300,000 stone blocks, each having a mass of nearly 2.5 t.

## Career Note

Architects are responsible for the design and visual appearance of buildings. Good architectural designs are appealing, safe, and functional. Architects must combine technical skills with a strong sense of style. Course work in mathematics, engineering, and art provide the necessary background for this profession.

# 10-1 Areas of Rectangles and Parallelograms

Earlier we measured lengths of segments and found perimeters of polygons. Now we will measure the part of the plane enclosed by a polygon. We call this measure the **area** of the polygon.

Just as we needed a unit length to measure segments, we now need a unit area. In the metric system a unit area often used is the **square centimeter (cm²).**

The rectangular regions shown in the diagrams have been divided into square centimeters to show the areas of the rectangles.

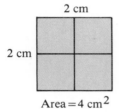

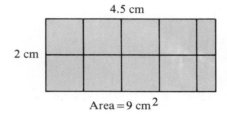

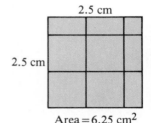

Notice that the area of each rectangle is the product of the lengths of two consecutive sides. These sides are called the **length** and the **width** of the rectangle. The length names the longer side and the width names the shorter side. We have the following formula for any rectangle.

---

### *Formula*

Area of rectangle = length × width

$$A = lw$$

---

The length and width of a rectangle are called its **dimensions.**

In the case of a parallelogram, we may consider either pair of parallel sides to be the **bases.** (The word *base* is also used to denote the length of the base.) The **height** is the perpendicular distance between the bases.

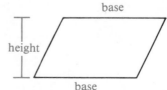

The colored region in the figure at the top of the next page can be moved to the right, as shown in the second figure, to form a rectangle having dimensions *b* and *h*. Thus, the area of the parallelogram is the same as the area of the rectangle, *bh.*

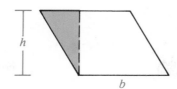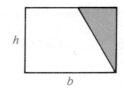

The area of any parallelogram can be found by using the following formula.

## Formula

Area of parallelogram = base × height

$$A = bh$$

**EXAMPLE 1**  Find the area of each parallelogram.

**a.**

30 m

25 m

**b.**

15 mm

20 mm

**Solution**

**a.** $A = bh$
$= 25 \times 30 = 750$
The area is 750 m².

**b.** $A = bh$
$= 15 \times 20 = 300$
The area is 300 mm².

**EXAMPLE 2**  A parallelogram has an area of 375 cm² and a height of 15 cm. Find the length of the base.

**Solution**

$$A = bh$$
$$375 = b \times 15$$
$$\frac{375}{15} = b$$
$$25 = b$$

The length of the base is 25 cm.

The unit areas used in the examples are **square meters (m²)**, **square millimeters (mm²)**, and **square centimeters (cm²)**. For very large regions, such as states or countries, we could use **square kilometers (km²)**.

Sometimes we may work with an unspecified unit of length. Then the unit of area is simply called a **square unit.** Thus, the area of the rectangle shown at the right is 250 square units.

10

25

## Class Exercises

**Find the area of each shaded region.**

**1.**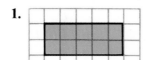

**2.**

**3.**

**4.**

**5.** Find the perimeters of the regions in Exercises 1 and 2.

**6.** Find the area of a square with sides 4 cm long.

**Complete.**

**7.** $1 \text{ cm} = \underline{\ ?\ } \text{ mm}$

$1 \text{ cm}^2 = \underline{\ ?\ } \text{ mm}^2$

**8.** $1 \text{ m} = \underline{\ ?\ } \text{ cm}$

$1 \text{ m}^2 = \underline{\ ?\ } \text{ cm}^2$

**9.** $1 \text{ km} = \underline{\ ?\ } \text{ m}$

$1 \text{ km}^2 = \underline{\ ?\ } \text{ m}^2$

**10.** Explain why the blue parallelogram and the gray one have equal areas.

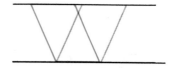

## Written Exercises

**Find the area and the perimeter of each rectangle or parallelogram.**

**A** **1.**

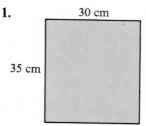

30 cm

35 cm

**2.**

11 m

4 m

5 m

**3.**

65 km

41 km

**4.**

5.6

1.3

1.2

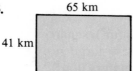

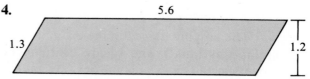

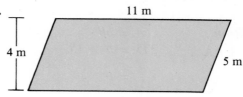

**Find the area and the perimeter of a rectangle having the given dimensions.**

5. 48 mm by 92 mm

6. 55 cm by 32 cm

7. 63.7 km by 39.1 km

8. 206.3 m by 33.15 m

**Copy and complete the tables.**

| Rectangle | 9. | 10. | 11. | 12. | 13. | 14. |
|---|---|---|---|---|---|---|
| length | 12.5 | 12 | ? | ? | 8.6 | ? |
| width | 7.5 | ? | 45 | 2.5 | ? | ? |
| perimeter | ? | ? | ? | 17.0 | 18.0 | 4 |
| area | ? | 60 | 3600 | ? | ? | 1 |

| Parallelogram | 15. | 16. | 17. | 18. | 19. | 20. |
|---|---|---|---|---|---|---|
| base | 18.5 | 2.7 | ? | 14.3 | 2.9 | ? |
| height | 4.0 | ? | 1.6 | 3.2 | ? | 0.4 |
| area | ? | 8.1 | 0.8 | ? | 11.6 | 3.6 |

**Find the area of each region.** (*Hint:* Subdivide each region into simpler ones if necessary.)

**B** 21.

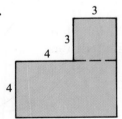

22.

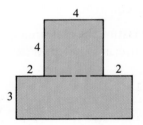

23.

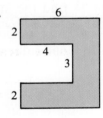

24.

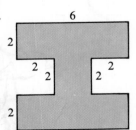

25.

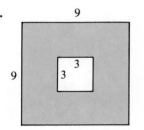

26.

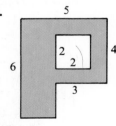

## Problems

**Solve.**

**A** **1.** How many square meters of wallpaper are needed to cover a wall 8 m long and 3 m high?

**2. a.** How many square yards of carpeting are needed to cover a floor that measures 8 yd by 5 yd?
**b.** How much will the carpeting cost at $24 per square yard?

**3. a.** How many square feet of vinyl floor covering are needed to cover a floor measuring 60 ft by 12 ft?
**b.** How much will the floor covering cost at $1.50 per square foot?

**4.** Jose wishes to line the open box shown at the right using five sheets of plastic. How many square centimeters will he need?

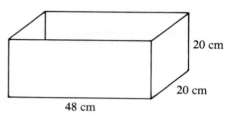

**B** **5.** Yoneko wishes to obtain 6 m² of plastic from a roll 40 cm wide. How many meters should she unroll?

**6.** A square pool 5 m on each side is surrounded by a brick walk 2 m wide. What is the area of the walk?

**7.** A construction site in the shape of a square is surrounded by a wooden wall 2 m high. If the length of one side of the wall is 45 m, find the area of the construction site.

**8.** A rectangular cow pasture has an area of 1925 m². If the length of one side of the pasture is 55 m, find the lengths of the other sides.

## Review Exercises

**Multiply.**

**1.** $\frac{1}{2} \times 8 \times 7$

**2.** $\frac{1}{3} \times 11 \times 9$

**3.** $\frac{1}{2} \times 1.3 \times 1.6$

**4.** $\frac{1}{2}(8 + 9)11$

**5.** $\frac{1}{4}(20 + 12)5$

**6.** $\frac{1}{2}(0.11 + 0.17)0.6$

**7.** $\frac{1}{3}(0.26 + 0.52)0.3$

**8.** $\frac{1}{4}(1.76 + 2.69)1.94$

# 10-2 Areas of Triangles and Trapezoids

Any side of a triangle can be considered to be the
**base.** The **height** is then the perpendicular distance
from the opposite vertex to the base line.

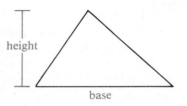

Let us find the area of a triangle having base $b$
and height $h$. The triangle and a congruent copy of
it can be put together to form a parallelogram as
shown in the diagrams.

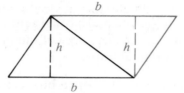

Since the area of the parallelogram is $bh$ and the area of the triangle is half the parallelogram, we have the following formula.

> ## Formula
>
> Area of triangle $= \frac{1}{2} \times$ base $\times$ height
>
> $$A = \frac{1}{2}bh$$

*EXAMPLE*   Find the area of each triangle.

**a.**

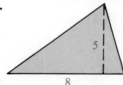

**b.**

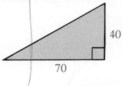

**c.**

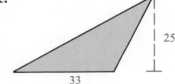

*Solution*

**a.** $A = \frac{1}{2}bh$

$\quad = \frac{1}{2} \times 8 \times 5$

$\quad = 20$

$A = 20$ square units

**b.** $A = \frac{1}{2}bh$

$\quad = \frac{1}{2} \times 70 \times 40$

$\quad = 1400$

$A = 1400$ square units

**c.** $A = \frac{1}{2}bh$

$\quad = \frac{1}{2} \times 33 \times 25$

$\quad = 4125$

$A = 412.5$ square units

Note in part (b) of Example 1 that the lengths of the sides of the *right angle* of the triangle were used as the base and the height. This can be done for any *right* triangle, even if the triangle is positioned so that it is "standing" on the side opposite the right angle.

The **height** of a trapezoid is the perpendicular distance between the parallel sides. These parallel sides are called the **bases** of the trapezoid. The method used to find the formula for the area of a triangle can be used to find the formula for the area of a trapezoid having bases $b_1$ and $b_2$ and height $h$. The trapezoid and a congruent copy of it can be put together to form a parallelogram. The area of the parallelogram is $(b_1 + b_2)h$ and the area of the original trapezoid is half of the area of the parallelogram.

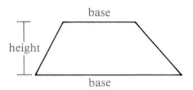

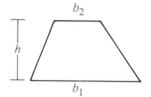 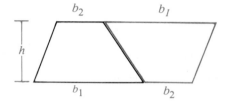

We, therefore, have the following formula.

## Formula

$$\text{Area of trapezoid} = \frac{1}{2} \times (\text{sum of bases}) \times \text{height}$$

$$A = \frac{1}{2}(b_1 + b_2)h$$

*EXAMPLE 2*   Find the area of each trapezoid.

**a.**
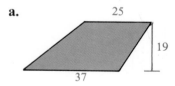

**b.**
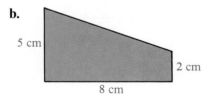

*Solution*      **a.** $A = \frac{1}{2}(b_1 + b_2)h$

$= \frac{1}{2} \times (37 + 25) \times 19 = 589$

$A = 589$ square units

**b.** $A = \frac{1}{2}(b_1 + b_2)h$

$= \frac{1}{2} \times (5 + 2) \times 8 = 28$

$A = 28$ m²

**EXAMPLE 3**   A trapezoid has an area of 200 cm² and bases of 15 cm and 25 cm. Find the height.

**Solution**

$$A = \frac{1}{2}(b_1 + b_2)h$$

$$200 = \frac{1}{2} \times (15 + 25) \times h$$

$$200 = \frac{1}{2} \times 40 \times h$$

$$200 = 20h$$
$$10 = h$$

The height is 10 cm.

## Class Exercises

**Find the area of each polygon.**

1.

8
10

2.

8
12

3.
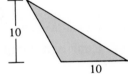
10
10

4.
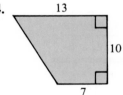
13
10
7

5.
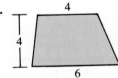
4
4
6

6.
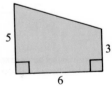
5
3
6

7.
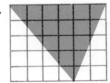

8.
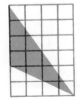

9.

10.
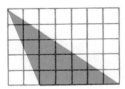

11.

12.
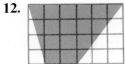

## Written Exercises

**Find the area of each polygon.**

**A** **1.**
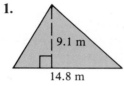
9.1 m
14.8 m

**2.**
6.6 km  6.0 km

**3.**
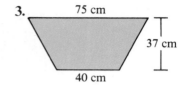
75 cm
37 cm
40 cm

**4.**
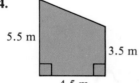
5.5 m
3.5 m
4.5 m

**5.**
29 mm
20 mm

**6.**
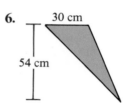
30 cm
54 cm

**7.** Triangle: base 122 km, height 30 km

**8.** Triangle: base 480 m, height 480 m

**9.** Trapezoid: bases 12.3 cm and 6.2 cm, height 4.8 cm

**10.** Trapezoid: bases 14.6 km and 22.4 km, height 14.0 km

**Copy and complete the tables.**

| Triangle | 11. | 12. | 13. | 14. |
|---|---|---|---|---|
| base | 6 cm | 16 mm | ? | ? |
| height | ? | ? | 2.4 m | 1.4 m |
| area | 72 cm² | 80 mm² | 4.8 m² | 0.42 m² |

**B**

| Trapezoid | 15. | 16. | 17. | 18. |
|---|---|---|---|---|
| base | 0.7 mm | 0.8 m | 3 | 2 |
| base | 1.7 mm | 1.2 m | ? | ? |
| height | ? | ? | 2 | 3 |
| area | 9.6 mm² | 1.0 m² | 18 | 18 |

**19.** In the figure at the right, $\overline{PQ}$ is parallel to $\overline{RT}$. Explain why triangles $PQR$, $PQS$, and $PQT$ all have the same area.

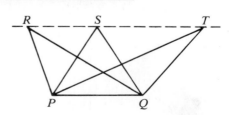

**In Exercises 20 and 21 find the area of (a) the blue part and (b) the red part of the pennant.**

**20.**

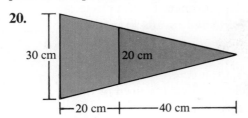

**21.**

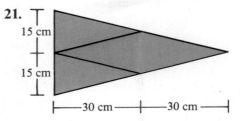

**C  22.** One triangle has a base and a height that are twice as large as the base and the height of another triangle. What is the ratio of their areas?

**23.** One rectangle has a base and a height that are 3 times the base and the height of another rectangle. What is the ratio of their areas? their perimeters?

## Review Exercises

**Simplify.**

**1.** $3 \times 5^2$

**2.** $1.2 \times 9^2$

**3.** $4 \times 1.3^2$

**4.** $3 \times 7^2 - 2 \times 4^2$

**5.** $4 \times 5^2 + 3 \times 8^2$

**6.** $11^2 \times 3 - 6^2 \times 4$

**7.** $5 \times 1.4^2 - 7 \times 1.1^2$

**8.** $2.5^2 \times 4 + 3.2^2 \times 7$

**9.** $17 \times 14^2 + 6 \times 16^2$

| | **Challenge** |
|---|---|

The ancient Egyptians worked primarily with fractions with a numerator of 1. They expressed fractions such as $\frac{2}{5}$ as the sum of these fractions: $\frac{2}{5} = \frac{1}{3} + \frac{1}{15}$. These sums were listed in a table. Find the following sums selected from the Egyptian fraction table.

**1.** $\frac{1}{8} + \frac{1}{52} + \frac{1}{104}$

**2.** $\frac{1}{12} + \frac{1}{51} + \frac{1}{68}$

**3.** $\frac{1}{12} + \frac{1}{76} + \frac{1}{114}$

**4.** $\frac{1}{24} + \frac{1}{58} + \frac{1}{174} + \frac{1}{232}$

**5.** $\frac{1}{20} + \frac{1}{124} + \frac{1}{155}$

**6.** $\frac{1}{24} + \frac{1}{111} + \frac{1}{296}$

# 10-3 Areas of Circles

Recall that there are two formulas for the circumference, $C$, of a circle. If the diameter of the circle is denoted by $d$ and the radius by $r$, then

$$C = \pi d \quad \text{and} \quad C = 2\pi r.$$

Two approximations for the number $\pi$ are 3.14 and $\frac{22}{7}$.

The part of the plane enclosed by a circle is called the **area of the circle.** This area is given by the following formula.

> ### Formula
>
> Area of circle $= \pi \times (\text{radius})^2$
>
> $$A = \pi r^2$$

*EXAMPLE 1*  Find the areas of the shaded regions.  Use $\pi \approx 3.14$.

**a.**

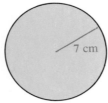

7 cm

**b.**

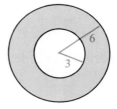

6

3

*Solution*

**a.** $A = \pi r^2$
$\quad \approx 3.14 \times 7^2$
$\quad \approx 3.14 \times 49 \approx 153.86$
$\quad A \approx 154 \text{ cm}^2$

**b.** $A = (\pi \times 6^2) - (\pi \times 3^2)$
$\quad = (\pi \times 36) - (\pi \times 9)$
$\quad = \pi(36 - 9) = \pi \times 27$
$\quad \approx 3.14 \times 27 \approx 84.78$
$\quad A \approx 84.8 \text{ square units}$

Recall that we give answers to only three digits when we use the approximation $\pi \approx 3.14$.  Sometimes, to avoid approximations, we give an answer in terms of $\pi$.

*EXAMPLE 2*  Find the area of the shaded region.  Leave your answer in terms of $\pi$.

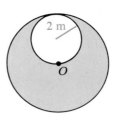

2 m

$O$

*Solution*

Area of shaded region
$= (\text{Area of large circle}) -$
$\qquad\qquad (\text{Area of small circle})$
We first find the area of the small circle.

$$A = \pi r^2 = \pi \times 2^2 = 4\pi$$

We then find the area of the large circle.
Since the radius of the large circle is the same as the diameter of the small circle, we know that the radius of the large circle is 4 m.

$$A = \pi r^2 = \pi \times 4^2 = 16\pi$$

Thus, area of shaded region is equal to $16\pi - 4\pi = 12\pi$
The area of the shaded region is $12\pi$ m².

To make the formula $A = \pi r^2$ seem reasonable to you, think of the circular region below cut like a pie.

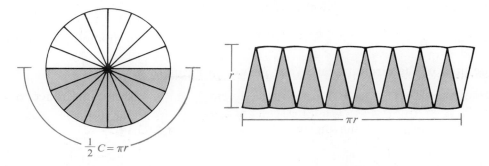

The pieces can be arranged to form a figure rather like a parallelogram with base $\pi r$ and height $r$. This suggests that the area is given by

$$\pi r \times r = \pi r^2.$$

**Reading Mathematics: *Diagrams***
To read and understand an explanation that is illustrated by a diagram, ask yourself questions about the diagram as you read. In the explanation above, for example, ask yourself, why is the figure "rather like" a parallelogram. How is it different? Why is the measure of the base $\pi r$?

## Class Exercises

**Solve.**

1. A circle has radius 10 cm. What is its area? Use $\pi \approx 3.14$.

2. A circle has diameter 14 units. What is its area? Use $\pi \approx \frac{22}{7}$.

3. A circle has diameter 6. What is its area? Give your answer in terms of $\pi$.

4. A circle has radius 5. What is its area? Give your answer in terms of $\pi$.

**5.** A circle has area $4\pi$ cm². What is its radius?

**6.** A circle has area $9\pi$. What is its diameter?

---

## Written Exercises

**Find the area of the circle. Use $\pi \approx 3.14$ and round the answer to three digits.**

**A** **1.** radius = 5 km          **2.** radius = 8 cm          **3.** diameter = 0.6 m

**Find the area of the circle. Use $\pi \approx \frac{22}{7}$.**

    **4.** radius = 14 cm          **5.** diameter = $3\frac{1}{2}$          **6.** diameter = 28 cm

**Find the area of the circle. Leave your answer in terms of $\pi$.**

    **7.** diameter = 20 m          **8.** radius = 15 cm          **9.** circumference = $4\pi$

**Find the radius of a circle having the given area. Use $\pi \approx 3.14$.**

**10.** $A = 78.5$ cm²                    **11.** $A = 314$ m²

**Find the diameter of a circle having the given area. Use $\pi \approx \frac{22}{7}$.**

**12.** $A = 154$ km²                    **13.** $A = 3\frac{1}{7}$ cm²

**Find the circumference of a circle with the given area in terms of $\pi$.**

**B** **14.** $A = 16\pi$                    **15.** $A = 4\pi$

**Find the area of the shaded region. Leave your answer in terms of $\pi$.**

**16.**

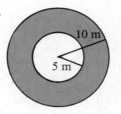

**17.**

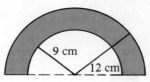

**18.**

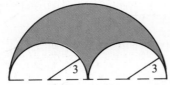

**19.**

**C 20.** Find a formula giving the area of a circle in terms of the diameter.

**21.** Find the formula giving the area of a circle in terms of the circumference.

## Problems

**Solve. Draw a sketch illustrating the problem if necessary. Use $\pi \approx 3.14$ and round the answer to three digits.**

**A 1.** A circular lawn 10 m in diameter is to be resodded at a cost of $14 per m². Find the total cost.

**2.** The Connaught Centre building in Hong Kong has 1748 circular plate glass windows, each 2.4 m in diameter. If glass costs $12 per m², what is the cost of the glass in a single window?

**3.** The circumference of a circular pond is 62.8 m. If its diameter is 20 m, find the area.

**B 4.** A circular pond 20 m in diameter is surrounded by a gravel path 5 m wide. The path is to be replaced by a brick walk costing $30 per square meter. How much will the walk cost?

**5.** The inner and outer radii of the grooved part of a phonograph record are 7 cm and 14 cm. What is the area of the grooved part?

**Hint for Problems 6 and 7: Let the radius of the circle be 1 unit.**

**C 6.** A circle is inscribed in a square. What fraction of the area of the square is taken up by the circle?

**7.** A square is inscribed in a circle. What fraction of the area of the circle is taken up by the square?

## Review Exercises

**Estimate to the nearest whole number.**

**1.** $(3.9)^2$ **2.** $(5.3)^2$ **3.** $(2.72)^2$ **4.** $(6.18)^2$

**5.** $2.7 \times 6$ **6.** $3.1 \times 4.9$ **7.** $3.14 \times 5.3$ **8.** $4.92 \times 5.13$

# 10-4 Using Symmetry to Find Areas

We say that the figure at the right is **symmetric with respect to a line,** $\overleftrightarrow{AB}$, because if it were folded along $\overleftrightarrow{AB}$, the lower half would fall exactly on the upper half. $\overleftrightarrow{AB}$ is called a **line of symmetry.**

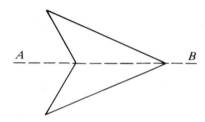

The diagrams below show that figures may have more than one line of symmetry.

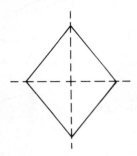 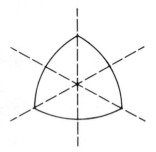 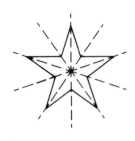

Symmetry can be helpful in finding areas.

**EXAMPLE 1**   Find the area of the symmetric figure.

**Solution**       The symmetry of this figure is such that the four triangles are congruent. Therefore,

$$A = 4 \times \tfrac{1}{2}bh$$

$$= 4 \times \left(\tfrac{1}{2} \times 7 \times 6\right) = 84$$

Thus, the area is 84 square units.

Figures can also be **symmetric with re-spect to a point.** Although the figure at the right has no line of symmetry, it is symmetric with respect to point $O$. A figure is symmetric with respect to a point, $O$, if for every point $P$ on the figure there corresponds an opposite point $Q$ on the figure such that $O$ is the mid-point of the segment $\overline{PQ}$. Every line through the point of symmetry divides the figure into two congruent figures.

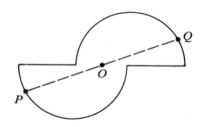

**EXAMPLE 2** Find the area of the symmetric figure.

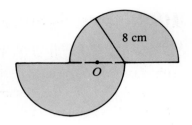

8 cm

**Solution** The dashed line divides the region into two semicircles of radius 8 cm.

$$A = 2 \times \left(\frac{1}{2}\pi r^2\right)$$

$$= 2 \times \left(\frac{1}{2}\pi \times 8^2\right)$$

$$\approx 3.14 \times 64 = 200.96$$

The area is approximately 201 cm².

## Class Exercises

**Copy each figure and show on your drawing any lines or points of symmetry.**

1.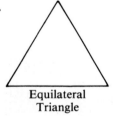

Equilateral
Triangle

2.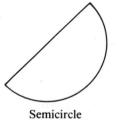

Semicircle

3.

Square

4.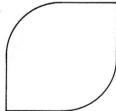

5.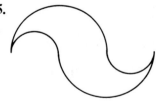

6.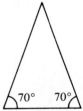

70°    70°

7.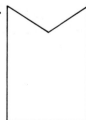

8.

9.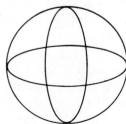

## Written Exercises

Copy each figure and show on your drawing any lines or points of symmetry.

**A**  **1.**

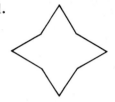

**2.**

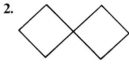

**3.**

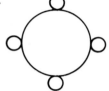

**4.**

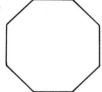

**5.**

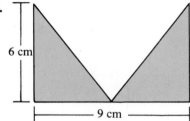

**6.**

Find the areas of the following symmetric figures. Give your answers in terms of $\pi$ if necessary.

**7.**

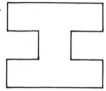

12 m

14 m

**8.**

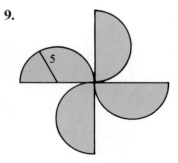

6 cm

9 cm

**B**  **9.**

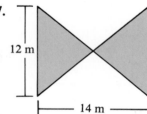

5

**10.**

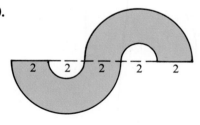

2  2  2  2  2

**11.**

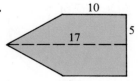

10

17

5

**12.**

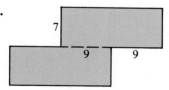

7

9  9

**13.**

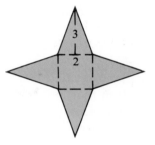

**14.**

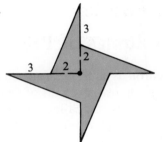

**15.**

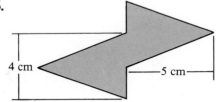

**16.**

---

## Self-Test A

**Find the area. Round the answer to three digits if necessary.**

**1.** rectangle
length: 21 cm
width: 13 cm

**2.** parallelogram
base: 7 m
height: 12 m

[10–1]

**3.** triangle
base: 18 cm
height: 8 cm

**4.** trapezoid
bases: 11 m and 7 m
height: 4 m

[10–2]

**5.** circle: Use $\pi \approx 3.14$.

radius: 15 mm

**6.** circle: Use $\pi \approx \frac{22}{7}$.

diameter: 28 cm

[10–3]

**7.** Copy the figure and show any lines or points of
symmetry.

**8.** Find the area of the symmetric figure.

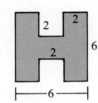

[10–4]

---

*Self-Test answers and Extra Practice are at the back of the book.*

# 10-5 Volumes of Prisms and Cylinders

A **polyhedron** is a figure formed of polygonal parts of planes, called **faces,** that enclose a region of space. A **prism** is a polyhedron that has two congruent faces, called **bases,** that are parallel. The other faces are regions bounded by parallelograms. The bases may also be parallelograms. Prisms are named according to their bases. Unless otherwise stated, we will only consider prisms whose other faces are rectangles.

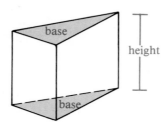

**Triangular Prism**

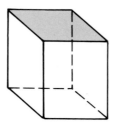

**Square Prism**

In each figure above, the bases are shaded. The perpendicular distance between the bases is the **height** of the prism.

A polyhedron together with the region inside it is called a **solid.** The measure of the space occupied by a solid is called the **volume** of the solid. The prism at the right, filled with 3 layers of 8 unit cubes, has 24 unit cubes. In this case, each unit cube is a cubic centimeter ($cm^3$). Thus the volume of the cube is 24 $cm^3$.

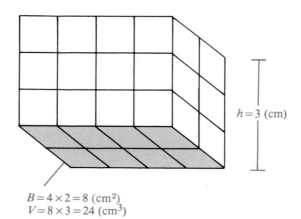

$h = 3$ (cm)

$B = 4 \times 2 = 8$ ($cm^2$)
$V = 8 \times 3 = 24$ ($cm^3$)

This example suggests a formula for finding the volume of any prism.

> ### *Formula*
> Volume of prism = base area × height
> $$V = Bh$$

**EXAMPLE 1**   A watering trough is in the form of a trapezoidal prism. Its ends have the dimensions shown. How long is the trough if it holds 12 m³?

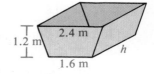

**Solution**   Examine the diagram carefully to determine which regions are the bases. In this diagram, one of the bases is at the front. You know that the volume is 12 m³. Find the area of the base.

$$B = \frac{1}{2}(1.6 + 2.4) \times 1.2 = 2.4 \ (\text{m}^2)$$

Then, use the formula.

$$V = Bh$$
$$12 = 2.4h$$
$$5 = h$$

The length of the trough is 5 m.

A **cylinder** is like a prism except that its bases are circles instead of polygons. We will only consider cylinders with congruent bases. The area of the base, $B$, is $\pi r^2$. Thus:

---

### Formula

Volume of cylinder = base area × height

$$V = \pi r^2 h$$

---

The volume of a container is often called its **capacity.** The capacity of containers of fluids is usually measured in **liters** (L) or **milliliters** (mL).

$$1 \text{ L} = 1000 \text{ cm}^3 \qquad 1 \text{ mL} = 1 \text{ cm}^3$$

It is easy to show that 1 m³ = 1000 L.

**EXAMPLE 2**   A cylindrical storage tank 1 m in diameter is 1.2 m high. Find its capacity in liters. Use $\pi \approx 3.14$.

**Solution**   Use the formula $V = \pi r^2 h$. The height is 1.2 m and, because the diameter is 1 m, the radius is 0.5 m.

$$V \approx 3.14 \times (0.5)^2 \times 1.2 = 0.942 \ (\text{m}^3)$$

Since 1 m³ = 1000 L, 0.942 m³ = 942 L. The capacity of the tank is approximately 942 L.

## Class Exercises

**Find the volume of the solid. In Exercises 5 and 6, leave your answer in terms of $\pi$.**

**1.**
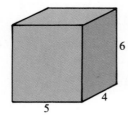
6

5

4

**2.**
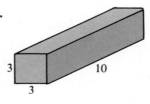
3

3

10

**3.**
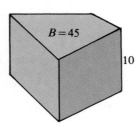
$B = 45$

10

**4.**
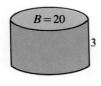
$B = 20$

3

**5.**
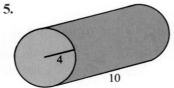
4

10

**6.**

5

4

**7. a.** 1 m = _?_ cm
   **b.** 1 m³ = _?_ cm³
   **c.** 1000 L = _?_ cm³

**8. a.** 1 cm = _?_ mm
   **b.** 1 cm³ = _?_ mm³
   **c.** 1 mL = _?_ mm³

## Written Exercises

**In this exercise set, use $\pi \approx 3.14$ and round the answer to three digits.**

**Find the volume of the solid.**

**A**

**1.**
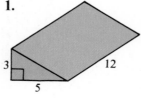
3

5

12

**2.**
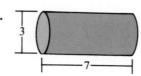
3

7

**3.**
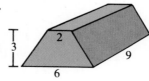
2

3

6

9

**4.**
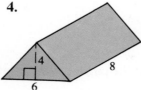
4

6

8

**5.**
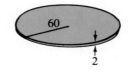
60

2

**6.**
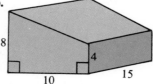
8

4

10

15

**366**   *Chapter 10*

**7.**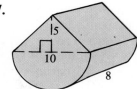

**8.**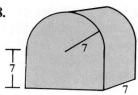

**Find the capacity in liters of the prism or cylinder.**

**9.** Square prism: 25 cm by 25 cm by 80 cm

**10.** Prism: base area = 1.2 m², height = 2.3 m

**11.** Cylinder: base radius = 1.2 m, height = 1.4 m

**12.** Cylinder: base diameter = 1 m, height = 75 cm

**The table below refers to cylinders. Copy and complete it. Leave your answers in terms of $\pi$.**

**B**

|  | **13.** | **14.** | **15.** | **16.** | **17.** | **18.** |
|---|---|---|---|---|---|---|
| **volume** | $12\pi$ | $150\pi$ | $100\pi$ | $20\pi$ | $18\pi$ | $100\pi$ |
| **base radius** | 2 | 5 | ? | ? | ? | ? |
| **base area** | ? | ? | $25\pi$ | $4\pi$ | ? | ? |
| **height** | ? | ? | ? | ? | 2 | 25 |

**Find the volume of the prism if the pattern were folded.**

**19.**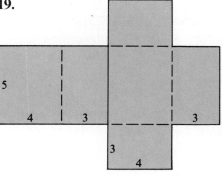

**20.**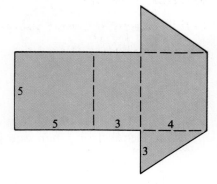

**C** **21.** What happens to the volume of a cylinder if this change is made?
   **a.** The radius is doubled.          **b.** The height is halved.
   **c.** The radius is doubled and the height is halved.

**22.** What happens to the volume of a cylinder if this change is made?
   **a.** The radius is halved.          **b.** The height is doubled.
   **c.** The radius is halved and the height is doubled.

In an *oblique prism,* the bases are *not* perpendicular to the other faces. The volume formula $V = Bh$ still applies, where $h$ is the perpendicular distance between the bases. Find the volume of each figure shown in red.

**23.**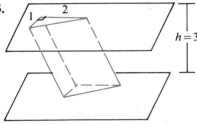

$h = 3$

**24.**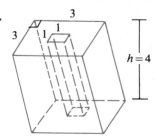

$h = 4$

## Problems

**Solve. Use $\pi \approx 3.14$ and round the answer to three digits.**

**A** **1.** Find the capacity in liters of the V-shaped trough shown below.

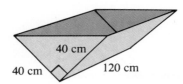

40 cm
40 cm
120 cm

**2.** Find the capacity in liters of the half-cylinder trough shown below.

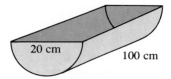

20 cm
100 cm

**B** **3.** Find the volume of metal in the copper pipe shown below.

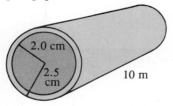

2.0 cm
2.5 cm
10 m

**4.** Find the volume of concrete in the construction block shown below.

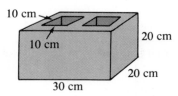

10 cm
10 cm
20 cm
20 cm
30 cm

**C** **5.** A cylindrical water bottle is 28 cm in diameter. How many centimeters does the water level drop when one liter is drawn off?

## Review Exercises

**Solve for $x$.**

**1.** $x = \frac{1}{3} \times 6.4 \times 15$     **2.** $x = \frac{1}{2}(7.2) \times 3$     **3.** $3.38 = 4x$     **4.** $\frac{1}{3}x = 11$

**5.** $3.14 \times 20^2 = x$     **6.** $(0.8)^2x = 5.12$     **7.** $314 = 3.14x^2$     **8.** $x^2 = 169$

# 10-6 Volumes of Pyramids and Cones

If we shrink one base of a prism to a point, we obtain a **pyramid** with that point as its **vertex.** A **cone** is obtained in the same way from a cylinder. In each case, the **height** of the solid is the perpendicular distance from its vertex to its base. As for a prism, the shape of the base of a pyramid determines its name. All other faces of a pyramid are triangles. A cone is like a pyramid except that its base is a circle instead of a polygon.

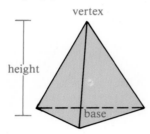

**Triangular Pyramid
or Tetrahedron**

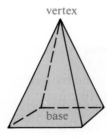

**Square Pyramid**

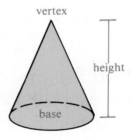

**Cone**

The volume of a pyramid can be found using this formula:

> ### *Formula*
>
> Volume of pyramid $= \frac{1}{3} \times$ base area $\times$ height
>
> $$V = \frac{1}{3}Bh$$

**EXAMPLE 1**    Find the volume of the square pyramid shown at the right.

**Solution**    First find the base area. Since the base is a square,

$$B = 12^2 = 144 \,(\text{cm}^2).$$

Then use the volume formula.

$$V = \frac{1}{3}Bh$$

$$= \frac{1}{3} \times 144 \times 25 = 1200 \,(\text{cm}^3)$$

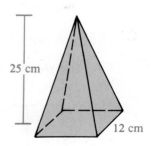

For a cone, the base area is given by the formula $B = \pi r^2$.

**Formula**

Volume of cone $= \frac{1}{3} \times$ base area $\times$ height

$$V = \frac{1}{3}\pi r^2 h$$

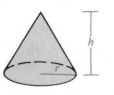

**EXAMPLE 2**　A conical container is 20 cm across the top and 21 cm deep. Find its capacity in liters. Use $\pi \approx 3.14$ and round to three digits.

**Solution**　The diameter of the base of the cone is 20 cm, so the radius is 10 cm.

$$V = \frac{1}{3}\pi \times 10^2 \times 21$$

$$\approx \frac{1}{3} \times 3.14 \times 100 \times 21 = 2198$$

Rounding the product to three digits, the volume is approximately 2200 cm³. Because 1000 cm³ equals 1 L, the capacity of the conical container is approximately 2.2 L.

**Reading Mathematics: *Vocabulary***

Words that we use in everyday speech may have different meanings in mathematics. For example, the everyday word *base* often refers to the part of an object that it is resting on. The geometrical term *base* refers to a particular face of a figure that may appear at the top, the side, or the end of the figure in a drawing. In Example 2, above, the base appears at the top of the container.

## Class Exercises

**Find the volume of the solid. In Exercise 3, the base is a square. In Exercise 4, leave your answer in terms of $\pi$.**

**1.**

$h = 10$

$B = 30$

**2.**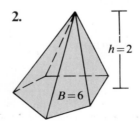

$h = 2$

$B = 6$

**3.**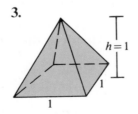

$h = 1$

1

1

**4.**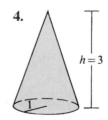

$h = 3$

1

**Complete.**

5. Except for the base, the shapes of all the faces of a pyramid are __?__ .

6. The shapes of all the faces of a tetrahedron are __?__ .

7. A cone has base radius 3 and height 6. The volume of the cone is __?__ π.

---

## Written Exercises

**For Exercises 1–10, use π ≈ 3.14 and round the answer to three digits if necessary.**

**Find the volume of the solid pictured.**

**A** 1.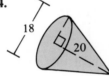

2.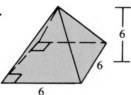

3.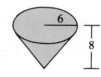

4.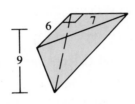

5.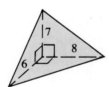

6.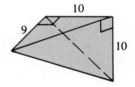

**Find the capacity in liters of the cone or pyramid described.**

7. A pyramid of height 12 cm having a 5 cm by 8 cm rectangle as base

8. A pyramid having height 10 m and a square base 15 m on a side

9. A cone having base radius 0.8 m and height 1.5 m

10. A cone of height 24 cm and base diameter 11 cm

**Copy and complete the table for pyramids.**

|              | 11. | 12. | 13. | 14. | 15. | 16. |
|--------------|-----|-----|-----|-----|-----|-----|
| base area, *B* | 5   | 33  | 4   | 13  | ?   | ?   |
| height, *h*    | 6   | 12  | ?   | ?   | 10  | 15  |
| volume, *V*    | ?   | ?   | 8   | 13  | 10  | 20  |

**Copy and complete the table for cones.**

**B**

| | 17. | 18. | 19. | 20. |
|---|---|---|---|---|
| radius, $r$ | 1 | 2 | ? | ? |
| height, $h$ | ? | ? | 4 | 15 |
| volume, $V$ | $5\pi$ | $12\pi$ | $12\pi$ | $5\pi$ |

**Complete.**

**C** **21.** A cone has volume 6 cm³. The volume of a cylinder having the same base and same height is __?__ cm³.

**22.** A cylinder of height 2 m has the same base and same volume as a cone. The cone's height is __?__ m.

## Problems

**Solve. Use $\pi \approx 3.14$ and round the answer to three digits.**

**A** **1.** The Great Pyramid in Egypt has a square base approximately 230 m on a side. Its original height was approximately 147 m. What was its approximate volume originally?

**2.** A volcano is in the form of a cone approximately 1.8 km high and 12 km in diameter. Find its volume.

**B** **3.** Find the volume of this buoy.

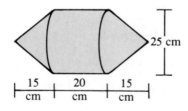

25 cm

15 cm   20 cm   15 cm

**4.** Find the volume of this tent. The floor of the tent is square.

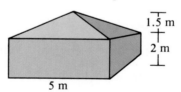

1.5 m
2 m
5 m

## Review Exercises

**Find the perimeter of the figure whose sides have the given lengths.**

**1.** rhombus: 9 m per side

**2.** trapezoid: 4 cm, 4 cm, 5 cm, 7 cm

**3.** triangle: 35 cm per side

**4.** parallelogram: 15 m, 7 m, 15 m, 7 m

**5.** square: 12.8 km per side

**6.** rectangle: 24.63 km by 37.40 km

**7.** square: 145.2 km per side

**8.** triangle: 3.70 m, 12.90 m, 14.85 m

# 10-7 Surface Areas of Prisms and Cylinders

The **surface area** of every prism and cylinder is made up of its two **bases** and its **lateral surface** as illustrated in the figures below. The lateral surface of a prism is made up of its **lateral faces.** Each lateral face is a rectangle.

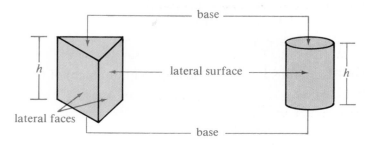

If we were to cut open and flatten out the prism and cylinder shown above, the bases and the lateral surfaces of the figures would look like this:

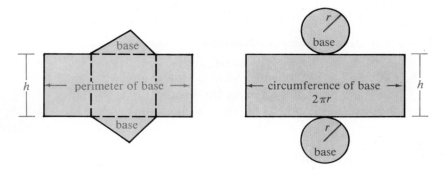

In each of the figures above, the area of the lateral surface, called the **lateral area,** is the product of the perimeter of the base and the height of the figure. To find the **total surface area** of a prism or cylinder, we simply add the area of the two bases to the lateral area of the figure.

---

### *Formulas*

For a prism or cylinder,

lateral area = perimeter of base × height

total surface area = lateral area + area of bases

---

**EXAMPLE 1**  Find (a) the lateral area, (b) the area of the bases, and (c) the total surface area of the prism shown.

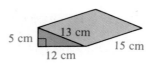

**Solution**
  **a.** perimeter of base = 5 + 12 + 13 = 30 (cm)
  height = 15 cm
  lateral area = 30 × 15 = 450 (cm²)

  **b.** area of bases = $2 \times \left(\frac{1}{2} \times 12 \times 5\right)$ = 60 (cm²)

  **c.** total surface area = 450 + 60 = 510 (cm²)

In a cylinder of base radius $r$, the perimeter (circumference) of the base is $2\pi r$. Thus:

---

### Formulas

For a cylinder,

  lateral area = $2\pi rh$

  area of bases = $2\pi r^2$

  total surface area = $2\pi rh + 2\pi r^2$

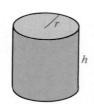

---

**EXAMPLE 2**  A can of paint will cover 50 m². How many cans are necessary to paint the inside (top, bottom, and sides) of the storage tank shown? Use $\pi \approx 3.14$.

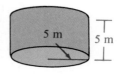

**Solution**
  Since $r = 5$ and $h = 5$,
  total surface area = $(2\pi \times 5 \times 5) + (2\pi \times 5^2)$
  $\approx (3.14 \times 50) + (3.14 \times 50) = 314$

  The total surface area is approximately 314 m².

  The number of cans of paint is 314 ÷ 50, or 6.28. Rounding *up* to the nearest whole number, the answer is 7 cans.

---

**Problem Solving Reminder**

For problems whose answers are the result of division, take time to consider whether it is reasonable *to round up or round down*. In Example 2, the answer to the division was 6.28 cans of paint. Because paint cannot be purchased in hundredths of a can, the answer was rounded up to the nearest whole number. If the answer had been rounded down, there would not have been enough paint to cover the interior of the tank.

## Class Exercises

**1.** If a prism has hexagonal bases, how many lateral faces does it have? How many faces does it have in all?

**Find the lateral area and the total surface area. Use $\pi \approx 3.14$ and round the answer to three digits.**

**2.**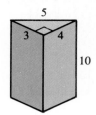

**3.**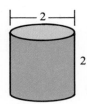

**4.**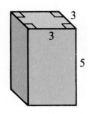

**5.** If each edge of a cube is one unit long, what is the total surface area in square units?

## Written Exercises

**Find the total surface area of a rectangular prism having the given dimensions.**

**A**

**1.** 45 cm by 30 cm by 20 cm

**2.** 5 cm by 18 cm by 25 cm

**3.** 2.5 m by 1.6 m by 0.8 m

**4.** 1.8 m by 0.6 m by 2.0 m

**Find (a) the lateral area and (b) the total surface area of the cylinder or prism. Use $\pi \approx 3.14$ and round the answer to three digits.**

**5.** A cylinder having base radius 12 cm and height 22 cm

**6.** A cylinder having base diameter 6 m and height 4.5 m

**7.**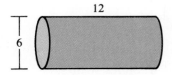

**8.**

**9.**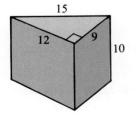

**10.**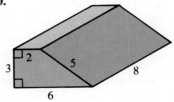

**If the patterns below were drawn on cardboard, they could be folded along the dotted lines to form prisms. Find (a) the volume and (b) the total surface area of each.**

**B 11.**

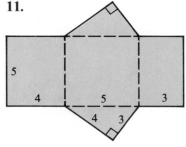

**12.**

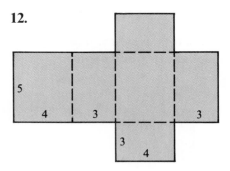

**13.** Find the length of the edge of a cube whose total area is 150 cm².

**14.** Three faces of a box have a common vertex and their combined area is 20 cm². What is the total surface area of the box?

**15.** One can of varnish will cover 64 m² of wood. If you want to put one coat of varnish on each of 24 wooden cubes with the dimensions shown at the right, how many cans of varnish should you buy?

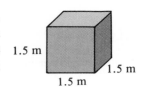

**16.** You have two cans of red paint, each of which will cover 100 m². Which of the two cylinders pictured below can you paint completely using just the paint that you have?

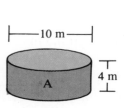
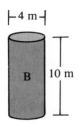

**17.** A prism of height 3 m has bases that are right triangles with sides 6 m, 8 m, and 10 m. Find the lateral area and the total surface area. (Be sure to draw and label a sketch.)

**18.** A prism of height 10 cm has bases that are right triangles with sides 5 cm, 12 cm, and 13 cm. Find the lateral area and the surface area. (Be sure to draw and label a sketch.)

**19.** If the number of faces of a prism is represented by *n*, write an algebraic expression to represent the number of lateral faces.

**C** **20.** The lateral surface of a cone is made up of many small wedges like the one shown in color. Thinking of the wedge as a triangle of base $b$ and height $s$, we see that its area is $\frac{1}{2}bs$. When we add all these areas together, we obtain $\frac{1}{2}(2\pi r)s$, or $\pi rs$, because the sum of all the $b$'s is $2\pi r$, the circumference of the base. Thus, for a cone:

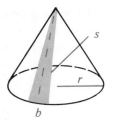

$$\text{lateral area} = \pi rs$$

$$\text{area of base} = \pi r^2$$

$$\text{total surface area} = \pi rs + \pi r^2$$

The length $s$ is called the **slant height** of the cone.

**Find the total surface area of each figure.**

**a.**

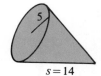

$s = 14$

**b.**

$s = 4$

$3$

$s = 6$

**c.**

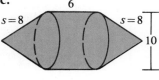

$6$

$s = 8$ $s = 8$ $10$

---

## Review Exercises

**Find the radius of the circle with the given circumference $C$ or diameter $d$. Use $\pi \approx 3.14$ and round the answer to three digits.**

**1.** $d = 12$ **2.** $C = 62.8$ **3.** $d = 210$ **4.** $C = 83.6$

**5.** $C = 220$ **6.** $d = 25$ **7.** $d = 76$ **8.** $C = 2.42$

---

### ▮▮▮ Challenge

**1.** Copy the square at the right and cut it into four pieces along the dashed lines. Re-form the four pieces into a larger square with a "hollow" square at the center.

**2.** Repeat step 1 several times, assigning different values to $a$ and $b$ each time. In each case, what is the area of the "hollow" square?

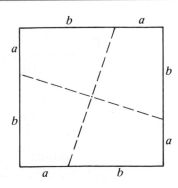

# 10-8 Volumes and Surface Areas of Spheres

The **sphere** with **radius** $r$ and **center** at $C$ consists of all points at the distance $r$ from point $C$. The word *radius* also is used for any segment having $C$ as one endpoint and a point of the sphere as another (for example, $\overline{CP}$ in the figure). The word **diameter** is also used in two ways: for a *segment* through $C$ having its endpoints on the sphere (for example, $\overline{AB}$), and for the *length* of such a segment.

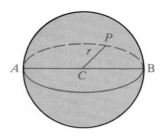

The following formulas for the surface area and volume of a sphere can be proved using higher mathematics.

---

### *Formulas*

For any sphere of radius $r$,

Area $= 4\pi \times (\text{radius})^2$                Volume $= \frac{4}{3}\pi \times (\text{radius})^3$

$A = 4\pi r^2$                              $V = \frac{4}{3}\pi r^3$

---

The surface area of a sphere is usually referred to simply as the *area* of the sphere.

---

**Reading Mathematics: *Diagrams***
Reading diagrams correctly is an important reading skill in mathematics. When three-dimensional objects are pictured on a two-dimensional page, the lines used to draw congruent segments may be of different lengths. For example, in the drawing above, you understand that all radii of a sphere are congruent and that $CP = CB$. Yet these radii cannot be drawn to be of equal length if the picture is to look realistic.

*EXAMPLE 1*  Find (a) the surface area and (b) the volume of a sphere having diameter 18. Leave your answer in terms of $\pi$.

*Solution*  The radius equals $\frac{1}{2}$ the diameter, so $r = 9$.
**a.** $A = 4\pi r^2 = 4\pi \times 9^2 = 4\pi \times 81 = 324\pi$

**b.** $V = \frac{4}{3}\pi r^3 = \frac{4}{3}\pi \times 9^3 = \frac{4}{3}\pi \times 729 = 972\pi$

**EXAMPLE 2**   The area of a sphere is $1600\pi$ cm². What is the radius?

**Solution**      Substitute $1600\pi$ for $A$ in the formula $A = 4\pi r^2$.

$$1600\pi = 4\pi r^2$$

$$\frac{1600\pi}{4\pi} = \frac{4\pi r^2}{4\pi}$$

$$400 = r^2$$

Since $20^2 = 400$, $r = 20$ cm.

## Class Exercises

**Complete. In Exercises 1 and 2, give your answers in terms of $\pi$.**

1. The area of a sphere of radius 1 is __?__, and the volume is __?__.

2. The area of a sphere of radius 2 is __?__, and the volume is __?__.

3. If the radius of a sphere is doubled, the area is multiplied by __?__.

4. If the radius of a sphere is doubled, the volume is multiplied by __?__.

5. If $\overline{AB}$ is a diameter of a sphere having center $C$, then $\overline{AC}$ and $\overline{BC}$ are __?__ of the sphere.

**Complete the following analogies with choice a, b, c, or d.**

6. Sphere: Circle = Cube: __?__
   **a.** Pyramid          **b.** Prism          **c.** Square          **d.** Cylinder

7. Cone: Pyramid = Cylinder: __?__
   **a.** Sphere          **b.** Prism          **c.** Circle          **d.** Triangle

## Written Exercises

**Copy and complete the table below. Leave your answers in terms of $\pi$.**

**A**

|                  | **1.** | **2.** | **3.** | **4.** | **5.** | **6.** |
|------------------|--------|--------|--------|--------|--------|--------|
| radius of sphere | 3      | 5      | 6      | 9      | 10     | 12     |
| surface area     | ?      | ?      | ?      | ?      | ?      | ?      |
| volume           | ?      | ?      | ?      | ?      | ?      | ?      |

**Half of a sphere is called a *hemisphere*. Solve. Leave your answers in terms of π.**

7. Find the volume of a hemisphere of radius 2.

8. Find the area of the curved surface of a hemisphere of radius 8.

9. Find the area of the curved surface of a hemisphere of radius 2.2.

10. Find the volume of a hemisphere of radius 3.3.

**Copy and complete the table below. Leave your answers in terms of π.**

B

|  | 11. | 12. | 13. | 14. | 15. | 16. |
|---|---|---|---|---|---|---|
| radius of sphere | 3.6 | ? | ? | 4.5 | ? | ? |
| surface area | ? | $64\pi$ | ? | ? | $256\pi$ | ? |
| volume | ? | ? | $972\pi$ | ? | ? | $288\pi$ |

**Solve. Leave your answers in terms of π.**

17. The observatory building shown at the right consists of a cylinder surmounted by a hemisphere. Find its volume.

18. The water tank at the right consists of a cone surmounted by a cylinder surmounted by a hemisphere. Find its volume.

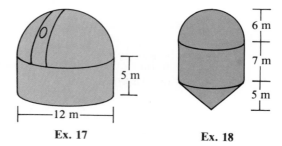

Ex. 17          Ex. 18

19. Earth's diameter is about $3\frac{2}{3}$ times that of the moon. How do their volumes compare?

20. The sun's diameter is about 110 times that of Earth. How do their volumes compare?

If a sphere fits snugly inside a cylinder it is said to be *inscribed* in the cylinder. Use the diagram at the right for Exercises 21 and 22.

C 21. a. A sphere of radius 4 is inscribed in a cylinder of height 8. What is the ratio of the volume of the sphere to the volume of the cylinder?

   b. A sphere of radius 5 is inscribed in a cylinder of height 10. What is the ratio of the volume of the sphere to the volume of the cylinder?

   c. A sphere of radius *r* is inscribed in a cylinder. What is the ratio of the volume of the sphere to the volume of the cylinder?

22. a. A sphere of radius 6 is inscribed in a cylinder of height 12. What is the ratio of the surface area of the sphere to the lateral area of the cylinder?

   b. A sphere of radius 7 is inscribed in a cylinder of height 14. What is the ratio of the surface area of the sphere to the lateral area of the cylinder?

   c. A sphere of radius *r* is inscribed in a cylinder. What is the ratio of the surface area of the sphere to the lateral area of the cylinder?

## Review Exercises

**Simplify these in your head if you can. Write down the answers.**

1. $7 \times 0.3$
2. $4 \times 5.1$
3. $0.8 \times 0.8$
4. $23.3 \times 100$
5. $3500 \div 1000$
6. $10.75 \div 100$
7. $300 \times 2.7$
8. $0.0004 \times 0.2$

### ▌▌▌ Calculator Key-In

The surface area of Earth is approximately 510,070,000 km². The surface area of Jupiter is approximately 64,017,000,000 km². About how many times greater than the surface area of Earth is the surface area of Jupiter?

If you try to enter the numbers as they are shown above on your calculator, you may find it will not accept more than eight digits. Many calculators have a key marked *EXP* that allows you to use scientific notation to express very large (or very small) numbers. For example, you can enter 510,070,000 by thinking of the number as $5.1007 \times 10^8$ and entering 5.1007 [EXP] 8. Your calculator may show 5.1007   08.

Try to find a calculator that will accept scientific notation and solve the problem above.

# 10-9 Mass and Density

The **mass** of an object is a measure of the amount of matter it contains. In the metric system, units of mass are the **gram (g)**, the **kilogram (kg)**, and the **metric ton (t)**.

> 1 g = mass of 1 cm³ of water under standard conditions. (Standard conditions are 4°C at sea-level pressure.)
>
> $$1 \text{ kg} = 1000 \text{ g}$$
>
> $$1 \text{ t} = 1000 \text{ kg}$$

The *weight* of an object is the force of gravity acting on it. While the mass of an object remains constant, its weight would be less on a mountaintop or on the moon than at sea level. In a given region, mass and weight are proportional (so that mass can be found by weighing).

The tables below give the masses of unit volumes (1 cm³) of several substances.

**TABLE 1**

| Substance | Mass of 1 cm³ | Mass of 1 m³ |
|-----------|---------------|--------------|
| Pine | 0.56 g | 0.56 t |
| Ice | 0.92 g | 0.92 t |
| Water | 1.00 g | 1.00 t |
| Aluminum | 2.70 g | 2.70 t |
| Steel | 7.82 g | 7.82 t |
| Gold | 19.3 g | 19.3 t |

**TABLE 2**

| Substance | Mass of 1 cm³ | Mass of 1 L |
|-----------|---------------|-------------|
| Helium | 0.00018 g | 0.00018 kg |
| Air | 0.0012 g | 0.0012 kg |
| Gasoline | 0.66 g | 0.66 kg |
| Water | 1.00 g | 1.00 kg |
| Milk | 1.03 g | 1.03 kg |
| Mercury | 13.6 g | 13.6 kg |

The mass per unit volume of a substance is its **density.** In describing densities we use a slash (/) for the word *per*. For example, the density of gasoline is 0.66 kg/L.

**EXAMPLE**  Find the mass of a block 3 cm by 5 cm by 10 cm made of wood having density 0.8 g/cm³.

**Solution**  The volume of the block is 3 × 5 × 10, or 150, cm³.

Since 1 cm³ of the wood has mass 0.8 g, 150 cm³ has mass 0.8 × 150, or 120, g.

The example illustrates the following formula.

> ## Formula
> Mass = Density × Volume

In Table 2, the number of grams per cubic centimeter is the same as the number of kilograms per liter. This is because there are 1000 g in a kilogram and 1000 cm$^3$ in a liter. Similarly in Table 1, the number of grams per cubic centimeter is the same as the number of metric tons per cubic meter.

## Class Exercises

**Complete.**

**1.** 1 t = __?__ kg        **2.** 1 kg = __?__ g        **3.** 1 t = __?__ g

**4.** 2400 kg = __?__ t     **5.** 0.62 kg = __?__ g     **6.** 400 g = __?__ kg

**Use the tables in this lesson to give the mass of the following.**

**7.** 2 L of water

**8.** 10 cm$^3$ of gasoline

**9.** 100 cm$^3$ of milk

**10.** 4 L of gasoline

**11.** helium filling a 1000 L tank

**12.** a cube of ice 2 cm on each edge

**13.** the block of steel shown

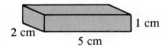

2 cm    5 cm    1 cm

## Written Exercises

**Complete.**

**A**  **1.** 6.3 kg = __?__ g     **2.** 2500 kg = __?__ t     **3.** 4.3 t = __?__ kg

**Use the tables in this lesson to find the mass of the following.**

**4.** 50 cm$^3$ of gold        **5.** 300 cm$^3$ of aluminum     **6.** 25 L of gasoline

**7.** 2.5 L of mercury        **8.** 500 m$^3$ of water         **9.** 500 m$^3$ of ice

**10.** 1 m$^3$ of air          **11.** 1 m$^3$ of helium         **12.** 4 L of milk

In Exercises 13–18, use $\pi \approx 3.14$ and round the answer to three digits.
In Exercises 13–15, the solid pictured is made of the specified material.
Use the tables in this lesson to find the mass of the solid.

**13.**

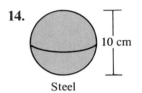

3 cm
3 cm
4 cm
Ice

**14.**

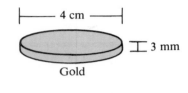

10 cm
Steel

**15.**

4 cm
3 mm
Gold

In Exercises 16–18, use the tables in this lesson to find the mass of the
named content of the container pictured.

**16.**

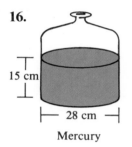

15 cm
28 cm
Mercury

**17.**

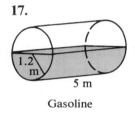

1.2 m
5 m
Gasoline

**18.**

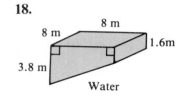

8 m
8 m
1.6 m
3.8 m
Water

---

## Problems

Solve. Use the tables in this lesson if no density is given. Use $\pi \approx 3.14$
and round the answer to three digits.

**A**  **1.** A solid gold bar has dimensions of
approximately 17 cm by 9 cm by
4.5 cm. Find its mass in kilograms.

**2.** A solid pine board has dimensions of
approximately 100 cm by 30 cm by
1.5 cm. Find its mass in grams.

**3.** The optical prism shown at the right
is made of glass having density
4.8 g/cm³. Find its mass.

**4.** A piece of ice is in the form of a
half-cylinder 5 cm in radius and
3 cm high. What is its mass?

**B**  **5.** What is the mass in metric tons of the
steel I-beam shown at the right?

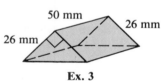

50 mm
26 mm
26 mm
Ex. 3

Ex. 4

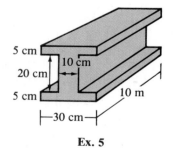

5 cm
10 cm
20 cm
5 cm
10 m
30 cm
Ex. 5

**6.** The drawing at the right shows the cross section of a two-kilometer-long tunnel that is to be dug through a mountain. How many metric tons of earth must be removed if its density is 2.8 t/m³?

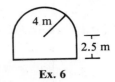

**Ex. 6**

---

## Self-Test B

**Use $\pi \approx 3.14$ and round answers involving $\pi$ to three digits.**

**Complete.**

**1.** Triangular prism: base area = 7.6 m², height = 5.5 m, volume = __?__ m³                                                                 [10–5]

**2.** Cylinder: base diameter = 38 cm, height = 24 cm, volume = __?__ cm³

**3.** Square prism: 17 cm by 17 cm by 32 cm, volume = __?__ cm³, capacity = __?__ L

**Find the volume of the solid.**

**4.** A cone with base radius 0.5 m and height 1.2 m                                                                 [10–6]

**5.** A pyramid with base of 180 m² and height 20 m

**6.** A square pyramid with base 26 cm on a side and height 48 cm

**Find (a) the lateral area and (b) the total surface area of the solid.**

**7.** A rectangular prism with base 4 by 3 and height 7                                                                 [10–7]

**8.** A cylinder with base diameter 10 and height 3

**9.** A square prism with height 42 and base 18 per side

**Find (a) the surface area and (b) the volume of a sphere with the given dimensions. Leave your answers in terms of $\pi$.**

**10.** radius = 18 cm                          **11.** diameter = 12 m                                                                 [10–8]

**Find the mass of the solid.**

**12.** A sphere of aluminum with diameter 18 cm, density 2.70 g/cm³                                                                 [10–9]

**13.** A 4 cm by 4 cm by 20 cm block of pine, density of 0.568 g/cm³

---

*Self-Test answers and Extra Practice are at the back of the book.*

# Locating Points on Earth

We can think of Earth as a sphere rotating on an axis that is a diameter with endpoints at the North Pole (N) and South Pole (S). A **great circle** on a sphere is the intersection of the sphere with a plane that contains the center of the sphere. The great circle whose plane is perpendicular to Earth's axis is called the **equator.** The equator divides Earth into the Northern and Southern Hemispheres.

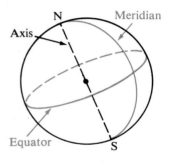

Semicircles with endpoints at the North and South Poles are called **meridians.** The meridian that passes through Greenwich, England, is called the **prime meridian.** The great circle on which the prime meridian lies divides Earth into the Eastern and Western Hemispheres.

The prime meridian and the equator are important parts of a degree-coordinate system that we use to describe the location of points on Earth. A series of circles whose planes are parallel to the equator, called **parallels of latitude,** identify the **latitude** of a point as a number of degrees between 0° and 90° *north* or *south of the equator.* The meridians identify the **longitude** of a point as the number of degrees between 0° and 180° *east* or *west of the prime meridian.*

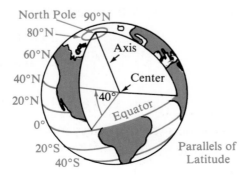

Parallels of Latitude

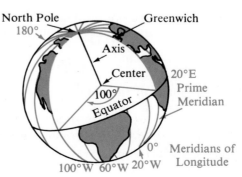

Meridians of Longitude

Thus, each point on the surface of Earth can be assigned an ordered pair of degree-coordinates: (latitude, longitude).

The flat map below, called a **Mercator projection,** shows the meridians and parallels of latitude marked off in 10° intervals as perpendicular lines. Notice that the city of Paris, France, is located at about 48°N, 2°E.

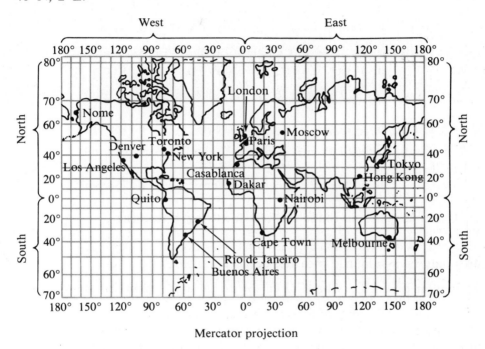

Mercator projection

**Use the map above to name the major city at the location specified.**

1. 40°N, 74°W
2. 34°S, 58°W
3. 23°S, 43°W
4. 34°S, 18°E

5. 55°N, 37°E
6. 37°S, 145°E
7. 33°N, 7°W
8. 34°N, 118°W

**Give the latitude and longitude of the following cities to the nearest 5°.**

9. Denver
10. Nairobi
11. London
12. Quito

13. Dakar
14. Toronto
15. Nome
16. Hong Kong

**Career Activity**

Ancient navigators determined their position by observing the sun and the stars. Modern navigators use much more sophisticated methods. If you were a navigator today what are some of the instruments and methods you might use?

# Chapter Review

**Write the letter of the correct answer.**

1. Find the area of a rectangle 17 cm long and 5 cm wide. [10–1]
   **a.** 44 cm²      **b.** 85 cm²      **c.** 42.5 cm²      **d.** 22 cm²

2. A parallelogram has a base of 16 m and a height of 4 m. Find the area.
   **a.** 32 m²      **b.** 40 m²      **c.** 20 m²      **d.** 64 m²

3. A triangle has a base of 14 m and a height of 10 m. Find the area. [10–2]
   **a.** 140 m²      **b.** 34 m²      **c.** 70 m²      **d.** 35 m²

4. A trapezoid has bases of 9 cm and 15 cm and a height of 6 cm. Find the area.
   **a.** 72 cm²      **b.** 144 cm²      **c.** 810 cm²      **d.** 189 cm²

5. A circle has a diameter of 18 cm. Find the area. Use $\pi \approx 3.14$ and round to three digits. [10–3]
   **a.** 28.3 cm²      **b.** 1020 cm²      **c.** 56.5 cm²      **d.** 254 cm²

**Complete. Use $\pi \approx 3.14$ and round answers involving $\pi$ to three digits.**

6. Figures may be symmetric with respect to a __?__ or a __?__. [10–4]

7. A rectangular prism has a base 12 cm by 14 cm and height 23 cm. Its volume is __?__. [10–5]

8. A cylinder has base radius 4 cm and height 15 cm. Its capacity is __?__ liters.

9. A cone with base diameter 6 and height 4.8 has volume __?__. [10–6]

10. A pyramid with base area 160 cm² and height 24 cm has volume __?__.

11. A cylinder with base radius 7 and height 12 has lateral area __?__ $\pi$ and surface area __?__ $\pi$. [10–7]

12. A prism has height 16. Its bases are triangles with base 15, height 12, and remaining side 18. Its total surface area is __?__.

13. A sphere with radius 24 has surface area __?__ $\pi$ and volume __?__ $\pi$. [10–8]

14. A sphere with radius 32 has surface area __?__ $\pi$ and volume __?__ $\pi$.

15. A rectangular storage tank has dimensions 14 m by 6 m by 4 m. It is filled with helium of density 0.00018 kg/L. The mass of the helium is __?__. [10–9]

# Chapter Test

**Find the area. Use $\pi \approx 3.14$ and round the answer to three digits.**

1. rectangle:
   length $= 14$ cm
   width $= 9$ m

2. parallelogram:
   base $= 23$ cm
   height $= 11$ cm
   [10–1]

3. triangle:
   base $= 9$ mm
   height $= 16$ mm

4. trapezoid:
   bases $= 8$ m and 14 m
   height $= 7$ m
   [10–2]

5. circle:
   radius $= 17$ mm

6. circle:
   diameter $= 42$ cm
   [10–3]

7. Copy the figure at the right and show any lines or points of symmetry.

8. Find the area of the symmetric figure at the right.

[10–4]

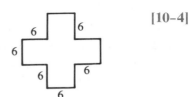

**Find the volume. Use $\pi \approx 3.14$ and round the answer to three digits.**

9. Cylinder: base radius $= 7.2$ m, height $= 5.8$ m
   [10–5]

10. Prism: base area $= 45$ cm$^2$, height $= 33$ cm

11. Cone: base diameter $= 6$ cm, height $= 14$ cm
    [10–6]

12. Pyramid: square base 17 cm on a side, height $= 21$ cm

**Find (a) the lateral area and (b) the total surface area of each. Use $\pi \approx 3.14$ and round the answer to three digits.**

13.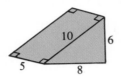

14.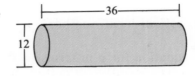

[10–7]

**Find (a) the surface area and (b) the volume of a sphere with the given dimensions. Leave your answer in terms of $\pi$.**

15. diameter $= 18$ cm

16. radius $= 45$ m
    [10–8]

17. Find the mass of a sphere with diameter 20 cm and density 8.9 g/cm$^3$. Use $\pi \approx 3.14$ and round the answer to three digits.
    [10–9]

18. Find the mass of a cube 9 cm on each edge with density 0.56 g/cm$^3$.

# Cumulative Review (Chapters 1–10)

## Exercises

Evaluate the expression if $a = 0.5$, $b = 5$, and $c = 2$.

**1.** $abc$    **2.** $a + b + c$    **3.** $b \div a$    **4.** $3a - 3b$

**5.** $ab^2$    **6.** $c^2 + b^3$    **7.** $6(a^2 - c^2)$    **8.** $10a - \dfrac{b}{c}$

Evaluate the expression if $x = \dfrac{1}{3}$, $y = 2\dfrac{3}{4}$, and $z = -\dfrac{3}{5}$.

**9.** $x + y$    **10.** $xz$    **11.** $3x + z$    **12.** $5z - x$

**13.** $x + 4y - 10z$    **14.** $6x + 8y + z$    **15.** $20z - (-y)$    **16.** $12x \div z$

Solve.

**17.** $15n + 7 > 52$    **18.** $6 - 2x < 21$    **19.** $38 \geq x - 2$

**20.** $\dfrac{n}{8} > 3$    **21.** $27 \leq 3x$    **22.** $3b + 5 < 56$

**23.** $4x + 7 = 35$    **24.** $-y + 10 = 26$    **25.** $\dfrac{3w}{7} = 9$

**26.** Find the measure of each angle of an equilateral triangle.

**27.** The measure of one angle in an isosceles triangle is 90°. Find the measures of the other angles.

**28.** An equilateral triangle has perimeter 37.5 cm. Find the length of each side.

**29.** What is 18% of 45?    **30.** 8 is what percent of 4000?

**31.** 6.48 is what percent of 27?    **32.** 105 is 42% of what number?

**33.** 316 is 0.5% of what number?    **34.** What is $34\frac{1}{2}\%$ of 1100?

Solve the equation for $y$ in terms of $x$.

**35.** $4x + y = 18$    **36.** $x + 2y = 15$    **37.** $6x - 6y = 50$

**38.** $\dfrac{x}{2} + y = 17$    **39.** $xy = 20$    **40.** $x^2 + y = 100$

Find the area of the circle described. Leave your answer in terms of $\pi$.

**41.** radius 14 cm    **42.** diameter 46 mm    **43.** radius 125 km

**44.** diameter 286 m    **45.** radius 1805 cm    **46.** diameter 305.6 m

**Find the surface area and volume for the sphere described. Leave your answer in terms of $\pi$.**

**47.** radius 8          **48.** radius 15          **49.** radius 36

---

## Problems

> **Problem Solving Reminders**
> Here are some reminders that may help you solve some of the problems on this page.
> • Determine what information is necessary to solve the problem.
> • Consider whether drawing a sketch will help.
> • If more than one method can be used to solve a problem, use one method to solve and one to check.

**Solve.**

**1.** The Aleutian Trench in the Pacific Ocean is 8100 m deep. Each story of an average skyscraper is about 4.2 m high. How many stories would a skyscraper as tall as the Aleutian Trench have? Round your answer to the nearest whole number.

**2.** These announcements were heard at a rocket launch: "Minus 45 seconds and counting," and "We have second-stage ignition at plus 110 seconds." How much time elapsed between the announcements?

**3.** A rancher bought 15 fence sections at $58 each to complete one side of a corral. If the length of the side is 375 m and each section is the same length, what is the length of a section?

**4.** George had 15 flat tires last year. The cost of repairing each tire was $6.35 plus 5% tax. How much did he spend on repairs for the year?

**5.** Fred bought a bag of nuts. He gave $\frac{1}{4}$ of it to one brother, $\frac{1}{6}$ to another, $\frac{1}{3}$ to his sister, and kept the rest. How much did he keep?

**6.** Gus is canning tomato sauce. He has 20 jars that have a diameter of 9 cm and height of 9 cm. If he must leave one centimeter of air space at the top of each jar, what volume of sauce can a jar hold? Use $\pi \approx 3.14$ and round the answer to three digits.

**7.** The cost of an average basket of groceries rose from $78.80 in April to $82.74 in June. What was the percent of increase?

**8.** A roll of 36-exposure film costs $5.85. Processing for color prints costs $.42 per print. What is the total cost for film and processing of 36 prints?

# 11

# Applying Algebra to Right Triangles

One of the basic shapes used in building hang-gliders such as the one shown at the right is the triangle. Triangles are especially well-suited for use in building because they are the simplest rigid forms. A rigid form is a figure that preserves its shape under pressure. Because of their rigidity, triangles are used in the construction of many large-scale projects, such as bridges, towers, and statues.

One important type of triangle is the right triangle. The ratios of the lengths of the sides of a right triangle are called trigonometric ratios. In this chapter, you will extend your knowledge of triangles and learn about trigonometric ratios. You will also study some methods for finding angles and lengths where triangles are involved.

## Career Note

The job of surveyor is a career in which knowledge of right triangles and trigonometry plays an important role. Surveyors are responsible for establishing legal land boundaries. About half of their time is spent on location measuring sites and collecting data for maps and charts. The remaining part of their time is spent preparing reports, drawing maps, and planning future surveys.

# 11-1 Square Roots

Recall that we can write $b \times b$ as $b^2$ and call it the *square* of $b$. The factor $b$ is a **square root** of $b^2$. A given number $a$ has $b$ as a square root if

$$b^2 = a.$$

Thus 9 has 3 as a square root because $3^2 = 9$.

Every positive number has two square roots, and these are opposites of each other. For example, the square roots of 25 are 5 and $-5$ because

$$5^2 = 5 \times 5 = 25 \quad \text{and} \quad (-5)^2 = (-5) \times (-5) = 25.$$

The only square root of 0 is 0 because $b \times b = 0$ only when $b = 0$.

In this chapter we will work mostly with positive square roots. We use $\sqrt{a}$ to denote the *positive* square root of $a$. Thus $\sqrt{25} = 5$, not $-5$. A symbol such as $2\sqrt{25}$ means *2 times the positive square root of 25*. The negative square root of 25 is $-\sqrt{25}$, or $-5$.

Negative numbers have no real-number square roots because no real number has a square that is negative.

If $\sqrt{a}$ is an integer, we call $a$ a **perfect square.** For example, 36 is a perfect square because $\sqrt{36}$ is the integer 6. Also, 144 is a perfect square because $\sqrt{144} = 12$.

If $a$ is not a perfect square, we can estimate $\sqrt{a}$ by finding the two consecutive integers between which the square root lies. In the process, we use the fact that the smaller of two positive numbers has the smaller positive square root.

**EXAMPLE** Between which two consecutive integers does $\sqrt{40}$ lie?

**Solution** 40 lies between the consecutive perfect squares 36 and 49.

$$36 < 40 < 49$$
$$\sqrt{36} < \sqrt{40} < \sqrt{49}$$
$$\text{Thus} \quad 6 < \sqrt{40} < 7.$$

## Class Exercises

**Read each symbol.**

1. $\sqrt{7}$     2. $3\sqrt{10}$     3. $-\sqrt{81}$     4. $\sqrt{64}$     5. $2\sqrt{14}$

If the given symbol names an integer, state the integer. If not, name the two consecutive integers between which the number lies.

**6.** $\sqrt{16}$     **7.** $-\sqrt{36}$     **8.** $\sqrt{21}$     **9.** $\sqrt{70}$     **10.** $\sqrt{50}$

**11.** $-\sqrt{49}$     **12.** $\sqrt{81}$     **13.** $\sqrt{69}$     **14.** $-\sqrt{144}$     **15.** $\sqrt{169}$

## Written Exercises

If the given symbol names an integer, state the integer. If not, name the two consecutive integers between which the number lies.

**A**   **1.** $\sqrt{43}$     **2.** $\sqrt{64}$     **3.** $-\sqrt{16}$     **4.** $\sqrt{24}$     **5.** $\sqrt{1}$

**6.** $\sqrt{0}$     **7.** $-\sqrt{6^2}$     **8.** $\sqrt{13}$     **9.** $\sqrt{54}$     **10.** $\sqrt{9}$

**11.** $\sqrt{30}$     **12.** $\sqrt{48}$     **13.** $\sqrt{15}$     **14.** $\sqrt{8^2}$     **15.** $\sqrt{2}$

**16.** $\sqrt{25} + \sqrt{16}$     **17.** $\sqrt{100} - \sqrt{49}$     **18.** $\sqrt{144} + \sqrt{25}$

**19.** $\sqrt{79 - 61}$     **20.** $-\sqrt{66 - 2}$     **21.** $\sqrt{100 - 19}$

Replace the __?__ with <, >, or = to make a true statement.

*EXAMPLE*   $\sqrt{9} + \sqrt{25}$ __?__ $\sqrt{9 + 25}$

*Solution*   $\sqrt{9} + \sqrt{25} = 3 + 5 = 8;\ \sqrt{9 + 25} = \sqrt{34} < 8.$
Thus $\sqrt{9} + \sqrt{25} > \sqrt{9 + 25}.$

**B**   **22.** $\sqrt{9} + \sqrt{16}$ __?__ $\sqrt{9 + 16}$         **23.** $\sqrt{16} + \sqrt{4}$ __?__ $\sqrt{16 + 4}$

**24.** $\sqrt{16} - \sqrt{9}$ __?__ $\sqrt{16 - 9}$         **25.** $\sqrt{25} - \sqrt{9}$ __?__ $\sqrt{25 - 9}$

**26.** $\sqrt{4} \times \sqrt{9}$ __?__ $\sqrt{4 \times 9}$         **27.** $\sqrt{25} \times \sqrt{4}$ __?__ $\sqrt{25 \times 4}$

**28.** $2\sqrt{2}$ __?__ $\sqrt{2 \times 2}$         **29.** $3\sqrt{12}$ __?__ $\sqrt{3 \times 12}$

Evaluate the expression.

**C**   **30.** $(\sqrt{25})^2$     **31.** $(\sqrt{81})^2$     **32.** $(\sqrt{49})^2$     **33.** $(\sqrt{11})^2$     **34.** $(\sqrt{2})^2$

## Review Exercises

Divide. Round the answer to the nearest hundredth.

**1.** $44 \div 6.7$         **2.** $35 \div 5.9$         **3.** $72 \div 8.3$         **4.** $96 \div 9.5$

**5.** $147 \div 12.3$         **6.** $230 \div 14.7$         **7.** $0.0165 \div 0.13$         **8.** $0.68 \div 0.81$

# 11-2 Approximating Square Roots

To get a close approximation of a square root, we can use the *divide-and-average* method. This method is based on the fact that

$$\text{if } \sqrt{a} = b, \text{ then } a = b \times b \text{ and } \frac{a}{b} = b.$$

In other words, when we divide a number $a$ by its square root $b$, the quotient is $b$. When we use an estimate for $b$ that is *less* than $b$ for a divisor, the quotient is then *greater* than $b$. The average of the divisor and quotient can be used as a new estimate for $b$.

**EXAMPLE 1**    Approximate $\sqrt{55}$ to the tenths' place.

**Solution**

Step 1    To estimate $\sqrt{55}$, first find the two integers between which $\sqrt{55}$ lies.

$$49 < \quad 55 < \quad 64$$
$$\sqrt{49} < \sqrt{55} < \sqrt{64}$$
$$7 < \sqrt{55} < \quad 8$$

Since 55 is closer to 49 than to 64, you might try 7.3 as an estimate of $\sqrt{55}$.

Step 2    Divide 55 by the estimate, 7.3. Compute to one more place than you want in the final answer.

```
       7.53
7.3)55.000
    51 1
     3 90
     3 65
       250
       219
        31
```

Step 3    Average the divisor and quotient.

$$\frac{7.3 + 7.53}{2} \approx 7.42$$

Use the average as the next estimate and repeat Steps 2 and 3 until your divisor and quotient agree in the tenths' place.

In this case, use 7.42 as your next estimate and divide.

```
        7.41
7.42)55.0000
     51 94
      3 060
      2 968
         920
         742
         178
```

As you can see, $\sqrt{55} \approx 7.4$ to the tenths' place.

We can approximate $\sqrt{55}$ to whatever decimal place we wish by repeating the steps shown in Example 1 on the previous page. Because $\sqrt{55}$ is a nonterminating, nonrepeating decimal, we say that $\sqrt{55}$ is an **irrational number.** Together, irrational and rational numbers form the set of **real numbers.** Other irrational numbers are discussed on page 119.

To find the square root of a number that is not an integer by the divide-and-average method, we use the same steps as in Example 1. The first step, finding the two integers between which a square root lies, is shown below.

**EXAMPLE 2**  Complete the first step in estimating each square root.

  **a.** $\sqrt{12.3}$ **b.** $\sqrt{0.7}$

**Solution**  The first step in estimating a square root is to find the two integers between which it lies.

**a.** $\begin{matrix} 9 < & 12.3 < & 16 \\ \sqrt{9} < & \sqrt{12.3} < & \sqrt{16} \\ 3 < & \sqrt{12.3} < & 4 \end{matrix}$  **b.** $\begin{matrix} 0 < & 0.7 < & 1 \\ \sqrt{0} < & \sqrt{0.7} < & \sqrt{1} \\ 0 < & \sqrt{0.7} < & 1 \end{matrix}$

---

## Class Exercises

**Give the first digit of the square root.**

**1.** $\sqrt{7}$  **2.** $\sqrt{11}$  **3.** $\sqrt{30}$  **4.** $\sqrt{50}$  **5.** $\sqrt{94}$

**Give the next estimate for the square root of the dividend.**

**EXAMPLE**  $3.8)\overline{15.000}$ with quotient $3.95$

**Solution**  $\dfrac{3.8 + 3.95}{2} \approx 3.88$

Thus 3.88 is the next estimate.

**6.** $2)\overline{4.8}$ with quotient $2.4$  **7.** $3)\overline{11}$ with quotient $3.7$  **8.** $5)\overline{28.5}$ with quotient $5.7$  **9.** $4.3)\overline{21.000}$ with quotient $4.88$

**10.** $14)\overline{225}$ with quotient $16$  **11.** $0.7)\overline{0.53}$ with quotient $0.76$  **12.** $0.6)\overline{0.3}$ with quotient $0.5$  **13.** $0.9)\overline{0.83}$ with quotient $0.92$

## Written Exercises

**Approximate to the tenths' place.**

**A**  1. $\sqrt{11}$        2. $\sqrt{13}$        3. $\sqrt{33}$        4. $\sqrt{23}$

   5. $\sqrt{8}$        6. $\sqrt{5}$        7. $\sqrt{26}$        8. $\sqrt{42}$

   9. $\sqrt{57}$        10. $\sqrt{75}$        11. $\sqrt{91}$        12. $\sqrt{69}$

   13. $\sqrt{5.7}$        14. $\sqrt{7.5}$        15. $\sqrt{9.1}$        16. $\sqrt{6.9}$

   17. $\sqrt{8.2}$        18. $\sqrt{4.6}$        19. $\sqrt{3.5}$        20. $\sqrt{1.6}$

**B**  21. $\sqrt{152}$        22. $\sqrt{285}$        23. $\sqrt{705}$        24. $\sqrt{328}$

   25. $\sqrt{15.2}$        26. $\sqrt{28.5}$        27. $\sqrt{70.5}$        28. $\sqrt{32.4}$

   29. $\sqrt{0.4}$        30. $\sqrt{0.6}$        31. $\sqrt{0.05}$        32. $\sqrt{0.21}$

**Approximate to the hundredths' place.**

   33. $\sqrt{2}$        34. $\sqrt{3}$        35. $\sqrt{5}$        36. $\sqrt{10}$

---

## Review Exercises

**Simplify.**

   1. $4(11 - 2) + 3(2) + 2(5 + 3)$          2. $5(81 \div 9) - 36 + 11$

   3. $7.3 + (9.14 - 6.91) - 2.81$          4. $11.32 - 67(8.01 - 7.92)$

   5. $3.617 + 0.7(5.301 - 4.911)$          6. $14.95 + 5(3.2 + 14) - 90.84$

   7. $(72 - 65) + 11(8.76 - 0.89)$          8. $5.143 + 0.3(8.914 - 7.126)$

### ▮▮▮ Calculator Key-In

On many calculators there is a square-root key. If you have access to such a calculator, use it to find the square roots below.

   1. a. $\sqrt{1}$        b. $\sqrt{100}$        c. $\sqrt{10{,}000}$

   2. a. $\sqrt{7}$        b. $\sqrt{700}$        c. $\sqrt{70{,}000}$

   3. a. $\sqrt{70}$        b. $\sqrt{7000}$        c. $\sqrt{700{,}000}$

   4. a. $\sqrt{0.08}$        b. $\sqrt{8}$        c. $\sqrt{800}$

   5. a. $\sqrt{0.54}$        b. $\sqrt{54}$        c. $\sqrt{5400}$

# 11-3 Using a Square-Root Table

Part of the Table of Square Roots on page 508 is shown below. Square roots of integers are given to the nearest thousandth.

| Number | Positive Square Root | Number | Positive Square Root | Number | Positive Square Root | Number | Positive Square Root |
|--------|--------|--------|--------|--------|--------|--------|--------|
| $N$ | $\sqrt{N}$ | $N$ | $\sqrt{N}$ | $N$ | $\sqrt{N}$ | $N$ | $\sqrt{N}$ |
| 1 | 1 | 26 | 5.099 | 51 | 7.141 | 76 | 8.718 |
| 2 | 1.414 | 27 | 5.196 | 52 | 7.211 | 77 | 8.775 |
| 3 | 1.732 | 28 | 5.292 | 53 | 7.280 | 78 | 8.832 |
| 4 | 2 | 29 | 5.385 | 54 | 7.348 | 79 | 8.888 |

We can use the table to approximate the square roots of integers from 1 to 100. For example, to find $\sqrt{78}$, first locate 78 under a column headed "Number." Then read off the value beside 78 in the "Square Root" column.

$$\sqrt{78} \approx 8.832$$

---

**Reading Mathematics:** *Tables*
Some tables may have many columns of information. When you are reading a table, use a ruler to help guide your eyes across the page or down a column. This way you can be sure to find the correct entry.

---

To find an approximate square root of a number that lies between two entries in the column headed "Number," we can use a process called **interpolation.** The interpolation process may cause an error in the last digit of the approximation.

**EXAMPLE 1**   Approximate $\sqrt{3.8}$ to the nearest thousandth by interpolation.

**Solution**   On a number line $\sqrt{3.8}$ lies between $\sqrt{3}$ and $\sqrt{4}$. We can assume that $\sqrt{3.8}$ is about 0.8 of the distance between $\sqrt{3}$ and $\sqrt{4}$.

$$\sqrt{3.8} \approx \sqrt{3} + 0.8(\sqrt{4} - \sqrt{3})$$
$$\approx 1.732 + 0.8(2.000 - 1.732)$$
$$\approx 1.732 + 0.2144 = 1.9464$$

Thus, rounded to the thousandths' place, $\sqrt{3.8} = 1.946$.

**EXAMPLE 2**   The area of a square display room is 71 m². Find the length of a side to the nearest hundredth of a meter.

**Solution**   Recall that the formula for the area of a square is $A = s^2$.

$$s^2 = 71$$
$$s = \sqrt{71}$$

Using the table on page 508, we see that $\sqrt{71} = 8.426$.
To the nearest hundredth of a meter, the length of a side is 8.43 m.

## Class Exercises

**Use the table on page 508 to find the approximate square root.**

**1.** $\sqrt{39}$   **2.** $\sqrt{83}$   **3.** $\sqrt{20}$   **4.** $\sqrt{95}$   **5.** $\sqrt{34}$

**6.** $\sqrt{49}$   **7.** $\sqrt{11}$   **8.** $\sqrt{52}$   **9.** $10\sqrt{87}$   **10.** $\frac{1}{10}\sqrt{22}$

**Find two decimals in the table on page 508 between which the square root lies.**

**11.** $\sqrt{54.3}$   **12.** $\sqrt{1.7}$   **13.** $\sqrt{60.7}$   **14.** $\sqrt{4.5}$   **15.** $\sqrt{28.6}$

## Written Exercises

**For Exercises 1–37, refer to the table on page 508.**

**Approximate to the nearest hundredth.**

**A**   **1.** $\sqrt{65}$   **2.** $\sqrt{31}$   **3.** $\sqrt{56}$   **4.** $\sqrt{13}$   **5.** $\sqrt{97}$

**6.** $10\sqrt{37}$   **7.** $10\sqrt{41}$   **8.** $\frac{1}{10}\sqrt{83}$   **9.** $\frac{1}{10}\sqrt{75}$   **10.** $\frac{1}{10}\sqrt{24}$

**11.** $3\sqrt{55}$   **12.** $2\sqrt{30}$   **13.** $6\sqrt{19}$   **14.** $4\sqrt{18}$   **15.** $7\sqrt{72}$

**B**   **16.** $\sqrt{39} + \sqrt{16}$   **17.** $\sqrt{28} + \sqrt{53}$   **18.** $\sqrt{71} - \sqrt{25}$

**19.** $\sqrt{91} - \sqrt{84}$   **20.** $2\sqrt{56} - \sqrt{4}$   **21.** $\sqrt{37} - 2\sqrt{8}$

**Approximate to the nearest hundredth by interpolation.**

**22.** $\sqrt{13.2}$   **23.** $\sqrt{5.9}$   **24.** $\sqrt{14.7}$   **25.** $\sqrt{81.6}$   **26.** $\sqrt{24.3}$

**27.** $\sqrt{8.7}$   **28.** $\sqrt{69.2}$   **29.** $\sqrt{50.9}$   **30.** $5\sqrt{84.4}$   **31.** $3\sqrt{81.9}$

**Approximate to the nearest hundredth.**

**C** **32.** $\sqrt{700}$   (*Hint:* $\sqrt{100 \times 7} = \sqrt{100} \times \sqrt{7} = 10\sqrt{7}$)     **33.** $\sqrt{500}$   **34.** $\sqrt{1100}$

**35.** $\sqrt{380}$   (*Hint:* $\sqrt{100 \times 3.8} = \sqrt{100} \times \sqrt{3.8} = 10\sqrt{3.8}$)   **36.** $\sqrt{420}$   **37.** $\sqrt{3470}$

## Problems

**Solve.**

**A** **1.** The area of a square is 85 m². Find the length of a side to the nearest hundredth of a meter.

**2.** A square floor has an area of 32 m². Find the length of a side to the nearest tenth of a meter.

**3.** The area of a square room measures 225 ft². How much will it cost to put molding around the ceiling at $.35 per foot?

**B** **4.** An isosceles right triangle has an area of 13.5 cm². Find the length of the equal sides to the nearest tenth of a centimeter.

**5.** A circle has an area of 47.1 cm². Find its radius to the nearest hundredth of a centimeter. (Use $\pi \approx 3.14$.)

**6.** The height of a parallelogram is half the length of its base. The parallelogram has an area of 67 cm². Find the height to the nearest tenth of a centimeter.

**C** **7.** The total surface area of a cube is 210 m². Find the length of an edge to the nearest hundredth of a meter.

**8.** One base of a trapezoid is three times as long as the other base. The height of the trapezoid is the same as the shorter base. The trapezoid has an area of 83 cm². Find the height to the nearest tenth of a centimeter.

## Review Exercises

**Write the value of the product.**

**1.** $0.9^2$    **2.** $0.05^2$    **3.** $0.06^2$    **4.** $0.7^2$    **5.** $3.7^2$    **6.** $1.9^2$    **7.** $24.3^2$    **8.** $5.03^2$

# 11-4 The Pythagorean Theorem

The longest side of a right triangle is opposite the right angle and is called the **hypotenuse.** The two shorter sides are called **legs.**

About 2500 years ago, the Greek mathematician Pythagoras proved the following useful fact about right triangles.

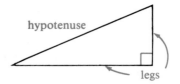

hypotenuse

legs

---

### The Pythagorean Theorem

If the hypotenuse of a right triangle has length $c$, and the legs have lengths $a$ and $b$, then

$$c^2 = a^2 + b^2.$$

---

The figure at the right illustrates the Pythagorean theorem. We see that the area of the square on the hypotenuse equals the sum of the areas of the squares on the legs:

$$25 = 9 + 16,$$

$$\text{or } 5^2 = 3^2 + 4^2.$$

The converse of the Pythagorean theorem is also true. It can be used to test whether a triangle is a right triangle.

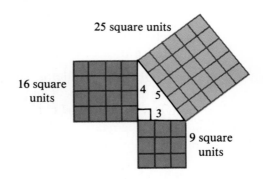

25 square units

16 square units

9 square units

---

### Converse of the Pythagorean Theorem

If the sides of a triangle have lengths $a$, $b$, and $c$, such that $c^2 = a^2 + b^2$, then the triangle is a right triangle.

---

**EXAMPLE 1**   Is the triangle with sides of the given lengths a right triangle?
  **a.** 4, 5, 7          **b.** 5, 12, 13

**Solution**   **a.** $4^2 = 16$, $5^2 = 25$, $7^2 = 49$
No, since $16 + 25 \neq 49$. The triangle with sides of lengths 4, 5, and 7 *cannot* be a right triangle.

**b.** $5^2 = 25$, $12^2 = 144$, $13^2 = 169$

Yes, since $25 + 144 = 169$. The triangle with sides of lengths 5, 12, and 13 is a right triangle.

Sometimes it may be necessary to solve for the length of a missing side of a right triangle. The example below illustrates the steps involved.

**EXAMPLE 2**   For right triangle $ABC$, find the length of the missing side to the nearest hundredth. Use the table on page 508.

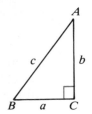

**a.** $a = 3$, $b = 7$         **b.** $a = 4$, $c = 9$

**Solution**    Using the equation $c^2 = a^2 + b^2$:

**a.** $c^2 = 3^2 + 7^2$
$\quad = 9 + 49 = 58$
$\quad c = \sqrt{58}$
$\quad c \approx 7.62$

**b.** $9^2 = 4^2 + b^2$
$\quad 9^2 - 4^2 = b^2$
$\quad 81 - 16 = b^2$
$\quad 65 = b^2$
$\quad \sqrt{65} = b$
$\quad b \approx 8.06$

---

## Class Exercises

**Without actually counting them, tell how many unit squares there are in the shaded square.**

**1.**

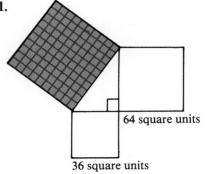

64 square units

36 square units

**2.**

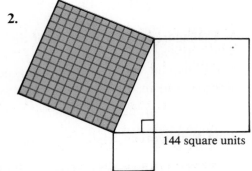

144 square units

25 square units

**Replace**  **with = or ≠ to make a true statement.**

**3.** $6^2 \underline{\ ?\ } 4^2 + 5^2$         **4.** $5^2 \underline{\ ?\ } 3^2 + 4^2$         **5.** $10^2 \underline{\ ?\ } 6^2 + 8^2$

**The lengths of the sides of a triangle are given. Is it a right triangle?**

**6.** 3, 4, 5         **7.** 7, 24, 25         **8.** 5, 10, 12         **9.** 10, 24, 26

## Written Exercises

**Find the area of the square.**

A   **1.** 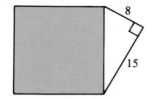    **2.** 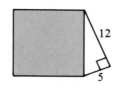    **3.**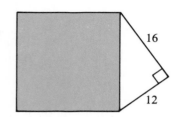

**Is the triangle with sides of the given lengths a right triangle?**

  **4.** 6, 8, 10                              **5.** 8, 15, 17

  **6.** 16 cm, 30 cm, 34 cm            **7.** 9 m, 12 m, 15 m

  **8.** 1.5 mm, 2.0 mm, 2.5 mm       **9.** 0.6 km, 0.8 km, 1.0 km

  **10.** 9 m, 21 m, 23 m                **11.** 20 cm, 21 cm, 29 cm

  **12.** 9 km, 40 km, 41 km            **13.** 8 m, 37 m, 39 m

**A right triangle has sides of lengths $a$, $b$, and $c$, with $c$ the length of the hypotenuse. Find the length of the missing side. If necessary, use the table on page 508 for the square-root values and round answers to the nearest hundredth.**

B   **14.** $a = 2, b = 1$   **15.** $a = 8, b = 6$   **16.** $a = 4, c = 9$   **17.** $b = 5, c = 6$

    **18.** $a = 5, b = 12$   **19.** $a = 9, b = 7$   **20.** $b = 11, c = 19$   **21.** $a = 24, c = 74$

A **Pythagorean triple** consists of three positive integers $a$, $b$, and $c$ that satisfy the equation $a^2 + b^2 = c^2$. You can find as many Pythagorean triples as you wish by substituting positive integers for $m$ and $n$ (such that $m > n$) in the following expressions for $a$, $b$, and $c$.

$$a = m^2 - n^2 \qquad b = 2mn \qquad c = m^2 + n^2$$

**Find Pythagorean triples using the given values of $m$ and $n$.**

C   **22.** $m = 5, n = 1$           **23.** $m = 6, n = 3$           **24.** $m = 4, n = 2$

---

## Problems

**Solve. Round your answer to the nearest tenth.**

A   **1.** A plane flies 90 km due east and then 60 km due north. How far is it then from its starting point?

**2.** The foot of a 6 m ladder is 2.5 m from the base of a wall. How high up the wall does the ladder reach?

**3.** The figure shows two cables bracing a television tower. What is the distance between the points where the cables touch the ground?

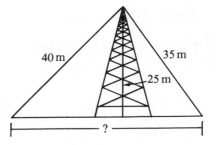

**4.** Find the length of a diagonal of a 10 cm by 10 cm square.

**B** **5.** A square has diagonals 10 cm long. Find the length of a side.

**6.** The diagonals of a rhombus are perpendicular and bisect each other. Find the length of a side of a rhombus whose diagonals have lengths 12 m and 18 m.

**C** **7.** Find the length of a diagonal of a 10 by 10 by 10 cube. (*Hint:* First find *AQ* using right triangle *PQA*. Then find the required length *AB* using right triangle *AQB*.)

**8.** Find the length of a diagonal of a box having dimensions 2 by 3 by 6. (See the *Hint* for Problem 7.)

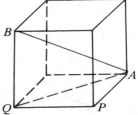

---

## Self-Test A

**If the given symbol names an integer, state the integer. If not, name the two consecutive integers between which the number lies.**

**1.** $\sqrt{56}$      **2.** $-\sqrt{81}$      **3.** $\sqrt{9} + \sqrt{25}$      **4.** $\sqrt{106 - 57}$      [11–1]

**5.** Use the divide-and-average method to approximate $\sqrt{86.4}$ to the nearest tenth.      [11–2]

**Use the table on page 508 and interpolation to approximate each square root to the nearest tenth.**

**6.** $\sqrt{18}$       **7.** $\sqrt{50}$       **8.** $\sqrt{9.8}$       **9.** $\sqrt{5.6}$       [11–3]

**A right triangle has sides of lengths *a*, *b*, and *c*, with *c* the length of the hypotenuse. Find the length of the missing side.**

**10.** $a = 3, b = 4$       **11.** $b = 96, c = 100$       **12.** $c = 20, a = 12$       [11–4]

*Self-Test answers and Extra Practice are at the back of the book.*

# 11-5 Similar Triangles

As you know, we say that two figures are congruent when they are identical in both shape and size. When two figures have the same shape, but do not necessarily have the same size, we say that the figures are **similar.**

For two *triangles* to be similar it is enough that the measures of their corresponding angles are equal.

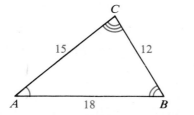 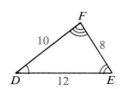

To indicate that the triangles shown above are similar, we can write the following:

$$\triangle ABC \sim \triangle DEF$$

The symbol $\sim$ means *is similar to*. Note that when we write expressions such as the one above, we list corresponding vertices in the same order.

In triangles *ABC* and *DEF*, we see that the lengths of the corresponding sides have the same ratio:

$$\frac{AB}{DE} = \frac{18}{12} = \frac{3}{2} \qquad \frac{BC}{EF} = \frac{12}{8} = \frac{3}{2} \qquad \frac{CA}{FD} = \frac{15}{10} = \frac{3}{2}$$

Therefore,

$$\frac{AB}{DE} = \frac{BC}{EF} = \frac{CA}{FD}.$$

Because the ratios are equal, we say that the lengths of the corresponding sides are *proportional.*

In general, we can say the following:

> For two similar triangles,
>
> corresponding angles are congruent
>
> and
>
> lengths of corresponding sides are proportional.

If △RUN ∼ △JOG, find the lengths marked $x$ and $y$.

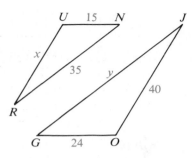

**Solution**

Since the corresponding vertices are listed in the same order, we know the following.

$$\angle R \cong \angle J$$

$$\angle U \cong \angle O$$

$$\angle N \cong \angle G$$

$$\frac{RU}{JO} = \frac{UN}{OG} = \frac{NR}{GJ}$$

Substituting the values that are given in the diagram, we obtain

$$\frac{x}{40} = \frac{15}{24} = \frac{35}{y}.$$

Therefore, we can set up one proportion involving $x$ and another involving $y$.

$$\frac{x}{40} = \frac{15}{24} \qquad\qquad \frac{15}{24} = \frac{35}{y}$$

$$24x = 15 \times 40 \qquad\qquad 15y = 35 \times 24$$

$$x = \frac{15 \times 40}{24} \qquad\qquad y = \frac{35 \times 24}{15}$$

$$x = \frac{600}{24} = 25 \qquad\qquad y = \frac{840}{15} = 56$$

The length marked $x$ is 25 and the length marked $y$ is 56.

In the two *right triangles* shown at the right, the measures of two acute angles are equal. Since all right angles have equal measure, 90°, the two remaining acute angles must also have equal measure. (Recall that the sum of the angle measures of any triangle is 180°.) Thus the two right triangles are similar.

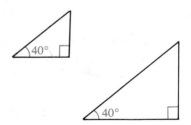

If an acute angle of one *right* triangle is congruent to an angle of a second *right* triangle, then the triangles are similar.

**EXAMPLE 2**  At the same time a tree on level ground casts a shadow 48 m long, a 2 m pole casts a shadow 5 m long. Find the height, $h$, of the tree.

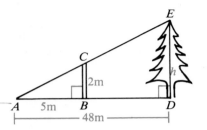

**Solution**  In the diagram, right triangles $ABC$ and $ADE$ share acute $\angle A$ and thus are similar. Since $\frac{AD}{AB} = \frac{DE}{BC} = \frac{EA}{CA}$, we can set up a proportion to solve for $h$.

$$\frac{h}{2} = \frac{48}{5}$$

$$5h = 48 \times 2$$

$$h = \frac{48 \times 2}{5} = 19.2$$

The height of the tree is 19.2 m.

---

**Reading Mathematics: *Diagrams***

To help you read a diagram that has overlapping triangles, you can redraw the diagram, pulling apart the individual triangles. For example, $\triangle ADE$ above can be separated as shown below.

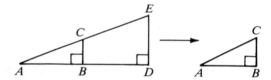 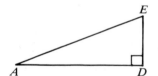

---

## Class Exercises

**In Exercises 1–4, $\triangle LOG \sim \triangle RIT$.**

1. Name all pairs of corresponding angles.

2. Name all pairs of corresponding sides.

3. $\frac{OL}{IR} = \frac{OG}{?}$

4. $\frac{LG}{?} = \frac{GO}{TI}$

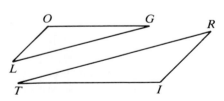

**True or false?**

5. $\triangle BAE$ and $\triangle DCE$ are right triangles.

6. $m \angle ABE = m \angle CDE$    7. $\triangle BAE \sim \triangle DCE$

8. $\frac{CE}{AE} = \frac{BA}{DE}$    9. $\frac{BA}{DC} = \frac{BE}{DE}$

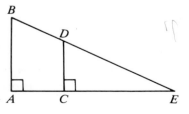

## Written Exercises

**Exercises 1–5 refer to the diagram at the right.** $\triangle ABC \sim \triangle PQR$.

**A**   **1.** $\dfrac{PR}{AC} = \dfrac{?}{CB}$      **2.** $\dfrac{BA}{?} = \dfrac{CB}{RQ}$

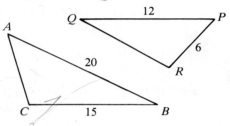

**3.** Find the length of $\overline{RQ}$.

**4.** Find the length of $\overline{AC}$.

**5.** $m \angle A = 47°$ and $m \angle Q = 29°$,
     $m \angle C = \underline{\quad?\quad}°$.

**Exercises 6–10 refer to the diagram at the right.** $\triangle AED \sim \triangle ACB$.

**6.** The length of $\overline{AE} = \underline{\quad?\quad}$.

**7.** The length of $\overline{AB} = \underline{\quad?\quad}$.

**8.** The length of $\overline{AD} = \underline{\quad?\quad}$.

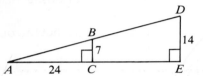

**9.** If $m \angle A = 25°$, then $m \angle ABC = \underline{\quad?\quad}°$
     and $m \angle ADE = \underline{\quad?\quad}°$.

**10.** If $\dfrac{BC}{DE} = \dfrac{1}{2}$, then $\dfrac{AB}{AD} = \underline{\quad?\quad}$.

**Find the lengths marked $x$ and $y$.**

**11.** $\triangle MAB \sim \triangle SRO$     **12.** $\triangle DIP \sim \triangle MON$     **13.** $\triangle RST \sim \triangle RUV$

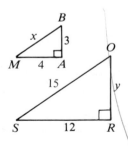    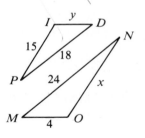   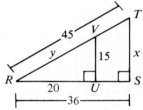

**B**   **14.** $\triangle GEM \sim \triangle TIM$     **15.** $\triangle PAR \sim \triangle PBT$     **16.** $\triangle ART \sim \triangle ABC$

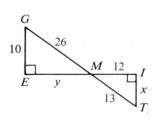    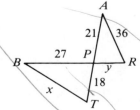   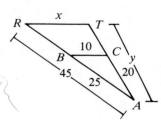

**17.** One day Jessica, who is 150 cm tall, cast a shadow that was 200 cm long while her father's shadow was 240 cm long. How tall is her father?

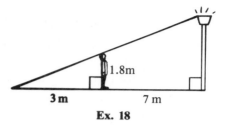

**18.** A person 1.8 m tall standing 7 m from a streetlight casts a shadow 3 m long. How high is the light?

**Ex. 18**

**19.** Find (a) the perimeter and (b) the area of the shaded trapezoid in the figure at the left below.

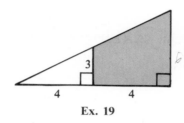

**Ex. 19**

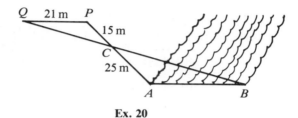

**Ex. 20**

**20.** To find the distance $AB$ across a river, surveyors laid off $\overline{PQ}$ parallel to $\overline{AB}$ and created 2 similar triangles. They made the measurements shown in the figure at the right above. Find the length of $\overline{AB}$.

**Exercises 21–22 refer to the diagram at the right.**

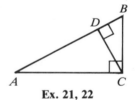

**C** **21.** Explain why $\triangle ADC \sim \triangle ACB$.

**22.** Explain why $\triangle ADC \sim \triangle CDB$.

**Ex. 21, 22**

**23.** If $\triangle MNO \sim \triangle PQR$ and $\triangle PQR \sim \triangle STU$, is $\triangle MNO \sim \triangle STU$? Explain your answer.

**24.** If $\triangle MNO \cong \triangle PQR$, is $\triangle MNO \sim \triangle PQR$? Explain your answer.

**25.** If $\triangle MNO \sim \triangle PQR$, is $\triangle MNO \cong \triangle PQR$? Explain your answer.

---

## Review Exercises

**Multiply.**

**1.** $300 \times 1.73$  **2.** $500 \times 3.9$  **3.** $7.81 \times 200$

**4.** $16.34 \times 120$  **5.** $2.18 \times 1.03$  **6.** $3.17 \times 6.01$

**7.** $54.2 \times 0.09$  **8.** $683 \times 2.48$  **9.** $71.4 \times 52.4$

# 11-6 Special Right Triangles

In an *isosceles right triangle* the two acute angles are congruent. Since the sum of the measures of these two angles is 90°, each angle measures 45°. For this reason, an isosceles right triangle is often called a **45° right triangle.**

In the diagram each leg is 1 unit long. If the hypotenuse is $c$ units long, by the Pythagorean theorem we know that $c^2 = 1^2 + 1^2 = 2$ and thus

$$c = \sqrt{2}.$$

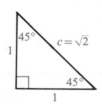

Every 45° right triangle is similar to the one shown. Since corresponding sides of similar triangles are proportional, we have the following property.

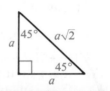

> If each leg of a 45° right triangle is $a$ units long, then the hypotenuse is $a\sqrt{2}$ units long.

**EXAMPLE 1** A square park measures 200 m on each edge. Find the length, $d$, of a path extending diagonally from one corner to the opposite corner. Use $\sqrt{2} \approx 1.414$.

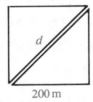

**Solution** Since the park is square, we can apply the property of 45° right triangles to solve for $d$. We use the fact that $d$ is the hypotenuse of the right triangle and that each side measures 200 m.

$$d = 200\sqrt{2} \approx 200 \times 1.414 = 282.8$$

Thus the path measures approximately 282.8 m.

A **30°–60° right triangle,** such as $\triangle ACB$, may be thought of as half an equilateral triangle. If hypotenuse $\overline{AB}$ is 2 units long, then the shorter leg $\overline{AC}$ (half of $\overline{AD}$) is 1 unit long. To find $BC$, we use the Pythagorean theorem:

$$(AC)^2 + (BC)^2 = (AB)^2$$
$$1^2 + (BC)^2 = 2^2$$
$$(BC)^2 = 2^2 - 1^2 = 3$$
$$BC = \sqrt{3}$$

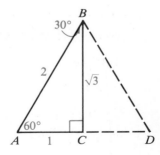

Every 30°–60° right triangle is similar to the one shown on the preceding page, and since corresponding sides of similar triangles are proportional, we have the following property.

If the shorter leg of a 30°–60° right triangle is *a* units long, then the longer leg is $a\sqrt{3}$ units long, and the hypotenuse is 2*a* units long.

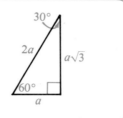

*EXAMPLE 2*    The hypotenuse of a 30°–60° right triangle is 8 cm long. Find the lengths of the legs.

**Solution**    Using the 30°–60° right triangle property, we know that

$$2a = 8, \ a = 4, \ \text{and} \ a\sqrt{3} = 4\sqrt{3}.$$

Thus the lengths of the legs are 4 cm and $4\sqrt{3}$ cm.

The symbol $\sqrt{\phantom{x}}$ is called the **radical sign.** An expression such as $\sqrt{3}$ or $\sqrt{x}$ is called a **radical.** We often leave answers *in terms of radicals* with the radical in the numerator. To rewrite expressions such as $\dfrac{6}{\sqrt{3}}$ so that the radical appears in the numerator, we may use the fact that $\sqrt{x} \times \sqrt{x} = x$.

*EXAMPLE 3*    Rewrite $\dfrac{6}{\sqrt{3}}$ in lowest terms with the radical in the numerator.

**Solution**    If we multiply the numerator and denominator by $\sqrt{3}$, we will find an equivalent fraction with the radical in the numerator.

$$\frac{6}{\sqrt{3}} = \frac{6 \times \sqrt{3}}{\sqrt{3} \times \sqrt{3}}$$

$$= \frac{6\sqrt{3}}{3}$$

$$= 2\sqrt{3}$$

Thus, $2\sqrt{3}$ is equivalent to $\dfrac{6}{\sqrt{3}}$ in lowest terms with the radical in the numerator.

## Class Exercises

**Rewrite the expression in lowest terms with the radical in the numerator.**

**1.** $\dfrac{1}{\sqrt{3}}$ **2.** $\dfrac{2}{\sqrt{3}}$ **3.** $\dfrac{2}{\sqrt{2}}$ **4.** $\dfrac{6}{\sqrt{2}}$ **5.** $\dfrac{1}{\sqrt{x}}$ **6.** $\dfrac{x}{\sqrt{x}}$

**Find the lengths marked $x$ and $y$ in the triangle. Give your answer in terms of radicals when radicals occur.**

**7.**

**8.**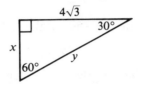

**9.**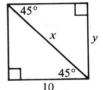

**10.**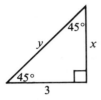

**11.**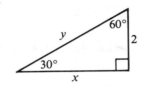

**12.**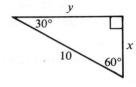

---

## Written Exercises

**Rewrite the expression in lowest terms with the radical in the numerator.**

**A** **1.** $\dfrac{6}{\sqrt{10}}$ **2.** $\dfrac{12}{\sqrt{13}}$ **3.** $\dfrac{3}{\sqrt{3}}$ **4.** $\dfrac{1}{\sqrt{2}}$ **5.** $\dfrac{2}{\sqrt{x}}$ **6.** $\dfrac{3x}{\sqrt{x}}$

**Approximate the lengths marked $x$ and $y$ to the nearest tenth. Use $\sqrt{2} \approx 1.414$ and $\sqrt{3} \approx 1.732$.**

**7.**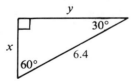

**8.**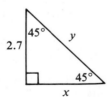

**9.**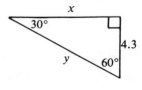

**10.**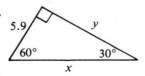

**11.**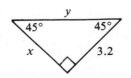

**12.**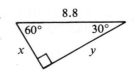

**In Exercises 13–16, give answers in terms of radicals with the radical in the numerator.**

**B** **13.** The hypotenuse of an isosceles right triangle has length 4. How long is each leg?

**14.** The hypotenuse of a 45° right triangle has length 10. How long is each leg?

**15.** The longer leg of a 30°–60° right triangle has length 15. How long are the other sides?

**16.** The side opposite the 60° angle of a right triangle has length 3. How long are the other sides?

**In Exercises 17–22, $\angle C$ is a right angle in $\triangle ABC$. Find the length of the missing side to the nearest tenth.**

**17.** $AC = 2$, $BC = 5$          **18.** $AB = 18$, $AC = 9$

**19.** $AB = 6\sqrt{2}$, $AC = 6$          **20.** $AC = CB = x$

**C** **21.** $\frac{1}{2}BA = CA = y$          **22.** $2BC = BA = z$

**Approximate the lengths marked $x$ and $y$ to the nearest tenth.**

**23.** 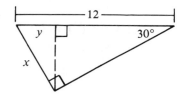          **24.**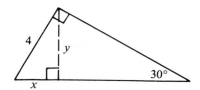

---

## Problems

**Solve. Round answers to the nearest tenth.**

**A** **1.** A ladder 10 m long resting against a wall makes a 60° angle with the ground. How far up the wall does it reach?

**2.** A baseball diamond is a square 90 ft on each side. How far is it diagonally from home plate to second base?

**3.** A hillside is inclined at an angle of 30° with the horizontal. How much altitude has Mary gained after hiking 40 m up the hill?

**4.** The diagram on the right shows the roof of a house. Find the dimensions marked $x$ and $y$.

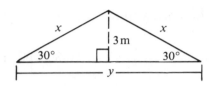

**B** **5.** A checkerboard has 8 squares on each side. If one side of a square is 5 cm long, how far is it from one corner of the board to the opposite corner?

**6.** Find the height of an equilateral triangle with sides 12 cm long.

**7.** Find the perimeters of the two squares shown in the diagram.

**8.** An equilateral triangle has sides 10 units long. Find (a) the height and (b) the area.

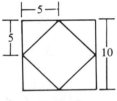

**Prob. 7**

**C** **9.** An equilateral triangle has sides $r$ units long. Find both (a) the height and (b) the area in terms of $r$.

**10.** The area of a square pan is 900 cm². What is the length of a diagonal of the pan?

**11.** A 10 m pole is supported in a vertical position by three 6 m guy wires. If one end of each wire is fastened to the ground at a 60° angle, how high on the pole is the other end fastened?

**12.** A rhombus has angles of 60° and 120°. Each side of the rhombus is 8 cm long. What are the lengths of the diagonals? (*Hint:* The diagonals bisect the angles of the rhombus.)

---

## Review Exercises

**Write the fraction as a decimal. Use a bar to show a repeating decimal.**

**1.** $\frac{5}{4}$     **2.** $\frac{3}{5}$     **3.** $\frac{2}{9}$     **4.** $\frac{7}{12}$     **5.** $\frac{11}{10}$     **6.** $\frac{3}{16}$     **7.** $\frac{1}{6}$     **8.** $\frac{4}{11}$

# 11-7 Trigonometric Ratios

Since each of the three triangles in the diagram below contains $\angle A$ and a right angle, the triangles are *similar* and the lengths of their sides are *proportional.* Thus, the ratios written below for the smallest triangle apply to all three triangles.

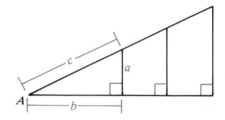

$$\frac{\text{length of side opposite } \angle A}{\text{length of hypotenuse}} = \frac{a}{c}$$

$$\frac{\text{length of side adjacent to } \angle A}{\text{length of hypotenuse}} = \frac{b}{c}$$

$$\frac{\text{length of side opposite } \angle A}{\text{length of side adjacent to } \angle A} = \frac{a}{b}$$

These ratios, called the **trigonometric ratios,** are so useful that each has been given a special name.

$\frac{a}{c}$ is called the **sine** of $\angle A$, or **sin A.**

$\frac{b}{c}$ is called the **cosine** of $\angle A$, or **cos A.**

$\frac{a}{b}$ is called the **tangent** of $\angle A$, or **tan A.**

It is important to understand that each trigonometric ratio depends only on the measure of $\angle A$ and *not* on the size of the right triangle.

The following shortened forms of the definitions may help you remember the trigonometric ratios.

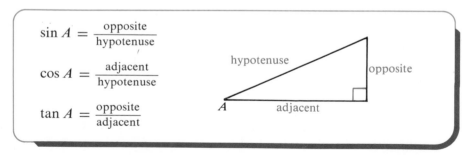

$$\sin A = \frac{\text{opposite}}{\text{hypotenuse}}$$

$$\cos A = \frac{\text{adjacent}}{\text{hypotenuse}}$$

$$\tan A = \frac{\text{opposite}}{\text{adjacent}}$$

*EXAMPLE 1*  For $\angle A$ find the value of each trigonometric ratio in lowest terms.

**a.** sin $A$ **b.** cos $A$ **c.** tan $A$

**Solution**    To find the value of the trigonometric ratios for $\angle A$, first find the value of $x$. By the Pythagorean theorem:

$$x^2 + 5^2 = 6^2$$
$$x^2 + 25 = 36$$
$$x^2 = 36 - 25$$
$$x^2 = 11$$
$$x = \sqrt{11}$$

**a.** $\sin A = \dfrac{\text{opposite}}{\text{hypotenuse}} = \dfrac{5}{6}$

**b.** $\cos A = \dfrac{\text{adjacent}}{\text{hypotenuse}} = \dfrac{\sqrt{11}}{6}$

**c.** $\tan A = \dfrac{\text{opposite}}{\text{adjacent}} = \dfrac{5}{\sqrt{11}} = \dfrac{5 \times \sqrt{11}}{\sqrt{11} \times \sqrt{11}} = \dfrac{5\sqrt{11}}{11}$

**EXAMPLE 2**    Find the sine, cosine, and tangent of a 30° angle to the nearest thousandth.

**Solution**    To find the values of these trigonometric ratios, first draw a 30°–60° right triangle. Let the shorter leg be 1 unit long and write in the lengths of the other sides according to the property of 30°–60° triangles that you learned in the previous lesson.

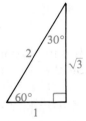

$$\sin 30° = \frac{1}{2} = 0.500$$

$$\cos 30° = \frac{\sqrt{3}}{2} \approx 0.866$$

$$\tan 30° = \frac{1}{\sqrt{3}} = \frac{\sqrt{3}}{3} \approx 0.577$$

---

## Class Exercises

**Find the value of the trigonometric ratio.**

**1.** $\sin A$        **2.** $\cos A$

**3.** $\tan A$        **4.** $\sin B$

**5.** $\cos B$        **6.** $\tan B$

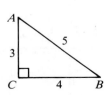

**Find the value of the trigonometric ratio.**

**7.** sin *P*          **8.** cos *P*

**9.** tan *P*          **10.** sin *R*

**11.** cos *R*          **12.** tan *R*

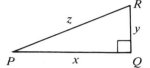

---

## Written Exercises

**Give the value of the sine, cosine, and tangent of ∠*A* and ∠*B*. Give all ratios in lowest terms.**

**A 1.**

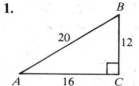

**2.**

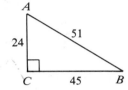

**3.**

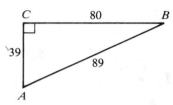

**4.**

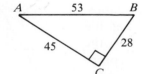

**5.**

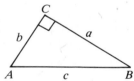

**6.**

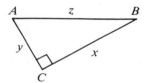

**B** In Exercises 7–12, give all ratios in lowest terms and with the radical in the numerator.

**Find the value of *x*. Then find tan *A*.**

**7.**

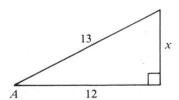

**8.**

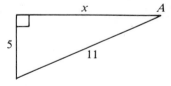

**Find the value of *x*. Then find sin *A*.**

**9.**

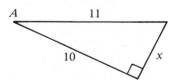

**10.**

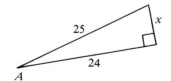

**Find the value of *x*. Then find cos *A*.**

**11.**

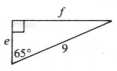

A ___x___

26    |10

**12.**

A ___10___

x    7

**Find the sine, cosine, and tangent. Write the answer both as a fraction and as a decimal to the nearest thousandth.** (*Hint:* See Example 2.)

**13. a.** $\sin 60°$  **b.** $\cos 60°$  **c.** $\tan 60°$       **14. a.** $\sin 45°$  **b.** $\cos 45°$  **c.** $\tan 45°$

**Give answers to the nearest tenth.**

**C  15.** $\sin 67° = 0.9205$
    **a.** $x =$ ___?___
    **b.** $y =$ ___?___
    **c.** $\sin 23° =$ ___?___

y |67°    7

x

**16.** $\tan 40° = 0.8391$
    **a.** $m =$ ___?___
    **b.** $n =$ ___?___
    **c.** $\cos 50° =$ ___?___

m    n

40°

15

**17.** $\cos 65° = 0.4226$
    **a.** $e =$ ___?___
    **b.** $f =$ ___?___
    **c.** $\tan 25° =$ ___?___

f

e

65°    9

## Review Exercises

**Select the number that is closest to the one given.**

**1.** 0.3765
    **a.** 0.3774        **b.** 0.3759

**2.** 8.1443
    **a.** 8.1430        **b.** 8.1451

**3.** 0.9004
    **a.** 0.9019        **b.** 0.8954

**4.** 1.8040
    **a.** 1.829         **b.** 1.788

**5.** 0.2126
    **a.** 0.2666        **b.** 0.1986

**6.** 11.4301
    **a.** 12.0801     **b.** 10.6801

**7.** 0.1758
    **a.** 0.1744        **b.** 0.1764

**8.** 3.2709
    **a.** 3.2712        **b.** 3.2705

  **Challenge**

Leslie is considering two job offers. Alloid Metals pays an hourly wage of $7.30. Acme Steel Company pays an annual salary of $14,040. Both jobs have a 40-hour work week. Which job offers a better salary?

# 11-8 Solving Right Triangles

The table on page 509 gives approximate values of the sine, the cosine, and the tangent of angles with measure 1°, 2°, 3°, . . . , 90°. To find sin 45°, look down the column headed "Angle" to 45°. To the right of it in the column headed "Sine," you see that sin 45° ≈ 0.7071.

The values in the table on page 509 are, in general, accurate to only four decimal places. However, in computational work with sine, cosine, and tangent, it is customary to use = instead of ≈. In this lesson, we will write *sin 45° = 0.7071* instead of *sin 45° ≈ 0.7071*.

We can use the values in the table to **solve right triangles,** that is, to find approximate measures of all the sides and all the angles of any right triangle.

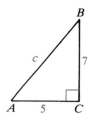

**EXAMPLE 1**  Solve $\triangle ABC$ by finding each measure.
    **a.** $c$ to the nearest tenth
    **b.** m $\angle A$
    **c.** m $\angle B$

**Solution**  **a.** The diagram indicates that $\angle C$ is a right triangle, therefore by the Pythagorean theorem:

$$c^2 = 7^2 + 5^2 = 74$$
$$c = \sqrt{74} = 8.6$$

To the nearest tenth, $c = 8.6$.

**b.** $\tan A = \frac{7}{5} = 1.4$

In the tangent column in the table, the closest entry to 1.4 is 1.3764, for angle measure 54°. Thus, to the nearest degree, m $\angle A = 54°$.

**c.** m $\angle A$ + m $\angle B$ + m $\angle C$ = 180°
    54° + m $\angle B$ + 90° = 180°
               m $\angle B$ = 180° − 90° − 54°
               m $\angle B$ = 36°

---

**Problem Solving Reminder**

To solve some problems, you might need to *use previously obtained solutions* in order to complete the answer. In Example 1, it was convenient to use the m $\angle A$ found in part *b* to solve for the m $\angle B$.

**EXAMPLE 2**  Solve $\triangle ABC$ by finding each measure.

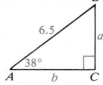

a. $m \angle B$
b. $a$ to the nearest tenth
c. $b$ to the nearest tenth

**Solution**

a. $m \angle A + m \angle B + m \angle C = 180°$
$38° + m \angle B + 90° = 180°$
$m \angle B = 180° - 90° - 38° = 52°$

b. $\sin 38° = \dfrac{a}{6.5}$

$a = \sin 38° \times 6.5$
$= 0.6157 \times 6.5 = 4.00205$

To the nearest tenth, $a = 4.0$.

c. $\cos 38° = \dfrac{b}{6.5}$

$b = \cos 38° \times 6.5$
$= 0.7880 \times 6.5 = 5.122$

To the nearest tenth, $b = 5.1$.

## Class Exercises

**State whether you would use the sine, cosine, or tangent ratio to find $x$ in each diagram.**

**1.**

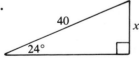

**2.**

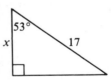

**3.**

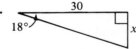

**4.**

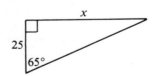

**For Exercises 5–18, use the table on page 509.**

**Find a value for the trigonometric ratio.**

**5.** $\cos 8°$      **6.** $\sin 67°$      **7.** $\tan 36°$      **8.** $\tan 82°$

**9.** $\sin 15°$      **10.** $\cos 75°$      **11.** $\cos 20°$      **12.** $\sin 70°$

**Find the measure of $\angle A$ to the nearest degree.**

**13.** $\sin A = 0.4$        **14.** $\tan A = 1.6$        **15.** $\cos A = 0.85$

**16.** $\cos A = 0.19$        **17.** $\tan A = 0.819$        **18.** $\sin A = 0.208$

## Written Exercises

**For Exercises 1–40, use the tables on page 508 and page 509.**

**Find $\sin A$, $\cos A$, and $\tan A$ for the given measure of $\angle A$.**

A  **1.** $25°$        **2.** $76°$        **3.** $88°$        **4.** $11°$        **5.** $39°$        **6.** $42°$

**7.** $74°$        **8.** $13°$        **9.** $40°$        **10.** $52°$        **11.** $65°$        **12.** $81°$

**Find the measure of $\angle A$ to the nearest degree.**

**13.** $\sin A = 0.9877$        **14.** $\cos A = 0.9205$        **15.** $\tan A = 0.0175$

**16.** $\cos A = 0.8572$        **17.** $\tan A = 4.0108$        **18.** $\sin A = 0.2250$

**Find the measure of the angle to the nearest degree or the length of the side to the nearest whole number.**

**19.** m $\angle A$        **20.** $x$

**21.** m $\angle B$        **22.** m $\angle L$

**23.** $t$        **24.** $u$

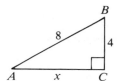

**Find the measure of $\angle A$ to the nearest degree.**

B  **25.** $\sin A = 0.8483$        **26.** $\cos A = 0.2758$        **27.** $\tan A = 0.4560$

**28.** $\sin A = 0.6559$        **29.** $\tan A = 2.7500$        **30.** $\cos A = 0.5148$

**Solve $\triangle ABC$. Round angle measures to the nearest degree and lengths to the nearest tenth.**

**31.** $a = 5$, $b = 8$        **32.** $a = 4$, $b = 7$

**33.** m $\angle A = 72°$, $c = 10$        **34.** m $\angle B = 26°$, $c = 8$

**35.** $b = 4$, $c = 9$        **36.** $a = 5$, $c = 7$

**37.** m $\angle B = 20°$, $b = 15$        **38.** m $\angle A = 80°$, $a = 9$

**39.** m $\angle A = 58°$, $b = 12$        **40.** m $\angle B = 39°$, $a = 20$

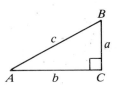

## Problems

**Give angle measures to the nearest degree and lengths to the nearest tenth.**

**A**  **1.** How tall is the tree in the diagram below?

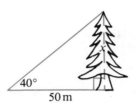
50 m

**2.** How tall is the flagpole in the diagram below?

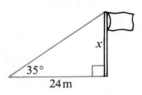

**3.** In the diagram below, the road rises 20 m for every 100 m traveled horizontally. What angle does it make with the horizontal?

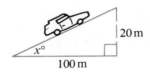

**4.** What angle does the rope make with the horizontal in the diagram below?

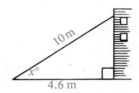

**B**  **5. a.** How tall is the building in the diagram below?
 **b.** How tall is the antenna?

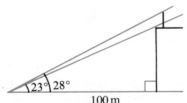

**6.** What is the height of the child in the diagram below?

**7.** In △ABC, $\overline{AC}$ is 8 cm long. The length of the altitude to $\overline{AB}$ is 5 cm. Find the measure of ∠A.

**8.** Triangle MNO is an isosceles triangle with $\overline{MO}$ congruent to $\overline{NO}$. The third side of the triangle, $\overline{MN}$, is 36 cm long. The perimeter of the triangle is 96 cm.
 **a.** Find the lengths of $\overline{MO}$ and $\overline{NO}$.
 **b.** The altitude from O to $\overline{MN}$ bisects $\overline{MN}$. Find the measure of ∠OMN.

**Give angle measures to the nearest degree and lengths to the nearest tenth.**

9. A surveyor is determining the direction in which tunnel $\overline{AB}$ is to be dug through a mountain. She locates point $C$ so that $\angle C$ is a right angle, the length of $\overline{AC}$ is 1.5 km, and the length of $\overline{BC}$ is 3.5 km. Find the measure of $\angle A$.

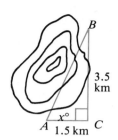

**C** 10. In $\triangle RST$, the measure of $\angle S$ is 142°. The length of $\overline{RS}$ is 10. Find the length of the altitude from vertex $R$.

## Self-Test B

**Exercises 1–4 refer to the diagram below. Complete.**

1. $\triangle TOY \underline{\ ?\ } \triangle TIN$                                 [11–5]

2. $\dfrac{TI}{TO} = \dfrac{IN}{OY} = \dfrac{?}{?}$

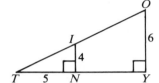

3. $TY = \underline{\ ?\ }$

4. $m \angle TIN \underline{\ ?\ } m \angle TOY$

**In Exercises 5 and 6, give answers in terms of radicals with the radical in the numerator.**

5. The hypotenuse of a 45° right triangle is $6\sqrt{2}$ cm long. What is the length of each leg?           [11–6]

6. The longer leg of a 30°–60° right triangle has length 12. What are the lengths of the shorter leg and the hypotenuse?

**Exercises 7–13 refer to the diagram at the right.**

**Complete in terms of $p$, $q$, and $r$.**

7. $\sin P = \underline{\ ?\ }$        8. $\cos P = \underline{\ ?\ }$          [11–7]

9. $\tan P = \underline{\ ?\ }$

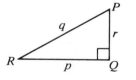

**Find the measure of the angles to the nearest degree and the lengths to the nearest tenth. Use the tables on page 508 and page 509.**

10. $r = 3$, $q = 6$             11. $m \angle P = 78°$, $r = 5$       [11–8]

12. $m \angle R = 23°$, $p = 8$        13. $m \angle R = 57°$, $q = 24$

*Self-Test answers and Extra Practice are at the back of the book.*

The following program produces a Pythagorean triple using any counting number greater than 2.

```
10   PRINT "INPUT THE DESIRED NUMBER";
20   INPUT N
30   IF N * (N - 1) * (N - 2) < > 0 THEN 60
40   PRINT "THERE IS NO SUCH TRIPLE."
50   GOTO 150
60   IF INT (N / 2) < > N / 2 THEN 120
70   IF N = 4 THEN 110
80   LET N = N / 2
90   LET C = C + 1
100   GOTO 60
110   LET N = 3
120   PRINT "A PYTHAGOREAN TRIPLE IS:"
130   LET B = INT (N ↑ 2 / 2)
140   PRINT 2 ↑ C * N,2 ↑ C * B,2 ↑ C * (B + 1)
150   END
```

1. Use the program to find a triple with the number 3. Does it give you the triple you had expected?

2. Use the program to find a triple with the number 4. Does it give you the triple you had expected?

3. Now try 5. Is the output what you had expected?

4. Input each of these numbers to produce Pythagorean triples:

   6, 7, 8, 11, 17, 24, 101

   Check three of your answers by multiplying.

5. Try 1 or 2 in the program. Can you explain why this output is true?

Trace the two squares as they are shown at the right. Can you draw one line that will divide each of the squares into two parts of equal area?

# Other Roots and Fractions as Exponents

As you learned earlier, $\sqrt{49} = 7$ because $7^2 = 49$. Recall that we chose 7 and not $-7$ since 7 is the positive square root of 49.

We can extend the idea of square roots to other roots. The fourth root of 16, denoted $\sqrt[4]{16}$, is 2 since $2^4 = 16$. In general,

> For any positive *even* integer $n$, and positive integer $c$,
> $$\sqrt[n]{a} = c, \text{ if } c^n = a.$$

The third root, or cube root, of a number may be positive or negative depending on the sign of the number. For example,

$$\sqrt[3]{64} = 4 \text{ because } 4^3 = 64$$

and $\sqrt[3]{-27} = -3$ because $(-3)^3 = -27.$

In general,

> For any positive *odd* integer $n$, and any integers $a$ and $c$,
> $$\sqrt[n]{a} = c, \text{ if } c^n = a.$$

The following cases will complete our discussion of roots.

If $a = 0$, then $\sqrt[n]{a} = 0$ because $a^n = 0^n = 0$.

If $a$ is a negative number and $n$ is *even*, there is no real $n$th root of $a$.

Recall that you add exponents when multiplying powers of the same base.

$$9^1 \times 9^2 = 9^{1+2} = 9^3$$

When this rule is applied to fractional exponents, we have

$$9^{\frac{1}{2}} \times 9^{\frac{1}{2}} = 9^{\frac{1}{2} + \frac{1}{2}} = 9^1 = 9.$$

This example suggests that we define $9^{\frac{1}{2}}$ as $\sqrt{9}$, since $(\sqrt{9})^2 = 9$.
In general,

$$a^{\frac{1}{n}} = \sqrt[n]{a}.$$

**Find the root. If the root does not exist, explain why.**

1. $\sqrt[3]{27}$      2. $\sqrt[6]{0}$      3. $\sqrt[4]{16}$      4. $\sqrt[3]{1000}$

5. $\sqrt[3]{64}$      6. $\sqrt[3]{-27}$      7. $\sqrt{-49}$      8. $\sqrt[4]{-16}$

9. $\sqrt[3]{-1000}$      10. $\sqrt[5]{-32}$      11. $\sqrt[3]{125}$      12. $\sqrt[7]{-1}$

**Write the expression without exponents.**

13. $36^{\frac{1}{2}}$    14. $27^{\frac{1}{3}}$    15. $64^{\frac{1}{3}}$    16. $64^{\frac{1}{6}}$    17. $16^{\frac{1}{4}}$    18. $1000^{\frac{1}{3}}$    19. $81^{\frac{1}{4}}$

**Calculator Activity**

Most scientific calculators have a $\boxed{\sqrt[x]{y}}$ key that can be used to approximate roots that are not integers. For example, to obtain $\sqrt[4]{7}$, enter $\boxed{7}$ $\boxed{\sqrt[x]{y}}$ $\boxed{4}$ $\boxed{=}$ to get 1.6265766. On some calculators, the $\boxed{\sqrt[x]{y}}$ key may be a second function. In this case, push the $\boxed{\text{2nd F}}$ key to activate the root function.

Other calculators have a $\boxed{y^x}$ and a $\boxed{1/x}$ key. In this case, we can obtain $\sqrt[4]{7}$ by calculating $7^{\frac{1}{4}}$ in the following way.

$$\boxed{7} \quad \boxed{y^x} \quad \underbrace{\boxed{4} \quad \boxed{1/x}} \quad \boxed{=}$$

This is $\frac{1}{4}$.

**Use a calculator to approximate the following.**

1. $\sqrt[3]{10}$    2. $\sqrt[3]{4}$    3. $\sqrt[5]{50}$    4. $\sqrt[4]{44}$    5. $5^{\frac{1}{4}}$    6. $11^{\frac{1}{3}}$

7. The volume of a sphere is related to its radius by the formula $V = \frac{4}{3}\pi r^3$. Use the formula to find the radius of a balloon that has a volume of 3500 ft³. Use $\pi \approx 3.14$.

# Chapter Review

**Complete.**

1. A positive number has exactly __?__ different square root(s). [11–1]

2. $\sqrt{169}$ is 13, therefore 169 is a __?__ square.

3. If 5.2 is used as an estimate for $\sqrt{28.6}$ in the divide-and-average [11–2]
   method, the next estimate will be __?__.

4. Using the divide-and-average method, $\sqrt{53} = $ __?__ to the tenths'
   place.

5. In the table on page 508, $\sqrt{24.6}$ lies between __?__ and __?__. [11–3]

6. Using interpolation and the table on page 498, $\sqrt{5.7} = $ __?__ to the
   nearest hundredth.

**True or false?**

7. The Pythagorean theorem applies to all triangles. [11–4]

8. The hypotenuse is the longest side of a right triangle and is opposite
   the right angle.

9. The measure of the diagonal of a 5 cm by 5 cm square is 50 cm.

**Exercises 10–12 refer to the diagram below.**
$\triangle ABC \sim \triangle DEF$. **Complete.**

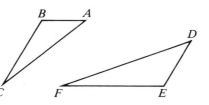

10. $\dfrac{BA}{?} = \dfrac{AC}{DF}$     11. $\angle C \cong \angle$ __?__ [11–5]

12. $\overline{BC}$ corresponds to __?__.

13. An isosceles right triangle is also called a __?__° right triangle. [11–6]

14. An equilateral triangle with sides 16 cm long has an altitude of
    __?__ cm.

15. The legs of an isosceles right triangle are 7 cm long. The length of
    the hypotenuse is __?__ cm.

**Exercises 16–21 refer to the diagram at the right. Match.**

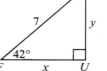

16. $\sin F$     17. $\cos F$     **A.** $\dfrac{x}{7}$     **B.** $\cos 42° \times 7$ [11–7]

18. $\tan N$     19. $y$     **C.** $\dfrac{y}{7}$     **D.** $\sin 42° \times 7$ [11–8]

20. $x$     21. $m \angle N$     **E.** $\dfrac{x}{y}$     **F.** 48°

# Chapter Test

If the given symbol names an integer, state the integer. If not, name the two consecutive integers between which the number lies.

1. $\sqrt{16}$   2. $-\sqrt{144}$   3. $\sqrt{169} - \sqrt{121}$   4. $\sqrt{38 + 43}$   [11–1]

Solve. Round your answer to the nearest tenth.

5. Use the divide-and-average method to approximate $\sqrt{13.7}$.   [11–2]

6. Using interpolation and the table on page 508, $\sqrt{12.6} \approx$ ___?___.   [11–3]

7. The area of a square deck is 65.6 m². Find the length of the side.

Is the triangle with sides of the given lengths a right triangle?

8. 6, 8, 10            9. 7, 11, 19            10. 8, 15, 17   [11–4]

Exercises 11–13 refer to the diagram at the right.
$\triangle MNO \sim \triangle XYZ$.

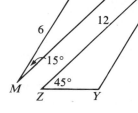

11. If $\frac{MN}{XY} = \frac{3}{4}$, then $\frac{NO}{YZ} =$ ___?___.   [11–5]

12. Find the length of $\overline{MO}$.

13. Find the measure of $\angle N$.

Give answers in terms of radicals with the radical in the numerator.

14. In a 30°–60° right triangle, the shorter leg has length 5. How long is (a) the longer leg and (b) the hypotenuse?   [11–6]

15. The hypotenuse of a 45° right triangle has a length of 36. How long is each leg?

Exercises 16–19 refer to the diagram at the right. Find a value for the trigonometric ratio.

16. $\cos A$     17. $\tan A$     18. $\cos B$     19. $\sin B$   [11–7]

$\triangle KLM$ is an isosceles triangle with $\overline{KL} \cong \overline{LM}$. The third side, $\overline{KM}$, is 42 cm long. The perimeter of $\triangle KLM$ is 112 cm.

20. Find the lengths of $\overline{KL}$ and $\overline{LM}$.     21. Find the height of $\triangle KLM$.   [11–8]

22. Find m $\angle K$ and m $\angle M$ to the nearest degree. Use the table on page 509.

# Cumulative Review (Chapters 1–11)

## Exercises

**Simplify the expression.**

**1.** $(8 + 2) \div (3 \times 9)$      **2.** $15 \div 8 + 7 \div 9$      **3.** $[48(2 + 5) - 6] \times 10$

**4.** $-26 - 7 + 9 \times 3$      **5.** $2(10 + 5) \div (-3 - 6)$      **6.** $[(7 - 5)(3 \times 5)] - 5$

**Evaluate the expression when $a = 10$, $b = -10$, and $c = 0.5$.**

**7.** $ab$      **8.** $ac$      **9.** $a + b$      **10.** $a - b$      **11.** $2ab$

**12.** $b^2$      **13.** $ac^2$      **14.** $a^2c$      **15.** $\dfrac{b}{a}$      **16.** $\dfrac{ab}{c}$

**Replace __?__ with $<$, $>$, or $=$ to make a true statement.**

**17.** $\dfrac{1}{2}$ __?__ $\dfrac{3}{8}$      **18.** $\dfrac{1}{5}$ __?__ $\dfrac{2}{8}$      **19.** $-\dfrac{4}{5}$ __?__ $\dfrac{3}{4}$      **20.** $-\dfrac{1}{3}$ __?__ $-\dfrac{7}{9}$

**Solve the equation.**

**21.** $12 + a = 50$      **22.** $x - 27 = 56$      **23.** $43m = 107.5$

**24.** $\dfrac{n}{38.6} = 15$      **25.** $a + \dfrac{2}{3} = 8$      **26.** $w - \dfrac{1}{5} = 12$

**27.** $2x + 4x = 48$      **28.** $5c - 12c = 35$      **29.** $3(t - 4) = 24$

**30.** $6z = 32 + z$      **31.** $2 + 5a = 30$      **32.** $7(x - 6) = 3x$

**Find the perimeter of the given polygon.**

**33.** square: sides of 2.9 ft      **34.** triangle: 7.11 in., 8.11 in., 12.45 in.

**Solve the proportion.**

**35.** $\dfrac{n}{15} = \dfrac{12}{45}$      **36.** $\dfrac{6}{7} = \dfrac{n}{35}$      **37.** $\dfrac{8}{n} = \dfrac{32}{40}$      **38.** $\dfrac{3}{5} = \dfrac{24}{n}$

**Graph each equation on a separate coordinate plane.**

**39.** $x + 2y = 7$      **40.** $2x + 2y = 9$      **41.** $x + \dfrac{1}{5}y = 1$

**Find the volume of a cylinder with the given dimensions. Leave your answer in terms of $\pi$.**

**42.** radius: 7    height: 12      **43.** radius: 2.3    height: 10.5

**Rewrite the expression in lowest terms with the radical in the numerator.**

**44.** $\dfrac{4}{\sqrt{10}}$      **45.** $\dfrac{5}{\sqrt{2}}$      **46.** $\dfrac{18}{\sqrt{37}}$      **47.** $\dfrac{2n}{\sqrt{n}}$      **48.** $\dfrac{6m}{\sqrt{m}}$

## Problems

> **Problem Solving Reminders**
> Here are some problem solving reminders that may help you solve some of the problems on this page.
> - Sometimes more than one method can be used to solve.
> - Supply additional information if necessary.
> - Check your results with the facts given in the problem.

**Solve.**

1. Evan is a parking lot attendant at the Lonestar Garage. When counting his tips from Monday he discovered he had 12 more quarters than dimes and 3 fewer nickels than quarters. If Evan earned a total of $19.45 in tips, how many of each type of coin did he have?

2. A regular pentagon has a perimeter of 378.5 m. Find the length of each side.

3. A discount store has an automatic markdown policy. Every 7 days, the price of an item is marked down 25% until the item is sold or 4 weeks have elapsed. If the first price of an item is $40 on September 17, what will the price be on October 1?

4. A window washer uses a vinegar and water solution in the ratio of one-half cup of vinegar to three cups of water. How much vinegar will be in two gallons of solution?

5. Lavender soap is sold in boxes of 3 bars for $6.50. To the nearest cent, what is the cost of one bar?

6. Bertha Magnuson had purchased 2500 shares of ABC stock for $3.50 per share. When she sold the stock, its value had gone down to $\frac{5}{8}$ of the total purchase price. How much did Bertha lose?

7. The area of a square table is 1936 in.². What will be the dimensions of a square tablecloth that drops 4 in. over each side of the table?

8. Mary has $300 to spend on new boots and a winter coat. She expects to spend $\frac{2}{5}$ as much for boots as for a coat. What is the most she can spend on boots?

# 12

# Statistics and Probability

For ecological and other reasons, it is sometimes important to determine the size and geographical location of a group of migrating birds such as those shown in the photograph. A total count of a particular kind of bird obviously poses a difficult problem. By using a sample count, however, only a small part of the bird population needs to be counted. The total number of birds in the entire area can then be estimated from the number in the sampled area.

The simplest method of counting birds and studying migration is direct observation. Because of the disadvantages of this method, more sophisticated methods are being developed and used. These include banding, radio tracking, and radar observation. With the help of computers, this data can be quickly collected and examined.

In this chapter, you will learn some methods for gathering and analyzing data.

## Career Note

Statisticians collect, analyze, and interpret numerical results of surveys and experiments. They may use the information that is gathered to determine suitable choices, evaluate a report, or redesign an existing program. Statisticians are usually employed in manufacturing, finance, or government positions. A thorough knowledge of mathematics and a background in economics or natural science is needed.

# 12-1 Picturing Numerical Data

Many scientific, social, and economic studies produce numerical facts. Such numerical information is called **data.** At the right are some data about the number of automobiles sold in the United States.

These data can be pictured by using a **bar graph** (below left) or a **broken-line graph** (below right). On the bar graph the height of each bar is drawn to the scale marked at the left and so is proportional to the data it represents. All bars have the same width. The broken-line graph can be made by joining the midpoints of the tops of the bars.

We can see from either graph that the most rapid 10-year increase in auto sales occurred between 1940 and 1950.

| Car Sales | |
|---|---|
| Year | Number Sold (nearest 100,000) |
| 1920 | 2,200,000 |
| 1930 | 3,400,000 |
| 1940 | 4,500,000 |
| 1950 | 8,000,000 |
| 1960 | 7,900,000 |
| 1970 | 8,200,000 |
| 1980 | 8,000,000 |

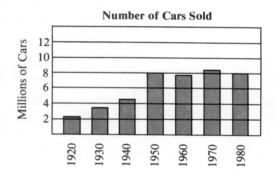

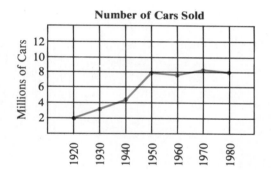

To draw a graph we must choose a **data unit** to mark off one of the axes. If the data are small numbers, the data unit can be a small number, such as 1 or 5. If the data are large numbers, a larger data unit should be chosen so that the graph will be a reasonable size. For example, the data unit in the graphs above is 2,000,000 automobiles.

*EXAMPLE*    The following table gives the average monthly temperatures in Minneapolis, Minnesota. Construct a bar graph to illustrate the data.

| Month | J | F | M | A | M | J | J | A | S | O | N | D |
|---|---|---|---|---|---|---|---|---|---|---|---|---|
| °C | −11.1 | −8.3 | −2.2 | 7.2 | 13.9 | 19.4 | 22.2 | 21.1 | 15.6 | 10.0 | 0 | −7.2 |

***Solution***   Label the horizontal axis with symbols for the months.   Label the vertical axis using a data unit of 5°. Then draw bars of equal widths and proper lengths. Draw the bars downward for negative data. Finally, give the graph a title.

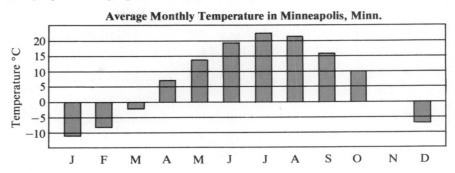

**Average Monthly Temperature in Minneapolis, Minn.**

Sometimes it is more convenient to arrange the bars in a graph horizontally. In the graph at the right, the data unit for the horizontal axis is 10,000 km² and the vertical axis is labeled with the names of the lakes.

We can estimate data from graphs. For example, we see from the graph that the area of Lake Erie is approximately 25,000 km².

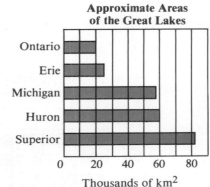

**Approximate Areas of the Great Lakes**

Thousands of km²

---

## Class Exercises

**Exercises 1–5 refer to the bar graph at the right.**

1. What data unit is used on the vertical axis?

2. Which mountain has the lowest elevation?  What is its elevation?

3. Which mountain has the highest elevation?  What is its elevation?

4. Which two mountains have nearly the same elevation?

5. Find the ratio of the elevation of Mount Vinson to the elevation of Mount McKinley.

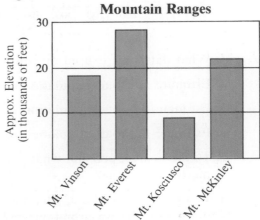

**Mountain Ranges**

**Exercises 6–10 refer to the broken-line graph at the right.**

6. What data unit is used on the vertical axis?

7. What was the approximate population in 1900? In 1950?

8. In approximately what year did the population pass 100 million? 150 million? 200 million?

9. In which 20-year period did the population increase the most?

10. Find the approximate total increase in population during the twentieth century.

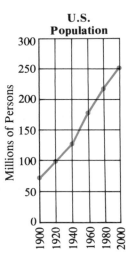

U.S. Population

## Written Exercises

The bar graph below shows the seven nations having populations over 100 million. Use the graph for Exercises 1–6.

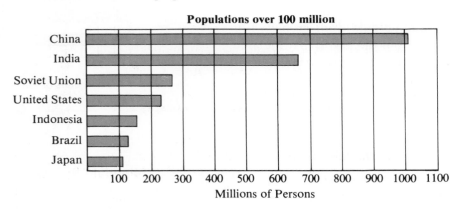

Populations over 100 million

A  1. What data unit is used on the horizontal axis?

2. Estimate the total population of the three largest nations.

3. Estimate the total population of the three smallest nations.

4. Estimate the total population of the seven nations.

5. To the nearest percent, what percent of the world's 4.5 billion people live in China?

6. To the nearest percent, what percent of the world's 4.5 billion people live in the seven nations shown on the graph?

**In Exercises 7-9, make a bar graph to illustrate the given data.**

7. The six longest highway tunnels are: Saint Gotthard, 16.2 km; Arlberg, 14 km; Frejus, 12.8 km; Mont Blanc, 11.7 km; Enasson, 8.5 km; and San Bernadino, 6.6 km.

8. The areas of the world's five largest islands in thousands of square kilometers are: Greenland, 2175; New Guinea, 792; Borneo, 725; Madagascar, 587; and Baffin, 507.

9. The table below gives the length of each continent's longest river.

| Continent | River | Length (km) |
|-----------|-------|-------------|
| Africa | Nile | 6632 |
| Asia | Yangtze | 6342 |
| Australia | Murray-Darling | 3693 |
| Europe | Volga | 3510 |
| North America | Mississippi-Missouri | 5936 |
| South America | Amazon | 6400 |

10. Make a broken-line graph to illustrate the given data.

**Number of U.S. High School Graduates (in thousands)**

| 1920 | 1930 | 1940 | 1950 | 1960 | 1970 | 1980 |
|------|------|------|------|------|------|------|
| 311 | 667 | 1221 | 1200 | 1864 | 2896 | 3078 |

B 11. Make a bar graph to illustrate the data. Net profits of the XYZ Company in thousands of dollars were: 30 in 1981; −20 (loss) in 1982; −5 (loss) in 1983; 45 in 1984; and 60 in 1985.

12. Make a broken-line graph to illustrate the average monthly temperatures in Minneapolis. (Use the table in the example in this lesson.)

## Review Exercises

**Perform the indicated operation.**

1. $\frac{47}{60} \times 360$    2. $\frac{74}{83} \times 249$    3. $\frac{118}{37} \times 111$    4. $\frac{85}{156} \times 312$

5. 25% of 540    6. 47% of 360    7. 73% of 180    8. 81% of 360

# 12-2 Pictographs and Circle Graphs

Nontechnical magazines often present data using pictures. These **pictographs** take the form of bar graphs with the bars replaced by rows or columns of symbols. Each symbol represents an assigned quantity. This amount must be clearly indicated on the pictograph. For example, the pictograph below illustrates the data on automobile sales given in the previous lesson.

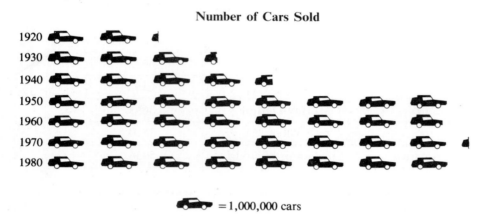

**Number of Cars Sold**

= 1,000,000 cars

*EXAMPLE 1*  The approximate numbers of different book titles published in the United States in selected years are: 11,000 in 1950; 15,000 in 1960; 36,000 in 1970; and 42,000 in 1980. Illustrate the data with a pictograph.

*Solution*  Stacks of books are appropriate symbols. We let one thick book represent 5000 titles and one thin book represent 1000 titles.

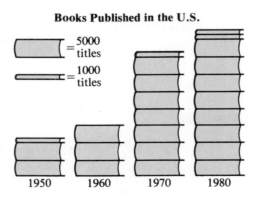

**Books Published in the U.S.**

The **circle graph** at the right shows how the world's water is distributed. Circle graphs are often more effective than other graphs in picturing how a total amount is divided into parts.

The following example illustrates how to make a circle graph.

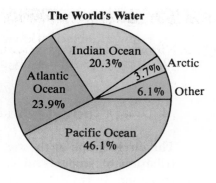

The World's Water

Indian Ocean 20.3%

Atlantic Ocean 23.9%

Arctic 3.7%

Other 6.1%

Pacific Ocean 46.1%

**EXAMPLE 2**   The seventh-grade class voted to decide where to have their year-end picnic. The results were as follows: Mountain Park, 62 votes; State Beach, 96 votes; City Zoo, 82 votes. Draw a circle graph to illustrate this distribution.

**Solution**   The whole circle represents the total number of votes. We plan to divide the circle into wedges to represent the distribution of the votes. Since the sum of all the adjacent angles around a point is 360°, the sum of the angle measures of all the wedges is 360°.

First find the total of all votes cast.

$$62 + 96 + 82 = 240$$

Then find the fraction of the vote cast for each place and the corresponding angle measure.

For Mountain Park:

$$\frac{62}{240} \times 360° = 93°$$

For State Beach:

$$\frac{96}{240} \times 360° = 144°$$

For City Zoo:

$$\frac{82}{240} \times 360° = 123°$$

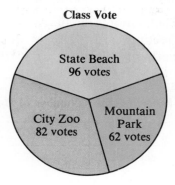

Class Vote

State Beach 96 votes

City Zoo 82 votes

Mountain Park 62 votes

Now use a compass to draw a circle and a protractor to draw three radii forming the angles found above. Finally, label the wedges and give the graph a title.

## Class Exercises

1. Draw a circle and divide it into 4 equal parts.

2. Draw a circle and divide it into 6 equal parts.

3. Draw a circle and divide it into 9 equal parts.

4. Draw a circle and divide it into 12 equal parts.

**The circle graph at the right pictures the distribution of students playing the various instruments in the school orchestra.**

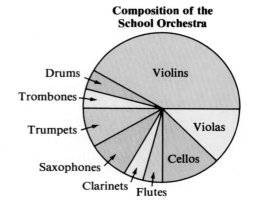

Composition of the School Orchestra

5. Which instrument is played by the greatest number of students in the orchestra?

6. How does the number of students playing stringed instruments (violins, violas, and cellos) compare with the number of students playing all the other instruments?

7. Which four instruments are played by the fewest students?

8. How does the number of saxophone players compare with the number of trumpet players?

## Written Exercises

**Complete the following tables and construct a circle graph for each.**

**A** 1. The Junior Athletic Association decided to raise money by selling greeting cards. Orders were obtained for the following kinds of cards.

| Kind of Card | Number of Boxes | Fraction of the Whole | Number of Degrees in the Angle |
|---|---|---|---|
| birthday | 60 | ? | ? |
| get well | 72 | ? | ? |
| friendship | 48 | ? | ? |
| thank you | 60 | ? | ? |
| Total | ? | ? | ? |

**2.** The winter issue of the school magazine contained the following kinds of material.

| Kind of Material | Number of Pages | Fraction of the Whole | Number of Degrees in the Angle |
|---|---|---|---|
| fiction | 32 | ? | ? |
| essays | 8 | ? | ? |
| sports | 16 | ? | ? |
| advertisements | 8 | ? | ? |
| Total | ? | ? | ? |

**Illustrate, using a circle graph.**

**3.** The surface of Earth is 30% land and 70% water.

**4.** Earth's atmosphere is 78% nitrogen, 21% oxygen, and 1% other gases.

**Illustrate, using a pictograph. The parentheses contain a suggestion as to what symbol to use.**

**5.** The XYZ Car Rental Company rented 520 cars in July, 350 cars in August, 350 cars in September, 400 cars in October, and 110 cars in November. (cars)

**6.** The Meadowbrook School ordered 200 cartons of milk the first week, 220 the second week, 240 the third week, and 180 the fourth week. (milk cartons)

**7.** The number of fish caught in Clear Lake: 4500 in 1970; 3500 in 1975; 4200 in 1980; 4800 in 1985. (fish)

**8.** The cost of higher education in the United States in billions of dollars: 1965 — $13; 1970 — $23; 1975 — $39; 1980 — $55. (dollars)

**Illustrate, using (a) a circle graph and (b) a pictograph. The parentheses contain a suggestion as to what symbol to use in the pictograph.**

**B** **9.** In the United States about 167 million people live in cities and about 59 million live in rural areas. (people)

**10.** A fund-raising event made $420 from the sale of antiques, $280 from crafts items, and $240 from food. (dollars)

**Illustrate, using (a) a circle graph and (b) a pictograph. The parentheses contain a suggestion as to what symbol to use in the pictograph.**

**11.** The average seasonal rainfall in Honolulu is 26 cm in winter, 7 cm in spring, 5 cm in summer, and 21 cm in autumn. (raindrops)

**12.** Of each dollar the United States government takes in, 47¢ comes from individual income taxes, 27¢ from Social Security, 12¢ from corporation taxes, and 14¢ from other sources. (piles of coins)

**13.** A family spends $440.75 of the monthly budget on food, $530.00 on rent, $617.00 total on clothes, medicine, and other items, and $175.25 on transportation. (dollars)

**14.** A total of 387 people were polled on Proposition Q, 46% favored it, 33% opposed it, and 21% had no opinion. (people)

**15.** The library received a $1300 grant. The librarian plans to spend 10% of the grant to extend magazine subscriptions, 35% to buy new books, 15% to repair damaged books, 30% to buy new furniture, and 10% to locate missing books. (books)

**C**   **16.** Which pictograph correctly shows that the production of a certain oil field doubled between 1975 and 1985? Explain your answer.

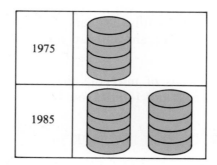

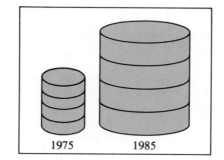

---

## Review Exercises

**Arrange in order from least to greatest.**

**1.** 4, 2.6, 7, 0.84, 3

**2.** 4, −8, −3, 1, 10, −5

**3.** 7, $3\frac{1}{2}$, −5, $4\frac{1}{4}$, −6

**4.** $\frac{7}{8}$, $\frac{5}{6}$, $\frac{11}{9}$, $\frac{7}{12}$, $\frac{12}{7}$

**5.** −6, −10, −2, −12, −9

**6.** $3\frac{7}{8}$, $2\frac{9}{10}$, $2\frac{8}{9}$, $3\frac{11}{16}$

**7.** 1.09, 1.1, 0.9, 0.09

**8.** $\frac{17}{4}$, $\frac{11}{2}$, $\frac{15}{8}$, $\frac{9}{4}$, $\frac{7}{5}$

# 12-3 Mean, Median, and Range

The daytime high temperatures in °C for five days in May were

$$23°, \ 14°, \ 18°, \ 28°, \ \text{and } 25°.$$

Here are two ways to summarize this information.

(1) The **mean** of the temperatures is their sum divided by the number of temperatures.

$$\frac{23° + 14° + 18° + 28° + 25°}{5} = \frac{108°}{5} = 21.6°$$

(2) The **median** of the temperatures is the middle temperature when they are arranged in order of size.

$$14°, \ 18°, \ 23°, \ 25°, \ 28° \qquad \text{Median} = 23°$$

For an even number of data items the median is midway between the two middle numbers.

The **range** of a set of data is the difference between the greatest and least numbers in the set. For example, the range of the above temperatures is $28° - 14°$, or $14°$.

**EXAMPLE** Find the mean, median, and range of each set of numbers. Round to the nearest tenth.

      **a.** 11, 19, 7, 45, 22, 38       **b.** 0, 1, 3, −4, 4, −6, 7

**Solution**     **a.** Mean $= \dfrac{11 + 19 + 7 + 45 + 22 + 38}{6} = \dfrac{142}{6} \approx 23.7$

        Arrange the numbers in order of size: 7, 11, 19, 22, 38, 45

        Median $= \dfrac{19 + 22}{2} = 20.5$       Range $= 45 - 7 = 38$

    **b.** Mean $= \dfrac{0 + 1 + 3 + (-4) + 4 + (-6) + 7}{7} = \dfrac{5}{7} \approx 0.7$

        Arrange the numbers in order of size: −6, −4, 0, 1, 3, 4, 7

        Median $= 1$       Range $= 7 - (-6) = 13$

In everyday conversation the word *average* is usually used for mean.

Analyzing data as we have been doing is part of the branch of mathematics called **statistics.**

## Class Exercises

**Find the mean, median, and range of each set of data.**

**1.** 1, 2, 3, 4, 5

**2.** 5, 3, 1

**3.** 10, 10, 6, 2

**4.** 3, 4, 5, 6, 8, 10

**5.** −4, −2, 0, 2, 4

**6.** 3, 2, 1, 0, −1, −2, −3

**7.** −3, −1, 0, 4, 5

**8.** 6, 5, 5, 0, −1

**9.** −15, −11, −7, −7, −7, 0, 2, 5, 9, 15, 18, 22

**10.** −37, −28, −15, −15, −6, 1, 13, 26, 34

**11.** 11, 17, 31, 43, 58, 61, 58, 89, 94, 58, 94, 107, 215

## Written Exercises

**Find the mean, median, and range of each set of data. If necessary, round to the nearest tenth.**

**A**

**1.** 30, 18, 21, 28, 23

**2.** 42, 58, 55, 61, 39

**3.** 7, 16, 20, 13, 26, 14

**4.** 85, 70, 93, 101, 116, 111

**5.** 47, 61, 53, 69, 45, 58

**6.** 17, 11, 9, 13, 7, 21, 8, 18

**7.** 3.6, 2.7, 2.9, 3.4, 3.4

**8.** 8.1, 9.2, 6.8, 7.3, 7.9, 6.9

**B**

**9.** −3, 2, −2, −5, 3

**10.** 8, −8, −12, 16, −8, 7

**11.** 1.3, −0.8, −0.1, 0.2, 0.9

**12.** 4.1, −3.2, −0.8, −1.5, 2.7, −0.1

**13.** Low Temperatures in February (°C)

| | | | | | | |
|---|---|---|---|---|---|---|
| −13° | −8° | −10° | −4° | 1° | 0° | −2° |
| −5° | −7° | −12° | −8° | −7° | −5° | 0° |
| −2° | −3° | −5° | 1° | 2° | 3° | 1° |
| 2° | 4° | 2° | 4° | 4° | 5° | 6° |

**14.** Elevations Along the Salton Sea Railway (meters)

| | | | | |
|---|---|---|---|---|
| 6.82 | 2.55 | 1.60 | −0.21 | −1.35 |
| −2.68 | −1.95 | −2.06 | −0.88 | −0.02 |
| 0.41 | 1.15 | 3.15 | 6.51 | 5.86 |

**444**  *Chapter 12*

**15.** Insert another number in the list 15, 23, 11, 17 in such a way that the median is not changed.

**16.** Replace one of the numbers in the list 15, 23, 11, 17 so that the median becomes 17.

**Find the value of $x$ such that the mean of the given list is the specified number.**

**17.** 6, 9, 13, $x$; mean $= 11$    **18.** 8, 14, 16, 12, $x$; mean $= 15$

**19.** Janet's scores on her first four mathematics tests were 98, 78, 84, and 96. What score must she make on the fifth test to have the mean of the five tests equal 90?

**20.** The heights of the starting guards and forwards on the basketball team are 178 cm, 185 cm, 165 cm, and 188 cm. How tall is the center if the mean height of the starting five is 182 cm?

C **21.** If each number in a list is increased by 5, how is the median affected?

**22.** If each number in a list is increased by 5, how is the mean affected?

## Review Exercises

**Perform the indicated operations. Round to the nearest tenth if necessary.**

**1.** $(4 \times 2 + 7 \times 5 + 6 \times 8) \div 15$    **2.** $(9 \times 4 + 6 \times 7 + 11 \times 5) \div 16$

**3.** $(14 \times 6 + 3 \times 0 + 16 \times 4) \div 10$    **4.** $(12 \times 1 + 8 \times 11 + 5 \times 6) \div 18$

**5.** $(18 \times 3 + 20 \times 5 + 7 \times 6) \div 14$    **6.** $(7 \times 15 + 4 \times 9 + 7 \times 2) \div 26$

**7.** $(11 \times 3 \times 2 + 8 \times 3 + 9) \div 33$    **8.** $(12 \times 2 \times 8 + 9 \times 3 \times 4) \div 15$

 **Challenge**

Cellini and Valdez were each appointed to a new job. Cellini is paid a starting salary of $16,500 a year with a $1000 raise at the end of each year. Valdez is paid a starting salary of $8000 for the first 6 months with a raise of $500 at the end of every 6 months. How much does each make at the end of the first year? The third year? Who has the better pay plan?

# 12-4 Frequency Distributions

Forty students took a quiz and received the following scores.

| | | | | | | | | | |
|---|---|---|---|---|---|---|---|---|---|
| 8 | 7 | 10 | 6 | 8 | 7 | 9 | 7 | 8 | 5 |
| 9 | 7 | 8 | 8 | 7 | 9 | 3 | 8 | 8 | 6 |
| 7 | 5 | 6 | 8 | 10 | 10 | 7 | 10 | 8 | 9 |
| 7 | 7 | 9 | 7 | 8 | 7 | 9 | 10 | 10 | 7 |

| Score | Frequency |
|---|---|
| 3 | 1 |
| 4 | 0 |
| 5 | 2 |
| 6 | 3 |
| 7 | 12 |
| 8 | 10 |
| 9 | 6 |
| 10 | 6 |

In order to analyze these data, we can arrange them in a **frequency table** as shown at the right. The number of times a score occurs is its **frequency**. Since a score of 9 was received by 6 students, we can say that the frequency of 9 is 6. The pairing of the scores with their frequencies is called a **frequency distribution.**

**EXAMPLE**  Find the range, mean, and median for the data above.

**Solution**  Since the lowest score is 3 and the highest is 10, the range is $10 - 3$, or 7.

To find the *mean* of the data, first multiply each possible score by its frequency. Enter the product in a third column as shown. The sum of the numbers in this column equals the sum of the 40 scores. Therefore:

$$\text{mean} = \frac{309}{40} = 7.725$$

| Score $x$ | Frequency $f$ | $x \times f$ |
|---|---|---|
| 3 | 1 | 3 |
| 4 | 0 | 0 |
| 5 | 2 | 10 |
| 6 | 3 | 18 |
| 7 | 12 | 84 |
| 8 | 10 | 80 |
| 9 | 6 | 54 |
| 10 | 6 | 60 |
| Total | 40 | 309 |

To find the median, first rearrange the data in order from least to greatest. Since there are 40 scores, the median is the average of the twentieth and twenty-first scores, both of which are 8.

$$\text{median} = \frac{8 + 8}{2} = 8$$

**Reading Mathematics:** *Tables*

As you use a table, reread the headings to remind yourself what each number represents. For example, the frequency table above shows that the *score* of 8 had a *frequency* of 10; that is, there are 10 scores of 8, not 8 scores of 10.

The **mode** of a set of data is the item that occurs with the greatest frequency. In the example on the preceding page, the mode is 7 since it has the greatest frequency, 12. Sometimes there is more than one mode. The mode usually is used with nonnumerical data, such as to determine the popularity of car colors.

## Class Exercises

**Find the range, the mean, the median, and the mode(s) of each student's test score, $x$.**

**1.** Evelyn

| $x$ | $f$ |
|---|---|
| 70 | 1 |
| 80 | 4 |
| 90 | 3 |
| 100 | 2 |

**2.** Bruce

| $x$ | $f$ |
|---|---|
| 75 | 2 |
| 85 | 3 |
| 90 | 1 |
| 100 | 4 |

**3.** Shana

| $x$ | $f$ |
|---|---|
| 70 | 4 |
| 85 | 3 |
| 95 | 2 |
| 100 | 1 |

**4.** Elroy

| $x$ | $f$ |
|---|---|
| 75 | 1 |
| 80 | 2 |
| 95 | 3 |
| 100 | 4 |

## Written Exercises

**In Exercises 1–6, find the range, the mean, the median, and the mode(s) of the data in each frequency table.**

**A**  **1.**

| $x$ | $f$ |
|---|---|
| 5 | 2 |
| 6 | 4 |
| 7 | 8 |
| 8 | 5 |
| 9 | 1 |

**2.**

| $x$ | $f$ |
|---|---|
| 0 | 3 |
| 1 | 2 |
| 2 | 7 |
| 3 | 8 |
| 4 | 5 |

**3.**

| $x$ | $f$ |
|---|---|
| 25 | 1 |
| 20 | 0 |
| 15 | 3 |
| 10 | 5 |
| 5 | 6 |

**4.**

| $x$ | $f$ |
|---|---|
| 18 | 1 |
| 15 | 0 |
| 12 | 3 |
| 9 | 6 |
| 6 | 7 |
| 3 | 1 |
| 0 | 2 |

**5.**

| $x$ | $f$ |
|---|---|
| 14 | 2 |
| 15 | 5 |
| 16 | 11 |
| 17 | 11 |
| 18 | 8 |
| 19 | 0 |
| 20 | 3 |

**6.**

| $x$ | $f$ |
|---|---|
| 28 | 2 |
| 27 | 4 |
| 26 | 11 |
| 25 | 11 |
| 24 | 4 |
| 23 | 2 |
| 22 | 1 |

In Exercises 7–12, make a frequency table for the given data, and then find the range, the mean, the median, and the mode(s) of the data. Round the mean to the nearest tenth if necessary.

**B** 7. The number of runs scored by Jan's team in recent softball games: 5, 1, 4, 4, 8, 6, 4, 1, 5, 0, 1, 5, 3, 8, 4, 5, 3, 4

8. Jim's 200 m dash practice times: 28 s, 29 s, 27 s, 27 s, 28 s, 29 s, 28 s, 26 s, 27 s, 26 s, 28 s, 27 s, 27 s, 25 s, 26 s

9.
| April in Pleasantville Average Temperatures (°C) | | | | |
|---|---|---|---|---|
| | 16 | 15 | 14 | 16 | 17 |
| 17 | 18 | 18 | 18 | 19 | 19 | 18 |
| 17 | 18 | 17 | 18 | 19 | 20 | 20 |
| 19 | 19 | 18 | 17 | 18 | 19 | 20 |
| 20 | 20 | 19 | 20 |

10.
| Class Test Scores | | | | | |
|---|---|---|---|---|---|
| 16 | 14 | 20 | 15 | 15 | 17 |
| 15 | 17 | 17 | 16 | 14 | 18 |
| 17 | 15 | 14 | 14 | 20 | 15 |
| 14 | 16 | 15 | 18 | 17 | 16 |
| 15 | 16 | 18 | 14 | 18 | 17 |

11.
| Samples of Steel Rods (diameters in millimeters) | | | | | | | | | | | |
|---|---|---|---|---|---|---|---|---|---|---|---|
| 101 | 102 | 100 | 97 | 101 | 103 | 100 | 100 | 101 | 100 | 99 | 100 |
| 99 | 100 | 101 | 103 | 102 | 100 | 101 | 99 | 100 | 100 | 101 |

12.
| Ages of Members of the Moose Hill Hiking Club | | | | | | | | | | | | | |
|---|---|---|---|---|---|---|---|---|---|---|---|---|---|
| 15 | 14 | 15 | 13 | 16 | 14 | 18 | 16 | 15 | 14 | 14 | 15 | 18 | 15 |
| 13 | 13 | 14 | 15 | 14 | 16 | 16 | 17 | 14 | 15 | 13 | 12 | 14 | 16 |
| 14 | 16 | 15 | 13 | 16 | 14 | 15 | 17 | 12 | 13 | 16 | 15 |

Exercises 13 and 14 refer to a class of 24 boys and 16 girls. (*Hint:* If you know the number of scores and their mean, you can find the sum of the scores.)

**C** 13. On test A, the mean of the boys' scores was 70 and the mean of the girls' scores was 75. What was the class mean?

14. On test B, the class mean was 75 and the mean of the girls' scores was 72. What was the mean of the boys' scores?

## *Review Exercises*

**Explain the meaning of each term.**

1. mean      2. median      3. mode      4. range

5. odd number      6. even number      7. prime number      8. multiple of 7

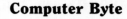

## Computer Byte

The following computer program will find the average, or mean, of several numbers. The program finds the sum of the numbers and then divides this sum by the number of numbers. For example, if you input 2, 4, 7, 8, and 11, the computer would find their sum, 32, and divide it by 5, the number of items, resulting in an answer of 6.4.

```
10   PRINT "TO FIND THE AVERAGE,"
20   PRINT "INPUT THE NUMBERS ONE AT A TIME."
30   PRINT "TYPE -1 AT END OF LIST."
40   PRINT
50   LET N = 0
60   LET S = 0
70   INPUT A
80   IF A = - 1 THEN 120
90   LET N = N + 1
100   LET S = S + A
110   GOTO 70
120   PRINT
130   PRINT "N = ";N
140   PRINT "SUM = ";S
150   PRINT "AVERAGE = ";S/N
160   END
```

**Use the program to find the average of the following. Be sure to type −1 at the end of each list.**

**1.** 12, 19, 23, 8, 17, 31

**2.** 57, 3, 86, 79, 101, 9

**3.** 542, 863, 921, 254, 378, 511

**4.** 1649, 15,241, 8463, 11,684

**5.** 887, 3105, 6324, 7048, 2103, 1298, 5541, 7201, 4961, 1114, 2260, 1954, 7322, 5665, 5321, 3742, 4457, 8916, 9923, 4309, 2385, 3342, 4187

**6.**

| | | | |
|---|---|---|---|
| 21,542 | 11,372 | 63,982 | 14,320 |
| 78,864 | 24,953 | 21,419 | 17,160 |
| 47,922 | 72,315 | 18,560 | 21,000 |
| 91,254 | 94,456 | 41,215 | 35,743 |
| 20,964 | 34,290 | 38,730 | 44,287 |

**7.**

| | | | |
|---|---|---|---|
| 502,784 | 339,065 | 764,290 | 274,415 |
| 114,902 | 765,329 | 314,675 | 441,823 |
| 345,245 | 465,367 | 468,901 | 687,390 |
| 113,823 | 589,001 | 892,030 | 389,210 |
| 992,084 | 314,675 | 440,871 | 572,903 |

# 12-5 Histograms and Frequency Polygons

When a bar graph is used to picture a frequency distribution, it is called a **histogram.** No spaces are left between the bars. The histogram for the quiz-score data of the table is shown at the left below.

The broken-line graph shown at the right below is the **frequency polygon** for the same distribution. The broken-line graph of the frequencies is connected to the horizontal axis at each end to form a polygon.

We can find the range, the mode, the mean, and the median from a histogram and from a frequency polygon.

| Score $x$ | Frequency $f$ |
|:---:|:---:|
| 3 | 1 |
| 4 | 0 |
| 5 | 2 |
| 6 | 3 |
| 7 | 12 |
| 8 | 10 |
| 9 | 6 |
| 10 | 6 |

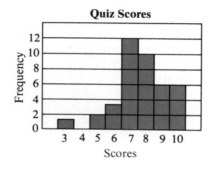

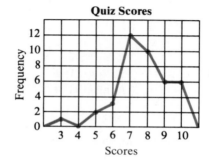

**EXAMPLE**   Find the range, the mode, the mean, and the median for the data represented in the graphs above.

**Solution**   The range is the difference between the least number representing data on the horizontal axis and the greatest number.

The range is $10 - 3$, or 7.

The tallest bar or highest point represents the mode.

The mode is 7.

To find the mean, we first multiply each data item by its frequency.

$3 \times 1 = 3$     $4 \times 0 = 0$     $5 \times 2 = 10$     $6 \times 3 = 18$
$7 \times 12 = 84$     $8 \times 10 = 80$     $9 \times 6 = 54$     $10 \times 6 = 60$

We then add the products.

$3 + 0 + 10 + 18 + 84 + 80 + 54 + 60 = 309$

We then add the frequencies.

$$1 + 0 + 2 + 9 + 12 + 8 + 5 + 3 = 40$$

We then divide the sum of the products by the sum of the frequencies.

$$309 \div 40 = 7.725$$

The mean is 7.725.

Since there are 40 data items, the median is the average of the middle two items, both of which are 8. Thus, their average is 8.

The median is 8.

## Class Exercises

**Refer to the histogram at the right.**

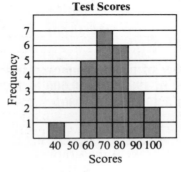

1. What is the mode of the scores?

2. What is the range of the scores?

3. How many students received scores of 90? 60? 50? 40?

4. How many students took the test?

5. What is the median score?

6. What is the mean of the scores? Round to the nearest tenth.

## Written Exercises

**Exercises 1–10 refer to the frequency polygon below.**

**A**   1. What is the range of the data?   2. What is the mode of the data?

**How many students did the following numbers of pushups?**

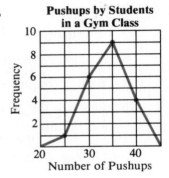

3. 25          4. 30          5. 35          6. 40

7. How many students are in the class?

8. What is the median of the data?

9. What is the mean of the data?

10. Draw a histogram for the data.

**Draw a histogram and a frequency polygon for the data in the following exercises of Lesson 12–4.**

**B** **11.** Exercise 7    **12.** Exercise 8    **13.** Exercise 9

**14.** Exercise 10    **15.** Exercise 11    **16.** Exercise 12

**C** **17.** Arrange the data listed below into the following intervals.

10–15, 15–20, 20–25, 25–30, 30–35, 35–40, 40–45

Then draw a histogram for the seven intervals.

23, 43, 27, 41, 12, 13, 19, 32, 43, 22, 36, 28, 44, 29, 31, 24, 34, 44, 37, 23, 17, 14, 23, 32, 36, 33, 43, 28, 39, 19

## Self-Test A

**The number of people employed in farming in the United States was: 11,100,000 in 1900; 10,400,000 in 1920; 9,500,000 in 1940; 5,400,000 in 1960; and 3,400,000 in 1980.**

**1.** Draw a bar graph.                                                    [12–1]

**2.** Draw a broken-line graph.

**3.** Draw a pictograph.  Use people as symbols.                          [12–2]

**4.** The sign-up records at the school's computer room showed the following usage: students, 40%; teachers, 30%; administration, 25%; other, 5%.  Draw a circle graph for the given data.

**5.** Find the mean, median, and range of 13, 6, 20, 20, 9, 15, 11, 10.    [12–3]

**6.** Make a frequency table for 12, 18, 15, 12, 9, 15, 15, 16, 9, 12, 18, 12.   [12–4]

**Exercises 7–10 refer to the histogram.**

**7.** How many runners had a time of 16 s?                                [12–5]

**8.** How many runners were there in all?

**9.** Draw a frequency polygon for the data.

**10.** What are the range, the mode, the mean, and the median of the data?

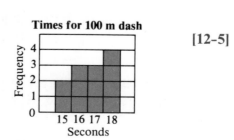

Times for 100 m dash

---

*Self-Test answers and Extra Practice are at the back of the book.*

## 12-6 Permutations

After school, Vilma plans to go to the music store and then to the pool. She can take any one of 3 routes from school to the music store and then take either of 2 routes from the store to the pool. In how many different ways can Vilma go from school to the pool?

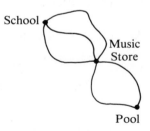

To answer, consider that for each route Vilma can take from school to the music store, she has a choice of either of 2 routes to continue from there. Thus, she has a choice of $3 \times 2$, or 6, possible different ways to go from school to the pool.

This example illustrates a general counting principle:

> If there are *m* ways to do **one thing** and *n* ways to do another, then there are $m \times n$ **ways to do both** things.

In mathematics, an arrangement **of a group** of things in a particular order is called a **permutation.** The **counting** principle can help you count the number of different **permutations of any** group of items.

*EXAMPLE 1* In how many different ways **can you** arrange the three cards shown at the right if you arrange them side by side in a row?

*Solution* Notice that there are 3 possibilities for the first card: A, B, or C. After the first card is selected, there are 2 possibilities for the second card. After the first and second cards have been selected, there is just 1 possibility for the third card.

Applying the counting principle, there are $3 \times 2 \times 1$, or 6, possible ways to arrange the three cards.

To check, list the permutations:

| | | |
|---|---|---|
| A B C | B A C | C A B |
| A C B | B C A | C B A |

Notice that the number of permutations of 3 things is $3 \times 2 \times 1$. We can write 3! (read *3 factorial*) to represent the expression $3 \times 2 \times 1$. In general, the number of permutations of *n* things is

$$n \times (n - 1) \times (n - 2) \times \cdots \times 3 \times 2 \times 1.$$

We can write this expression as *n*!.

If we let $_nP_n$ represent the number of permutations of a group of $n$ things when using all $n$ things, we can write the following formula.

**Formula**

For the number of permutations of $n$ things taken $n$ at a time,

$$_nP_n = n \times (n-1) \times (n-2) \times \cdots \times 3 \times 2 \times 1 = n!$$

**EXAMPLE 2**   How many four-digit whole numbers can you write using the digits 1, 2, 3, and 4 if no digit appears more than once in each number?

**Solution**   You need to find the number of permutations of 4 things taken 4 at a time.
Using the formula,

$$_4P_4 = 4! = 4 \times 3 \times 2 \times 1 = 24.$$

Therefore, 24 four-digit whole numbers can be written using the given digits.

Sometimes we work with arrangements that involve just a portion of the group at one time.

**EXAMPLE 3**   This year, 7 dogs are entered in the collie competition at the annual Ridgedale Kennel Club show. In how many different ways can first, second, and third prizes be awarded in the competition?

**Solution**   You want to find the number of permutations of 7 things taken 3 at a time. There are 7 choices for first prize, 6 for second, and 5 for third. Thus,

$$7 \times 6 \times 5 = 210.$$

There are 210 possible ways to award the prizes.

If we let $_nP_r$ represent the number of permutations of $n$ objects taken $r$ at a time, we can write the following formula.

**Formula**

For the number of permutations of $n$ things taken $r$ at a time, we use the following formula carried out to $r$ factors:

$$_nP_r = n \times (n-1) \times (n-2) \times \cdots$$

Using the formula in Example 3 above, we find $_7P_3 = 7 \times 6 \times 5 = 210$.

## Class Exercises

**Find the value of each.**

**1.** 4!      **2.** 3!      **3.** 2!      **4.** 1!

**5.** $_5P_5$      **6.** $_3P_3$      **7.** $_5P_4$      **8.** $_6P_3$

**Solve.**

**9.** In wrapping a gift, you have a choice of 3 different boxes and 4 different wrapping papers. In how many different ways can you wrap the gift?

**10.** There are 3 roads from Craig to Hartsdale, 2 roads from Hartsdale to Lee, and 4 roads from Lee to Trumbull. In how many different ways can you travel from Craig to Trumbull by way of Hartsdale and Lee?

**Use the formula to answer. Then list all of the permutations to check.**

**11.** In how many different ways can you arrange the letters in the word CAR?

**12.** In how many different ways can you arrange the 4 cards shown at the right if you take 3 at a time and arrange them side by side?

R    Y    G    B

---

## Written Exercises

**Find the value of each.**

**A**

**1.** 5!      **2.** 6!      **3.** 7!      **4.** 8!

**5.** $_6P_6$      **6.** $_7P_7$      **7.** $_8P_8$      **8.** $_5P_5$

**9.** $_6P_4$      **10.** $_8P_4$      **11.** $_{43}P_3$      **12.** $_{23}P_2$

---

## Problems

Exercises 1–4 refer to the map at the right. Tell how many different ways you can travel from one city to the other.

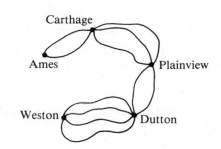

**A**

**1.** Ames to Plainview     **2.** Carthage to Dutton

**3.** Carthage to Weston     **4.** Ames to Dutton

5. A furniture store sells couches that are available in 3 different styles, 7 different colors, and 2 different sizes. How many different couches are available?

6. How many different sandwiches can you make using one kind of bread, one kind of meat, and one kind of cheese with these choices?
   Bread: white, rye, whole wheat
   Cheese: Swiss, cheddar
   Meat: turkey, chicken, roast beef

**Use a formula to solve.  List the permutations to check your answer.**

7. In how many different ways can you arrange the 4 books shown side by side on a shelf?

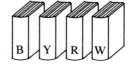

8. In how many different ways can Carla, Dean, and Ellen be seated in a row of 3 chairs?

9. How many different two-digit whole numbers can you make with the digits 1, 3, 5, and 7 if no digit appears more than once in each number?

10. In how many different ways can you arrange the letters in the word RING if you take the letters 3 at a time?

**Solve.**

**B** 11. In how many different ways can 7 books be arranged side by side on a shelf?

12. In how many different ways can 6 students stand in a row of 6?

13. In how many different ways can you arrange the letters in the word ANSWER if you take the letters 5 at a time?

14. How many different four-digit numbers can you make using the digits 1, 2, 3, 5, 7, 8, and 9 if no digit appears more than once in a number?

**Using the digits 1, 2, 4, 5, 7, and 8, how many different three-digit numbers can you form according to each of the following rules?**

**C** 15. Each digit may be repeated any number of times in a number.

16. The numbers are even numbers and no digit appears more than once in a number.

17. There is a 5 in the ones' place and no digit appears more than once in a number.

**18.** In how many different ways can you arrange the letters in the word ROOT? (*Hint:* Notice that two of the letters are indistinguishable.)

**19.** In how many different ways can you arrange the letters in the word NOON?

The diagram at the right, called a **Venn diagram,** illustrates the following statement.

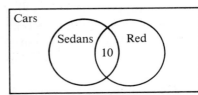

> Of 100 cars, 35 are sedans, 25 are red, and 10 are red sedans.

The rectangular region represents the number of cars. One circular region represents the number of sedans; the other represents the number of red cars. The overlapping portion of the two circular regions represents the number of cars that are red and sedans. The number of red cars that are not sedans is 25 − 10, or 15.

**Draw a Venn diagram to illustrate the statement. Then answer the questions.**

**20.** Of 240 knit caps, 110 are striped, 65 are blue, and 45 are striped and blue.
   **a.** How many are striped but not blue?
   **b.** How many are blue but not striped?
   **c.** How many are striped or blue (striped, or blue, or both)?
   **d.** How many are neither striped nor blue?

## Review Exercises

**Simplify.**

**1.** $\frac{70}{5} \times \frac{1}{2}$

**2.** $\frac{15}{12} \times \frac{36}{60}$

**3.** $\frac{21}{5} \times \frac{10}{14}$

**4.** $\frac{27}{51} \times \frac{17}{3}$

**5.** $\frac{9 \times 8}{3 \times 2}$

**6.** $\frac{7 \times 6}{4 \times 3}$

**7.** $\frac{11 \times 10 \times 9}{3 \times 2 \times 1}$

**8.** $\frac{23 \times 22 \times 21 \times 20}{4 \times 3 \times 2 \times 1}$

---

**■■■■ Challenge**

You and a friend have decided to jog through Peachtree Park. The park has five entrance gates and several paths as shown in the diagram at the right. To jog along all of the paths without covering any path more than once, through which gate would you enter?

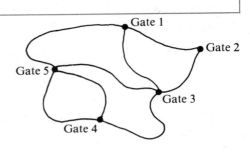

# 12-7 Combinations

Sometimes we select groups of objects from a larger group without regard to the order of the objects selected. Such groups in which the order is not considered are called **combinations.**

Suppose there is a group of 4 students—Jon, Lin, Meg, and Ray—who wish to go to the computer center today. If only 3 of them may go today, what are the possible combinations of these students who may be selected to go? To find any one combination, just leave out 1 student from the group of 4. The possible combinations are listed below.

Jon, Lin, Meg    Jon, Lin, Ray    Jon, Meg, Ray    Lin, Meg, Ray

The number of groups of 3 students that can be selected from a group of 4 students is 4.

Here is another way to approach the problem.

$$\begin{bmatrix} \text{number of groups} \\ \text{of 3 students you} \\ \text{can select from 4} \end{bmatrix} \times \begin{bmatrix} \text{number of ways of} \\ \text{arranging 3 stu-} \\ \text{dents in a group} \end{bmatrix} = \begin{bmatrix} \text{number of ways of} \\ \text{arranging 3 out} \\ \text{of 4 students} \end{bmatrix}$$

$$N \qquad\qquad \times \qquad\qquad {}_3P_3 \qquad\qquad = \qquad\qquad {}_4P_3$$

Solving for $N$, we find that

$$N = \frac{{}_4P_3}{{}_3P_3} = \frac{4 \times 3 \times 2}{3 \times 2 \times 1} = 4.$$

This formula gives us the same answer that we found by showing and counting the number of combinations of 3 students that may be selected from a group of 4 students.

**EXAMPLE 1**    How many combinations of 4 cards can be chosen from the cards shown at the right?

**Solution**    Use the formula to find the number of combinations of 6 cards taken 4 at a time.

$$N = \frac{{}_6P_4}{{}_4P_4} = \frac{\overset{3}{\cancel{6}} \times 5 \times \overset{1}{\cancel{4}} \times \overset{1}{\cancel{3}}}{\underset{1}{\cancel{4}} \times \underset{1}{\cancel{3}} \times \underset{1}{\cancel{2}} \times 1} = \frac{15}{1} = 15$$

List the combinations to check.

A B C D    A B D E    A C D E    A D E F    B C E F

A B C E    A B D F    A C D F    B C D E    B D E F

A B C F    A B E F    A C E F    B C D F    C D E F

In general, if we let $_nC_r$ represent the number of combinations of $n$ things taken $r$ at a time, we can use the following formula.

## Formula

For the number of combinations of $n$ things taken $r$ at a time,

$$_nC_r = \frac{_nP_r}{_rP_r} = \frac{n \times (n-1) \times \cdots \text{(to } r \text{ factors)}}{r!}$$

**EXAMPLE 2**  Four of 7 students who have volunteered will be chosen to hand out programs at the drama club performance. How many combinations of these students can be selected?

**Solution**  We wish to find the number of combinations of 7 students taken 4 at a time.
Using the formula, we find

$$_7C_4 = \frac{_7P_4}{_4P_4} = \frac{7 \times 6 \times 5 \times 4}{4 \times 3 \times 2 \times 1} = \frac{35}{1} = 35.$$

## Class Exercises

**Find the value of each.**

**1.** $_5C_3$    **2.** $_5C_4$    **3.** $_6C_2$    **4.** $_7C_2$    **5.** $_8C_4$    **6.** $_8C_6$    **7.** $_{12}C_3$    **8.** $_{20}C_2$

**Use the formula to solve. List the combinations to check your answer.**

**9.** How many combinations of 2 cards can be chosen from the cards shown at the right?

**10.** How many combinations of 3 letters can be chosen from the letters A, B, C, D, and E?

## Problems

**A**  **1.** How many combinations of 4 books can you choose from 6 books?

**2.** How many groups of 3 types of plants can be selected from 6 types?

**3.** How many straight lines can be formed by connecting any 2 of 6 points, no 3 of which are on a straight line?

**4.** How many groups of 4 fabrics can be selected from 7 fabrics?

B **5.** How many ways can a class of 21 students select 2 of its members as class representatives for student government?

**6.** The school photographer wants to photograph 3 students from a club with 14 members. How many combinations can be made?

**7.** Kristen must answer 5 of 10 questions on her quiz. How many combinations of questions are possible?

**8.** Philip wishes to check 2 books out of his school library. If the library contains 800 books, in how many ways might Philip make his choice of books?

**9.** You have a total of 4 coins: a penny, a nickel, a dime, and a quarter. How many different amounts of money can you form using the given number of these coins?
   **a.** 1 coin   **b.** 2 coins   **c.** 3 coins   **d.** 4 coins   **e.** one or more coins

**10.** There are three books left in the sale rack: a book about sailing, a cookbook, and a book about baseball. How many combinations can be formed using the given number of these books?
   **a.** 1 book   **b.** 2 books   **c.** 3 books   **d.** one or more books

C **11.** In how many ways can a five-member committee of 3 seniors and 2 juniors be selected from a group of 14 seniors and 8 juniors? (*Hint:* Find the number of combinations of each type and determine their product.)

**12.** A basketball squad has 17 members. The coach has designated 3 members to play center, 6 to play guard, and 8 to play forward. How many ways can the coach select a starting team of 1 center, 2 guards, and 2 forwards? (*Hint:* See the hint in Exercise 11.)

## Review Exercises

**Explain the meaning of each term.**

   **1.** factor          **2.** multiple          **3.** even number          **4.** odd number

   **5.** cube          **6.** at least one          **7.** at most one          **8.** exactly one

# 12-8 The Probability of an Event

In a simple game, the five cards shown are turned face down and mixed so that all choices are *equally likely*.

You then draw a card. If the card is a heart, you win a prize. To find your chance of winning, notice that there are 5 possible **outcomes.** Notice, too, that 2 of the outcomes are hearts. We say that 2 of the outcomes **favor** the event of drawing a heart. The **probability,** or chance, of drawing a heart is $\frac{2}{5}$. If we let $H$ stand for the event of drawing a heart, we may write $P(H) = \frac{2}{5}$. This statement is read as *the probability of event H is $\frac{2}{5}$.*

In general we have the following:

---

### *Formula*

The probability of an event $E$ is

$$P(E) = \frac{\text{number of outcomes favoring event } E}{\text{number of possible outcomes}}$$

for equally likely outcomes.

---

When all outcomes are equally likely, as in drawing one of the cards in the game described above, we say that the outcomes occur *at random,* or *randomly.*

In this chapter, we shall often refer to *experiments* such as drawing cards, drawing marbles from a bag, or rolling game cubes. We shall always assume that the outcomes of these experiments occur at random. When we refer to spinners like the one shown at the right, we shall always assume that the pointer stops at random but not on a division line.

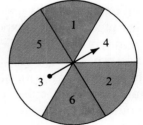

Often it is helpful to list all possible outcomes of an experiment. We could show the possible outcomes for the spinner shown above by using letters to represent the colors and by writing the following:

| | | |
|---|---|---|
| *R*1 | *B*2 | *W*3 |
| *R*6 | *B*5 | *W*4 |

**EXAMPLE 1**  Find the probability that the pointer of the spinner shown on the preceding page stops on a wedge of the type described.

a. even-numbered                     b. odd-numbered
c. red                               d. not red
e. even-numbered *and* red           f. green
g. not green

*Solution*      The number of possible outcomes is 6.

a. $P(\text{even-numbered}) = \frac{3}{6} = \frac{1}{2}$

b. $P(\text{odd-numbered}) = \frac{3}{6} = \frac{1}{2}$

c. $P(\text{red}) = \frac{2}{6} = \frac{1}{3}$

d. $P(\text{not red}) = \frac{4}{6} = \frac{2}{3}$

e. $P(\text{even-numbered and red}) = \frac{1}{6}$

f. $P(\text{green}) = \frac{0}{6} = 0$

g. $P(\text{not green}) = \frac{6}{6} = 1$

The events in parts (f) and (g) of the example illustrate these facts:

> The probability of an impossible event is 0.
> The probability of a certain event is 1.

Sometimes it is useful to picture the possible outcomes of an experiment. Consider the experiment of rolling the two game cubes that are shown at the right. One cube is blue, one cube is red. The numbers 1 through 6 are printed on each cube, one number per face. An outcome can be represented by an ordered pair of numbers. The array at the right shows the 36 possible outcomes when the two cubes are rolled. The encircled dot stands for the outcome of a 5 on the top face of the red cube and a 3 on the top face of the blue cube, or the ordered pair (5, 3).

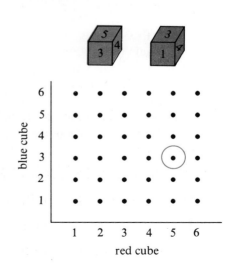

**EXAMPLE 2** Two game cubes are rolled.
  **a.** Find the probability that one cube *or* the other cube shows a 5 (that is, that the number 5 is on the top face of either or both cubes).
  **b.** Find the probability that the sum of the top faces is 5.

**Solution** First, make a sketch to show the possible outcomes. Then circle the event.

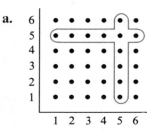

$$P(\text{a } 5) = \frac{11}{36}$$

$$P(\text{sum} = 5) = \frac{4}{36} = \frac{1}{9}$$

---

**Problem Solving Reminder**
Sometimes *making a sketch* can help you solve a problem. In Example 2, above, picturing an outcome as the graph of an ordered pair simplifies the problem.

## Class Exercises

**Two red, one white, and three blue marbles are put into a bag. Find each probability for a marble chosen at random.**

**1.** $P(\text{red})$     **2.** $P(\text{white})$     **3.** $P(\text{blue})$     **4.** $P(\text{green})$

**5.** $P(\text{not green})$     **6.** $P(\text{red or white})$

**7.** $P(\text{white or blue})$     **8.** $P(\text{red or blue})$

**Exercises 9–12 refer to the spinner at the right. Find the probability that the pointer stops on a wedge of the type described.**

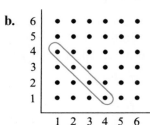

 **9.** numbered with a factor of 6

**10.** numbered with a multiple of 3

**11.** even-numbered or blue

**12.** even-numbered and blue

## Written Exercises

**Find the probability of each roll if you use a single game cube.**

**A**   **1.** a 5                         **2.** an odd number

     **3.** a number less than 5         **4.** a number greater than 3

     **5.** a 7                         **6.** a number less than 7

**Each of the 20 cards shown at the right has a letter, a number, and a color. Each card is equally likely to be drawn. Find each probability.**

     **7.** $P(C)$                  **8.** $P(A)$

     **9.** $P(1)$                 **10.** $P(2)$

    **11.** $P(\text{red})$             **12.** $P(\text{blue})$

    **13.** $P(\text{not A})$          **14.** $P(\text{not D})$

    **15.** $P(1 \text{ or } 2)$         **16.** $P(1, 2, 3, \text{ or } 4)$

    **17.** $P(\text{neither 1 nor 2})$    **18.** $P(\text{not 1, or 2, or 3, or 4})$

**Exercises 19–28 refer to the spinner below. Find the probability that the pointer stops on a wedge of the type described.**

**19.** red

**20.** white or blue

**21.** numbered with a factor of 12

**22.** even-numbered

**23.** numbered with a multiple of 3

**24.** numbered with a multiple of 4

**25.** odd-numbered or red

**26.** odd-numbered and red

**27.** red and a factor of 6

**28.** blue and a multiple of 5

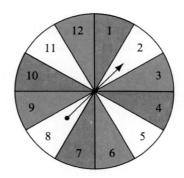

**Two game cubes are rolled. Find the probability that this sum shows.**

**B**   **29.** 7              **30.** 12             **31.** 3             **32.** 9

    **33.** 7 or 11        **34.** 2 or 12       **35.** less than 7     **36.** 6, 7, or 8

For Exercises 37 and 38, use a *tree diagram* to picture the possible outcomes of the random experiment.

*EXAMPLE*  Two coins are tossed. Find the probability of obtaining at least one head.

*Solution*  Let H stand for heads and T stand for tails.

| First Coin | Second Coin | Outcomes |
|---|---|---|
| H | H | H, H |
|  | T | H, T |
| T | H | T, H |
|  | T | T, T |

There are 4 possible outcomes. Three of the outcomes have one or more heads.

$$P(\text{at least one H}) = \frac{3}{4}$$

**37.** You have two bags of marbles, each of which holds one blue marble, one red marble, and one green marble. You choose one marble at random from each bag. Find the probability of obtaining the following.
  **a.** two red marbles   **b.** at least one blue marble   **c.** no green marbles
  **d.** at most one green marble   **e.** exactly one red marble

**38.** Three coins are tossed. Find the probability of obtaining the following.
  **a.** at least two heads   **b.** three heads   **c.** no heads
  **d.** at most two tails   **e.** exactly one head

**C**  **39.** You have one penny, one nickel, one dime, and one quarter in your pocket. You select two coins at random. What is the probability that you have taken at least 25¢ from your pocket?

## Review Exercises

**Evaluate each expression using the given values of the variable.**

$1 - x$

**1.** $x = 1$   **2.** $x = \frac{1}{2}$   **3.** $x = \frac{2}{3}$   **4.** $x = -1$

$\frac{x}{1 - x}$

**5.** $x = \frac{1}{3}$   **6.** $x = \frac{2}{3}$   **7.** $x = \frac{3}{7}$   **8.** $x = \frac{4}{7}$

# 12-9 Odds in Favor and Odds Against

**EXAMPLE 1**   A game is played with the spinner shown at the right.  To win the game, the pointer must stop on a wedge that shows a prime number.  Find each probability.
**a.** You win.          **b.** You do not win.

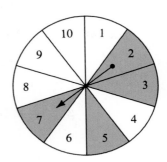

**Solution**   There are 10 possible outcomes.
For you to win, the pointer must stop on 2, 3, 5, or 7.  Therefore:

**a.** $P(\text{win}) = \frac{4}{10} = \frac{2}{5}$

**b.** $P(\text{not win}) = \frac{6}{10} = \frac{3}{5}$

Example 1 illustrates this general fact:

> If the probability that an event occurs is $p$, then the probability that the event does not occur is $1 - p$.

In Example 1 there are 4 ways of winning and 6 ways of not winning.  We therefore say:

**a.** The *odds in favor* of winning are 4 to 6, or 2 to 3.

**b.** The *odds against* winning are 6 to 4, or 3 to 2.

> ## *Formula*
> If the probability that an event occurs is $p$ (where $p \neq 0$ and $p \neq 1$), then:
>
> $$\text{Odds in favor of the event} = \frac{p}{1 - p}$$
>
> $$\text{Odds against the event} = \frac{1 - p}{p}$$

Odds are usually expressed in the form "$x$ to $y$," where $x$ and $y$ are integers having no common factor.

**EXAMPLE 2**  Find the odds (a) in favor of and (b) against rolling a sum of 6 with two game cubes.

**Solution**  From the array of possible outcomes, we see that:

$$P(\text{sum} = 6) = \frac{5}{36} = p$$

$$P(\text{sum} \neq 6) = 1 - p = 1 - \frac{5}{36} = \frac{31}{36}$$

**a.** Odds in favor $= \dfrac{p}{1 - p}$

$$\frac{\frac{5}{36}}{\frac{31}{36}} = \frac{5}{36} \div \frac{31}{36} = \frac{5}{36} \times \frac{36}{31} = \frac{5}{31}$$

Odds in favor are 5 to 31.

**b.** Odds against are 31 to 5.

**EXAMPLE 3**  The chance of rain tomorrow is 40%. What are the odds against its raining?

**Solution**  The probability of rain is $p = 40\% = 0.4$.

Odds against rain $= \dfrac{1 - p}{p} = \dfrac{1 - 0.4}{0.4} = \dfrac{0.6}{0.4} = \dfrac{3}{2}$

Odds against rain are 3 to 2.

## Class Exercises

1. The chance of rain tomorrow is 20%. Find the odds against rain.

2. Find the odds against rolling a 4 with one game cube.

**Exercises 3–6 refer to the spinner at the right. Find the odds (a) in favor of and (b) against the pointer stopping on a wedge of the type described.**

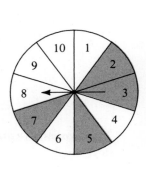

3. an odd number          4. a multiple of 3

5. a factor of 10          6. a number less than 7

7. The odds are 1 to 1 that an event will occur. What is the probability that the event will occur?

8. The probability that an event will occur is $\frac{1}{2}$. What are the odds against the event?

## Written Exercises

**Exercises 1–6 refer to a bag containing 6 red, 2 white, and 4 blue marbles. Find the odds in favor of drawing a marble of the type described.**

**A**  **1.** red               **2.** white              **3.** blue

     **4.** white or blue           **5.** blue or red         **6.** red or white

**7.** The chance of rain tomorrow is 60%. What are the odds against rain?

**8.** The probability that the Lions football team will win the next game is 0.6. What are the odds that it will win?

**9.** The Student Union Party has a 30% chance of winning in the next school election. What are the odds against its winning?

**In Exercises 10–15 a card has been drawn at random from the 20 cards that are shown. Find the odds against drawing a card of the type described.**

**10.** an A

**11.** a blue card

**12.** a red card

**13.** a C

**14.** a 1, 2, or 3

**15.** B2, B3, B4, or B5

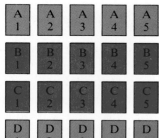

**Two game cubes are rolled. Find the odds against obtaining the sum described.**

**16.** 7            **17.** 11             **18.** 7 or 11

**19.** 2 or 12       **20.** greater than 7       **21.** less than 6

**B**  **22. a.** even            **b.** odd

     **23. a.** divisible by 3       **b.** not divisible by 3

**The two game cubes are rolled again. Find the odds in favor of the event described.**

**24.** Exactly one 5 shows.       **25.** At least one 5 shows.

**26.** Two even numbers show.       **27.** At least one odd number shows.

**C** **28.** The odds in favor of the Melodies winning the music competition are 5 to 3. What is the probability that the Melodies will win?

**29.** The odds in favor of drawing a red marble at random from a bag of marbles are 1 to 8. What is the probability of drawing a red marble?

## Review Exercises

**Perform the indicated operations. Simplify.**

**1.** $\dfrac{5}{6} + \dfrac{7}{12} - \dfrac{11}{24}$    **2.** $\dfrac{7}{8} + \dfrac{15}{16} - \dfrac{7}{24}$    **3.** $\dfrac{11}{20} + \dfrac{9}{10} - \dfrac{8}{15}$    **4.** $\dfrac{5}{9} + \dfrac{25}{27} - \dfrac{71}{81}$

**5.** $\dfrac{17}{18} - \dfrac{2}{3} + \dfrac{5}{12}$    **6.** $\dfrac{37}{40} - \dfrac{41}{60} + \dfrac{11}{12}$    **7.** $\dfrac{11}{14} - \dfrac{17}{28} + \dfrac{9}{49}$    **8.** $\dfrac{7}{9} - \dfrac{5}{12} + \dfrac{13}{18}$

 **Computer Byte**

If we toss a coin, there are two possible outcomes—a head or a tail. If these two outcomes are equally likely, then the probability of each is $\frac{1}{2}$. Does this mean that if we toss a coin 10 times, we will get 5 heads and 5 tails? (Not necessarily.)

The following program will simulate tossing a coin. The program is based on a list of "random numbers" that are decimals between 0 and 1. (Usage of the RND function varies. Check this program with the manual for the computer that you are using, and make any necessary changes.)

```
10   PRINT "HOW MANY TOSSES";
20   INPUT N
30   LET H = 0
40   FOR I = 1 to N
50   LET A = RND (1)
60   IF A < .5 THEN 90
70   PRINT TAB( 6);"TAIL"
80   GOTO 110
90   LET H = H + 1
100   PRINT "HEAD"
110   NEXT I
120   PRINT
130   PRINT "H = ";H; TAB( 10);"T = ";N − H;
140   PRINT  TAB( 20);"H/N = ";H/N
150   END
```

**1.** Run the program 10 times for N = 25.
   **a.** For how many runs was 0.45 < H/N < 0.55?
   **b.** What percent was this?

# 12-10 Mutually Exclusive Events

The pointer shown at the right stops at random but not on a division line. Consider the following events.

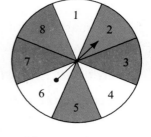

*R:* The pointer stops on a red wedge.

*B:* The pointer stops on a blue wedge.

*O:* The pointer stops on an odd-numbered wedge.

*E:* The pointer stops on an even-numbered wedge.

Events *R* and *B* cannot both occur at once. Such events are said to be **mutually exclusive.** Notice that events *B* and *E* are *not* mutually exclusive because they both occur when the pointer stops on a blue wedge that is even-numbered.

In the spinner above, five of the eight wedges are colored red or blue. Therefore,

$$P(R \text{ or } B) = \frac{5}{8}.$$

Notice that

$$P(R) = \frac{1}{4}, \ P(B) = \frac{3}{8}, \text{ and } P(R) + P(B) = \frac{1}{4} + \frac{3}{8} = \frac{5}{8}.$$

Thus, $P(R \text{ or } B) = P(R) + P(B)$.

---

If *A* and *B* are mutually exclusive events, then

$$P(A \text{ or } B) = P(A) + P(B)$$

---

**EXAMPLE** The probability that a randomly chosen car is green is 0.15 and that it is red is 0.25.

    **a.** Find the probability that the next car you see will be red or green.

    **b.** Find the odds against its being red or green.

*Solution*     **a.** The events described are mutually exclusive. Thus,

$$P(\text{red or green}) = P(\text{red}) + P(\text{green})$$
$$= 0.25 + 0.15 = 0.40$$

The probability that the next car will be red or green is 40%.

    **b.** Odds against red or green $= \dfrac{1 - 0.40}{0.40} = \dfrac{0.6}{0.4} = \dfrac{3}{2}$

The odds against the next car's being red or green are 3 to 2.

## Class Exercises

**Are events *A* and *B* mutually exclusive?**

1. You take a test.
   *A:* You pass it.
   *B:* You fail it.

2. You take a test.
   *A:* You score less than 9.
   *B:* You score more than 6.

3. Two coins are tossed.
   *A:* Two heads result.
   *B:* Two tails result.

4. Two game cubes are rolled.
   *A:* The sum is 5.
   *B:* A 5 shows on one cube.

5. Two game cubes are rolled.
   *A:* Cubes show the same number.
   *B:* The sum is 7.

6. Two game cubes are rolled.
   *A:* Cubes show the same number.
   *B:* The sum is 8.

**$A$ and $B$ are mutually exclusive events. Find $P(A$ or $B)$.**

7. $P(A) = \frac{1}{4}$, $P(B) = \frac{3}{8}$

8. $P(A) = 0.2$, $P(B) = 0.5$

9. $P(A) = \frac{1}{2}$ and $P(B) = \frac{2}{3}$. Are $A$ and $B$ mutually exclusive events?

## Written Exercises

**Solve. *A* and *B* are mutually exclusive events.**

A

1. $P(A) = \frac{1}{5}$, $P(B) = \frac{2}{3}$. Find $P(A$ or $B)$.

2. $P(A) = 0.32$, $P(B) = 0.45$. Find $P(A$ or $B)$.

3. $P(A) = 0.4$, $P(A$ or $B) = 0.7$. Find $P(B)$.

4. $P(B) = \frac{1}{3}$, $P(A$ or $B) = \frac{3}{4}$. Find $P(A)$.

**In Exercises 5–10, find the probability that the pointer stops on a wedge of the type described.**

5. **a.** red **b.** white **c.** red or white

6. **a.** blue **b.** red **c.** blue or red

7. **a.** blue **b.** not blue

8. **a.** white **b.** not white

9. **a.** odd-numbered **b.** red **c.** odd-numbered or red

10. **a.** odd-numbered **b.** blue **c.** odd-numbered or blue

**Exercises 11–15 refer to the cards at the right. Find the probability that a card drawn at random is of the type described.**

11. **a.** a 5          **b.** less than 3
    **c.** a 5 or less than 3

12. **a.** a 2          **b.** greater than 3
    **c.** a 2 or greater than 3

13. **a.** a D          **b.** a red card greater than 2
    **c.** a D or a red card greater than 2

14. **a.** a B          **b.** a blue card greater than 3
    **c.** a B or a blue card greater than 3

15. **a.** a C          **b.** a red card less than 4
    **c.** a C or a red card less than 4

**Two game cubes are rolled. Find the probability of the event described.**

**B** 16. **a.** The sum is 5.
      **b.** A 5 shows.
      **c.** Neither is the sum 5 nor does a 5 show.

17. **a.** The sum is 4.
    **b.** A 6 shows.
    **c.** Neither is the sum 4 nor does a 6 show.

18. **a.** The sum is 9.
    **b.** The roll is a double (for example, $\boxed{3}$ $\boxed{3}$ ).
    **c.** The sum is 9 or a double is rolled.

19. **a.** The sum is 7.
    **b.** The roll is a double.
    **c.** The sum is 7 or a double is rolled.

20. Joe's batting average (the probability of his getting a hit) is 0.350. What are the odds against his getting a hit?

21. The probability of Jan's team winning the next softball game is 55%. Find the odds that the team will lose.

**C** 22. The probability that the next soup ordered will be chicken is 0.45; that it will be tomato is 0.35. What are the odds against its being chicken or tomato?

23. At West High School the probability that a randomly chosen student is a senior is 0.20. The probability that the student is a junior is 0.25. Find the odds against the student's being a junior or senior.

## Challenge

A sightseer at Breathless Gorge dropped his sandwich from a gondola. The object fell at the rate of 9.8 meters per second (m/s) after the first second, 19.6 m/s after the second second, and 29.4 m/s after the third second. How fast will the sandwich be falling at 4 s?

## Self-Test B

1. In how many different ways can you arrange 4 boxes side by side on a shelf? **[12–6]**

2. How many different three-digit numbers can you make with the digits 1, 2, 3, 4, and 5 if no digit appears more than once in each number?

3. How many combinations of 3 letters can be chosen from the letters A, B, C, D, and E? **[12–7]**

4. In how many ways can a committee of 2 people be selected from a group of 15 people?

**Find the probability that the pointer on the spinner shown at the right stops on a wedge of the type described.**

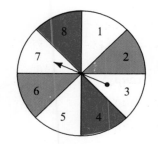

5. 3          6. odd-numbered          **[12–8]**

7. red          8. a number less than 9

9. What are the odds in favor of the pointer on the spinner shown stopping on a wedge with a number greater than 6?          **[12–9]**

10. What are the odds against the pointer on the spinner shown stopping on a blue wedge?

11. The probability of snow next week is 75%. What are the odds in favor of snow?

12. Events $A$ and $B$ are mutually exclusive. $P(A) = \frac{5}{12}$. $P(B) = \frac{1}{6}$. Find $P(A \text{ or } B)$.          **[12–10]**

13. Events $A$ and $B$ are mutually exclusive. $P(A) = 0.5$. $P(A \text{ or } B) = 0.7$. Find $P(B)$.

*Self-Test answers and Extra Practice are at the back of the book.*

# Standard Deviation

For many purposes the *range* of a set of data is a poor measure of its spread. Consider, for example, the frequency distributions and histograms below for the mass in kilograms of various rocks found by three groups of rock collectors.

| Group A | | Group B | | Group C | |
|---|---|---|---|---|---|
| Mass (kg) | Frequency | Mass (kg) | Frequency | Mass (kg) | Frequency |
| 1 | 4 | 1 | 1 | 1 | 1 |
| 3 | 1 | 3 | 3 | 3 | 2 |
| 7 | 1 | 5 | 2 | 5 | 4 |
| 9 | 4 | 7 | 3 | 7 | 2 |
|  |  | 9 | 1 | 9 | 1 |

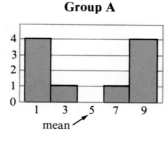

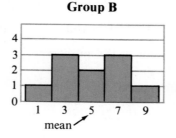

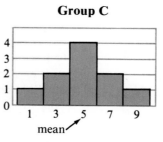

Each distribution has a range of 8. But a glance at each histogram reveals that distribution A is more spread out around its center, or mean, than are distributions B and C.

A more useful measure of the spread of a distribution is the **deviation** from the mean. Suppose a bowler has scores of 198, 210, and 156. The mean score is 188. The table at the right shows how much each score varies, or *deviates,* from the mean.

| Score | Deviation |
|---|---|
| 198 | +10 |
| 210 | +22 |
| 156 | −32 |

The **variance** and **standard deviation** are two commonly used measures of how data are scattered around the mean. The variance is computed by squaring each deviation from the mean, adding these squares, and dividing the sum by the number of entries in the distribution. The standard deviation is found by computing the positive square root of the variance. The standard deviation is a more useful statistic than the variance because comparisons are being made of common units. For example, when data are given in meters, the variance will be in square meters, but the standard deviation will be in meters.

The table below shows the calculation of the variance and standard deviation for each frequency distribution on the preceding page. In each case, $x$ is the mass and $m$ is the mean, which is 5. Notice that the more a distribution is spread out from its mean, the larger its standard deviation.

| Group A | | | Group B | | | Group C | | |
|---|---|---|---|---|---|---|---|---|
| | deviation | | | deviation | | | deviation | |
| $x$ | $(x - m)$ | $(x - m)^2$ | $x$ | $(x - m)$ | $(x - m)^2$ | $x$ | $(x - m)$ | $(x - m)^2$ |
| 1 | $-4$ | 16 | 1 | $-4$ | 16 | 1 | $-4$ | 16 |
| 1 | $-4$ | 16 | 3 | $-2$ | 4 | 3 | $-2$ | 4 |
| 1 | $-4$ | 16 | 3 | $-2$ | 4 | 3 | $-2$ | 4 |
| 1 | $-4$ | 16 | 3 | $-2$ | 4 | 5 | 0 | 0 |
| 3 | $-2$ | 4 | 5 | 0 | 0 | 5 | 0 | 0 |
| 7 | 2 | 4 | 5 | 0 | 0 | 5 | 0 | 0 |
| 9 | 4 | 16 | 7 | 2 | 4 | 5 | 0 | 0 |
| 9 | 4 | 16 | 7 | 2 | 4 | 7 | 2 | 4 |
| 9 | 4 | 16 | 7 | 2 | 4 | 7 | 2 | 4 |
| 9 | 4 | 16 | 9 | 4 | 16 | 9 | 4 | 16 |
| | | 136 kg² | | | 56 kg² | | | 48 kg² |

variance $= \frac{136}{10} = 13.6 \text{ kg}^2$   variance $= \frac{56}{10} = 5.6 \text{ kg}^2$   variance $= \frac{48}{10} = 4.8 \text{ kg}^2$

standard
deviation $= \sqrt{13.6}$
$\approx 3.7 \text{ kg}$

standard
deviation $= \sqrt{5.6}$
$\approx 2.4 \text{ kg}$

standard
deviation $= \sqrt{4.8}$
$\approx 2.2 \text{ kg}$

**Find (a) the mean, (b) the deviation, (c) the variance, and (d) the standard deviation of the given data. Use the table on page 508, or approximate the square root to the nearest hundredth by interpolation.**

1. From a sample of four dairy cows, a farmer recorded the following yields for one day: 11 gal, 13 gal, 9 gal, and 15 gal.

2. In a survey of local stores, the following prices were quoted for a World watch: $34, $27, $41, and $38.

3. The fuel efficiency ratings of five new cars were 20, 19, 20, 22, and 33 mi/gal.

# Chapter Review

**Complete.**
**Refer to the bar graph for Exercises 1–3.**

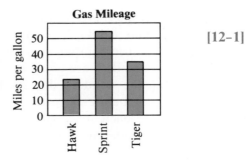

**Gas Mileage**

1. The car with the best gas mileage is the
   __?__ .                                                    [12–1]

2. The Tiger gets approximately __?__ miles
   per gallon.

3. If we join the midpoints of the tops of the
   bars in the bar graph, we obtain a __?__
   graph.

4. To show on a circle graph that 25% of a team is in the seventh grade,     [12–2]
   you would use a wedge of a circle that has an angle measure of __?__
   degrees.

**Use the data 65, 67, 67, 69, 70, 73 for Exercises 5–9.**

5. The range is __?__ .          6. The median is __?__ .           [12–3]

7. The mean is __?__ .           8. The mode is __?__ .             [12–4]

9. The frequency of 67 is __?__ .

**Refer to the frequency table for Exercises 10 and 11.**

10. If you drew a histogram                                          [12–5]
    for the data, the tallest bar
    would represent __?__ .

11. The range of the data is
    __?__ .

| Cost of a Quart of Milk | | | | |
|---|---|---|---|---|
| **Price** | 59¢ | 60¢ | 61¢ | 62¢ |
| **Frequency** | 4 | 5 | 2 | 1 |

**True or false?**

12. You can arrange the letters in the word DRAW in 4 different ways if      [12–6]
    you take the letters 3 at a time.

13. A group of 3 people can be selected from 8 people in 56 ways.           [12–7]

14. When a single game cube is rolled, the probability that the roll         [12–8]
    shows a number greater than 3 is $\frac{1}{2}$.

15. The odds in favor of rolling a 2 with a single game cube are 1 to 5.     [12–9]

16. The probability of rolling a 3 or a 6 with a single game cube is $\frac{1}{3}$.   [12–10]

# Chapter Test

**Refer to the bar graph for Exercises 1 and 2.**

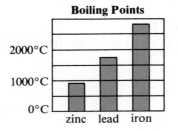

**Boiling Points**

1. Give the approximate boiling points of the three elements. [12–1]

2. Draw a broken-line graph for the data.

3. Draw a pictograph for the data below. [12–2]

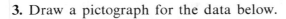

| Number of Shares of Stock Sold | | | |
|---|---|---|---|
| **1980** | 1,546,000 | **1981** | 2,341,000 |
| **1982** | 3,995,000 | **1983** | 4,784,000 |

4. Find the mean, the median, and the range of 28, 14, 19, 24, 30. [12–3]

**Refer to the frequency table for Exercises 5–7.**

5. Give the mean, the median, the range, and the mode of the data. Round the mean to the nearest tenth. [12–4]

6. Draw a histogram for the data. [12–5]

7. Draw a frequency polygon for the data.

| Number of Minutes | Frequency |
|---|---|
| 155 | 2 |
| 158 | 3 |
| 159 | 1 |
| 162 | 1 |
| 163 | 1 |
| 166 | 2 |
| 167 | 3 |
| 168 | 5 |

**Solve.**

8. In how many different ways can 3 people sit in a row of 5 seats? [12–6]

9. How many combinations of 3 colors can you choose from 7 colors? [12–7]

**Exercises 10–13 refer to the spinner shown below at the right. Find the probability that the pointer will stop on a wedge of the type described.**

10. blue          11. green [12–8]

12. odd-numbered and red

13. What are the odds in favor of the pointer stopping on a red wedge? [12–9]

14. Events $A$ and $B$ are mutually exclusive. $P(A) = \frac{1}{3}$. $P(B) = \frac{5}{12}$. Find $P(A \text{ or } B)$. [12–10]

# Cumulative Review (Chapters 1–12)

## Exercises

**Simplify.**

1. $5.4 + (7 \times 6.3)$

2. $\dfrac{16 + 9}{32 + 8}$

3. $13(9.2 \div 16)$

4. $4x - y + xy + 3y$

5. $2m + 2n - m + n$

6. $3p^2(p + 5)$

**Perform the indicated operation. Write the answer as a proper fraction in lowest terms or as a mixed number in simple form.**

7. $\dfrac{3}{5} + \dfrac{4}{7}$

8. $\dfrac{8}{11} - \dfrac{1}{3}$

9. $\dfrac{3}{4} \times \dfrac{7}{8}$

10. $-1\dfrac{1}{2} \div -2\dfrac{1}{2}$

**Solve.**

11. $6x = -4 + (-18)$

12. $-3 - 7 = x + 1$

13. $x(4 - 2) = 40$

14. $x + (-9) > -2$

15. $4 + (-12) \le x - 6$

16. $11 - x > 0$

**Find the radius of a circle with the given circumference. Use $\pi \approx \dfrac{22}{7}$.**

17. $C = 264$

18. $C = 343.2$

19. $C = 83.6$

20. $C = 897.6$

**Solve.**

21. What is 39% of 120?

22. What is 125% of 89?

23. What percent of 60 is 12?

24. What percent of 75 is 35?

25. 18 is 50% of what number?

26. 35 is 8% of what number?

**Use a straight-line graph to complete the ordered pairs, then find the slope.**

27. $A(7, 0)$, $B(5, 1)$, $C(-1, ?)$, $D(?, 7)$

28. $A(0, 0)$, $B(2, 1)$, $C(4, ?)$, $D(?, 3)$

**Find the perimeter and the area of the polygon described.**

29. A rectangle with sides 136.5 m and 97.5 m

30. A parallelogram with base 15 cm, one side 7 cm, and height 6 cm

31. A triangle with base 63 m, sides 51 m and 30 m, and height 24 m

**Is the triangle with sides of the given lengths a right triangle?**

32. 12, 16, 20

33. 4, 6, 9

34. 9, 12, 15

# Problems

**Solve.**

1. Rosalie has 2 ten-dollar bills. She would like to buy a book for
$6.95, some puzzles for a total cost of $12.50, and birthday cards for
a total cost of $3.60. Does she have enough money?

2. A faucet leaks at the rate of $\frac{1}{4}$ cup of water per hour. How much
water will leak in 24 hours?

3. A Rosebud rocking chair can be purchased for $245.70 in a furni-
ture store or for $189 at a factory store. What is the percent of
markup for the furniture store?

4. In order for two elements to combine, the sum of their valence
numbers must be equal to 0. One atom of carbon has a valence
number of 4, and one atom of fluorine has a valence number of
$-1$. How many atoms of fluorine are needed to combine with 1
atom of carbon?

5. Find the length of a diagonal of a square if each side of the square
has a length of 7 cm. Round to the nearest tenth.

6. Two cyclists depart at the same time from the same point, traveling
in opposite directions. One cyclist travels at a speed of 14 mi/h,
while the other cyclist travels at a speed of 16 mi/h. How long must
the cyclists travel to be at least 135 mi apart?

7. Find the area of a rectangular yard that has one side of length 12 m
and perimeter 54 m.

8. You can buy whole chickens at Sal's Market for $1.09/lb. For an
additional $.25/lb, the butcher will cut up the chicken and remove
the bones. If chicken is sold in tenths of a pound, how much bone-
less chicken can be purchased for $7.50?

9. A rectangular painting is 5 ft wide and has an area of 15 ft². The
frame around the painting is 3 in. wide. What are the dimensions of
the framed painting?

# Skill Review

## Addition

```
 236
   4
 108
+ 57
 405
```

```
  14.71
   3.009
+291.681
 309.400
```

**Add.**

**1.**  84
  + 15

**2.**  46
  + 32

**3.**  25
  + 18

**4.**  59
  + 16

**5.**  26
  + 95

**6.**  79
  + 56

**7.**  653
  + 88

**8.**  296
  + 35

**9.**  8.7
  + 0.5

**10.**  0.36
  + 2.44

**11.**  299.1
  + 68.33

**12.**  933.068
  + 724.3997

**13.**  5.4
  12.137
  1.25
  + 306.49

**14.**  1.6256
  6.006
  9.36
  + 1.49

**15.**  24.603
  18.4
  2.9
  + 0.216

**16.**  911.34
  0.52
  26.1
  + 83.192

**17.** $257 + 631$

**18.** $463 + 451$

**19.** $358 + 164$

**20.** $10.4 + 3.25$

**21.** $19.6 + 0.462$

**22.** $4.78 + 31.1$

**23.** $5.176 + 2.98$

**24.** $67.4 + 45.93$

**25.** $19.71 + 2.388$

**26.** $27.0564 + 281.4$

**27.** $3.51032 + 14.99$

**28.** $6.8 + 37.51 + 108.2$

**29.** $89.71 + 5.5 + 0.62$

**30.** $5326 + 703 + 9427 + 8$

**31.** $1027 + 349 + 8 + 12$

**32.** $16.24 + 5.6 + 18.09 + 6.7$

**33.** $2.55 + 0.34 + 0.42 + 3.57$

**34.** $61.476 + 14.1 + 0.59 + 366$

**35.** $90.072 + 32.4 + 24 + 8.6$

**36.** $115 + 20 + 9 + 7603$

**37.** $70 + 1328 + 94 + 5$

**38.** $193.7 + 4.08 + 11.5 + 1.9026$

**39.** $0.428 + 83.7 + 6.999 + 7.06$

**40.** $29 + 72 + 604 + 396$

**41.** $785 + 120 + 7 + 653$

**42.** $356 + 9 + 12 + 2301$

**43.** $6 + 24 + 315 + 1409$

**44.** $37.7 + 0.6 + 6.834 + 16$

**45.** $1165 + 0.08 + 17.1 + 94.028$

**46.** $570.2 + 74.4 + 6.553 + 9.2$

**47.** $7.8118 + 27.6 + 5.302 + 14$

# Skill Review

## Subtraction

$$\begin{array}{r} 630 \\ -\ 249 \\ \hline 381 \end{array}$$

$$\begin{array}{r} 300.710 \\ -\ 46.008 \\ \hline 254.702 \end{array}$$

**Subtract.**

**1.**
$$\begin{array}{r} 64 \\ -\ 51 \end{array}$$

**2.**
$$\begin{array}{r} 98 \\ -\ 37 \end{array}$$

**3.**
$$\begin{array}{r} 56 \\ -\ 30 \end{array}$$

**4.**
$$\begin{array}{r} 79 \\ -\ 42 \end{array}$$

**5.**
$$\begin{array}{r} 896 \\ -\ 241 \end{array}$$

**6.**
$$\begin{array}{r} 613 \\ -\ 402 \end{array}$$

**7.**
$$\begin{array}{r} 978 \\ -\ 365 \end{array}$$

**8.**
$$\begin{array}{r} 784 \\ -\ 463 \end{array}$$

**9.**
$$\begin{array}{r} 18.636 \\ -\ 13.435 \end{array}$$

**10.**
$$\begin{array}{r} 67.86 \\ -\ 54.22 \end{array}$$

**11.**
$$\begin{array}{r} 27.954 \\ -\ 16.21 \end{array}$$

**12.**
$$\begin{array}{r} 46.5822 \\ -\ 24.001 \end{array}$$

**13.**
$$\begin{array}{r} 8.434 \\ -\ 6.297 \end{array}$$

**14.**
$$\begin{array}{r} 35.061 \\ -\ 9.875 \end{array}$$

**15.**
$$\begin{array}{r} 6.952 \\ -\ 5.06 \end{array}$$

**16.**
$$\begin{array}{r} 583.86 \\ -\ 279.9 \end{array}$$

**17.**
$$\begin{array}{r} 453 \\ -\ 17.46 \end{array}$$

**18.**
$$\begin{array}{r} 55.7 \\ -\ 3.9 \end{array}$$

**19.**
$$\begin{array}{r} 0.081 \\ -\ 0.007 \end{array}$$

**20.**
$$\begin{array}{r} 2.0056 \\ -\ 0.918 \end{array}$$

**21.** $57 - 32$

**22.** $86 - 45$

**23.** $98 - 36$

**24.** $49 - 26$

**25.** $78 - 34$

**26.** $76 - 65$

**27.** $85 - 37$

**28.** $42 - 29$

**29.** $90 - 57$

**30.** $94.7 - 39$

**31.** $48.08 - 7.95$

**32.** $279.5 - 33.7$

**33.** $64.08 - 7.05$

**34.** $605.01 - 31.23$

**35.** $0.072 - 0.009$

**36.** $1314 - 197$

**37.** $6240 - 3078$

**38.** $8521 - 2364$

**39.** $7017 - 3468$

**40.** $4320 - 2006$

**41.** $9481 - 2556$

**42.** $22.916 - 17.4$

**43.** $25.006 - 3.98$

**44.** $518.2 - 327.41$

**45.** $2.7736 - 0.1531$

**46.** $800.6 - 315$

**47.** $0.9001 - 0.035$

**48.** $207.001 - 44.62$

**49.** $91.003 - 17.6$

**50.** $23,025 - 18,769$

**51.** $55,317 - 40,769$

# Skill Review

## Multiplication

```
  476
 ×15
 2380
  476
 7140
```

```
341.6  1 place
×0.27  2 places
23912
6832
92.232 3 places
```

**Multiply.**

**1.** 13
×2

**2.** 21
×4

**3.** 53
×3

**4.** 64
×2

**5.** 47
×3

**6.** 16
×5

**7.** 90
×4

**8.** 34
×6

**9.** 84
×12

**10.** 63
×23

**11.** 71
×65

**12.** 40
×31

**13.** 785
×1.2

**14.** 6.61
×3

**15.** 808
×17.2

**16.** 22.5
×8.9

**17.** 89.06
×0.5

**18.** 37.5
×0.28

**19.** 212.8
×0.67

**20.** 93.65
×4.11

**21.** $20 \times 30$

**22.** $60 \times 50$

**23.** $10 \times 90$

**24.** $50 \times 700$

**25.** $900 \times 80$

**26.** $600 \times 80$

**27.** $3400 \times 200$

**28.** $400 \times 1500$

**29.** $300 \times 470$

**30.** $18 \times 345$

**31.** $563 \times 27$

**32.** $99 \times 198$

**33.** $314 \times 16$

**34.** $408 \times 70$

**35.** $923 \times 60$

**36.** $11.6 \times 38.51$

**37.** $47.3 \times 6.05$

**38.** $86.9 \times 121.75$

**39.** $36.91 \times 0.51$

**40.** $8.05 \times 0.003$

**41.** $20.35 \times 3.7$

**42.** $854.6 \times 2.19$

**43.** $41.6 \times 212.5$

**44.** $5129.36 \times 0.008$

**45.** $85.004 \times 93.11$

**46.** $9.8413 \times 16.55$

**47.** $3.7509 \times 0.031$

**48.** $83.751 \times 98.230$

**49.** $251 \times 0.0074$

**50.** $326.11 \times 7.001$

**51.** $0.8612 \times 0.0101$

# Skill Review

## Division

$$6.8\overline{)85.068}$$

quotient: 12.51

```
  12.51
6.8)85.068
  68
  170
  136
   346
   340
    68
    68
     0
```

To round a quotient to a particular place, divide to one place beyond the place specified and then round.

```
       22.988
2.71)62.30000
     54 2
      8 10
      5 42
      2 680
      2 439
        2410
        2168
         2420
         2168
          252
```

Rounded to the nearest hundredth the quotient is 22.99.

**Divide.**

1. $5\overline{)85}$

2. $8\overline{)96}$

3. $2\overline{)92}$

4. $13\overline{)273}$

5. $32\overline{)512}$

6. $17\overline{)714}$

7. $4\overline{)140.8}$

8. $7\overline{)160.3}$

9. $18\overline{)13.68}$

10. $5.06\overline{)480.7}$

11. $3.8\overline{)59.66}$

12. $8.1\overline{)423.63}$

**Divide. Round to the nearest tenth if necessary.**

13. $8\overline{)86}$

14. $6\overline{)50}$

15. $4\overline{)63}$

16. $49\overline{)172}$

17. $32\overline{)424}$

18. $26\overline{)817}$

19. $7.3\overline{)4001}$

20. $6.9\overline{)117}$

21. $0.56\overline{)27.8}$

**Divide. Round to the nearest hundredth if necessary.**

22. $315 \div 9$

23. $513 \div 7$

24. $536 \div 8$

25. $405 \div 5$

26. $341 \div 6$

27. $11 \div 4$

28. $3775 \div 15.6$

29. $6.3 \div 2.4$

30. $11.9 \div 4.6$

31. $276 \div 37.5$

32. $88.01 \div 0.6$

33. $764 \div 0.4$

34. $21.6 \div 28$

35. $12.75 \div 51$

36. $1066 \div 27$

37. $17,063 \div 33$

38. $24,474 \div 60$

39. $19,412 \div 0.15$

40. $1.0472 \div 0.66$

41. $565.28 \div 11.8$

42. $47.611 \div 0.4$

43. $29,919 \div 56$

44. $15,960 \div 38$

45. $23,674 \div 77$

46. $11,500 \div 25$

47. $21,984 \div 36$

48. $87,048 \div 403$

49. $186.22 \div 0.038$

50. $28,400 \div 75$

51. $1234.5 \div 67$

52. $8.7553 \div 0.47$

53. $96.14 \div 0.026$

54. $0.054 \div 1.86$

55. $0.0269 \div 4.001$

56. $365.4 \div 74.1$

57. $11.99 \div 0.121$

# Extra Practice: Chapter 1

**Simplify the numerical expression.**

**1.** $44.23 - 1.6$      **2.** $83 \times 6$      **3.** $4.80 \div 30$      **4.** $74 + 116$

**Evaluate the expression when $d = 6$ and $y = 4$.**

**5.** $y - 3$      **6.** $d + d$      **7.** $2y \times 3$      **8.** $y + d + 2$

**Simplify the numerical expression.**

**9.** $18 \div (2 \times 3) + 6$      **10.** $5 \times (5 - 2) \div 3$      **11.** $\dfrac{8 - 2 \times 3}{(6 - 4)(5 + 1)}$

**Evaluate the expression when $x = 4.2$ and $y = 9$.**

**12.** $4(x + y)$      **13.** $5y - x$      **14.** $(y - x)(x + y)$

**Evaluate the expression when $a = 5$ and $b = 10$.**

**15.** $3a - \dfrac{a}{5}$      **16.** $a(b - 2) \div 3$      **17.** $\dfrac{ab - 20}{a + b}$

**Evaluate.**

**18.** $3^4$      **19.** $13^2$      **20.** $25^3$      **21.** $34^2$

**22.** $17^3$      **23.** $7^5$      **24.** $30^3$      **25.** $9^4$

**Multiply.**

**26.** $9^2 \times 3^4$      **27.** $10^3 \times 2^4$      **28.** $7^3 \times 13^2$      **29.** $15^3 \times 4^5$

**30.** $6^3 \times 100^2$      **31.** $4^3 \times 1^5$      **32.** $2^5 \times 11^3$      **33.** $3^3 \times 2^6$

**Write the decimal in expanded form.**

**34.** 734      **35.** 516.21      **36.** 0.024      **37.** 25.2

**38.** 2138      **39.** 91.9      **40.** 0.38      **41.** 307.009

**Write as a decimal.**

**42.** 18 and 21 hundredths      **43.** 5 and 4 thousandths      **44.** 242 and 6 tenths

**45.** 9 and 9 ten-thousandths      **46.** 85 thousandths      **47.** 6 ten-thousandths

**Round to the place specified.**

**48.** hundreds; 871.21      **49.** tenths; 113.93      **50.** thousandths; 1.00414

**51.** hundredths; 0.0577      **52.** tens; 382.45      **53.** hundredths; 45.552

**54.** thousandths; 3.4279      **55.** hundreds; 84      **56.** tenths; 74.991

**What value of the variable makes the statement true?**

**57.** $7.4 \times n = 1.84 \times 7.4$      **58.** $6.81 = r + 6.81$

**59.** $8.726z = z$      **60.** $3.4(5.12 + 9.4) = (5.12 + 9.4)p$

**61.** $(79 \times k) + (79 \times 4) = 79 \times 12$      **62.** $a \times 24.5 = 24.5$

**Solve for the given replacement set.**

**63.** $q + 9 = 17$; $\{8, 9, 10\}$      **64.** $b - 12 = 35$; $\{45, 46, 47\}$

**65.** $x \div 14 = 8$; $\{110, 112, 114\}$      **66.** $53 + m = 66$; $\{13, 15, 17\}$

**67.** $8f = 56$; $\{5, 6, 7\}$      **68.** $17t = 102$; $\{7, 8, 9\}$

**69.** $2g - 8 = 14$; $\{10, 11, 12\}$      **70.** $4(a + 6) = 36$; $\{1, 2, 3\}$

**Use inverse operations to solve.**

**71.** $k + 16 = 28$      **72.** $n \div 6 = 122$      **73.** $7p = 217$      **74.** $p - 11 = 41$

**75.** $m \div 12 = 204$      **76.** $17 + z = 53$      **77.** $6m = 240$      **78.** $a - 35 = 41$

**Solve, using the five-step plan.**

**79.** Recently, Ken ran a 400 m race in 52.3 s. This was 0.2 s slower than the school record. What is the school record?

**80.** On July 1, Kay had $542.07 in her savings account. On September 1, she had $671.82 in her account. How much did she save between July 1 and September 1?

**81.** A machine produces 3 plastic parts each minute that it runs. If the machine runs for 7 h, how many parts will it produce?

**82.** Exercise World is buying 4 new exercise bicycles for $129.95 each and 6 exercise mats for $47.85 each. What is the total cost?

**Solve and check.**

**83.** John owes Tom $18. John earns $5 an hour and already has $3. How long must John work to earn enough to pay Tom back?

**84.** Rosa sells newspapers at the bus stop. If each paper costs $.25, how many must Rosa sell to take in $10?

# Extra Practice: Chapter 2

**Express as an integer.**

**1.** $|^-3|$      **2.** $|5|$      **3.** $|1|$      **4.** $|0|$      **5.** $|^-6|$

**6.** $|4|$      **7.** $|^-2|$      **8.** $|^-7|$      **9.** $|9|$      **10.** $|^-8|$

**Graph the number and its opposite on the same number line.**

**11.** 6      **12.** $^-9$      **13.** 8      **14.** 5      **15.** $^-1$

**Write the integers in order from least to greatest.**

**16.** 0, 4, $^-4$, 3, $^-3$      **17.** 5, 0, $^-2$, $^-8$, 6      **18.** 7, $^-3$, $^-2$, 1, 9

**19.** $^-1$, 3, $^-7$, 5, $^-4$      **20.** $^-6$, $^-7$, 0, $^-5$, $^-2$      **21.** $^-4$, 5, 4, $^-5$, 0

**Replace** _?_ **with =, >, or < to make a true statement.**

**22.** 31 _?_ 61      **23.** $18 \times 4$ _?_ 82      **24.** 47 _?_ $49 - 9$

**List the integers that can replace $x$ to make the statement true.**

**25.** $|x| = 5$      **26.** $|x| = 2$      **27.** $|x| = 0$

**28.** $|x| = 4$      **29.** $|x| \leq 6$      **30.** $|x| \leq 3$

**Graph the numbers in each exercise on the same number line.**

**31.** 1, $^-1.5$, 0, $^-2$      **32.** $^-2$, $^-4$, 2.5, $^-3.5$      **33.** 3, 1.2, $^-2.4$, $^-5$

**34.** 7.6, $^-6.7$, 6, $^-7$      **35.** 1, 1.5, $^-2$, 2.5      **36.** 2, $^-1.8$, $^-2.1$, 0

**Draw an arrow to represent the decimal number described.**

**37.** The number 5, with starting point $^-4$

**38.** The number $^-8$, with starting point 6

**39.** The number 6, with starting point $^-5$

**40.** The number $^-6$, with starting point 3

**Find the sum.**

**41.** $2.7 + 7.2$      **42.** $^-4.9 + {}^-7.6$      **43.** $^-2.25 + 2.25$

**44.** $^-3.8 + {}^-3.8$      **45.** $^-148 + {}^-256$      **46.** $6.1 + {}^-2.3$

**47.** $^-0.6 + {}^-2.3$      **48.** $18.12 + 1.66$      **49.** $^-5.2 + 2.9$

**50.** $^-2.8 + {}^-3.9$      **51.** $1.9 + 19$      **52.** $^-14.75 + 9.94$

**Find the difference.**

**53.** $30.5 - 18$      **54.** $16 - 24.3$      **55.** $10.6 - 11.2$

**56.** $16.2 - (-7)$      **57.** $5 - (-15.3)$      **58.** $18 - (-9.7)$

**59.** $-8.3 - 5.2$      **60.** $-7 - 14.6$      **61.** $-12.3 - 8$

**62.** $43.4 - (-136)$      **63.** $-3 - (-8.2)$      **64.** $-12.9 - (-4.5)$

**Evaluate when $x = -3.2$ and $y = -5.7$.**

**65.** $-y$      **66.** $-x$      **67.** $-|x|$

**68.** $-|y|$      **69.** $y - x$      **70.** $-x - y$

**71.** $x - (-y)$      **72.** $-x - (-y)$      **73.** $-y - (-x)$

**Find the quotient.**

**74.** $-15.5 \div 0.5$      **75.** $-16.4 \div 4$      **76.** $-150 \div 7.5$

**77.** $36.6 \div -0.06$      **78.** $38 \div -0.4$      **79.** $8.19 \div (-9)$

**80.** $-3.038 \div (-7)$      **81.** $-63.7 \div (-0.007)$      **82.** $-246 \div (-0.6)$

**83.** $-0.003 \div (1)$      **84.** $0 \div (-0.85)$      **85.** $-0.9 \div (-1.8)$

**Find the product.**

**86.** $2.4(-1.2)$      **87.** $-1.4(3.4)$      **88.** $4.5(-1.7)$

**89.** $-8.2(-6.1)$      **90.** $-4.3(-3.4)$      **91.** $-6.2(-5.3)$

**92.** $-2.15(1.15)$      **93.** $3.14(-8.1)$      **94.** $-5.22(8.11)$

**95.** $4.25(-3.14)$      **96.** $1.11(-1.11)$      **97.** $-6.75(2.32)$

**Evaluate the expression if $a = 4$, $b = 2$, and $c = 6$.**

**98.** $5c^3$      **99.** $(4b)^3$      **100.** $2ab^4$      **101.** $2a^3$

**102.** $3(ab)^2$      **103.** $(3c)^b$      **104.** $a^2 - b^4$      **105.** $c^2a$

**Write the expression without exponents.**

**106.** $7^{-3}$      **107.** $(-4)^{-2}$      **108.** $2^{-5}$      **109.** $(-6)^{-4}$      **110.** $8^{-1}$

**111.** $(-9)^{-2}$      **112.** $5^{-4}$      **113.** $(-1)^{-7}$      **114.** $3^{-6}$      **115.** $(-5)^{-3}$

**116.** $8^5 \times 8^{-7}$      **117.** $10^4 \times 10^{-3}$      **118.** $5^9 \times 5^{-9}$      **119.** $4^{-2} \times 4^0$

**120.** $9^{17} \times 9^{-17}$      **121.** $3^{-6} \times 3^4$      **122.** $6^{-2} \times 6^5$      **123.** $2^{10} \times 2^{-8}$

# Extra Practice: Chapter 3

**List all the factors of each number.**

| | | | | | |
|---|---|---|---|---|---|
| **1.** 6 | **2.** 7 | **3.** 10 | **4.** 13 | **5.** 15 | **6.** 18 |
| **7.** 41 | **8.** 42 | **9.** 52 | **10.** 56 | **11.** 59 | **12.** 64 |

**Which of the numbers 2, 3, 4, 5, 9, and 10 are factors of the given number?**

| | | | | | |
|---|---|---|---|---|---|
| **13.** 155 | **14.** 168 | **15.** 189 | **16.** 210 | **17.** 272 | **18.** 305 |
| **19.** 1080 | **20.** 1100 | **21.** 1260 | **22.** 1362 | **23.** 1423 | **24.** 1485 |

**Determine whether each number is prime or composite. If the number is composite, give the prime factorization.**

| | | | | | |
|---|---|---|---|---|---|
| **25.** 10 | **26.** 34 | **27.** 54 | **28.** 41 | **29.** 30 | **30.** 18 |

**Write as a proper fraction in lowest terms or as a mixed number in simple form.**

**31.** $\frac{15}{45}$      **32.** $-\frac{22}{64}$      **33.** $\frac{7}{5}$      **34.** $\frac{180}{240}$

**35.** $-\frac{144}{96}$      **36.** $-\frac{58}{12}$      **37.** $\frac{75}{125}$      **38.** $\frac{-145}{95}$

**39.** $\frac{14}{-8}$      **40.** $\frac{32}{28}$      **41.** $\frac{-63}{81}$      **42.** $-\frac{48}{54}$

**Complete.**

**43.** $\frac{1}{4} + \frac{1}{4} + \frac{1}{4} + \frac{1}{4} = \underline{\ ?\ }$      **44.** $\underline{\ ?\ } \times \frac{1}{3} = -\frac{2}{3}$

**45.** $5 \times \underline{\ ?\ } = -1$      **46.** $\frac{3}{10} = 3 \div \underline{\ ?\ }$

**47.** $\left(-\frac{1}{7}\right) + \left(-\frac{1}{7}\right) = \underline{\ ?\ }$      **48.** $4 \times \underline{\ ?\ } = \frac{4}{9}$

**49.** $\frac{2}{5} = \frac{?}{20}$      **50.** $\frac{-8}{9} = \frac{?}{27}$      **51.** $-\frac{12}{30} = -\frac{2}{?}$

**52.** $\frac{20}{-36} = \frac{?}{-9}$      **53.** $3 = \frac{12}{?}$      **54.** $-4 = \frac{-28}{?}$

**Write as an improper fraction.**

**55.** $5\frac{2}{3}$      **56.** $3\frac{4}{5}$      **57.** $-4\frac{3}{10}$      **58.** $-15\frac{1}{6}$

**59.** $-7\frac{1}{8}$      **60.** $9\frac{3}{25}$      **61.** $6\frac{1}{4}$      **62.** $-5\frac{11}{12}$

Write the set of fractions as equivalent fractions with the least common denominator (LCD).

**63.** $\frac{5}{6}, \frac{3}{8}$

**64.** $\frac{4}{5}, -\frac{3}{10}$

**65.** $\frac{4}{8}, \frac{5}{12}$

**66.** $-\frac{8}{15}, -\frac{7}{20}$

**67.** $-\frac{11}{28}, \frac{17}{42}$

**68.** $\frac{7}{30}, \frac{19}{70}$

Add or subtract. Write the answer as a proper fraction in lowest terms or as a mixed number in simple form.

**69.** $\frac{5}{13} + \frac{4}{13}$

**70.** $-\frac{7}{8} - \frac{5}{8}$

**71.** $\frac{5}{12} - \frac{1}{12}$

**72.** $\frac{5}{9} + \left(-\frac{1}{4}\right)$

**73.** $-\frac{9}{14} - \frac{5}{32}$

**74.** $\frac{3}{7} - \left(-\frac{4}{5}\right)$

**75.** $-4\frac{1}{12} + 2\frac{4}{5}$

**76.** $5\frac{2}{3} - 2\frac{5}{8}$

**77.** $-3\frac{7}{12} + 7\frac{5}{16}$

**78.** $2\frac{3}{7} - \left(-4\frac{5}{6}\right)$

**79.** $-1\frac{3}{28} + \left(-4\frac{10}{21}\right)$

**80.** $-15\frac{1}{2} - \left(-8\frac{3}{4}\right)$

Multiply or divide. Write the answer as a proper fraction in lowest terms or as a mixed number in simple form.

**81.** $-\frac{3}{4} \times \frac{4}{5}$

**82.** $\frac{21}{56} \times \frac{20}{25}$

**83.** $\frac{7}{8} \times \left(-\frac{4}{7}\right)$

**84.** $6\frac{3}{4} \times \left(-1\frac{1}{3}\right)$

**85.** $-2\frac{5}{8} \times \left(-\frac{16}{19}\right)$

**86.** $-4\frac{2}{7} \times 2\frac{1}{4}$

**87.** $\frac{4}{9} \div \frac{2}{9}$

**88.** $-\frac{3}{8} \div \left(-\frac{5}{16}\right)$

**89.** $-\frac{8}{15} \div \left(-2\frac{2}{7}\right)$

**90.** $-3\frac{3}{5} \div \left(2\frac{4}{15}\right)$

**91.** $-\frac{5}{6} \div 4\frac{1}{2}$

**92.** $3\frac{5}{9} \div (-32)$

Write as a terminating or repeating decimal. Use a bar to show a repeating decimal.

**93.** $\frac{3}{8}$

**94.** $\frac{3}{5}$

**95.** $\frac{21}{22}$

**96.** $-\frac{5}{16}$

**97.** $\frac{1}{6}$

**98.** $-\frac{5}{9}$

**99.** $-\frac{8}{15}$

**100.** $-\frac{161}{189}$

**101.** $\frac{287}{385}$

**102.** $3\frac{5}{12}$

**103.** $-2\frac{3}{11}$

**104.** $4\frac{1}{18}$

Write as a proper fraction in lowest terms or as a mixed number in simple form.

**105.** $0.6$

**106.** $-0.04$

**107.** $1.34$

**108.** $-4.22$

**109.** $-3.025$

**110.** $0.\overline{6}$

**111.** $-2.\overline{09}$

**112.** $1.\overline{7}$

**113.** $8.2121\ldots$

**114.** $-2.1666\ldots$

# Extra Practice: *Chapter 4*

**Use transformations to solve each equation. Write down all the steps.**

**1.** $r + 30 = 80$

**2.** $31 + d = 47$

**3.** $\frac{1}{4}b = \frac{3}{4}$

**4.** $x - 21 = 19$

**5.** $79 - a = 17$

**6.** $14 + 12 = 17 + a$

**7.** $1.23 = 1.50 - a$

**8.** $\frac{7}{8} + a = 4$

**9.** $13t = 52$

**10.** $84 = 14k$

**11.** $\frac{n}{6} = 12$

**12.** $15 = \frac{x}{3}$

**13.** $28 = 7v$

**14.** $3 = \frac{p}{3}$

**15.** $\frac{t}{7} = 5(3 + 4)$

**16.** $9 = \frac{x}{10}$

**17.** $\frac{1}{7} = \frac{3}{7}k$

**18.** $\frac{12}{5}c = \frac{3}{10}$

**19.** $0.25a = 1.0$

**20.** $1.7 = 1.7x$

**21.** $\frac{5}{9} = \frac{12}{5}k$

**22.** $1.25 = 0.6f$

**23.** $6.25n = 1.25$

**24.** $\frac{11}{15} = \frac{3}{5}y$

**25.** $\frac{p}{6} - 12 = 3$

**26.** $72 = \frac{p}{3} + 3$

**27.** $3.6x - 2.5 = 15.5$

**28.** $1.24 - 1.2m = 1.0$

**29.** $17 = 5y - 3$

**30.** $\frac{8}{3} = \frac{2}{5}v - \frac{1}{3}$

**31.** $\frac{m}{3} = \frac{1}{3}(6 + 12)$

**32.** $2x - 3 = 15 - 6$

**Write a variable expression for the word phrase.**

**33.** The difference when a number $t$ is subtracted from eighteen

**34.** Five added to the product of a number $x$ and nine

**35.** Forty divided by a number $m$, decreased by sixteen

**36.** The remainder when a number $q$ is subtracted from two hundred

**37.** Eleven more than three times a number $y$

**38.** The sum of a number $r$ and six, divided by twelve

**Write an equation for each word sentence.**

**39.** The product of six and a number *n* is fifty-four.

**40.** Twelve less than two times a number *n* is seventy.

**41.** The sum of a number *n* and nineteen is sixty-one.

**Choose a variable and write an equation for each problem.**

**42.** The altitude of the Dead Sea is 1296 ft below sea level. How much greater is the altitude of Death Valley, California, at 282 ft below sea level?

**43.** An astronaut enters a space capsule 1 hr 40 min before launch time. How long has the astronaut been in the capsule 2 h 32 min after the launch?

**44.** The Torrance baseball team scored 10 runs in the first 3 innings. Gardena scored 2 in the first and 5 in the fourth. If Torrence does not score again, how many more runs does Gardena need to win?

**Write an equation for each problem. Solve the equation using the five-step method. Check your answer.**

**45.** Mercury melts at 38.87° below 0°C and boils at 356.9°C. What is the difference between these temperatures?

**46.** Yolanda rode the bus from a point 53 blocks south of Carroll Avenue to a point 41 blocks north of Carroll Avenue. How many blocks did she travel?

**47.** John sailed for 7 hours on Swan Lake. If he sailed for two more hours than he fished, how long did he fish?

**48.** In a local election 1584 people voted. The winner received 122 votes more than the loser. How many votes did each candidate receive?

**49.** It takes Fritz 55 min to commute to and from work. The ride home takes 7 min less than the ride to work. How long does it take each way?

**50.** Ann swam a total of 78 laps on Monday and Tuesday. She swam 12 more laps on Tuesday than on Monday. How far did she swim each day?

# Extra Practice: *Chapter 5*

**Draw a sketch to illustrate each of the following.**

1. Three points on a line
2. Two intersecting lines
3. Three noncollinear points
4. Two intersecting planes
5. Two rays with a common endpoint, *A*
6. Two segments with a common endpoint, *B*

**Complete.**

7. 10 km = __?__ m
8. 0.27 km = __?__ m
9. 9 m = __?__ cm
10. 0.5 km = __?__ m
11. 77 m = __?__ cm
12. 175 mm = __?__ m
13. 1500 m = __?__ km
14. 81 cm = __?__ mm
15. 0.1575 km = __?__ m
16. 900 mm = __?__ m = __?__ cm
17. 2 m = __?__ cm = __?__ mm
18. If the sum of the measures of two angles is 180°, the angles are __?__.
19. A small square is often used to indicate a(n) __?__ angle.
20. Perpendicular lines form __?__° angles.
21. The __?__ is the common endpoint of two rays that form an angle.
22. If m∠*A* = 40°, the complement of ∠*A* measures __?__°.
23. A(n) __?__ angle has a measure between 90° and 180°.
24. A triangle with at least two sides congruent is called a(n) __?__ triangle.
25. Triangles can be classified by their __?__ or their __?__.
26. The sum of the measures of the angles of a triangle is __?__°.
27. A(n) __?__ triangle has two perpendicular sides.
28. The sum of the lengths of any two sides of a triangle is __?__ than the length of the third side.
29. A triangle with three congruent sides is called a(n) __?__ triangle.
30. A triangle with all its angles less than 90° is called a(n) __?__ triangle.

**True or false?**

31. A regular polygon has all the sides equal.
32. All the angles of a rectangle have a measure of 90°.

33. A trapezoid always has one pair of congruent sides.

34. The opposite sides of a parallelogram are congruent.

35. Only two sides of a rhombus are congruent.

36. A square with a perimeter of 24 cm has 4 sides that measure 6 cm.

37. A regular decagon, 5 cm on a side, has a perimeter of 50 cm.

**Solve. Use $\pi \approx 3.14$ and round answers to three digits.**

38. The radius of a circle is 30 cm. Find the diameter.

39. The radius of a circle is 56 mm. Find the circumference.

40. The diameter of a circle is 3 m. Find the circumference.

41. The diameter of a circle is 16 mm. Find the circumference.

42. The circumference of a circle is 20 m. Find the diameter.

43. The circumference of a circle is 25.8 m. Find the radius.

**Complete the statements about the pair of congruent figures.**

44. $\overline{KL} \cong$ ___?___    45. $\overline{NK} \cong$ ___?___

46. $\overline{LM} \cong$ ___?___    47. $\overline{SP} \cong$ ___?___

48. $\angle N \cong$ ___?___    49. $\angle S \cong$ ___?___

50. $\angle L \cong$ ___?___    51. $\angle Q \cong$ ___?___

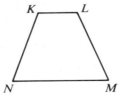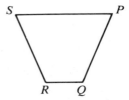

**Are the triangles in each pair congruent? If so, name the triangles that are congruent and explain why they are congruent.**

52.

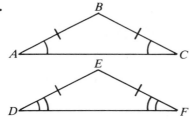

53.

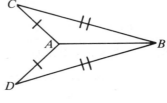

**Use a compass and a straightedge to make each construction.**

54. Draw a segment $AB$. Construct the perpendicular bisector.

55. Draw an acute angle $ABC$. Construct $\overrightarrow{BX}$ so that it bisects $\angle ABC$.

56. Construct an angle congruent to $\angle ABC$ of Exercise 55.

# Extra Practice: *Chapter 6*

**Express each ratio as a fraction in lowest terms.**

1. $\dfrac{10 \text{ min}}{1 \text{ h}}$
2. $\dfrac{6 \text{ cm}}{2 \text{ m}}$
3. $\dfrac{250 \text{ mL}}{2 \text{ L}}$
4. $\dfrac{6 \text{ kg}}{600 \text{ g}}$

5. $\dfrac{7 \text{ days}}{4 \text{ weeks}}$
6. $\dfrac{3 \text{ h}}{45 \text{ min}}$
7. $\dfrac{15 \text{ mm}}{5 \text{ cm}}$
8. $\dfrac{60 \text{ cm}}{150 \text{ mm}}$

**Solve. Express each ratio in lowest terms.**

9. A student walked 25 km in 4 h. What was the average speed in kilometers per hour?

10. A special typewriter has 46 keys. Of these 46 keys, 26 are used for letters of the alphabet. What is the ratio of the number of keys not used for letters to the total number of keys?

**Solve.**

11. In order to make 4 servings, a recipe calls for 3 eggs to be used. How many eggs will be needed to make 32 servings?

12. Charlie runs 5 mi in 37 min. If he could maintain the same speed, how long would it take him to run 15 mi?

13. A dozen apples cost $1.80. What is the unit price?

14. A car travels 234 km in 4.5 h. What is the car's average speed?

15. $\dfrac{19}{20} = \dfrac{n}{10}$
16. $\dfrac{8}{48} = \dfrac{n}{4}$
17. $\dfrac{4}{13} = \dfrac{12}{n}$
18. $\dfrac{n}{27} = \dfrac{12}{81}$

19. $\dfrac{40}{n} = \dfrac{5}{9}$
20. $\dfrac{n}{64} = \dfrac{12}{8}$
21. $\dfrac{100}{40} = \dfrac{20}{n}$
22. $\dfrac{7}{3} = \dfrac{n}{18}$

23. Joe bought 4 shock absorbers for the price of 3 at a clearance sale. He paid a total of $39. How much would the 4 shock absorbers cost at their regular price?

24. Billie runs 4 km in 25 min. How long will it take to run 10,000 m?

25. A supermarket sells 12 oranges for $1.30. How much would 18 oranges cost at the same rate?

**A map has a scale of 1 cm:8 km. What actual distance does each map length represent?**

26. 8 cm
27. 5.5 cm
28. 3 cm
29. 7.9 cm

30. 3.75 cm
31. 10.2 cm
32. 6.25 cm
33. 15.25 cm

**Express as a fraction in lowest terms or as a mixed number in simple form.**

**34.** 5%          **35.** 10%          **36.** 72%          **37.** 39%          **38.** 2%

**39.** 27%          **40.** 25%          **41.** 200%          **42.** 163%          **43.** 215%

**Express as a percent.**

**44.** $\frac{3}{4}$          **45.** $\frac{1}{10}$          **46.** $\frac{3}{5}$          **47.** $\frac{1}{8}$          **48.** 7

**49.** $\frac{7}{16}$          **50.** $\frac{43}{20}$          **51.** $\frac{36}{25}$          **52.** $\frac{23}{4}$          **53.** $\frac{22}{5}$

**Express each percent as a decimal.**

**54.** 15%          **55.** 71%          **56.** 5%          **57.** 15.2%          **58.** 98%

**59.** 2.4%          **60.** 7.5%          **61.** 625%          **62.** 0.42%          **63.** 200%

**Express each decimal as a percent.**

**64.** 0.45          **65.** 0.23          **66.** 1.33          **67.** 0.05          **68.** 12.5

**69.** 0.0025          **70.** 10.2          **71.** 0.53          **72.** 0.008          **73.** 0.125

**Express each fraction as a decimal, then as a percent.**

**74.** $\frac{2}{5}$          **75.** $\frac{1}{4}$          **76.** $\frac{7}{10}$          **77.** $\frac{5}{8}$          **78.** $\frac{1}{16}$

**79.** $1\frac{3}{8}$          **80.** $\frac{5}{16}$          **81.** $8\frac{3}{4}$          **82.** $\frac{13}{16}$          **83.** $5\frac{1}{5}$

**Answer each question by writing an equation and solving it. Round to the nearest tenth of a percent if necessary.**

**84.** 12% of 50 is what number?          **85.** 110% of 99 is what number?

**86.** What percent of 81 is 27?          **87.** 85% of what number is 425?

**88.** 15% of $30 is how much?          **89.** What percent of 250 is 75?

**90.** What percent of 256 is 32?          **91.** 60% of what number is 144?

**92.** 5% of 1000 is what number?          **93.** 140% of what number is 35?

**94.** 18% of $85 is how much?          **95.** What percent of 825 is 25?

**96.** 30% of what number is 23.7?          **97.** 6% of what number is 3.48?

**98.** What percent of 325 is 65?          **99.** 12% of 408 is what number?

# Extra Practice: Chapter 7

Find the percent increase or decrease from the first number to the second. Round to the nearest tenth of a percent if necessary.

**1.** 45 to 66　　　　**2.** 50 to 60　　　　**3.** 140 to 84　　　　**4.** 256 to 500

**5.** 15 to 75　　　　**6.** 144 to 96　　　　**7.** 360 to 234　　　　**8.** 196 to 49

Find the new number produced when the given number is increased or decreased by the given percent.

**9.** 64; 25% increase　　　　　　　　**10.** 80; 65% decrease

**11.** 78; 150% increase　　　　　　　**12.** 480; 35% increase

**Solve.**

**13.** Mervyn's is having a sale on batteries. The regular price of $2.40 is decreased by 20%. What is the sale price?

**14.** A bicycle that usually sells for $120 is on sale for $96. What is the percent of discount?

**15.** Great Hikes purchases backpacks from the manufacturer at $10 each. Then the backpacks are sold at the store for $25 each. What is the percent of markup?

**16.** If there is a sales tax of 5%, how much will it cost to buy a $57 handbag?

**17.** A wheelbarrow sells for $36. Next month the price will be marked up 6%. How much will the wheelbarrow cost next month?

**18.** A furniture store has a total income of $3600 per week and total operating costs of $3096 per week. To the nearest tenth, what percent of the income is profit?

**19.** The Thort Company made a profit of $32,175 on sales of $371,250. To the nearest tenth, what percent of the sales was profit?

**20.** Dora's commission last month was $621. Her total sales were $8280. What is her rate of commission?

**21.** Foster Realty Co. charges a 6% commission for the sale of property. If a homeowner wishes to clear $130,000 after paying the commission, for how much must the home be sold?

**22.** Ivan works for salary plus commission. He earns $500 per month plus 4% commission on all sales. What sales level is needed for Ivan to earn $675 in a given month?

23. Last Friday, 720 people at Data Tech drove cars to work. Of these, 585 bought gas on the way home. What percent of the drivers bought gas on the way home? Use a proportion to solve.

24. How much does a $53.00 iron cost if it is taxed at 5%? Use a proportion to solve.

25. Jill has $525 in a savings account that pays 6% simple interest. How much will she have in the account after 1 year, if she makes no deposits or withdrawals?

26. Holly has a savings account that pays 6.5% interest, compounded monthly. If she started the account with $750 and makes no deposits or withdrawals, how much will be in the account after 3 months?

27. Phil borrowed $5000 for one year. The interest on the loan came to $689.44. To the nearest tenth, what was the interest rate?

28. Victor had a two-year loan with a simple interest rate of 13% annually. At the end of the two years he had paid $1454.44 in interest. What was the original of Victor's loan?

29. Consider a principal of $2750 earning an interest rate of 10.5%. If compounded semiannually for one year, what is the total interest earned on $2750? (If you do not have a calculator, round to the nearest penny at every step.)

30. A $600 deposit is left in an account for 9 months. If the account earns an 8% interest compounded quarterly, what is the total interest earned at the end of the 9 months? (If you do not have a calculator, round to the nearest penny at every step.)

31. Carol has 60 boxes of light bulbs. If she sells 15% of the boxes today, how many will be left?

32. At a pet store, fish that usually cost $1.25 each are on sale for $.75 each. What is the percent of decrease?

33. A restaurant raises its price for the salad bar 20% to $1.50. What did the salad bar cost before?

34. The band concert drew an audience of 270 on Thursday. On Friday, attendance increased 30%. How many people attended on Friday?

35. The discount rate at an appliance store sale is 15%. What is the sale price of a dishwasher if the original price was $600?

36. Nancy Allen earns a 12% commission on her sales. Last month she earned $4800. What was the total value of her sales?

# Extra Practice: *Chapter 8*

**Solve.**

1. $2a + 7a = 36$
2. $12x - 8x = 4$
3. $-10y + 12y = 1$
4. $d - 7d = -72$
5. $-5t + 12t = 3$
6. $-3p - 9p = 6$
7. $x + 3x + 6 = 8$
8. $4c + 8c - 3 = 9$
9. $-n - 3 + 10n = 9$
10. $5z + 10 - 12z = -4$
11. $8 + y - 9y = -4$
12. $d + 12 - 7d = 48$
13. $4a = a + 3$
14. $16p = 12 + 8p$
15. $32 - 6x = 2x$
16. $50 + t = 6t$
17. $3x = -6x + 36$
18. $y = 12y - 44$
19. $3c + 4 = 5c + 5$
20. $2n - 5 = 7n + 20$
21. $d - 9 = 24 - 10d$
22. $-x + 7 = 6x + 21$
23. $3 - 4y = 13 + 2y$
24. $-3a + 6 = 5 - 2a$

**Solve.**

25. At Moor's Deli, a ham sandwich and a glass of milk cost a total of $3.25. If the sandwich costs four times as much as the milk, what is the cost of each?

26. The sum of two consecutive integers is 135. What are the two integers?

27. A 60 ft piece of rope is cut into two pieces, one three times as long as the other. What is the length of each piece?

28. The perimeter of an isosceles triangle is 75 cm. If the congruent sides of the triangle are each twice as long as the remaining side, what is the length of each side?

**Replace __?__ with < or >.**

29. $15 \underline{\phantom{?}} 23$
30. $57 \underline{\phantom{?}} 49$
31. $35 \underline{\phantom{?}} 18$
32. $7 + 11 \underline{\phantom{?}} 31$
33. $17 \underline{\phantom{?}} 12 + 4$
34. $29 - 15 \underline{\phantom{?}} 17$

**Write an inequality for each word sentence.**

35. Seventeen is greater than twelve.

36. A number plus twelve is less than twenty-two.

**37.** Twice a number is greater than sixty-nine.

**38.** Seven is smaller than two times five.

**39.** Eighty minus twenty-six is greater than two times twenty-one.

**40.** Thirty-three is less than six times a number.

**Use transformations to solve the inequality. Write down all the steps.**

**41.** $y + 5 \le -17$

**42.** $q - 8 > 24$

**43.** $5\frac{1}{6} - \frac{1}{3} < k$

**44.** $7.8 + 4.3 \ge m$

**45.** $f < (17 - 3)0.5$

**46.** $z \le \frac{1}{2}(25 - 7)$

**47.** $6e > 42$

**48.** $-5a \le 45$

**49.** $28 < -7u$

**50.** $52 \ge 4x$

**51.** $\frac{w}{-3} > 21$

**52.** $\frac{c}{4} \ge 17$

**53.** $-11 \le \frac{a}{5}$

**54.** $\frac{e}{-2} < -23$

**55.** $10 > \frac{v}{7}$

**56.** $-5x + 7 + 2x - 13 < 9$

**57.** $-22 \ge 10d - 4 - 3d + 12$

**58.** $2 > \frac{4}{5}n + 3 - \frac{1}{5}n + 6$

**59.** $5 + \frac{3}{7}u - 8 + \frac{2}{7}u \le 5$

**60.** $-w + 8 \le 4w - 17$

**61.** $-6s - 4 < s + 31$

**62.** $10 + \frac{2}{3}h > -\frac{5}{3}h + 6$

**63.** $8 - \frac{1}{6}p \ge 4 - \frac{1}{3}p$

**64.** $4(m + 3) \ge 48$

**65.** $35 < -5(x + 2)$

**66.** $34 < (18t - 6)\frac{1}{3}$

**67.** $\frac{-1}{7}(14 + 21u) \ge 7$

**Solve.**

**68.** Of all pairs of consecutive integers whose sum is greater than 250, find the pair whose sum is least.

**69.** Two cars start from the same point traveling in different directions. One car travels at a speed of 52 mi/h, and the other car travels at a speed of 48 mi/h. How long must they travel to be 300 mi apart?

**70.** A purse contains 20 coins, all either quarters or dimes. If the total value of the coins is greater than $3.50, at least how many are quarters?

# Extra Practice: Chapter 9

**For Exercises 1–6, give the coordinates of the point.**

1. $A$  2. $B$  3. $C$

4. $D$  5. $E$  6. $F$

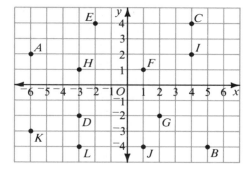

**Name the point for the ordered pair.**

7. $(2, -2)$  8. $(-3, -4)$

9. $(4, 2)$  10. $(-6, -3)$

11. $(1, -4)$  12. $(-3, 1)$

**a. Graph the given ordered pairs on a coordinate plane.**
**b. Connect all the points in the order listed by means of line segments to produce a closed figure.**
**c. Name the figure as specifically as you can.**

13. $(3, 3)$, $(3, 0)$, $(-2, 0)$, $(-5, 0)$, $(-5, 3)$, $(3, 3)$

14. $(2, -2)$, $(2, -6)$, $(-2, -6)$, $(-2, -2)$, $(0, -2)$, $(2, -2)$

15. $(1, 5)$, $(-2, 2)$, $(-5, -1)$, $(0, -1)$, $(6, -1)$, $(1, 5)$

16. $(0, 7)$, $(3, 5)$, $(2, 2)$, $(-2, 2)$, $(-3, 5)$, $(0, 7)$

17. $(7, -4)$, $(4, -1)$, $(2, 1)$, $(1, 2)$, $(1, -4)$, $(7, -4)$

**Tell whether the ordered pair is a solution of the given equation.**

$-x + 2y = 4$
18. $(4, 0)$  19. $(-4, 0)$  20. $(2, 3)$  21. $(-2, 3)$

$x + 3y = 6$
22. $(0, 2)$  23. $(6, 0)$  24. $(2, 1)$  25. $(3, 1)$

$2x + y = 5$
26. $(3, 1)$  27. $(2, 1)$  28. $(1, 3)$  29. $(0, 5)$

**a. Solve the equation for $y$ in terms of $x$.**
**b. Find the solutions of the equation for the given values of $x$.**

30. $2x + y = 8$
   values of $x$: $-1, 0, 1$

31. $y - 3x = 4$
   values of $x$: $-5, 0, 5$

32. $-x + y = 1$
   values of $x$: $-1, \frac{1}{2}, 2$

33. $3x - 2y = -6$
   values of $x$: $-2, 0, 2$

**34.** $2x + y = 6$
values of $x$: 0, 3, 1

**35.** $x - y = 3$
values of $x$: 6, $-1$, $-3$

**Graph the equation on a coordinate plane. Use a separate set of axes for each equation.**

**36.** $2x + y = 4$

**37.** $x - 3y = -3$

**38.** $x + 2y = -4$

**39.** $3x + 4y = 12$

**40.** $2x + 5y = 10$

**41.** $-x + y = -1$

**42.** $x + y = 2$

**43.** $3x - y = 3$

**44.** $4x - 2y = 3$

**45.** $2x = y - 2$

**46.** $y = x - 3$

**47.** $-3x + 4y = 12$

**Use a graph to solve the system. Do the lines intersect, coincide, or are they parallel?**

**48.** $\quad x - y = -2$
$\quad 2x + y = 5$

**49.** $\quad x - y = 6$
$\quad 2x + y = 0$

**50.** $2x + y = -2$
$\quad x - y = -4$

**51.** $2x + y = 2$
$\quad 2x + y = -3$

**52.** $6x + 3y = 6$
$\quad 2x + \;\; y = 2$

**53.** $3x + y = -6$
$\quad 2x - y = 1$

**54.** A parking lot attendant charges $3 for the first hour and $1 for each extra hour after that. Write an equation that relates the total cost of parking ($y$) to the number of extra hours a car is parked ($x$). Graph the equation. What is the slope of the graph? From your graph determine how much it would cost to park a car for 5 h after the initial hour.

**55.** The initial charge for a phone call to Jonesboro is 25¢ for the first 3 min. After that the charge is 5¢ for every additional minute. Write an equation that relates the total cost of a phone call ($y$) to the number of additional minutes on the phone ($x$). Graph the equation. What is the slope of the graph? From the graph determine the cost of a call requiring 7 min beyond the initial 3.

**56.** The temperature in a town rose at a constant rate from 5 A.M. to 12 noon. At 9 A.M. the temperature was $17°C$. At 11 A.M. the temperature was $21°C$. Use this information to draw a graph and determine what the temperature was at 5 A.M. and at 12 noon.

**Graph the inequality.**

**57.** $x + 2y < 4$

**58.** $2x - y < 1$

**59.** $6x + y \geq 7$

**60.** $-x + 3y < 0$

**61.** $2x + y \geq 1$

**62.** $3x + y \leq -2$

# Extra Practice: Chapter 10

**Find the area and perimeter of a rectangle with the given dimensions.**

**1.** 100 m by 50 m      **2.** 27 mm by 13 mm      **3.** 125 cm by 61 cm

**4.** If a rectangle has a width of 57 km and an area of 3648 km², what is
(a) the length and (b) the perimeter?

**Find the area of a parallelogram with the given dimensions.**

**5.** $b = 12$ cm, $h = 11$ cm      **6.** $b = 105$ m, $h = 28$ m      **7.** $b = 17$ km, $h = 7$ km

**8.** If a parallelogram has a height of 13 cm and an area of 201.5 cm²,
what is the base?

**Find the area of each polygon.**

**9.** Triangle: base 23 cm, height 7 cm      **10.** Triangle: base 35 m, height 41 m

**11.** Trapezoid: bases 11.5 m and 6.5 m, height 15 m

**12.** If a triangle has a height of 113 cm and an area of 4576.5 cm², what
is the base?

**13.** If a trapezoid has an area of 531 mm² and bases of 32.1 mm and
21 mm, what is the height?

**Solve. Use $\pi \approx 3.14$ and round the answer to three digits.**

**14.** A circle has a radius of 42 cm. What is the area?

**15.** A circle has a diameter of 22 m. What is the area?

**16.** A circle has an area of $12\frac{4}{7}$ ft². What is the diameter?

**17.** A circle has a radius of 56 m. What is the area?

**Copy each figure and show on your drawing any lines or points of symmetry.**

**18.**

**19.**

**20.**
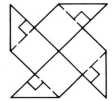

**21.** A prism with height 5 m has base area 173 m². Find the volume.

**22.** Find the volume of a cylinder with height 82 cm and base area 50 cm².

**23.** A right triangle with legs 15 mm and 7 mm is the base of a prism with height 24 mm. Find the volume of the prism.

**24.** Find the volume of a cylinder with height 300 cm and diameter 15 cm.

**25.** The volume of a triangular prism is 2340 cm³. Find the height of the prism if the base area is 78 cm².

**26.** A cone with base area 234 mm² is 82 mm high. Find its volume and its capacity in liters.

**27.** A pyramid with height 52 cm has a rectangular base measuring 12 cm by 17 cm. Find the volume.

**28.** The base of a cone has diameter 70 cm. If the height of the cone is 120 cm, what is the volume?

**For Exercises 29–32, find (a) the lateral surface area and (b) the total surface area of the figure described.**

**29.** A prism of height 15 cm whose bases are right triangles with sides 15 cm, 36 cm, and 39 cm

**30.** A cylinder with radius 4 cm and height 11 cm

**31.** A cube with edges that are 15 cm

**32.** A rectangular prism of height 27 cm and base 15 cm by 12 cm

**For Exercises 33–36, leave your answers in terms of $\pi$.**

**33.** Find the surface area of a sphere with radius 18 cm.

**34.** Find the volume of a sphere with diameter 30 cm.

**35.** Find the radius of a sphere whose surface area is $576\pi$ m².

**36.** Find the volume of a sphere whose radius is 16 cm.

**37.** If the mass of 1 L of gasoline is 0.66 kg, what is the mass of 4 L?

**38.** Each edge of a cube of steel measures 12 cm. Find the mass of the cube if the mass of 1 cm³ of steel is 7.82 g.

**39.** A rectangular prism of aluminum is 15 cm long, 7 cm wide, and 30 cm high. Find the mass of the prism if the mass of 1 cm³ of aluminum is 2.708 g.

# Extra Practice: Chapter 11

**If the given symbol names an integer, state the integer. If not, name the two consecutive integers between which the number lies.**

1. $\sqrt{9}$

2. $\sqrt{39}$

3. $-\sqrt{27}$

4. $\sqrt{10}$

5. $\sqrt{17}$

6. $\sqrt{5}$

7. $\sqrt{81}$

8. $-\sqrt{12}$

9. $\sqrt{60}$

10. $-\sqrt{78}$

11. $\sqrt{25}$

12. $\sqrt{19}$

13. $\sqrt{16}$

14. $\sqrt{43}$

15. $\sqrt{84}$

16. $\sqrt{64}$

**Approximate to the tenths' place, using the divide-and-average method.**

17. $\sqrt{24}$

18. $\sqrt{30}$

19. $\sqrt{92}$

20. $\sqrt{74}$

21. $\sqrt{51}$

22. $\sqrt{29}$

23. $\sqrt{10}$

24. $\sqrt{86}$

25. $\sqrt{37}$

26. $\sqrt{80}$

27. $\sqrt{5.6}$

28. $\sqrt{9.8}$

**For Exercises 29–53, refer to the table on page 508. Approximate the square root to the nearest hundredth.**

29. $\sqrt{8}$

30. $\sqrt{19}$

31. $\sqrt{5}$

32. $\sqrt{11}$

33. $\sqrt{24}$

34. $\sqrt{52}$

35. $\sqrt{39}$

36. $\sqrt{2}$

37. $\sqrt{31}$

38. $2\sqrt{18}$

39. $\sqrt{58}$

40. $5\sqrt{45}$

**Approximate the square root to the nearest hundredth by interpolation.**

41. $\sqrt{15.8}$

42. $\sqrt{6.3}$

43. $\sqrt{21.4}$

44. $\sqrt{83.6}$

**State whether or not a triangle with sides of the given lengths is a right triangle.**

45. 2, 3, 4

46. 3, 4, 5

47. 10, 20, 24

48. 8, 15, 17

49. 10, 12, 15

50. 6, 8, 10

51. 30, 40, 50

52. 7, 9, 12

53. 12, 16, 20

**A right triangle has sides of lengths $a$, $b$, and $c$, with $c$ the length of the hypotenuse. Find the length of the missing side. If necessary, use the table on page 508 for the square root values and round answers to the nearest hundredth.**

54. $a = 5, b = 8$

55. $a = 4, b = 5$

56. $a = 3, c = 5$

57. $b = 12, c = 15$

58. $b = 15, c = 17$

59. $a = 7, b = 7$

**60.** $a = 6$, $c = 10$           **61.** $b = 8$, $c = 17$          **62.** $a = 13$, $c = 14$

**Find the lengths marked $x$ and $y$. In each exercise, the triangles are similar.**

**63.**

 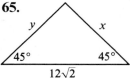

**64.**

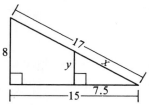

**Find the lengths marked $x$ and $y$ in the triangle. Give your answer in terms of radicals when radicals occur.**

**65.**

**66.**

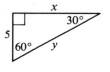

**67.**

**68.**

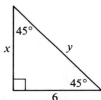

**Rewrite the expression in lowest terms with the radical in the numerator.**

**69.** $\dfrac{2}{\sqrt{3}}$       **70.** $\dfrac{5}{\sqrt{x}}$       **71.** $\dfrac{x}{\sqrt{2a}}$       **72.** $\dfrac{7}{\sqrt{4}}$

**Use the diagram to name each ratio.**

**73.** $\tan A$         **74.** $\sin A$

**75.** $\cos A$        **76.** $\tan B$

**77.** $\cos B$        **78.** $\sin B$

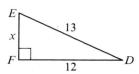

**For Exercises 79–88, use the tables on pages 508 and 509. Find $\sin A$, $\cos A$, and $\tan A$ for the given measure of $\angle A$.**

**79.** $27°$        **80.** $68°$        **81.** $89°$        **82.** $5°$

**Find the measure of $\angle A$ to the nearest degree.**

**83.** $\sin A = 0.436$        **84.** $\cos A = 0.224$        **85.** $\tan A = 1.35$

**Refer to the diagram at the right.**

**86.** Find the value of $x$.

**87.** Find $\angle D$ to the nearest degree.

**88.** Find $\angle E$ to the nearest degree.

# Extra Practice: Chapter 12

**Illustrate.**

1. Make a broken-line graph and a bar graph to illustrate the given data.

### Class Typing Speed

| Words per minute | 20 | 19 | 18 | 17 | 16 | 15 | 14 |
|---|---|---|---|---|---|---|---|
| Number of students | 1 | 5 | 1 | 12 | 6 | 4 | 1 |

2. During the baseball season, Tom hit 16 home runs, Fred hit 12, Jamie hit 10, Dale hit 6, and Corey hit 6. Illustrate the data using both a pictograph and a circle graph.

**Find the mean, the median, and the range of each set of data.**

3. 2, 4, 5, 8, 8, 10, 12

4. 6, 7, 7, 9, 11, 12, 12, 16

5. 1, 2, 3, 3, 3, 4, 4, 7, 8, 9

6. 310, 220, 300, 300, 240, 220, 300

7. 210, 250, 190, 180, 155

8. 1.4, 1.7, 2.7, 1.9, 2.1, 2.2

9. 8000, 6000, 3000, 8000, 7000, 1000

10. 1.9, 2.1, 2.1, 1.7, 1.5, 2.1, 1.5, 1.9

**Make a frequency table for the given data. Then find the mean, the median, the range, and the mode.**

11. 8, 9, 10, 11, 9, 12, 12, 16, 12, 14, 11, 10, 10, 11, 10

12. 9, 7, 9, 12, 10, 11, 5, 8, 8, 7, 12, 7, 11, 10

13. 4, 2, 1, 2, 4, 2, 5, 1, 3, 3, 4, 3, 5, 2, 1, 2, 5, 5

14. 16, 17, 15, 16, 17, 18, 18, 16, 15, 17

**Make a table, and arrange the data listed below into the following intervals: 4.5–34.5, 34.5–64.5, 64.5–94.5. Then use the table to make a histogram and a frequency polygon.**

15. 64 91 86 23 67

    16 42 63 25 46

    44 88 48 67 86

    68 25 46 86 67

    52 25 67 32 86

16. 61 84 88  8 17

    23 27 46 10 23

    65 42 65 46 65

    23 35 46 27 46

     5 42 27 46 84

**Find the value of each.**

**17.** 6!  **18.** 3!  **19.** 10!  **20.** 2!  **21.** 5!

**22.** In how many different ways can you arrange the letters in the word DIRECT if you take the letters 3 at a time?

**23.** $_4C_2$  **24.** $_{10}C_3$  **25.** $_6C_3$  **26.** $_{12}C_2$  **27.** $_{10}C_5$

**28.** How many combinations of 3 fish can you choose from 7 fish?

**29.** There are 52 basketball teams entered in a tournament. How many combinations can make it to the final game?

**A bag contains 3 green, 2 blue, 1 red, and 1 white marble. Find the probability for a marble chosen at random.**

**30.** $P(\text{green})$  **31.** $P(\text{not green})$  **32.** $P(\text{yellow})$

**33.** $P(\text{not yellow})$  **34.** $P(\text{green or red})$  **35.** $P(\text{red, white, or green})$

**Kate has 24 albums: 4 by the Deltas, 6 by the Squares, 6 by the Tuscon Band, 3 by Eliot Smith, 3 by the Marks, and 2 by the Deep River Quartet. She selects one at random.**

**36.** Find the odds in favor of selecting a record by the following.
  **a.** the Marks  **b.** the Deep River Quartet  **c.** the Squares

**37.** Find the odds against selecting a record by the following.
  **a.** Eliot Smith  **b.** the Tuscon Band  **c.** the Squares

**Two game cubes are rolled. Find the probability of each.**

**38. a.** The cubes show the same number.  **39. a.** The difference is 1.
  **b.** The sum is 3.  **b.** The sum is 12.
  **c.** The cubes show the same number  **c.** The difference is 1 or the sum is
  or the sum is 3.  12.

**40.** $P(A) = 0.40$, $P(B) = 0.60$, $P(A \text{ and } B) = 0.24$. Find $P(A \text{ or } B)$.

**41.** Find the probability that a month, chosen at random, begins with the letter J or has 30 days.

**42.** $P(A) = 0.25$, $P(B) = 0.20$, $P(A \text{ and } B) = 0.05$. Find $P(A \text{ or } B)$.

**43.** Two game cubes are rolled. Find the probability of each event.
  **a.** The cubes show the same number.
  **b.** The sum is 8.
  **c.** The cubes show the same number and the sum is 8.
  **d.** The cubes show the same number or the sum is 8.

# Table of Square Roots of Integers from 1 to 100

| Number | Positive Square Root | Number | Positive Square Root | Number | Positive Square Root | Number | Positive Square Root |
|---|---|---|---|---|---|---|---|
| $N$ | $\sqrt{N}$ | $N$ | $\sqrt{N}$ | $N$ | $\sqrt{N}$ | $N$ | $\sqrt{N}$ |
| 1 | 1 | 26 | 5.099 | 51 | 7.141 | 76 | 8.718 |
| 2 | 1.414 | 27 | 5.196 | 52 | 7.211 | 77 | 8.775 |
| 3 | 1.732 | 28 | 5.292 | 53 | 7.280 | 78 | 8.832 |
| 4 | 2 | 29 | 5.385 | 54 | 7.348 | 79 | 8.888 |
| 5 | 2.236 | 30 | 5.477 | 55 | 7.416 | 80 | 8.944 |
| 6 | 2.449 | 31 | 5.568 | 56 | 7.483 | 81 | 9 |
| 7 | 2.646 | 32 | 5.657 | 57 | 7.550 | 82 | 9.055 |
| 8 | 2.828 | 33 | 5.745 | 58 | 7.616 | 83 | 9.110 |
| 9 | 3 | 34 | 5.831 | 59 | 7.681 | 84 | 9.165 |
| 10 | 3.162 | 35 | 5.916 | 60 | 7.746 | 85 | 9.220 |
| 11 | 3.317 | 36 | 6 | 61 | 7.810 | 86 | 9.274 |
| 12 | 3.464 | 37 | 6.083 | 62 | 7.874 | 87 | 9.327 |
| 13 | 3.606 | 38 | 6.164 | 63 | 7.937 | 88 | 9.381 |
| 14 | 3.742 | 39 | 6.245 | 64 | 8 | 89 | 9.434 |
| 15 | 3.873 | 40 | 6.325 | 65 | 8.062 | 90 | 9.487 |
| 16 | 4 | 41 | 6.403 | 66 | 8.124 | 91 | 9.539 |
| 17 | 4.123 | 42 | 6.481 | 67 | 8.185 | 92 | 9.592 |
| 18 | 4.243 | 43 | 6.557 | 68 | 8.246 | 93 | 9.644 |
| 19 | 4.359 | 44 | 6.633 | 69 | 8.307 | 94 | 9.695 |
| 20 | 4.472 | 45 | 6.708 | 70 | 8.367 | 95 | 9.747 |
| 21 | 4.583 | 46 | 6.782 | 71 | 8.426 | 96 | 9.798 |
| 22 | 4.690 | 47 | 6.856 | 72 | 8.485 | 97 | 9.849 |
| 23 | 4.796 | 48 | 6.928 | 73 | 8.544 | 98 | 9.899 |
| 24 | 4.899 | 49 | 7 | 74 | 8.602 | 99 | 9.950 |
| 25 | 5 | 50 | 7.071 | 75 | 8.660 | 100 | 10 |

Exact square roots are shown in red. For the others, rational approximations are given correct to three decimal places.

# Table of Trigonometric Ratios

| Angle | Sine | Cosine | Tangent | Angle | Sine | Cosine | Tangent |
|-------|------|--------|---------|-------|------|--------|---------|
| 1° | .0175 | .9998 | .0175 | 46° | .7193 | .6947 | 1.0355 |
| 2° | .0349 | .9994 | .0349 | 47° | .7314 | .6820 | 1.0724 |
| 3° | .0523 | .9986 | .0524 | 48° | .7431 | .6691 | 1.1106 |
| 4° | .0698 | .9976 | .0699 | 49° | .7547 | .6561 | 1.1504 |
| 5° | .0872 | .9962 | .0875 | 50° | .7660 | .6428 | 1.1918 |
| 6° | .1045 | .9945 | .1051 | 51° | .7771 | .6293 | 1.2349 |
| 7° | .1219 | .9925 | .1228 | 52° | .7880 | .6157 | 1.2799 |
| 8° | .1392 | .9903 | .1405 | 53° | .7986 | .6018 | 1.3270 |
| 9° | .1564 | .9877 | .1584 | 54° | .8090 | .5878 | 1.3764 |
| 10° | .1736 | .9848 | .1763 | 55° | .8192 | .5736 | 1.4281 |
| 11° | .1908 | .9816 | .1944 | 56° | .8290 | .5592 | 1.4826 |
| 12° | .2079 | .9781 | .2126 | 57° | .8387 | .5446 | 1.5399 |
| 13° | .2250 | .9744 | .2309 | 58° | .8480 | .5299 | 1.6003 |
| 14° | .2419 | .9703 | .2493 | 59° | .8572 | .5150 | 1.6643 |
| 15° | .2588 | .9659 | .2679 | 60° | .8660 | .5000 | 1.7321 |
| 16° | .2756 | .9613 | .2867 | 61° | .8746 | .4848 | 1.8040 |
| 17° | .2924 | .9563 | .3057 | 62° | .8829 | .4695 | 1.8807 |
| 18° | .3090 | .9511 | .3249 | 63° | .8910 | .4540 | 1.9626 |
| 19° | .3256 | .9455 | .3443 | 64° | .8988 | .4384 | 2.0503 |
| 20° | .3420 | .9397 | .3640 | 65° | .9063 | .4226 | 2.1445 |
| 21° | .3584 | .9336 | .3839 | 66° | .9135 | .4067 | 2.2460 |
| 22° | .3746 | .9272 | .4040 | 67° | .9205 | .3907 | 2.3559 |
| 23° | .3907 | .9205 | .4245 | 68° | .9272 | .3746 | 2.4751 |
| 24° | .4067 | .9135 | .4452 | 69° | .9336 | .3584 | 2.6051 |
| 25° | .4226 | .9063 | .4663 | 70° | .9397 | .3420 | 2.7475 |
| 26° | .4384 | .8988 | .4877 | 71° | .9455 | .3256 | 2.9042 |
| 27° | .4540 | .8910 | .5095 | 72° | .9511 | .3090 | 3.0777 |
| 28° | .4695 | .8829 | .5317 | 73° | .9563 | .2924 | 3.2709 |
| 29° | .4848 | .8746 | .5543 | 74° | .9613 | .2756 | 3.4874 |
| 30° | .5000 | .8660 | .5774 | 75° | .9659 | .2588 | 3.7321 |
| 31° | .5150 | .8572 | .6009 | 76° | .9703 | .2419 | 4.0108 |
| 32° | .5299 | .8480 | .6249 | 77° | .9744 | .2250 | 4.3315 |
| 33° | .5446 | .8387 | .6494 | 78° | .9781 | .2079 | 4.7046 |
| 34° | .5592 | .8290 | .6745 | 79° | .9816 | .1908 | 5.1446 |
| 35° | .5736 | .8192 | .7002 | 80° | .9848 | .1736 | 5.6713 |
| 36° | .5878 | .8090 | .7265 | 81° | .9877 | .1564 | 6.3138 |
| 37° | .6018 | .7986 | .7536 | 82° | .9903 | .1392 | 7.1154 |
| 38° | .6157 | .7880 | .7813 | 83° | .9925 | .1219 | 8.1443 |
| 39° | .6293 | .7771 | .8098 | 84° | .9945 | .1045 | 9.5144 |
| 40° | .6428 | .7660 | .8391 | 85° | .9962 | .0872 | 11.4301 |
| 41° | .6561 | .7547 | .8693 | 86° | .9976 | .0698 | 14.3007 |
| 42° | .6691 | .7431 | .9004 | 87° | .9986 | .0523 | 19.0811 |
| 43° | .6820 | .7314 | .9325 | 88° | .9994 | .0349 | 28.6363 |
| 44° | .6947 | .7193 | .9657 | 89° | .9998 | .0175 | 57.2900 |
| 45° | .7071 | .7071 | 1.0000 | 90° | 1.0000 | .0000 | Undefined |

# Summary of Formulas

## Circumference

$C = \pi d$

$C = 2\pi r$

## Area

**Rectangle:** $A = lw$

**Triangle:** $A = \frac{1}{2}bh$

**Parallelogram:** $A = bh$

**Trapezoid:** $A = \frac{1}{2}(b_1 + b_2)h$

**Circle:** $A = \pi r^2$

## Volume

**Prism:** $V = Bh$

**Cylinder:** $V = \pi r^2 h$

**Pyramid:** $V = \frac{1}{3}Bh$

**Cone:** $V = \frac{1}{3}\pi r^2 h$

**Sphere:** $V = \frac{4}{3}\pi r^3$

## Lateral Area

**Prism:** lateral area = perimeter of base $\times$ height

**Cylinder:** lateral area = $2\pi rh$

## Surface Area

**Prism:** total surface area = lateral area + area of bases

**Cylinder:** total surface area = $2\pi rh + 2\pi r^2$

**Sphere:** $A = 4\pi r^2$

## Mass

Mass = Density $\times$ Volume

## Distance

distance = rate $\times$ time

## Percentage

percentage = rate $\times$ base

## Interest

Interest = Principal $\times$ rate $\times$ time

# Glossary

**Absolute value** (p. 46) The distance from 0 to the graph of a number on the number line.

**Acute angle** (p. 174) An angle with measure between 0° and 90°.

**Acute triangle** (p. 179) A triangle with three acute angles.

**Angle** (p. 173) A figure formed by two rays with a common endpoint.

**Annual rate of interest** (p. 262) A percent of the principal figured on a yearly basis.

**Arc** (p. 198) A part of a circle.

**Area** (p. 346) Amount of surface, measured in square units.

**Bar graph** (p. 434) A graph in which the length of each bar is proportional to the number it represents.

**Base of a geometric figure** (pp. 346, 351, 352, 364) A selected side or face.

**Base, numerical** (p. 9) A number that is raised to some power. In $5^2$, 5 is the base.

**Bisector** (p. 198) The line dividing a geometric figure into two congruent parts.

**Broken-line graph** (p. 434) A graph made by joining successive plotted points.

**Capacity** (p. 365) A measure of the volume of a container.

**Center** (pp. 188, 378) The point that is equidistant from all points on a circle or a sphere.

**Chord** (p. 188) A segment joining two points on a circle.

**Circle** (p. 188) A plane figure composed of all points measuring the same distance from a given point in the plane.

**Circle graph** (p. 260) A graph that uses the area of a circle to represent a sum of data. The area is divided into segments proportional to the data.

**Circumference** (p. 188) The perimeter of a circle.

**Collinear points** (p. 164) Two or more points that lie on the same line.

**Combination** (p. 458) An arrangement of a group of things in which order does not matter.

**Common denominator** (p. 100) A common multiple used as the denominator of two or more fractions that are equivalent to the given fractions.

**Common factor** (p. 97) A number that is a factor of two or more numbers.

**Common multiple** (p. 100) A number that is a multiple of two or more numbers.

**Compass** (p. 188) A tool used to draw a circle.

**Complementary angles** (p. 174) Two angles whose measures have a sum of 90°.

**Cone** (p. 369) A closed figure formed by a circular region and a curved surface that come to a point.

**Congruent figures** (pp. 192, 360) Figures that have the same size and shape.

**Coordinate** (p. 14) The number paired with a point on the number line.

**Coordinate plane** (p. 310) A plane marked with two perpendicular number lines, used to graph ordered pairs of numbers.

**Corresponding angles** (p. 192) The angles at matching vertices of congruent figures.

**Cosine of an angle** (p. 416) If $\angle A$ is one of the acute angles in a right triangle, the cosine of $\angle A$ is the ratio of the length of the side adjacent to $\angle A$ to the length of the hypotenuse.

**Counting numbers** (p. 89) The set of numbers 1, 2, 3, 4 . . . .

**Cross-multiplying** (p. 217) A method for solving and checking proportions.

**Cube** (p. 364) A rectangular prism having square faces.

**Cube root of a number** (p. 426) One of the three equal factors of a number.

**Cylinder** (p. 365) A geometric solid having two parallel bases that are congruent and one curved surface joining the bases.

**Data** (p. 434) Numerical information.

**Decimal system** (p. 13) The place-value numeration system that uses 10 as a base.

**Degree** (p. 173) Unit of angle measure.

**Density** (p. 382) The mass per unit volume of a substance.

**Diameter** (pp. 188, 378) A chord that contains the center of a circle or a sphere. Also, the length of such a chord.

**Dimensions** (p. 346) Length, width and height of a space figure.

**Endpoint** (p. 164) The point at the end of a line segment or ray.

**Equation** (p. 23) A mathematical sentence with an equals sign to indicate that two expressions name the same number.

**Equilateral triangle** (p. 179) A triangle in which all sides are congruent.

**Equivalent equations** (p. 130) Equations that have the same solution.

**Equivalent fractions** (p. 96) Fractions that name the same number.

**Equivalent inequalities** (p. 293) Inequalities that have the same solutions.

**Evaluate an expression** (p. 3) To replace variables in an expression with specified values and then complete the indicated arithmetic.

**Expanded form** (p. 13) The method of representing a number as the sum of products of each digit and powers of 10.

**Even number** (p. 84) Any multiple of 2.

**Exponent** (p. 9) A number indicating how many times the base is used as a factor.

**Extremes of a proportion** (p. 218) The first and last terms of a proportion.

**Factor** (p. 9) Any of two or more whole numbers that are multiplied to form a product.

**Fraction** (p. 97) An indicated quotient, for example $\frac{2}{5}$. The denominator, 5 in the example, tells the number of equal parts into which the whole has been divided. The numerator, 2, tells how many of these parts are being considered.

**Frequency distribution** (p. 446) A table that pairs each item in a set of data with its frequency.

**Frequency polygon** (p. 450) A line graph of frequencies connected to the horizontal axis at each end to form a polygon.

**Function** (p. 317) A set of ordered pairs in which no two different ordered pairs have the same $x$-coordinate.

**Geometric construction** (p. 198) A geometric drawing for which only a compass and a straightedge may be used.

**Graph of an equation** (p. 320) The line consisting of all points whose coordinates satisfy the equation.

**Graph of a number** (p. 14) The point on the number line paired with the number.

**Graphs** *See* Bar graph, Broken-line graph, Circle graph, Histogram, Pictograph.

**Greatest common factor (GCF)** (p. 97) The greatest whole number that is a factor of two or more given whole numbers.

**Grouping symbols** (p. 5) Symbols such as parentheses, ( ), and brackets, [ ], that are used to group expressions.

**Height** (p. 346) The perpendicular distance between the bases of a geometric figure. In triangles, cones, and pyramids, the perpendicular distance from the base to the opposite vertex.

**Histogram** (p. 450) A bar graph that shows a frequency distribution.

**Hypotenuse** (p. 402) The side opposite the right angle in a right triangle.

**Identity elements** (p. 19) 0 is the identity element for addition, because it can be added to any number without changing the value of the number. 1 is the identity element for multiplication, because it may be multiplied by any number without changing the value of the number.

**Improper fraction** (p. 98) A positive fraction whose numerator is greater than or equal to its denominator, or the opposite of such a fraction.

**Inequality** (p. 47) A mathematical sentence formed by placing an inequality sign between two expressions.

**Inscribed polygon** (p. 189) A polygon that has all of its vertices on the circle.

**Integers** (p. 46) The whole numbers and their opposites: . . . , −2, −1, 0, 1, 2, . . . .

**Interest** (p. 262) The amount of money paid for the use of money.

**Interpolation** (p. 399) A method of approximation.

**Inverse operations** (p. 26) Operations that undo each other. Addition and subtraction are inverse operations, as are multiplication and division.

**Irrational number** (p. 119) All real numbers that are not rational.

**Isosceles triangle** (p. 179) A triangle with at least two sides congruent.

**Lateral area** (p. 373) The surface area of a solid, not including the bases.

**Least common denominator (LCD)** (p. 100) The least common multiple of two or more denominators.

**Least common multiple (LCM)** (p. 100) The least number that is a multiple of two or more nonzero numbers.

**Legs of a right triangle** (p. 402) The two sides forming the right angle.

**Like terms** (p. 282) Terms in which the variable parts are the same.

**Line** (p. 164) A figure determined by two points and extending in both directions without end.

**Line segment** (p. 164) Two points on a line and all the points between them.

**Linear equation in two variables** (p. 321) An equation with two variables that can be written in the form $ax + by = c$, where $a$ and $b$ are both not 0.

**Lowest terms** (p. 97) A fraction is in lowest terms when the numerator and the denominator have no common factor but 1.

**Mass** (p. 382) The measure of the amount of matter an object contains.

**Mean** (p. 443) The value found by dividing the sum of a group of numbers by the number of numbers in the group. Also called *average*.

**Means of a proportion** (p. 218) The second and third terms of a proportion.

**Median** (p. 443) The number that falls in the middle when data are listed from least to greatest. If the number of data is even, the median is the mean of the two middle items.

**Midpoint** (p. 169) The point of a segment that divides it into two congruent segments.

**Mixed number** (p. 98) A whole number plus a proper fraction.

**Mode** (p. 447) The number that occurs most often in a set of data.

**Multiple** (p. 84) A product of a given number and any whole number.

**Mutually exclusive events** (p. 470) Events that cannot both occur at the same time.

**Noncollinear points** (p. 164) Points not on the same line.

**Number line** (p. 14) A line on which consecutive integers are assigned to equally spaced points on the line in increasing order from left to right.

**Number sentence** (p. 23) An equation or inequality indicating the relationship between two mathematical expressions.

**Numerical coefficient** (p. 3) In an expression such as $3ab$, the number 3 is the numerical coefficient of $ab$.

**Numerical expression** (p. 2) An expression that names a number, such as $2 + 3$.

**Obtuse angle** (p. 174) An angle with measure between 90° and 180°.

**Obtuse triangle** (p. 179) A triangle that has one obtuse angle.

**Odd number** (p. 84) A whole number that is not a multiple of 2.

**Odds of an event** (p. 466) A ratio that compares the probability of an event occurring and the probability of the event not occurring.

**Open sentence** (p. 23) A mathematical sentence that contains one or more variables.

**Opposites** (p. 46) A pair of numbers such as −4 and 4.

**Ordered pair of numbers** (p. 312) A pair of numbers whose order is important.

**Origin** (pp. 14, 312) The graph of zero on a number line, or (0, 0) in a rectangular coordinate plane.

**Outcome** (p. 461) The result of an event.

**Parallel lines** (p. 165) Lines in the same plane that do not intersect.

**Parallel planes** (p. 165) Planes that do not intersect.

**Parallelogram** (p. 183) A quadrilateral with both pairs of opposite sides parallel.

**Percent** (p. 227) A ratio of a number to 100, shown by the symbol %.

**Percent of change** (p. 246) The amount of change divided by the original amount.

**Perfect square** (p. 394) A number whose square root is an integer.

**Perimeter** (p. 184) The distance around a plane figure.

**Permutation** (p. 453) An arrangement of a group of things in a particular order.

**Perpendicular bisector** (p. 198) The line that is perpendicular to a segment at its midpoint.

**Perpendicular lines** (p. 173) Two lines that intersect to form 90° angles.

**Pi** (p. 188) The ratio of the circumference of a circle to its diameter.

**Pictograph** (p. 438) A form of bar graph with the bars replaced by rows or columns of symbols.

**Plane** (p. 165) A flat surface extending infinitely in all directions.

**Point** (p. 164) The simplest figure in geometry representing an exact location.

**Polygon** (p. 183) A closed plane figure made up of line segments.

**Polyhedron** (p. 364) A three-dimensional figure formed of polygonal parts of planes.

**Power of a number** (pp. 9, 70) A product in which all the factors, except 1, are the same. For example, $2^3 = 2 \times 2 \times 2 = 8$, so 8 is the third power of 2.

**Prime factorization** (p. 90) An expression showing a positive integer as the product of prime factors.

**Prime number** (p. 89) A whole number greater than 1 that has only two whole number factors, itself and 1.

**Principal** (p. 262) An amount of money on which interest is paid.

**Prism** (p. 364) A polyhedron that has two parallel, congruent faces called bases. The other faces are parallelograms.

**Probability** (p. 461) The ratio of the number of outcomes favoring an event to the total number of possible outcomes.

**Proper fraction** (p. 98) A positive fraction whose numerator is less than its denominator, or the opposite of such a fraction.

**Proportion** (p. 217) An equation stating that two ratios are equal.

**Protractor** (p. 173) A device used to measure angles.

**Pyramid** (p. 369) A polyhedron that has a polygonal base and three or more triangular faces.

**Quadrilateral** (p. 183) A polygon with four sides.

**Radical** (p. 412) An expression such as $\sqrt{5}$ or $\sqrt{a}$; the radical sign symbol used to denote the square root of a number.

**Radius** (pp. 188, 356) A line segment joining any point on a circle or sphere to the center. Also, the length of that segment.

**Random variable** (p. 461) A variable whose value is determined by the outcome of a random experiment.

**Range** (p. 443) The difference between the greatest and the least numbers in a set of data.

**Rate** (p. 214) A ratio that compares quantities of different kinds of units.

**Ratio** (p. 210) An indicated quotient of two numbers.

**Ray** (p. 164) A part of a line with one endpoint.

**Real number** (p. 119) Any number that is either a rational number or an irrational number.

**Reciprocals** (p. 113) Two numbers whose product is 1.

**Rectangle** (p. 184) A quadrilateral with four right angles.

**Regular polygon** (p. 183) A polygon in which all sides are congruent and all angles are congruent.

**Relatively prime numbers** (p. 97) Two or more numbers that have no common factor but 1.

**Replacement set** (p. 23) The given set of numbers that a variable may represent.

**Rhombus** (p. 184) A parallelogram in which all sides are congruent.

**Right angle** (p. 174) An angle with measure 90°.

**Right triangle** (p. 179) A triangle with a right angle.

**Rigid motions** (p. 193) Motions such as rotation, translation, and reflection, used to move a figure to a new position without changing its shape or size.

**Rounding** (p. 14) A method of approximating a number.

**Scalene triangle** (p. 179) A triangle with no two sides congruent.

**Scientific notation** (p. 76) A method of expressing a number as the product of a power of 10 and a number between 1 and 10.

**Segment** (p. 164) *See* line segment.

**Semicircle** (p. 188) Half of a circle.

**Sides of an equation** (p. 23) The mathematical expressions to the right and to the left of the equals sign.

**Sides of a figure** (pp. 178, 183, 212) The rays that form an angle or the segments that form a polygon.

**Similar figures** (p. 196) Figures that have the same shape but not necessarily the same size.

**Simple form** (p. 106) A mixed number is in simple form if its fractional part is expressed in lowest terms.

**Simplify an expression** (p. 2) To replace an expression with its simplest name.

**Sine of an angle** (p. 416) If $\angle A$ is an acute angle of a right triangle, the sine of $\angle A$ is the ratio of the length of the side opposite $\angle A$ to the length of the hypotenuse.

**Skew lines** (p. 166) Two nonparallel lines that do not intersect.

**Slope of a line** (p. 328) The steepness of a line; that is, the ratio of the change in the $y$-coordinate to the change in the $x$-coordinate when moving from one point on a line to another point.

**Solid** (p. 364) An enclosed region of space bounded by planes.

**Solution** (pp. 23, 130) A value of a variable that makes an equation or inequality a true sentence.

**Solving a right triangle** (p. 420) The process of finding the measures of the sides and angles of a right triangle.

**Sphere** (p. 378) A figure in space made up of all points equidistant from a given point.

**Square** (p. 184) A rectangle with all four sides congruent.

**Square root of a number** (p. 394) One of the two equal factors of the number.

**Statistical measures** (p. 443) Measures including the range, mean, median, and mode used to analyze numerical data.

**Supplementary angles** (p. 174) Two angles whose measures have a sum of 180°.

**Surface area** (p. 373) The total area of a solid.

**Symmetric** (p. 360) A figure is symmetric with respect to a line if it can be folded on that line so that every point on one side coincides exactly with a point on the other side. A figure is symmetric with respect to a point $O$ if for each point $A$ on the figure there is a point $B$ on the figure for which $O$ is the midpoint of $AB$.

**System of equations** (p. 324) A set of two or more equations in the same variables.

**Tangent of an angle** (p. 416) If $\angle A$ is an acute angle of a right triangle, the tangent of $\angle A$ is the ratio of the length of the side opposite $\angle A$ to the length of the side adjacent to $\angle A$.

**Terms of an expression** (p. 3) The parts of a mathematical expression that are separated by a $+$ sign.

**Terms of a proportion** (p. 217) The numbers in a proportion.

**Transformation** (pp. 130, 282) Rewriting an equation or inequality as an equivalent equation or inequality.

**Trapezoid** (p. 183) A quadrilateral with exactly one pair of parallel sides.

**Triangle** (p. 178) A polygon with three sides.

**Trigonometric ratios** (p. 416) Any of the sine, cosine, or tangent ratios.

**Value of a variable** (p. 2) Any number that a variable represents.

**Variable** (p. 2) A symbol used to represent one or more numbers.

**Variable expression** (p. 2) A mathematical expression that contains a variable.

**Vertex of an angle** (p. 173) The common endpoint of two intersecting rays.

**Vertex of a polygon or polyhedron** (p. 178) The point at which two sides of a polygon or three or more edges of a polyhedron intersect.

**Volume** (p. 364) A measure of the space occupied by a solid.

**$x$-axis** (p. 312) The horizontal number line on a coordinate plane.

**$x$-coordinate** (p. 312) The first number in an ordered pair of numbers that designates the location of a point on the coordinate plane. Also called the *abscissa*.

**$y$-axis** (p. 312) The vertical number line on a coordinate plane.

**$y$-coordinate** (p. 312) The second number in an ordered pair of numbers that designates the location of a point on the coordinate plane. Also called the *ordinate*.

# Index

Profit, 254
Programs in BASIC, 122
  for determining probability, 469
  for finding averages, 449
  for finding lowest common multiple, 121
  for plotting linear equations, 338
  to produce Pythagorean triple, 425
Programming languages, 39, 122
  See also BASIC
Proper fraction, 98
Properties, 18
  addition and subtraction, 19, 282
    of zero, 19
  associative, 18
  commutative, 18, 63, 282
  for decimals, 118, 119
  distributive, 20, 282
  of 45° right triangles, 411
  multiplication and division, 19, 282
    of one, 19
    of zero, 19
  for positive and negative fractions, 92–93, 103, 118, 119
  of proportions, 218
  of straight line, 328
  of 30–60 degree right triangles, 412
Proportions, 208, 217, 220
  of lengths and widths, 224
  and percent, 257, 260
  in problem solving, 257
  property of, 218
  in scale drawings, 224
Protractor, 173
Pyramids, 344, 369
PYTHAGORAS, 402
Pythagorean Theorem, 402, 411
  converse of, 402
Pythagorean triple, 404

Quadrants I–IV, 313
Quadrilateral, 183
Quantities, graphing relationships between, 328–329
Quotients, 67

Radical, 412
Radius, 188, 356, 378
Random outcomes, of probability, 461
Range, 443, 474
Rates, 214, 257
  of interest, 262
  See also Ratios
Rational numbers, 82, 94, 113, 118–119, 397
Ratio(s), 208, 210
  the Golden Ratio, 239
  percent, 227
  in proportions, 217
  and rates, 214
  scale, 224
  slope as, 328
  of sides in a triangle, 406
  trigonometric, 416
Ray, 164, 199
Reading Mathematics, vii, 3, 60, 71, 111, 132, 144, 184, 268, 288, 312, 357, 370, 378, 399, 408, 446
Real number line. See Number line
Real numbers, 119, 397
Reciprocals, 113, 136
Rectangle, 184, 239, 346
Rectangular coordinate system, 312
  See also coordinate plane
Reflection, 194

Relatively prime, 97
Repeating decimal, 117, 118
Replacement set, 23, 130
Research activity, 275, 305
Review Exercises, 4, 8, 12, 17, 25, 32, 53, 58, 66, 69, 72, 88, 91, 95, 99, 133, 135, 146, 151, 168, 172, 182, 187, 191, 197, 213, 216, 219, 223, 230, 233, 249, 253, 256, 266, 284, 292, 296, 300, 315, 319, 327, 333, 350, 355, 359, 368, 372, 377, 381, 395, 401, 410, 415, 419, 437, 442, 445, 448, 457, 460, 465, 469
  See also Chapter Review, Cumulative Review

Rhombus, 184
Right angle, 174
Right triangles, 179, 352, 402, 407
  isosceles, 411

solving, 420
Rigid motions, 193–194
Root(s), 394, 396, 399, 426
  square, 394, 396, 399
Rotation, 194
Rounding, to whole numbers, 14
Rulers, metric, 169
Rules
  for divisibility, 85–86
  for exponents and powers of ten, 9, 70, 73
  for fractions
    adding and subtracting, 103, 106
    dividing, 113
    multiplying, 109
  for order of operations, 6
  for percentage
    expressing a fraction as a percent, 227, 232
    expressing a percent as a fraction, 228
    expressing a percent as a decimal, 231
    expressing a decimal as a percent, 231
  for positive and negative numbers, 54, 55, 59, 63, 64, 67
  for rounding decimal numbers, 14
  for scientific notation, 76
  See also Properties and Formulas
RUN command, 123, 203, 338

Scale, 46, 208, 224
  drawings, 224
Scientific notation, 76, 203
Segment, 164
  congruent, 169
  diagonal, 183
  radius, 188, 378
  of a sphere, 378
Self-Test A, 22, 62, 102, 142, 177, 226, 261, 290, 323, 363, 405, 452
Self-Test B, 37, 75, 121, 155, 201, 237, 273, 303, 337, 385, 424, 473
  See also Chapter Review, Cumulative Review, Review Exercises

## Credits

Mechanical art: ANCO/Boston. Cover concept: Kirchoff/Wohlberg, Inc., cover photograph: Balthazar Korab. Page 1, Gregory Heisler/Gamma-Liaison; 32, Milt & Joan Mann/The Marilyn Gartman Agency; 45, Ralph Wetmore/Photo Researchers, Inc.; 58, Clyde H. Smith/Peter Arnold, Inc.; 62, Gary Milburn/Tom Stack & Associates; 76, NASA; 83, © Paulo Bonino 1982/Photo Researchers, Inc.; 88, © Dan McCoy/Rainbow; 116, Greig Cranna; 129, NASA; 151, VANSCAN™ Thermogram by Daedalus Enterprises; 156, © Julie Houck 1983; 163, Grant Heilman Photography; 186, © S.L.O.T.S./ Taurus Photos; 190, NASA; 200, Rainbow; 202, Dr. J. Lorre/Photo Researchers, Inc.; 209, Peter Arnold, Inc.; 216, Benn Mitchell © 1981/The Image Bank; 221, Greig Cranna; 230, Brian Parker/Tom Stack & Associates; 236, © Peter Menzel/Stock, Boston; 238, Greig Cranna; 245, Langridge/McCoy/Rainbow; 255, © Dick Luria 1982/The Stock Shop; 272, © John Lee/The Image Bank; 274, Thomas Hovland/Grant Heilman Photography; 281, Courtesy of French National Railroads; 289, Hank Morgan/Rainbow; 304, German Information Center; 311, Photo Courtesy of LEXIDATA; 333, © Lou Jones 1981/The Image Bank; 345; Owen Franken/Stock, Boston; 359, © Rick Smolan/Stock, Boston; 380, © Van Bucher 1982/Photo Researchers, Inc.; 386, © Dick Davis 1972/Photo Researchers, Inc.; 393, © Jerry Wachter Photography/Focus on Sports; 401, Milt & Joan Mann/The Marilyn Gartman Agency; 415, © 1979 Stuart Cohen/Stock, Boston; 426, © C. B. Jones 1982/Taurus Photos; 433, Robert C. Fields/Animals Animals; 444, Frederic Lewis Inc.; 460, Mike Mazzaschi/Stock, Boston; 474, Russ Kinne/Photo Researchers, Inc.

# Answers to Selected Exercises

## 1 Introduction to Algebra

**PAGE 4 WRITTEN EXERCISES** **1.** 48
**3.** 105 **5.** 14.15 **7.** 4.5 **9.** 25 **11.** 4 **13.** 12
**15.** 27 **17.** 2 **19.** 47 **21.** 26 **23.** 512 **25.** 9
**27.** 24 **29.** 2.5 **31.** 112.5 **33.** 680.805
**35.** 3.15 **37.** 2

**PAGE 4 REVIEW EXERCISES** **1.** 51.87
**3.** 45.93 **5.** 0.27 **7.** 6.3

**PAGES 7–8 WRITTEN EXERCISES** **1.** 92
**3.** 51 **5.** 2 **7.** 126 **9.** 7 **11.** 36 **13.** 10
**15.** 70 **17.** 36 **19.** 196 **21.** 3 **23.** 1512
**25.** 91 **27.** 84 **29.** 0.5 **31.** 4.1 **33.** 5827.248
**35.** 14.6 **37.** $2x \times (y - 4) + 2x$
or $2x \times y - (4 + 2)x$
**39.** $x \times [y + (z \div 3 + 1)] - z$

**PAGE 8 REVIEW EXERCISES** **1.** 5 **3.** 6
**5.** 3 **7.** 168

**PAGE 8 CALCULATOR KEY-IN** **1.** 288
**3.** 16.5 **5.** 6048

**PAGES 11–12 WRITTEN EXERCISES** **1.** $5^6$
**3.** $8^9$ **5.** $7^7$ **7.** $4^6$ **9.** $10^1$ **11.** $10^6$
**13.** 1,000,000 **15.** 10,000 **17.** 16 **19.** 256
**21.** 400 **23.** 3375 **25.** 6400 **27.** 256 **29.** 1296
**31.** 4096 **33.** 125 **35.** 169 **37.** 400
**39.** 1,680,700 **41.** 0 **43.** 961 **45.** 256
**47.** 6912 **49.** 72,000 **51.** 0 **53.** 27 **55.** 90,000
**57.** 27 **59.** 218 **61.** 25 **63.** 225 **65.** 3375
**67.** 16

**PAGE 12 REVIEW EXERCISES** **1.** 60 **3.** 57
**5.** 30 **7.** 24

**PAGES 16–17 WRITTEN EXERCISES**
**1.** $(3 \times 10) + 8$ **3.** $(8 \times 1000) + (9 \times 10) + 1$
**5.** $(4 \times 0.1) + (7 \times 0.01)$
**7.** $(6 \times 0.01) + (3 \times 0.001)$
**9.** $(1 \times 0.1) + (8 \times 0.01) + (7 \times 0.001)$
**11.** 54.57 **13.** 9002.146 **15.** 7.43 **17.** 19.005
**19.** 0.0048 **21.** 6.025 **23.** 30 **25.** 290 **27.** 650
**29.** 160 **31.** 72.46 **33.** 0.06 **35.** 0.01 **37.** 18.17
**39.** 0.001 **41.** 401.090 **43.** 250.341
**45.** 8.100 **47.** $(5 \times 10^3) + (2 \times 10^2) + (8 \times 10^1)$
**49.** $(1 \times 10^2) + (8 \times 10^1) +$
$$(3 \times 10^0) + \left(8 \times \frac{1}{10^2}\right)$$
**51.** $\left(9 \times \frac{1}{10^2}\right) + \left(1 \times \frac{1}{10^3}\right)$

**53.** $(7 \times 10^0) + \left(4 \times \frac{1}{10^1}\right) + \left(8 \times \frac{1}{10^2}\right) +$
$$\left(2 \times \frac{1}{10^3}\right)$$
**55.** $(2 \times 10^2) + (4 \times 10^0) + \left(5 \times \frac{1}{10^1}\right)$
**57.** $(3 \times 10^1) + (8 \times 10^0) + \left(3 \times \frac{1}{10^3}\right)$ **59.** b
**61.** a **63.** b **65.** c

**PAGE 17 REVIEW EXERCISES** **1.** 9.4
**3.** 960 **5.** 60 **7.** 14.8

**PAGES 21–22 WRITTEN EXERCISES**
**1.** 14.6; associative **3.** 14.24; distributive
**5.** 18.97; commutative and associative
**7.** 0.21; commutative and associative
**9.** False **11.** True **13.** $n = 0$ **15.** $t = 3.2$
**17.** $w = 0$ **19.** $g = 15.9$ **21.** $n = 3$
**23.** $d = 1565$ **25.** $t = 0$ **27.** 3.88 **29.** 400
**31.** 900 **33.** 32 For exercises 35 and 37,
answers will vary. An example is given.
**35.** $a = 2, b = 3, c = 4$
**37.** $a = 2, b = 4, c = 6$

**PAGE 22 SELF-TEST A** **1.** 46 **2.** 15 **3.** 168
**4.** 9 **5.** 4 **6.** 6 **7.** 16 **8.** 512 **9.** 9
**10.** 100,000 **11.** 100,000 **12.** 80 **13.** 3.18
**14.** 300 **15.** 102 **16.** 93 **17.** 21 **18.** 67.96

**PAGES 24–25 WRITTEN EXERCISES**
**1.** True **3.** False **5.** False **7.** True **9.** True
**11.** True **13.** 29 **15.** 8 **17.** No solution
**19.** No solution **21.** 1518 **23.** 1.26 **25.** 11
**27.** 12 **29.** 9 **31.** 2 **33.** No solution **35.** 4
For exercises 37 and 39, answers will vary. An
example is given. **37.** $3x + 5 = 35$
**39.** $(x + 20) \div 6 - 10 = 10$ **41.** $+, -$
**43.** $+, \div$ **45.** $\div, -, \times$ **47.** $-, +, \div$

**PAGE 25 REVIEW EXERCISES** **1.** 11.67
**3.** 27.14 **5.** 7.337 **7.** 2.84

**PAGE 28 WRITTEN EXERCISES**
**1.** $x = 15 - 8$; 7 **3.** $f = 74 - 38$; 36
**5.** $y = 14 + 9$; 23 **7.** $b = 32 + 25$; 57
**9.** $c = 27 \div 3$; 9 **11.** $m = 108 \div 9$; 12
**13.** $n = 9 \times 6$; 54 **15.** $g = 7 \times 16$; 112
**17.** $a = 17 - 17$; 0 **19.** $n = 42 \div 14$; 3
**21.** $h = 297 \times 11$; 3267 **23.** $d = 110 + 87$; 197
**25.** $q = 465 \div 31$; 15 **27.** $p = 358 - 208$; 150
**29.** $c = 536 - 511$; 25 **31.** $g = 401 + 19$; 420
**33.** $x = 31.9 - 1.6$; 30.3 **35.** $c = 2.56 \div 1.6$;
1.6 **37.** 6 **39.** 7 **41.** 13 **43.** 17 **45.** 24

**47.** 120 **49.** 0.54 **51.** 3.3 **53.** 12.84
**55.** ×, + **57.** ×, − **59.** ÷, +

**PAGE 28 REVIEW EXERCISES** **1.** 2 **3.** 52
**5.** 60 **7.** 31

**PAGES 31–32 PROBLEMS** **1.** 11.65 m
**3.** $41,520 **5.** $7.20 **7.** 936 people **9.** $1596
**11.** $4.25

**PAGE 32 REVIEW EXERCISES** **1.** 340
**3.** 175 **5.** 500 **7.** 400

**PAGES 35–37 PROBLEMS** **1.** $113.70
**3.** $24.79 **5.** 22 bonus points **7.** smaller:
4263 square ft; larger: 16,620 square ft
**9.** 41 points **11.** $488.25 **13.** 12,948 shares
**15.** $3.90

**PAGE 37 SELF-TEST B** **1.** 29 **2.** 8 **3.** No
solution **4.** 3 **5.** 44 **6.** 16 **7.** 9 **8.** $114.15
**9.** 49 passengers **10.** $182.24

**PAGE 40 CHAPTER REVIEW** **1.** F **3.** A
**5.** D **7.** False **9.** True **11.** True **13.** False
**15.** False **17.** False **19.** False **21.** d

**PAGE 42 CUMULATIVE REVIEW
EXERCISES** **1.** 809 **3.** 3.1 **5.** 351.9 **7.** 27
**9.** 25 **11.** 2 **13.** 43 **15.** 16.5 **17.** 6 **19.** 27
**21.** 64 **23.** 15,625 **25.** a **27.** a **29.** False
**31.** False **33.** 7 **35.** 3 **37.** 19 **39.** 2 **41.** 13
**43.** 2

**PAGE 43 CUMULATIVE REVIEW
PROBLEMS** **1.** $3010 **3.** $42.95 **5.** $20
**7.** $17

# 2 Positive and Negative Numbers

**PAGES 48–49 WRITTEN EXERCISES**

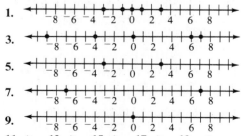

**11.** > **13.** > **15.** > **17.** > **19.** <
**21.** > **23.** > **25.** > **27.** <
**29.** 18 < 32 < 46, or 46 > 32 > 18
**31.** 103 < 130 < 310, or 310 > 130 > 103
**33.** 689 < 698 < 986, or 986 > 698 > 689

**35.** 3 **37.** 0 **39.** 9 **41.** 8 **43.** ⁻15, ⁻2, 0, 6
**45.** ⁻12, ⁻8, ⁻1, 1, 7 **47.** ⁻14, ⁻10, 4, 8, 14
**49.** ⁻6.4, ⁻2.7, 0.6, 3.1
**51.** 6, ⁻6

**53.** 4, ⁻4

**55.** 0

**57.** ⁻1, 0, 1

**59.** ⁻4, ⁻3, ⁻2, ⁻1, 0, 1, 2, 3, 4

**61.** ⁻7, ⁻6, ⁻5, ⁻4, ⁻3, 3, 4, 5, 6, 7

**63.** positive **65.** No number has a negative
absolute value.

**PAGE 49 REVIEW EXERCISES**

**PAGES 52–53 WRITTEN EXERCISES**

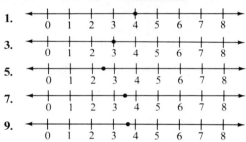

**11.**

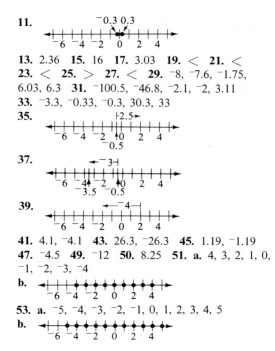

**13.** 2.36 **15.** 16 **17.** 3.03 **19.** $<$ **21.** $<$
**23.** $<$ **25.** $>$ **27.** $<$ **29.** ⁻8, ⁻7.6, ⁻1.75,
6.03, 6.3 **31.** ⁻100.5, ⁻46.8, ⁻2.1, ⁻2, 3.11
**33.** ⁻3.3, ⁻0.33, ⁻0.3, 30.3, 33
**35.**

**37.**

**39.**

**41.** 4.1, ⁻4.1 **43.** 26.3, ⁻26.3 **45.** 1.19, ⁻1.19
**47.** ⁻4.5 **49.** ⁻12 **50.** 8.25 **51. a.** 4, 3, 2, 1, 0,
⁻1, ⁻2, ⁻3, ⁻4
**b.**

**53. a.** ⁻5, ⁻4, ⁻3, ⁻2, ⁻1, 0, 1, 2, 3, 4, 5
**b.**

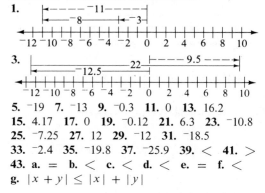

**PAGE 53 REVIEW EXERCISES 1.** 11.14;
distributive **3.** 27.18; multiplicative property
of one **5.** 132.3; distributive **7.** 1075;
commutative and associative **9.** 0;
multiplicative property of zero

**PAGE 57 WRITTEN EXERCISES**
**1.**

**3.**

**5.** ⁻19 **7.** ⁻13 **9.** ⁻0.3 **11.** 0 **13.** 16.2
**15.** 4.17 **17.** 0 **19.** ⁻0.12 **21.** 6.3 **23.** ⁻10.8
**25.** ⁻7.25 **27.** 12 **29.** ⁻12 **31.** ⁻18.5
**33.** ⁻2.4 **35.** ⁻19.8 **37.** ⁻25.9 **39.** $<$ **41.** $>$
**43. a.** $=$ **b.** $<$ **c.** $<$ **d.** $<$ **e.** $=$ **f.** $<$
**g.** $|x + y| \leq |x| + |y|$

**PAGE 58 PROBLEMS 1.** ⁻7 + 13 + ⁻3;
3°C **3.** 3.5 + ⁻11 + 2; lost; 5.5 yd lost
**5.** ⁻4.30 + 2.50 + 2.60; gain

**PAGE 58 REVIEW EXERCISES 1.** 156.1
**3.** 64.9 **5.** 44.89 **7.** 4.2 **9.** 285

**PAGES 60–61 WRITTEN EXERCISES**
**1.** 7 + (−19) **3.** 6.2 + 8.3 **5.** − 12 **7.** 14.5
**9.** −6 **11.** −27 **13.** 42 **15.** 15 **17.** −43
**19.** 20 **21.** 44 **23.** −1.1 **25.** 2.2 **27.** −24.6
**29.** −24.6 **31.** 3.37 **33.** 24.8 **35.** −41.3
**37.** 1.8 **39.** 4.5 **41.** −6.2 **43.** 1.7 **45.** 10.7
**47.** −10.7 **49.** 1.7 **51.** 8 **53.** 8 **55.** −19
**57.** −18 **59.** 7 **61.** −8 **63. a.** $=$ **b.** $>$
**c.** $>$ **d.** $>$ **e.** $|x − y| \geq |x| − |y|$

**PAGES 61–62 PROBLEMS 1.** −3 − (−11);
8°C change **3.** 1266 − 1455; 229 m
**5.** 18.75 − 1.00 − 1.75 − 1.50; $14.50
**7.** 78 − 5.8 − 7.5 − 12; 52.7 cm **9.** $44.22

**PAGE 62 SELF-TEST A 1.** $<$ **2.** $>$ **3.** $<$
**4.** $=$ **5.** $=$ **6.** ⁻54, ⁻4.52, ⁻0.25, 0, 5.4
**7.** ⁻7.3, ⁻3.79, ⁻0.37, ⁻0.09, 37 **8.** 33 **9.** 0
**10.** ⁻1.44 **11.** 35 **12.** 0 **13.** − 36 **14.** 11.6
**15.** − 11.6 **16.** − 11.6

**PAGES 65–66 WRITTEN EXERCISES**
**1.** 27 **3.** − 56 **5.** 96 **7.** 0 **9.** − 12
**11.** 5.1 **13.** − 2.9 **15.** 15.4 **17.** − 5 **19.** 459
**21.** − 66.69 **23.** 0 **25.** 3.42 **27.** − 111.24
**29.** 37.875 **31.** 61.38 **33.** − 1.054 **35.** 0
**37.** 361.9 **39.** − 40.81 **41.** − 8.2 **43.** − 100
**45.** − 2 **47.** − 2 **49.** 0 **51.** 2 **53.** − 30
**55.** 0

**PAGE 66 REVIEW EXERCISES 1.** 56
**3.** 154 **5.** 13 **7.** 26

**PAGE 66 CALCULATOR KEY–IN** 8192

**PAGES 68–69 WRITTEN EXERCISES**
**1.** − 6 **3.** 3 **5.** 0 **7.** − 12 **9.** − 7.5 **11.** 3.6
**13.** − 35 **15.** − 5.3 **17.** 8 **19.** 0 **21.** − 0.5
**23.** − 9.5 **25.** − 2.2 **27.** 13.8 **29.** 0.1545
**31.** − 6.3 **33.** 4.7 **35.** − 1.85 **37.** − 3.2
**39.** 0.625 **41.** − 4.2 **43.** − 1.2 **45.** − 2
**47.** − 72 **49.** − 8

**PAGE 69 REVIEW EXERCISES 1.** 49
**3.** 784 **5.** 112 **7.** 1 **9.** 37

**PAGE 69 CALCULATOR KEY–IN 1.** 1.1
**3.** − 28.201 **5.** 5101.015

**PAGES 71–72 WRITTEN EXERCISES**
**1.** 64 **3.** 100 **5.** 14 **7.** 225 **9.** 1 **11.** $3^2$
**13.** $9^2$ **15.** $(16 \times 4)^2$ **17.** $2^7$ **19.** $10^5$ **21.** $n^{11}$
**23.** 100 **25.** 81 **27.** 576 **29.** 1 **31.** 1875

**33.** 225 **35.** 1 **37.** 1000 **39.** 96 **41.** 52
**43.** 4 **45.** $\frac{1}{8}$ **47.** 1

PAGE 72 REVIEW EXERCISES **1.** $-2$
**3.** 2 **5.** 12 **7.** 1

PAGE 72 CALCULATOR KEY–IN **1.** 56,
56 **3.** 3243, 3243 $\qquad a^2 - b^2, (a + b)(a - b)$

PAGE 74 WRITTEN EXERCISES **1.** $-\frac{1}{32}$
**3.** $\frac{1}{1000}$ **5.** 1 **7.** $\frac{1}{25}$ **9.** $\frac{1}{64}$ **11.** $\frac{1}{49}$ **13.** 1
**15.** $\frac{1}{27}$ **17.** $\frac{1}{256}$ **19.** $-\frac{1}{8}$ **21.** 1 **23.** $-3$
**25.** 5 **27.** $-2$ **29.** $-2$ **31.** 8 **33.** 0 **35.** 1
**37.** $-4$ **39.** $-3$ **41.** $\frac{1}{x^5}$ **43.** $\frac{1}{a^5}$ **45.** $\frac{1}{w^8}$

PAGE 75 SELF-TEST B **1.** $-47.46$
**2.** $-136.68$ **3.** 378 **4.** $-11$ **5.** $-31$ **6.** 8
**7.** 8 **8.** 225 **9.** 1296 **10.** $\frac{1}{16}$ **11.** $-\frac{1}{216}$
**12.** $\frac{1}{343}$ **13.** $\frac{1}{81}$

PAGES 76–77 ENRICHMENT
**1.** $5.798 \times 10^3$ **3.** $8.915673 \times 10^6$ **5.** 1.75
**7.** $5.01 \times 10^{-2}$ **9.** 3790 **11.** 301,000
**13.** 0.056 **15.** 0.0000000399 **17.** $7.4 \times 10^{-4}$
**19.** $1.5 \times 10^{11}$

PAGE 78 CHAPTER REVIEW **1.** $<$ **3.** $>$
**5.** $=$ **7.** $<$ **9.** $>$ **11.** $>$ **13.** False
**15.** True **17.** True **19.** False **21.** True
**23.** False **25.** False **27.** True **29.** True
**31.** True **33.** True **35.** True **37.** c **39.** d
**41.** d

PAGES 80–81 CUMULATIVE REVIEW
EXERCISES **1.** 6 **3.** 180 **5.** 3 **7.** 17
**9.** True **11.** True **13.** False **15.** False
**17.** True **19.** ⁻6.18 **21.** $-8$ **23.** 4.3
**25.** $-0.12$ **27.** $-9$ **29.** $-1.6$ **31.** 4.009
**33.** 7700 **35.** 909.1 **37.** 103.6 **39.** $10^{10}$ or
10,000,000,000 **41.** 64 **43.** 6 **45.** 100
**47.** $-20, -7, 0, 6, 12, 15$ **49.** $-5, -3, 0, 1, 7,$
9

PAGE 81 CUMULATIVE REVIEW
PROBLEMS **1.** $59.46 **3.** 4 times **5.** Deli
Delights

# 3 Rational Numbers

PAGES 87–88 WRITTEN EXERCISES **1.** 1,
2, 3, 6, 7, 14, 21, 42 **3.** 1, 2, 4, 8, 16, 32 **5.** 1,
2, 4, 7, 8, 14, 28, 56 **7.** 1, 2, 3, 4, 6, 7, 12, 14,
21, 28, 42, 84 **9.** 1, 41 **11.** 2, 3, 4 **13.** 3, 5
**15.** 3, 9 **17.** 2, 3, 4, 5, 10 **19.** none

**21. a.** yes **b.** no **c.** no **23. a.** yes **b.** yes
**c.** yes **25. a.** yes **b.** yes **c.** yes **27.** 4
**29.** 5 **31.** 9 **33.** if number represented by last
three digits is a multiple of 8 **35.** if last two
digits are 00, 25, 50, or 75

PAGE 88 REVIEW EXERCISES **1.** $<$
**3.** $>$ **5.** $<$ **7.** $>$

PAGE 91 WRITTEN EXERCISES
**1.** composite **3.** composite **5.** composite
**7.** prime **9.** composite **11.** composite
**13.** $2^2 \cdot 3$ **15.** $2^3 \cdot 3$ **17.** $3 \cdot 13$ **19.** $2 \cdot 3 \cdot 11$
**21.** $2 \cdot 3^3$ **23.** $2^2 \cdot 3 \cdot 7$ **25.** $2^2 \cdot 7^2$
**27.** $2^2 \cdot 7 \cdot 11$ **29.** $2 \cdot 3 \cdot 19$ **31.** All other even
numbers are divisible by 2. **33.** All prime
numbers greater than 2 are odd numbers. The
sum of two odd numbers is an even number.
**35.** 1, 3, 7, 9 **37.** 1001; the three digit number

PAGE 91 REVIEW EXERCISES **1.** 2
**3.** $-1$ **5.** 240 **7.** 2

PAGE 94–95 WRITTEN EXERCISES
**1.**
**3.**
**5.**
**7.** $\frac{-1}{2}, \frac{1}{-2}$ **9.** $\frac{1}{-11}, -\frac{1}{11}$ **11.** $\frac{-2}{9}, -\frac{2}{9}$
**13.** $\frac{-13}{6}, \frac{13}{-6}$ **15.** $\frac{-3}{4}, \frac{3}{-4}$ **17.** 6 **19.** $\frac{1}{4}$
**21.** 3 **23.** $-3$ **25.** $-\frac{7}{8}$ **27.** $-1$ **29.** $-\frac{1}{3}$
**31.** $-3$ **33.** $-9$ **35.** $\frac{7}{6}$ **37.** $\frac{3}{7}$ **39.** $-\frac{1}{9}$
**41.** $-\frac{2}{7}$ **43.** $-\frac{3}{8}$

PAGE 95 REVIEW EXERCISES **1.** 3, 4, 8,
18, 36 **3.** 1, 5, 6, 10, 15, 30, 45 **5.** 6, $x$

PAGE 99 WRITTEN EXERCISES **1.** 4
**3.** 5 **5.** 7 **7.** 3 **9.** 125 **11.** $\frac{3}{5}$ **13.** $2\frac{3}{4}$
**15.** $-\frac{1}{9}$ **17.** $-5\frac{2}{3}$ **19.** $-\frac{2}{5}$ **21.** $10\frac{5}{12}$
**23.** $-6\frac{1}{13}$ **25.** $-42\frac{6}{7}$ **27.** $\frac{17}{8}$ **29.** $\frac{99}{16}$
**31.** $-\frac{35}{8}$ **33.** $-\frac{49}{3}$ **35.** $-\frac{83}{10}$ **37.** 3 **39.** $-1$
**41.** 0 **43.** 24 **45.** $\frac{b}{5}$ **47.** $\frac{d}{e}$ **49.** $\frac{n}{3}$ **51.** $\frac{v}{3u}$

PAGE 99 REVIEW EXERCISES **1.** 49
**3.** 1296 **5.** 1 **7.** 5184 **9.** 243

PAGES 101–102 WRITTEN EXERCISES
**1.** $\frac{4}{12}, \frac{1}{12}$ **3.** $\frac{6}{8}, \frac{5}{8}$ **5.** $\frac{6}{27}, -\frac{1}{27}$ **7.** $\frac{70}{147}, \frac{6}{147}$
**9.** $-\frac{20}{30}, \frac{7}{30}$ **11.** $\frac{16}{300}, \frac{21}{300}$ **13.** $-\frac{48}{112}, -\frac{7}{112}$
**15.** $\frac{35}{294}, \frac{30}{294}$ **17.** $\frac{70}{80}, \frac{25}{80}, \frac{42}{80}$ **19.** $\frac{108}{252}, \frac{441}{252},$
$\frac{112}{252}$ **21.** $\frac{2}{130}, \frac{78}{130}, \frac{45}{130}$ **23.** $-\frac{60}{72}, \frac{16}{72}, -\frac{63}{72}$
**25.** $\frac{2a}{6}, \frac{b}{6}$ **27.** $\frac{20h}{500}, \frac{5h}{500}, \frac{4h}{500}$ **29.** $\frac{3}{3c}, \frac{2}{3c}$
**31.** $\frac{3yz}{xyz}, \frac{xz}{xyz}, \frac{5xy}{xyz}$ **33.** $<$ **35.** $<$ **37.** $>$

PAGE 102 SELF-TEST A **1.** 2, 3, 4, 9 **2.** 3,
9 **3.** 2, 3, 4, 5, 9, 10 **4.** none **5.** composite;
$2^2 \cdot 3^3$ **6.** prime **7.** composite; $3 \cdot 29$
**8.** prime **9.** $-\frac{1}{3}$ **10.** $\frac{7}{9}$ **11.** $\frac{2}{-3}$ **12.** $\frac{1}{4}$
**13.** $-2\frac{2}{5}$ **14.** $\frac{16}{21}$ **15.** $-1\frac{23}{48}$ **16.** $4\frac{1}{4}$ **17.** $\frac{11}{5}$
**18.** $-\frac{11}{3}$ **19.** $\frac{94}{15}$ **20.** $\frac{19}{12}$ **21.** $-\frac{69}{8}$ **22.** $\frac{28}{40},$
$\frac{35}{40}$ **23.** $-\frac{30}{147}, \frac{14}{147}$ **24.** $\frac{27}{450}, \frac{12}{450}$ **25.** $-\frac{16}{30},$
$-\frac{1}{30}$

PAGES 104–105 WRITTEN EXERCISES
**1.** $\frac{2}{3}$ **3.** $-\frac{4}{17}$ **5.** 0 **7.** $-\frac{11}{5}$ **9.** $\frac{16}{21}$ **11.** $\frac{7}{2}$
**13.** $-\frac{3}{8}$ **15.** $-\frac{51}{10}$ **17.** $\frac{287}{156}$ **19.** $-\frac{41}{72}$
**21.** $\frac{7}{12}$ **23.** $\frac{3}{5}$ **25.** $-\frac{17}{18}$ **27.** $\frac{19}{12}$ **29.** $\frac{1}{10}$

PAGE 105 REVIEW EXERCISES **1.** $1\frac{3}{8}$
**3.** $3\frac{2}{7}$ **5.** $5\frac{11}{12}$ **7.** $9\frac{4}{13}$

PAGES 107–108 WRITTEN EXERCISES
**1.** $8\frac{4}{5}$ **3.** $\frac{10}{11}$ **5.** $6\frac{2}{3}$ **7.** $-20\frac{1}{7}$ **9.** $3\frac{5}{6}$
**11.** $2\frac{1}{2}$ **13.** $-2\frac{5}{24}$ **15.** $-3\frac{17}{30}$ **17.** $5\frac{3}{4}$
**19.** $-3\frac{7}{8}$ **21.** $11\frac{2}{3}$ **23.** $-19\frac{1}{5}$ **25.** $9\frac{11}{12}$
**27.** $-8\frac{1}{4}$ **29.** $-8\frac{2}{7}$ **31.** 4 **33.** $\frac{2}{15}$ **35.** $\frac{1}{4}$

PAGE 108 PROBLEMS **1.** $14\frac{11}{16}$ yd **3.** $\frac{5}{6}$ mi

**5.** $1\frac{5}{8}$ in. **7.** right : $1\frac{5}{16}$ in.; bottom : $\frac{11}{16}$ in.

PAGE 108 REVIEW EXERCISES **1.** $\frac{2}{-7}$
**3.** $\frac{5}{-9}$ **5.** $-\frac{1}{7}$ **7.** 4 **9.** 3

PAGES 111–112 WRITTEN EXERCISES
**1.** 7 **3.** $-2$ **5.** $\frac{1}{28}$ **7.** $-\frac{1}{60}$ **9.** $\frac{10}{27}$ **11.** $-\frac{1}{4}$
**13.** $-\frac{2}{27}$ **15.** $\frac{5}{8}$ **17.** 0 **19.** $\frac{15}{64}$ **21.** 1
**23.** $43\frac{11}{12}$ **25.** $-22\frac{1}{7}$ **27.** $3\frac{10}{27}$ **29.** $\frac{1}{24}$
**31.** $\frac{1}{72}$ **33.** $-4\frac{1}{2}$ **35.** $6n$ **37.** $-\frac{y}{2}$ **39.** $\frac{2}{5}$
**41.** 3

PAGE 112 REVIEW EXERCISES **1.** 60
**3.** $-36$ **5.** 8 **7.** $-3\frac{1}{9}$

PAGE 115 WRITTEN EXERCISES **1.** $\frac{2}{3}$
**3.** 25 **5.** $\frac{25}{72}$ **7.** $3\frac{8}{9}$ **9.** 6 **11.** $5\frac{1}{5}$ **13.** $-\frac{9}{16}$
**15.** 7 **17.** $5\frac{23}{64}$ **19.** $-2$ **21.** $-9$ **23.** $\frac{1}{3}$
**25.** $-\frac{1}{2}$

PAGES 115–116 PROBLEMS **1.** $\frac{1}{19}$
**3.** \$7,500 **5.** \$4 per hour **7.** \$6.50 **9.** $\frac{1}{6}$ of
the class **11.** $\frac{3}{20}$ of the workers

PAGE 116 REVIEW EXERCISES
**1.** $n = 240 - 135$; 105 **3.** $y = 156 \div 4$; 39
**5.** $x = 432 \div 12$; 36 **7.** $m = 418 \div 38$; 11

PAGE 120 WRITTEN EXERCISES **1.** 0.25
**3.** $0.\overline{2}$ **5.** 0.9 **7.** $-0.\overline{6}$ **9.** $-0.375$
**11.** $-0.12$ **13.** 1.1 **15.** $0.58\overline{3}$ **17.** $0.2\overline{6}$
**19.** $-1.3\overline{8}$ **21.** 0.15 **23.** $3.\overline{285714}$ **25.** $\frac{1}{20}$
**27.** $-\frac{3}{5}$ **29.** $2\frac{7}{100}$ **31.** $5\frac{1}{8}$ **33.** $-1\frac{3}{8}$
**35.** $12\frac{5}{8}$ **37.** $\frac{9}{40}$ **39.** $-1\frac{413}{500}$ **41.** $-\frac{5}{9}$
**43.** $-1\frac{1}{90}$ **45.** $\frac{5}{33}$ **47.** $\frac{35}{99}$ **49.** $-1\frac{4}{33}$
**51.** $2\frac{121}{900}$ **53.** $\frac{41}{333}$ **55.** rational **57.** rational
**59.** 3.0, 3.00$\overline{9}$, 3.0$\overline{9}$, 3.1 **61. a.** 1 **b.** $=$
**63. a.** $-1\frac{1}{4}$ **b.** $=$

PAGE 121 SELF-TEST B  **1.** $\frac{7}{12}$  **2.** $-\frac{21}{30}$
**3.** $-\frac{2}{15}$  **4.** $22\frac{4}{9}$  **5.** $-10\frac{1}{24}$  **6.** $17\frac{3}{8}$  **7.** $3\frac{3}{4}$
**8.** $-\frac{1}{24}$  **9.** $-8\frac{1}{7}$  **10.** $1\frac{1}{2}$  **11.** $-\frac{1}{8}$
**12.** $-4\frac{1}{2}$  **13.** 0.625  **14.** $0.\overline{18}$  **15.** $-0.0125$
**16.** $1.1\overline{6}$  **17.** $\frac{7}{8}$  **18.** $1\frac{2}{3}$  **19.** $-2\frac{213}{1000}$  **20.** $\frac{7}{30}$

PAGE 121 COMPUTER BYTE  **1.** 300
**3.** 510  **5.** 2640

PAGE 123 ENRICHMENT  **1.** Answers will vary.

PAGE 124 CHAPTER REVIEW  **1.** 4, 6, 8, 12, 24  **3.** c  **5.** False  **7.** True  **9.** False
**11.** False  **13.** F  **15.** G  **17.** K  **19.** L  **21.** I
**23.** J

PAGE 126 CUMULATIVE REVIEW EXERCISES  **1.** 58  **3.** 7  **5.** 5  **7.** 12
**9.** 225  **11.** 8  **13.** 25  **15.** 250
**17.** $75.70 > 75.40 > 75.06$
**19.** $0.33 > 0.30 > 0.03$  **21.** $-7, 7$  **23.** $-5$, $-6, -7$ and so on; also 5, 6, 7, and so on
**25.** $-5, 5$  **27.** $-32, -33, -34, -35, -36$, $-37, -38, -39, 32, 33, 34, 35, 36, 37, 38, 39$
**29.** 18  **31.** 12  **33.** $-4$  **35.** $-\frac{4}{5}$  **37.** $\frac{2}{5}$
**39.** $-\frac{10}{7}$  **41.** $-\frac{2}{7}$  **43.** $\frac{2}{5}$  **45.** $-5\frac{5}{8}$
**47.** $8\frac{23}{136}$

PAGE 127 CUMULATIVE REVIEW PROBLEMS  **1.** \$196.88  **3.** No  **5.** \$787.50
**7.** \$444.75

## Chapter 4 Solving Equations
PAGE 133 WRITTEN EXERCISES  **1.** 12
**3.** $-1$  **5.** $-4$  **7.** 23  **9.** 17  **11.** 7  **13.** 12
**15.** 6  **17.** 8  **19.** $-\frac{2}{5}$  **21.** $3\frac{2}{3}$  **23.** 1.1
**25.** 0.253  **27.** 1.196  **29.** $9\frac{3}{4}$  **31.** $2\frac{7}{12}$
**33.** 0.358  **35.** $3\frac{4}{5}$  **37.** 1.73

PAGE 133 REVIEW EXERCISES  **1.** 5  **3.** 3
**5.** 9

PAGE 135 WRITTEN EXERCISES  **1.** 15
**3.** 18  **5.** 6  **7.** 54  **9.** $-84$  **11.** 28  **13.** 5
**15.** $-10$  **17.** 91  **19.** 11  **21.** $7\frac{1}{2}$  **23.** $-221$

**25.** $-252$  **27.** $-\frac{5}{6}$  **29.** $9\frac{2}{6}$  **31.** 364  **33.** 247
**35.** $\frac{2}{19}$  **37.** $10\frac{1}{2}$

PAGE 135 REVIEW EXERCISES  **1.** $\frac{3}{14}$
**3.** $\frac{2}{15}$  **5.** $3\frac{1}{19}$

PAGE 138 WRITTEN EXERCISES  **1.** 88
**3.** 20  **5.** $-1.5$  **7.** $-24$  **9.** 60  **11.** 19.5
**13.** 18  **15.** 20  **17.** $-30$  **19.** $-49.5$  **21.** $13\frac{1}{2}$
**23.** $10\frac{1}{2}$  **25.** 6.5  **27.** $-0.8$  **29.** $-\frac{1}{9}$  **31.** $-\frac{5}{6}$
**33.** 7.595  **35.** 0.02148  **37.** 5.7  **39.** 15.6
**41.** 3.3

PAGE 138 REVIEW EXERCISES  **1.** 30
**3.** 55  **5.** 103  **7.** 17  **9.** 19

PAGE 138 CALCULATOR KEY-IN  **1.** 0.3
**3.** 2.7203791  **5.** 3.3807

PAGES 141–142 WRITTEN EXERCISES
**1.** 11  **3.** 7  **5.** 11  **7.** 10  **9.** 3  **11.** 80
**13.** 30  **15.** 55  **17.** 24  **19.** $4\frac{1}{3}$  **21.** $22\frac{2}{5}$
**23.** $7\frac{1}{2}$  **25.** $3\frac{1}{3}$  **27.** $1\frac{7}{9}$  **29.** $1\frac{17}{21}$  **31.** $4\frac{2}{7}$  **33.** $\frac{1}{4}$
**35.** $\frac{25}{66}$  **37.** $\frac{5}{11}$  **39.** $\frac{2}{15}$  **41.** 27

PAGE 142 SELF-TEST A  **1.** $-24$  **2.** 10
**3.** 9  **4.** 56  **5.** $-13$  **6.** $-68$  **7.** 17  **8.** $-33$
**9.** 150  **10.** $-2\frac{4}{13}$  **11.** $-80$  **12.** $-1.74$
**13.** $27\frac{1}{2}$  **14.** 120  **15.** 6  **16.** $-3$

PAGE 142 CALCULATOR KEY-IN  852

PAGES 145–146 WRITTEN EXERCISES
**1.** $8b$  **3.** $53 - d$  **5.** $30 + t$  **7.** $g + 9$
**9.** $78 - m$  **11.** $n + 19$  **13.** $d \div 11$
**15.** $12 - z$  **17.** $11t + 15$  **19.** $91(m + n)$
**21.** $r \div (83 - 10)$  **23.** $c[(12 + 9) + 3]$
**25.** $(60 + 40 + 10) \div d$  **27.** $b + 3$  **29.** $x + 6$
**31.** $25q$  **33.** $x \div 60$  **35.** $x - 10$

PAGE 146 REVIEW EXERCISES  **1.** 27
**3.** 576  **5.** $-15$  **7.** 19

PAGE 148 WRITTEN EXERCISES
**1.** $5d = 20$  **3.** $3w - 7 = 8$  **5.** $5 \div r = 42$
**7.** $n - 1 = 5$  **9.** $2n \div 3 = 15$
**11.** $(4 + x) \div 2 = 34$  **13.** $59 - x = 3 + 2x$
**15.** $(x - 5) \div 3 = 2$

PAGE 148 REVIEW EXERCISES  **1.** $n - 4$
**3.** $n \div 7$  **5.** $40 - n$  **7.** $2n$

PAGES 150–151 PROBLEMS  **1.** $9n = 1170$
**3.** $18n = 13.50$  **5.** $144 - n = 116$
**7.** $n = (12 + 18) - 19$  **9.** $2000 + n = 2650$

**11.** $\frac{4}{5}n = 180$  **13.** $2n - 30 = 20$  **15.** $\frac{1}{9}n = 5$

PAGE 151 REVIEW EXERCISES
**1.** $n - 8 = 43$  **3.** $n + 14 = 70$
**5.** $n - 17 = 34$

PAGES 154–155 PROBLEMS  **1.** 250 lb
**3.** 480 books  **5.** 24 shirts  **7.** 25 people
**9.** 150 lb  **11.** 24 gal

PAGE 155 SELF-TEST B  **1.** $12x$  **2.** $60 - d$
**3.** d  **4.** c  **5.** a  **6.** $21n = 189$
**7.** $n - 450 = 1845$  **8.** 412 tennis balls

PAGE 157 ENRICHMENT  **1.** 3  **3.** 3
**5.** 2; 7

PAGE 158 CHAPTER REVIEW  **1.** 20
**3.** divide  **5.** a  **7.** c  **9.** b  **11.** b

PAGE 160 CUMULATIVE REVIEW
EXERCISES  **1.** 5  **3.** 5  **5.** $-1$  **7.** 32
**9.** 16  **11.** $<$  **13.** $<$  **15.** $<$  **17.** $>$
**19.** $x = 8$  **21.** $x = -20$  **23.** $x = -18.5$
**25.** $x = 5$  **27.** $x = -12$  **29.** $-3\frac{7}{12}$  **31.** $-1\frac{3}{8}$
**33.** $1\frac{3}{20}$  **35.** $-21$  **37.** $-\frac{1}{9}$  **39.** $3\frac{11}{48}$  **41.** 20
**43.** 24  **45.** $-14$  **47.** 0.75  **49.** 11  **51.** $-6$

PAGE 161 CUMULATIVE REVIEW
PROBLEMS  **1.** $2.97  **3.** 220,000 readers
**5.** $418.50  **7.** $656.38  **9.** 100 inquiries

# 5 Geometric Figures

PAGES 167–168 WRITTEN EXERCISES
**1.** $\overrightarrow{YX}$  **3.** $\overrightarrow{PQ}$  **5.** $\overleftrightarrow{XY}$, $\overleftrightarrow{XZ}$, $\overrightarrow{XY}$, $\overrightarrow{YX}$  **7.** S, P,
O or T, Q, O  **9.** $\overline{PS}$, $\overline{PO}$, $\overline{PQ}$  **11.** $\overline{ST}$ and
$\overline{PQ}$  **13.** Answers may vary; for example, $\overrightarrow{PS}$
and $\overrightarrow{OT}$  **15.** Answers may vary; for example,
$\overrightarrow{OS}$ and $\overrightarrow{OT}$  **17.** $\overrightarrow{OS}$  **19.** $\overrightarrow{OP}$, $\overrightarrow{OQ}$, $\overrightarrow{OS}$, $\overrightarrow{OT}$
**21.** Answers may vary; for example, $\overleftrightarrow{AX}$ and
$\overleftrightarrow{XY}$; X  **23.** Answers may vary; for example,
$\overleftrightarrow{XW}$ and $\overleftrightarrow{AD}$, plane $XWD$  **25.** true  **27.** false
**29.** false  **31.** true  **33.** one  **35.** Two
nonparallel lines in a plane must intersect, and
their intersection is a point.

PAGE 168 REVIEW EXERCISES  **1.** 2.9
**3.** 4.6  **5.** 3.0  **7.** 9.3

PAGES 171–172 WRITTEN EXERCISES
**1. a.** 8 cm  **b.** 80 mm  **3. a.** 9 cm  **b.** 92 mm
**5. a.** 8 cm  **b.** 82 mm  **7.** $\overline{AE}$ and $\overline{PT}$; $\overline{VZ}$ and
$\overline{GJ}$  **9.** 450; 4500  **11.** 2.5, 2500  **13.** 60; 6000

**15.** 2  **17.** 0.625  **19.** 4500  **21.** 3.74 m by
5.2 m  **23.** 4.675 m by 7.05 m  **25.** 27 mm
**27.** 65 cm  **29.** 2.1875  **31.** 35

PAGE 172 REVIEW EXERCISES  **1.** 90
**3.** 40  **5.** 140  **7.** 105  **9.** 65

PAGES 175–177 WRITTEN EXERCISES
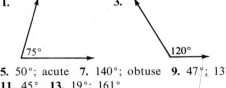
**1.** 75°  **3.** 120°

**5.** 50°; acute  **7.** 140°; obtuse  **9.** 47°; 137°
**11.** 45°  **13.** 19°; 161°
**15.**    **17.**

**19.** $\overleftrightarrow{BC}$ and $\overleftrightarrow{AB}$; $\overleftrightarrow{AD}$ and $\overleftrightarrow{AB}$  **21.** $\angle CED$ and
$\angle DEA$; $\angle CEB$ and $\angle AEB$
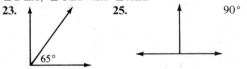
**23.** 65°  **25.** 90°

**27.** true  **29.** $m\angle 1 = m\angle 3 = 60°$,
$m\angle 2 = m\angle 4 = 120°$

PAGE 177 SELF-TEST A  **1.** C ———— D

**2.** A —— X ——  **3.** R —— S ——
**4.** Answers may vary; for example, $\overrightarrow{XW}$ and
$\overrightarrow{YZ}$  **5.** Answers may vary; for example, plane
$WXY$ and plane $DAB$  **6.** Answers may vary;
for example, $\overleftrightarrow{BC}$ and $\overleftrightarrow{ZC}$  **7.** Answers may
vary; for example, plane $DCB$ and plane
$ZCD$  **8.** 4  **9.** 0.87  **10.** 0.785  **11.** 10.9
**12.** $MB$; $\overline{MB}$  **13.** 90°  **14.** congruent  **15.** is
perpendicular to  **16.** acute  **17.** 48  **18.** 73

PAGES 181–182 WRITTEN EXERCISES
**1.** 80°  **3.** 90°  **5.** isosceles  **7.** equilateral
**9.** isosceles, obtuse  **11.** scalene, acute
**13.** $\triangle ACB$, $\triangle DEC$, $\triangle DEA$  **15.** $\triangle BDC$
**17. a.**    **b.**    **19.** 60°
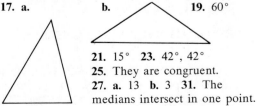
**21.** 15°  **23.** 42°, 42°
**25.** They are congruent.
**27. a.** 13  **b.** 3  **31.** The
medians intersect in one point.

PAGE 182 REVIEW EXERCISES  **1.** 15.08
**3.** 15.52  **5.** 20.33  **7.** 21.873  **9.** 24.768

PAGES 185–187 WRITTEN EXERCISES
**1.** pentagon  **3.** hexagon  **5.** triangle  **7.** square
**9.** 14.1  **11.** 312 cm  **13.** 72.6 mm  **15.** 1356 m
**17.** 108°  **19.** 2.56 m  **21.** 12 m  **23.** 3
**25.**

**27.** 22

**29. a.** 360°  **b.** 1440°  **c.** 360°  **d.** 1080°
**31.** 5

PAGE 187 REVIEW EXERCISES  **1.** 22.4
**3.** 49  **5.** 10.9  **7.** 44.8

PAGES 190–191 WRITTEN EXERCISES
**1.** 25.1 cm  **3.** 2830 mm  **5.** 134 m  **7.** 90.1 m
**9.** 143 km  **11.** 199 m  **13.** 1.59 mm
**15.** 3.74 km  **17.** 2.79 cm  **19.** 40,074 km
**21.** 47.1 km  **23.** 94.2 m  **25.** 35.7  **27.** 41.4
**29.** 22.7  **31.** $S = \frac{\pi d}{2}$  **33.** Draw a circle and
label three points $A$, $B$, $C$ on the circle. Let
point $D$ be a point not on the circle.
Quadrilateral $ABCD$ cannot be inscribed in the
circle.  **35.** An angle inscribed in a semicircle is
a right angle.

PAGE 191 REVIEW EXERCISES  **1.** 18
**3.** 27  **5.** 8  **7.** 14

PAGES 196–197 WRITTEN EXERCISES
**1.** c  **3.** SSS  **5.** rotation or reflection  **7.** a
rotation, two reflections, or a translation and a
reflection  **9.** $\triangle GHK \cong \triangle FHK$; SAS
**11. a.** $\angle N$  **b.** $\angle Y$  **c.** $\overline{NS}$  **d.** $\overline{RL}$  **13. a.** $\overline{FG}$
**b.** $\overline{EH}$  **c.** $\angle EFG$  **d.** $\overline{CD}$  **15.** no

PAGE 197 REVIEW EXERCISES  **1.** 7
**3.** 11  **5.** 72  **7.** 4

PAGE 197 CALCULATOR KEY-IN
**1.** 3.1604938  **3.** 3.141$\overline{6}$  **5.** 3.1416  The closest
approximation is $\frac{355}{113}$.

PAGES 200–201 WRITTEN EXERCISES
**1.**

**5.**

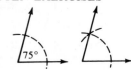

**7.** The perpendicular bisector appears to
pass through the vertex of the angle
formed by the two congruent sides.  **9.** yes
**11.**

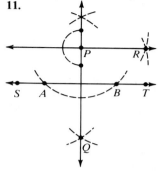

**13.** yes

PAGE 201 SELF-TEST B  **1.** equilateral
**2.** 180  **3.** 3  **4.** octagon  **5.** parallelogram
**6.** 33 cm  **7.** 100 cm  **8.** true  **9.** false
**10.** $\overline{FG}$, $J$, $IJF$
**11.**

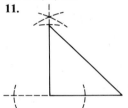

**12.**

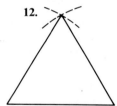

PAGE 203 ENRICHMENT  **1.** 57,960,000
**3.** 149,730,000  **5.** 12  **7.** 32,186,000,000
**9.**
```
10  FOR I=1 TO 5
20  PRINT "HOW MANY HOURS";
30  INPUT T
40  LET R=760
50  PRINT "DISTANCE TRAVELED"
60  PRINT "IS ";R*T;" MILES."
70  NEXT I
80  END
```

PAGE 204 CHAPTER REVIEW  **1.** collinear
points  **3.** midpoint  **5.** false  **7.** true
**9.** false  **11.** d  **13.** c
**15.**

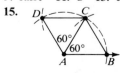

$\triangle ABC$ is
equilateral, thus
$\angle CAB = 60°$.
$\triangle ABC \cong DAC$,
thus $\angle DAC = 60°$,
$\angle DAB = 120°$.

PAGE 206 CUMULATIVE REVIEW
EXERCISES  **1.** 8  **3.** 7  **5.** 12  **7.** −19

**9.** 9 **11.** 45 **13.** 14 **15.** $-18\frac{3}{4}$ **17.** 72
**19.** 24 **21.** 30 **23.** 15 **25.** $\frac{4}{5}$ **27.** $-\frac{1}{9}$ **29.** $\frac{2}{3}$
**31.** 0.4 **33.** 0.4375 **35.** $-0.2\overline{7}$ **37.** $\frac{7}{100}$ **39.** 2
**41.** $-1\frac{8}{33}$ **43.** true

**PAGE 207 CUMULATIVE REVIEW**
**PROBLEMS 1.** $31.11 **3.** $124 **5.** 52 m
**7.** 127 boxes **9.** 22 m

# 6 Ratio, Proportion, and Percent

**PAGES 211–212 WRITTEN EXERCISES**
**1. a.** $\frac{5}{7}$ **b.** $\frac{5}{12}$ **c.** $\frac{12}{7}$ **3. a.** $\frac{9}{4}$ **b.** $\frac{9}{13}$ **c.** $\frac{13}{4}$
**5.** $\frac{3}{8}$ **7.** $\frac{7}{7}$ **9.** $\frac{15}{4}$ **11. a.** $\frac{5}{21}$ **b.** $\frac{21}{26}$ **c.** $\frac{26}{5}$
**13. a.** $\frac{1}{3}$ **b.** $\frac{2}{1}$ **c.** $\frac{3}{2}$ **15.** $\frac{3}{5}$ **17.** $\frac{14}{9}$

**PAGES 212–213 PROBLEMS 1.** $\frac{80}{3}$
**3.** $62\frac{1}{2}$ lb/ft³ **5.** $\frac{11}{2}$ **7. a.** 2:5 **b.** 2:5 **c.** $\frac{2}{5}$
**9.** $\frac{3}{7}$ **11.** $\frac{5}{18}$

**PAGE 213 REVIEW EXERCISES 1.** 9 **3.** 9
**5.** 5 **7.** 12

**PAGES 215–216 PROBLEMS 1.** 17¢ **3.** $16
**5.** $5.25 **7.** 195 km **9.** 240 km **11.** Best
Brand Pear Tomatoes **13.** Sparkle Window
Cleaner **15.** 5 gal **17.** 62 km/h; 74 km/h

**PAGE 216 REVIEW EXERCISES 1.** 20 cm
**3.** 84 cm **5.** 142.4 m **7.** 57.4 m **9.** 49.44 cm

**PAGE 219 WRITTEN EXERCISES 1.** 4
**3.** 56 **5.** 35 **7.** 60 **9.** 72 **11.** 45 **13.** $7\frac{1}{2}$
**15.** $8\frac{1}{2}$ **17.** 10 **19.** 12 **21.** 20 **23.** 15
**25.** $5\frac{3}{5}$ **27.** $4\frac{1}{2}$ **29.** $\frac{1}{3}$ **31.** $\frac{4}{1}$

**PAGE 219 REVIEW EXERCISES 1.** $13\frac{1}{3}$
**3.** $22\frac{7}{9}$ **5.** $8\frac{11}{18}$ **7.** $43\frac{1}{2}$ **9.** $60\frac{3}{10}$

**PAGES 221–223 PROBLEMS 1. a.** 880 km
**b.** 14 h **3. a.** $3.75 **b.** 28 oranges
**5. a.** 5.3 cm³ **b.** 337.5 cm³ **7.** $1.20 **9.** 8 cans
**11.** $11\frac{1}{4}$ oz; $10\frac{2}{3}$ servings **13.** 6 min
**15.** 12 wins **17.** 12 vests **19.** $1333\frac{1}{3}$ ft²
**21.** 6955 votes for; 2782 votes against
**23.** 76,800 km

**PAGE 223 REVIEW EXERCISES 1.** $\dfrac{200\ cm}{15\ cm}$
**3.** $\dfrac{30\ ft}{5\ ft}$ **5.** $\dfrac{2000\ m}{450\ m}$ **7.** $\dfrac{36\ in.}{20\ in.}$

**PAGE 223 CALCULATOR KEY-IN**
**1.** 1.4925373 **3.** 3.9385584 **5.** 0.208982

**PAGES 225–226 WRITTEN EXERCISES**
**1.** 1104 in. **3.** 576 in. **5.** 960 in. **7. a.** 2 cm

**b.** 800 km **9. a.** 10.5 cm **b.** 4200 km
**11. a.** 3 cm **b.** 1200 km **13.** 400 km
**15.** 1,384,300 km **17.** 139,700 km

**PAGE 226 SELF-TEST A 1.** $\frac{3}{2}$ **2.** $\frac{17}{9}$ **3.** $\frac{2}{11}$
**4.** $1.15 **5.** $.61 **6.** 2 **7.** 4 **8.** 135 **9.** $.67
**10.** $110 **11.** 1 in. **12.** 1 cm:16 cm

**PAGE 229 WRITTEN EXERCISES 1.** $\frac{3}{4}$
**3.** $\frac{9}{20}$ **5.** $\frac{3}{25}$ **7.** $1\frac{1}{4}$ **9.** $\frac{31}{200}$ **11.** $\frac{43}{400}$ **13.** 80%
**15.** 30% **17.** 48% **19.** 62% **21.** 220%
**23.** 102% **25.** $87\frac{1}{2}$% **27.** $\frac{1}{2}$% **29.** $\frac{3}{4}$%
**31.** $302\frac{1}{2}$% **33.** $\frac{1}{6}$ **35.** $\frac{5}{12}$

**PAGE 230 PROBLEMS 1.** 38% **3.** 60%
**5.** 5% **7.** $62\frac{1}{2}$ **9.** 10%; 70%; 80%

**PAGE 230 REVIEW EXERCISES 1.** 0.45
**3.** 0.275 **5.** 0.04 **7.** $0.\overline{72}$ **9.** $0.57\overline{3}$

**PAGE 233 WRITTEN EXERCISES 1.** 0.93
**3.** 1.14 **5.** 2.6 **7.** 0.495 **9.** 0.006 **11.** 0.0005
**13.** 59% **15.** 9% **17.** 260% **19.** 1283%
**21.** 0.7% **23.** 8.67% **25.** 0.375; 37.5%
**27.** 0.006; 0.6% **29.** 1.625; 162.5% **31.** 0.5875;
58.8% **33.** $0.708\overline{3}$; 70.8% **35.** 0.2857143;
28.6% **37.** $16\frac{2}{3}$% **39.** $88\frac{8}{9}$% **41.** $83\frac{1}{3}$%

**PAGE 233 REVIEW EXERCISES 1.** 2.95
**3.** 15.438 **5.** 23.8576

**PAGES 235–236 WRITTEN EXERCISES**
**1.** 40% **3.** 270 **5.** 1300 **7.** 5.4 **9.** 83.3%
**11.** 104.5 **13.** 87 **15.** 0.896 **17.** 114.3%
**19.** 224 **21.** 54 **23.** 252

**PAGES 236–237 PROBLEMS 1.** $12.50
**3.** $77.5% **5.** 10% **7.** 11.5% **9.** 4 billion
**11. a.** 10% **b.** 40%

**PAGE 237 SELF-TEST B 1.** $\frac{27}{100}$ **2.** $\frac{83}{100}$
**3.** $1\frac{16}{25}$ **4.** $2\frac{9}{10}$ **5.** 5% **6.** 37.5% **7.** 400%
**8.** 325% **9.** 0.45 **10.** 0.78 **11.** 3.48
**12.** 0.008 **13.** 64% **14.** 81% **15.** 785%
**16.** 6.8% **17.** 25% **18.** 205% **19.** 11 **20.** 120

**PAGE 239 ENRICHMENT 1.** 3, 21, 144, and
987 are divisible by 3; 5, 55, and 610 are
divisible by 5. **3.** The new pattern consists of
every other Fibonacci number, starting with 3.
**5.** The new pattern consists of every other
Fibonacci number, starting with 5.

**PAGE 240 CHAPTER REVIEW 1. a 3.** b
**5.** c **7.** a **9.** b **11.** c **13.** e **15.** b **17.** c

PAGE 242 CUMULATIVE REVIEW
EXERCISES  **1.** 5  **3.** 2  **5.** 6  **7.** 3  **9.** 4
**11.** 36  **13.** false  **15.** false  **17.** $\frac{10}{15}, \frac{3}{15}$
**19.** $-\frac{3}{24}, \frac{8}{24}$  **21.** $\frac{24}{60}, \frac{10}{60}, \frac{15}{60}$  **23.** 5  **25.** 3
**27.** 16  **29.** 1  **31.** 64  **33.** $2\frac{7}{13}$  **35.** rhombus
**37.** diameter

PAGE 243 CUMULATIVE REVIEW
PROBLEMS  **1.** $79.38  **3.** 23,500 ft
**5.** 1230 lb  **7.** l: 15 m; w: 5 m  **9.** $616.13

# 7 Percents and Problem Solving

PAGES 247–248 WRITTEN EXERCISES
**1.** 15%  **3.** 40%  **5.** 82.5%  **7.** 46.9%  **9.** 0.8%
**11.** 27.5%  **13.** 132  **15.** 57.2  **17.** 205.8
**19.** 124.5  **21.** 60  **23.** 336  **25.** 338.8  **27.** 64
**29.** 120

PAGES 248–249 PROBLEMS  **1.** 1,134
employees  **3.** 25%  **5.** 7511 books  **7.** 16.7%
**9.** 246 homes  **11.** $1.25

PAGE 249 REVIEW EXERCISES  **1.** 9.72
**3.** 17  **5.** 520  **7.** 220.5  **9.** 0.04

PAGES 252–253 PROBLEMS  **1.** $67.15
**3.** 14% discount  **5.** $26.46  **7.** $30  **9.** 25%
markup  **11.** $12.50  **13.** 72% of the original
price  **15.** The final prices will be equal.

PAGE 253 REVIEW EXERCISES  **1.** radius
**3.** diameter

PAGES 255–256 PROBLEMS  **1.** $129
**3.** $978  **5.** $10,900  **7.** $64,893.62
**9.** $58,510.64  **11.** $7,380  **13.** $4,250

PAGE 256 REVIEW EXERCISES  **1.** 10
**3.** 6  **5.** 128  **7.** 0.4

PAGE 259 WRITTEN EXERCISES  **1.** 5.12
**3.** 80%  **5.** 40

PAGES 259–261 PROBLEMS  **1.** 82.5%
**3.** 49 books  **5.** $34.80; $27.84  **7.** $36.50
**9. a.** 108°  **b.** $3.6 million
**11. b.** food $450; clothing $270; housing $540;
medical $180; other $360
**13. b.** salaries $489,600;
maintenance/repair $54,400; books/supplies
$34,000; recreation $34,000; after-school
programs $34,000; teacher training $34,000

PAGE 261 SELF-TEST A  **1.** 25%  **2.** 35%
**3.** 23%  **4.** $103.70  **5.** 30% discount

**6.** $6,360  **7.** $72,688  **8.** 437.4  **9.** 80%  **10.** 75
violations

PAGES 264–265 WRITTEN EXERCISES
**1.** $384; $1664  **3.** $745.20; $3505.20  **5.** $1692;
$7332  **7.** $5550.60; $11,930.60  **9.** 14%
**11.** 8.5%  **13.** 16%  **15.** $\frac{1}{2}$ yr  **17.** $4450
**19.** $6720  **21.** $1850  **23.** $102.12

PAGES 265–266 PROBLEMS  **1.** $577.50
**3.** $96.25  **5.** 7.5%  **7.** 8 yr

PAGE 266 REVIEW EXERCISES  **1.** 0.375
**3.** 0.05  **5.** 1.496  **7.** 0.833  **9.** 2.396

PAGE 266 CALCULATOR KEY-IN  **1.** 16
**3.** 48%  **5.** 86.$\overline{6}$%

PAGES 268–269 WRITTEN EXERCISES
**1.** $7056  **3.** $1560.60  **5.** $3149.28
**7.** $2137.84  **9.** $3975.35  **11.** $9724.05
**13.** $7.70  **15.** 9 mo

PAGE 269 PROBLEMS  **1.** $1852.20  **3.** 12%
simple interest will earn more.  **5.** 13.2%

PAGES 271–273 PROBLEMS  **1.** $2.50
**3.** $21,942  **5.** $165  **7.** $7000  **9.** $56,400
**11.** 10%

PAGE 273 SELF-TEST B  **1.** $10,000
**2.** 14%  **3.** $3,149.28  **4.** 9 mo  **5.** $2,169.65

PAGE 275 ENRICHMENT  **1.** 125.86; 162.75;
20.00  **3.** 249.75; 323.80; 35.00  **5.** 665.61;
676.09; 90.00

PAGE 276 CHAPTER REVIEW  **1.** 25%
**3.** 95  **5.** c  **7.** true  **9.** c

PAGE 278 CUMULATIVE REVIEW
EXERCISES  **1.** 8  **3.** 7  **5.** 12  **7.** $-19$
**9.** 9  **11.** 45  **13.** $-7, -2.5, -2.2, 0, 3, 6.4$
**15.** $-8.4, -3, -1.6, -1.0, 0.5$  **17.** True
**19.** True  **21.** True  **23.** $\frac{6}{9}, \frac{5}{9}$  **25.** $-\frac{88}{33}, \frac{27}{33}$
**27.** $\frac{28}{12}, \frac{9}{12}$  **29.** $\frac{68}{128}, -\frac{3}{128}$  **31.** 12  **33.** 180
**35.** 63  **37.** rotation and translation; ASA
**39.** rotation and translation; SSS

PAGE 279 CUMULATIVE REVIEW
PROBLEMS  **1.** 15 cm  **3.** 4.4 km  **5.** up $4\frac{5}{8}$
points  **7.** 58.125 L  **9.** 10 students

# 8 Equations and Inequalities

PAGES 283–284 WRITTEN EXERCISES
**1.** $2m$  **3.** $-7c$  **5.** $c-5$  **7.** $-7y+77$

**9.** $-4a - 30$  **11.** $6x - 14$  **13.** $\frac{1}{2}$  **15.** 9
**17.** $-1\frac{1}{4}$  **19.** 5  **21.** $-\frac{1}{2}$  **23.** $-1$  **25.** $\frac{1}{2}$
**27.** $-1$  **29.** $2\frac{1}{2}$  **31.** 1  **33.** $-1$  **35.** 6
**37.** $-\frac{3}{4}$  **39.** $-1\frac{2}{3}$  **41.** $1\frac{1}{2}$  **43.** 5  **45.** $-3$
**47.** $-14$

PAGE 284 REVIEW EXERCISES  **1.** 1
**3.** 15  **5.** 12  **7.** $10\frac{1}{2}$  **9.** 4

PAGE 286 WRITTEN EXERCISES  **1.** 5
**3.** 7  **5.** $-6$  **7.** $-\frac{1}{2}$  **9.** $-4$  **11.** $-\frac{3}{5}$  **13.** 4
**15.** $-2\frac{1}{2}$  **17.** 3  **19.** 1  **21.** $\frac{1}{3}$  **23.** 6  **25.** 1
**27.** $-3$  **29.** $-8\frac{1}{2}$  **31.** $-16$  **33.** $6\frac{2}{3}$  **35.** 80

PAGE 286 REVIEW EXERCISES  **1.** 4  **3.** 6
**5.** 12  **7.** 24

PAGES 288–289 PROBLEMS  **1.** 16 ft, 24 ft
**3.** 35 min, 45 min  **5.** 10 cm, 20 cm, 20 cm
**7.** 74 points  **9.** 34 ft  **11.** 10 yr  **13.** 1000 m

PAGE 290 SELF-TEST A  **1.** 4  **2.** $-\frac{1}{2}$
**3.** $-2$  **4.** 2  **5.** $-6$  **6.** $\frac{1}{5}$  **7.** 16  **8.** $-9$
**9.** 5  **10.** $-4$  **11.** $-2$  **12.** 2  **13.** $-4$  **14.** $\frac{7}{9}$
**15.** $-11$  **16.** $7\frac{1}{2}$  **17.** 20 ft  **18.** 106

PAGE 291–292 WRITTEN EXERCISES
**1.** $12 < 22$  **3.** $6 > 0$  **5.** $0 < 8 < 10$
**7.**  

$$\leftarrow\!+\!+\!+\!+\!+\!+\!\bullet\!+\!\uparrow\!+\!+\!+\!\rightarrow$$
$$m \quad m+1$$

**11.** $6 > t$  **13.** $p > q$  **15.** $10d < 5n$
**17.** $6 < 2n < 8$  **19.** $x < a < y$
**21.** $2 < 6 < 10 < 20 < 50$  **23.** $a < 5 < 8 < b$

PAGE 292 REVIEW EXERCISES  **1.** 5  **3.** 5
**5.** 9  **7.** 77

PAGES 295–296 WRITTEN EXERCISES  All
the numbers: **1.** less than 12  **3.** greater than
42  **5.** less than 3  **7.** greater than or equal
to 20  **9.** less than $-18$  **11.** less than 7
**13.** less than $-7$  **15.** less than or equal to 12
**17.** less than or equal to 15  **19.** less than
$-26$  **21.** less than or equal to 60  **23.** less
than or equal to $-11$  **25.** less than $-168$
**27.** greater than 21.3  **29.** less than $5\frac{3}{7}$
**31.** less than or equal to $-3.87$  **33.** less than
or equal to 47  **35.** less than or equal to $-10.8$
**37.** greater than or equal to 3  **39.** greater
than 21  **41.** greater than $-4.1$  **43.** less than
or equal to $-92.4$  **45.** greater than $22\frac{13}{15}$
**47.** greater than or equal to $-9$  **49.** greater
than $\frac{1}{5}$  **51.** less than $-1.7$

PAGE 296 REVIEW EXERCISES  **1.** $-2.6$,
$-1.4$, 1.2, 3.2  **3.** $-12.2$, $-12.09$, $-11.2$, 12,
112  **5.** $-30.05$, $-5.3$, $-5.03$, 0.35, 3.05
**7.** $-2.89$, $-2.8$, 2.089, 2.89, 28.9

PAGES 299–300 WRITTEN EXERCISES  All
the numbers: **1.** less than $-6$

$$\leftarrow\!+\!+\!+\!\bullet\!+\!+\!+\!+\!+\!+\!+\!+\!+\!+\!+\!+\!+\!\rightarrow$$
$$-10 \ -8 \ -6 \ -4 \ -2 \ \ 0 \ \ 2 \ \ 4 \ \ 6 \ \ 8$$

**3.** greater than or equal to $-9$

$$\leftarrow\!+\!\bullet\!+\!+\!+\!+\!+\!+\!+\!+\!+\!+\!+\!+\!+\!+\!\rightarrow$$
$$-10 \ -8 \ -6 \ -4 \ -2 \ \ 0 \ \ 2 \ \ 4 \ \ 6 \ \ 8$$

**5.** less than $-30$

$$\leftarrow\!+\!+\!+\!+\!+\!+\!+\!+\!\bullet\!+\!+\!+\!+\!+\!+\!\rightarrow$$
$$-38 \qquad -34 \qquad -30 \qquad -26$$

**7.** greater than or equal to $-42$

$$\leftarrow\!+\!+\!+\!+\!\bullet\!+\!+\!+\!+\!+\!+\!+\!+\!+\!+\!\rightarrow$$
$$-46 \qquad -42 \qquad -38 \qquad -34 \qquad -30$$

**9.** greater than $5\frac{2}{5}$  **11.** greater than or equal to
$-10$  **13.** greater than 2  **15.** greater than or
equal to 3  **17.** greater than $-4$  **19.** less than
or equal to 5  **21.** less than 8  **23.** less than or
equal to $-3\frac{1}{13}$  **25.** less than or equal to $-3$
**27.** greater than 1  **29.** less than or equal
to $-2$  **31.** greater than 15  **33.** less than 2
**35.** greater than 24  **37.** less than $-7.2$
**39.** greater than or equal to $-10.3$

PAGE 300 REVIEW EXERCISES  **1.** $n + 9$
**3.** $n - 16$  **5.** $n + 3$  **7.** $\frac{n}{5}$

PAGES 302–303 PROBLEMS  **1.** 36, 37
**3.** 4 hr  **5.** \$22,714.29  **7.** 3, 4, 5  **9.** 400
employees  **11.** 51 checks

PAGE 303 SELF-TEST B  **1.** $a < b$
**2.** $35 > 4z$  All the numbers: **3.** less than or
equal to 28  **4.** greater than 3  **5.** greater than
or equal to 7  **6.** greater than $4\frac{2}{3}$  **7.** greater
than $-14$  **8.** greater than 5  **9.** less than 2
**10.** greater than 2  **11.** less than or equal to 0
**12.** greater than or equal to $-6$  **13.** 74, 75

PAGE 306 CHAPTER REVIEW  **1.** False
**3.** True  **5.** True  **7.** False  **9.** b  **11.** c
**13.** D  **15.** B  **17.** E  **19.** G  **21.** b

PAGE 308 CUMULATIVE REVIEW
EXERCISES  **1.** $-5$  **3.** 18  **5.** 2  **7.** 2  **9.** 4
**11.** 64  **13.** 3  **15.** 7  **17.** 5  **19.** 4  **21.** $-\frac{1}{4}$
**23.** 3  **25.** $\frac{4}{8}$, $\frac{7}{8}$  **27.** $\frac{4}{20}$, $\frac{15}{20}$  **29.** $\frac{8}{12}$, $\frac{3}{12}$, $\frac{5}{12}$
**31.** skew  **33.** diameter  **35.** vertex  **37.** $91\frac{2}{3}\%$
**39.** 200

PROBLEMS **1.** $79.38 **3.** 23,500 ft
**5.** 1230 lb **7.** 52 cm **9.** $616.13

# 9 The Coordinate Plane

## PAGES 314–315 WRITTEN EXERCISES
**1.** $(-5, -5)$ **3.** $(5, 1)$ **5.** $(-3, -2)$
**7.** $(-1, 4)$ **9.** $I$ **11.** $V$ **13.** $D$ **15.** $W$ **17.** $M$
**19. c.** rectangle **21. c.** parallelogram
**23. b.** $(-5, 1), (-2, -2), (2, -2), (2, 1)$
**d.** Translation **25. b.** $(-2, 3), (2, 3), (4, 1),$
$(4, -1), (1, 1), (-1, 1), (-4, -1), (-4, 1)$
**d.** Translation **27. b.** $(-6, 5), (-2, 1), (-9, 3)$
**d.** Reflection

## PAGE 315 REVIEW EXERCISES **1.** 8
**3.** 27 **5.** $-5$ **7.** $-15$

## PAGES 318–319 WRITTEN EXERCISES
**1.** Yes **3.** Yes **5.** No **7.** No **9.** No **11.** No
**13.** Yes **15.** Yes **17.** $y = -2x + 7$
**19.** $y = 4x + 16$ **21.** $y = \frac{2}{3}x - 2$
**23. a.** $y = x + 7$ **b.** $(2, 9), (-5, 2), (7, 14)$
**25. a.** $y = \frac{1}{2}x + 5$ **b.** $(4, 7), (0, 5), (6, 8)$
**27. a.** $y = -\frac{1}{4}x + 5$ **b.** $(4, 4), (-8, 7), (0, 5)$
**29. a.** $y = \frac{3}{2}x - 3$ **b.** $(4, 3), (-2, -6),$
$(-8, -15)$ **31. a.** $y = \frac{1}{2}x + \frac{5}{2}$ **b.** $(3, 4), (5, 5),$
$(-1, 2)$ **33. a.** $y = \frac{5}{2}x + 3$ **b.** $(2, 8), (1, 5\frac{1}{2}),$
$(-3, -4\frac{1}{2})$ In Exercises 35–43, answers will
vary. Examples are given. **35.** $(-1, 0), (0, -1),$
$(1, -2)$ **37.** $(-1, 1), (0, 0), (1, -1)$
**39.** $(-1, 0), (0, 5), (1, 10)$ **41.** $(-3, -6),$
$(0, -4), (3, -2)$ **43.** $(-1, 4), (0, 6), (1, 8)$
**45.** $x + y = 0$ **47.** $x + y = -3$

## PAGE 319 REVIEW EXERCISES **1.** 3
**3.** 28 **5.** 3 **7.** $-56$ **9.** $-4$

## PAGES 321–323 WRITTEN EXERCISES

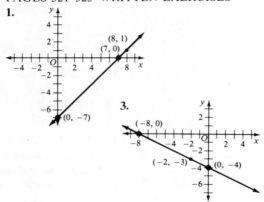

**1.**

**3.**

**5.–33.** Each graph is a straight line through the
points listed: **5.** $(0, 10), (1, 8), (5, 0)$ **7.** $(0, 0),$
$(1, 1), (2, 2)$ **9.** $(-2, -1), (0, 0)$ **11.** $(0, -6),$
$(6, 0)$ **13.** $(-3, 0), (0, -3)$ **15.** $(0, 4), (1, 0)$
**17.** $(-2, 0), (0, 6)$ **19.** $(-8, 0), (0, 2)$
**21.** $(-4, 0), (0, 3)$ **23.** $(-\frac{5}{2}, 0), (0, 1)$
**25.** $(0, 6), (6, 0)$ **27.** $(0, -6), (2, 0)$ **29.** $(-12, 0), (0, 6)$
**31.** $(0, -\frac{1}{2}), (\frac{1}{3}, 0)$ **33.** $(0, -8), (6, 0)$

**35.**

**37.**

**39.**

**47.** $-\frac{3}{4}$

## PAGE 323 SELF-TEST A **1.** $(-3, 6)$
**2.** $(4, -4)$
**3.** $(6, 6)$
**4.** $C$
**5.** $T$
**6.** $P$ **7.–12.**

**13.** $(0, \frac{1}{2}), (-4, -\frac{1}{2}), (9, 2\frac{3}{4})$ **14.** $(-3, 12),$
$(\frac{1}{2}, 5), (11, -16)$
**15.**

**16.**

**17.**

**18.**

PAGES 326–327 WRITTEN EXERCISES
**1.** Intersect **3.** Parallel **5.** Coincide
**7.** Parallel **9.** Intersect **11.** Coincide
**13.** Intersect **15.** Parallel **17.** Intersect **19.** 2

PAGE 327 REVIEW EXERCISES **1.** $1\frac{1}{8}$
**3.** $\frac{7}{8}$ **5.** $\frac{4}{5}$ **7.** $\frac{1}{13}$

PAGE 327 CALCULATOR KEY-IN
$13,107.20

PAGES 331–333 PROBLEMS **1.** $\frac{3}{2}$ **3.** 4°C
**5.** 245 mi **7. a.** $y = 4.5x + 2$ **c.** 7 lb
**9. a.** $y = \frac{7}{12}x + \frac{1}{6}$ **c.** $10\frac{2}{3}$ min **11.** 5.5 s

PAGE 333 REVIEW EXERCISES All the
numbers: **1.** greater than $1\frac{1}{3}$ **3.** greater than or
equal to $-1$ **5.** less than or equal to 0
**7.** greater than or equal to $-1$ **9.** greater
than 41

PAGE 333 CALCULATOR KEY-IN 1; 121;
12,321; 1,234,321; 123,454,321; and so on.

PAGES 336–337 WRITTEN EXERCISES
**1.** $y = -3x - 6$ **3.** $y = x - 2$
**5.** $y = -x + 3$
**7.**

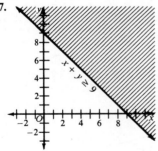

**9.** Solid line through $(-1, 0)$ and $(0, 2)$, shaded
region at left above line **11.** Solid line
through $(0, 4)$ and $(2, 0)$, shaded region at right
above line **13.** Solid line through $(-\frac{5}{2}, 0)$ and
$(0, 5)$, shaded region at left above line
**15.** Dashed line through $(0, -\frac{1}{3})$ and $(\frac{1}{2}, 0)$,
shaded region at left above line **17.** Solid line
through $(0, 8)$ parallel to $x$-axis, shaded region
below line **19.** Solid line through $(-1, 0)$ and
$(0, -1)$, shaded region at left below line.
**21.** Solid line through $(-3, 0)$ parallel to $y$-
axis, shaded region to left of line. **23.** Dashed
line through $(-\frac{5}{2}, 0)$ and $(0, 5)$, shaded region
at right below line **25.** $x < 4$ **27.** $y \geq -3$

PAGE 337 SELF-TEST B **1.** intersect
**2.** coincide **3.** parallel **4.** 4 wk

**5.**

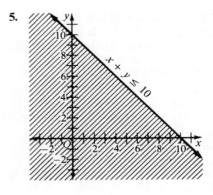

**6.**

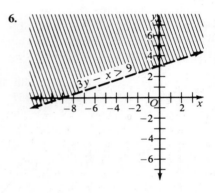

**7.**

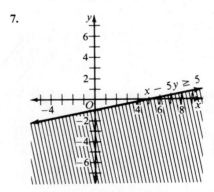

PAGES 338–339 ENRICHMENT
**1.** $y = 0.5x + 1.32$ **3.** 2.9 kg
**5.** $y = -3.3x + 20.8$ **a.** 15.85°C **b.** 6.3 h
**7.** $y = 0.59x + 7.39$; $10,930; $11,520

PAGE 340 CHAPTER REVIEW **1.** False
**3.** True **5.** C **7.** B **9.** Straight line through
$(0, 0)$ and $(1, -1)$ **11.** Straight line through
$(0, 2)$ and $(-1, 0)$ **13.** C **15.** constant
**17.** dashed

PAGE 342 CUMULATIVE REVIEW
EXERCISES  1. $5b$  3. $y + z - 12$
5. $3p + 7$  7. 288.01  9. 100.76  11. 4.8911
13. 554.09  15. 3.0088  17. $-425.0612$
19. $-8.174$  21. $-\frac{3}{14}$  23. $1\frac{1}{8}$  25. $-4\frac{47}{50}$
27. $t = -\frac{3}{4}(\frac{1}{5})$; $-\frac{3}{20}$  29. $x + 12 > 2x$; the
solutions are all the numbers less than 12
31. 20.7 m  33. 1.6  35. 42  37. no  39. yes

PAGE 343 CUMULATIVE REVIEW
PROBLEMS  1. $84.88  3. 7.5 times
5. $9799; $10,038  7. 44.1%  9. 0.5 kW·h

## 10 Areas and Volumes

PAGES 348–349 WRITTEN EXERCISES
1. 1050 cm²; 130 cm  3. 2665 km²; 212 km
5. 4416 mm²; 280 mm  7. 2490.67 km²;
205.6 km  9. perimeter: 40; area: 93.75
11. length: 80; perimeter: 250  13. width: 0.4;
area: 3.44  15. 74  17. 0.5  19. 4  21. 37
square units  23. 30 square units  25. 72
square units

PAGE 350 PROBLEMS  1. 24 m²
3. a. 720 ft²  b. $1080  5. 15 m  7. 2025 m²

PAGE 350 REVIEW EXERCISES  1. 28
3. 1.04  5. 40  7. 0.078

PAGES 354–355 WRITTEN EXERCISES
1. 67.34 m²  3. 2127.5 cm²  5. 290 mm²
7. 1830 km²  9. 44.4 cm²  11. 24 cm  13. 4 m
15. 8 mm  17. 15  19. The base and the height
are the same for each triangle.  21. 450 cm²;
450 cm²  23. 9 : 1; 3 : 1

PAGE 355 REVIEW EXERCISES  1. 75
3. 6.76  5. 292  7. 1.33  9. 4868

PAGES 358–359 WRITTEN EXERCISES
1. 78.5 km²  3. 0.283 m²  5. $9\frac{5}{8}$ square units
7. $100\pi$ m²  9. $4\pi$ square units  11. 10 m
13. 2 cm  15. $4\pi$  17. $31\frac{1}{2}\pi$ cm²

19. $(25\pi - 50)$m²  21. $A = \dfrac{C^2}{4\pi}$

PAGE 359 PROBLEMS  1. $1100  3. 314 m²
5. 462 cm²  7. $\frac{100}{157}$

PAGE 359 REVIEW EXERCISES  1. 16
3. 9  5. 18  7. 15

PAGES 362–363 WRITTEN EXERCISES
1.

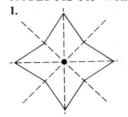

3.

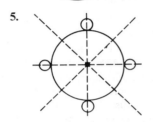

5.

7. 84 m²  9. $50\pi$ square units  11. 135
square units  13. 16 square units
15. $(200 + 50\pi)$cm²

PAGE 363 SELF-TEST A  1. 273 cm²
2. 84 m²  3. 72 cm²  4. 36 m²  5. 707 mm²
6. 616 cm²  7.                8. 28 square
units

PAGES 366–368 WRITTEN EXERCISES
1. 90 cubic units  3. 108 cubic units  5. 22,608
cubic units  7. 514 cubic units  9. 50 L
11. 6330 L  13. $4\pi$; 3  15. 5; 4  17. 3; $9\pi$
19. 60 cubic units  21. a. quadrupled
b. halved  c. doubled  23. 3

PAGE 368 PROBLEMS  1. 96 L
3. 7070 cm³  5. 1.62 cm

PAGE 368 REVIEW EXERCISES  1. 32
3. 0.845  5. 1256  7. 10; $-10$

PAGES 371–372 WRITTEN EXERCISES
**1.** 72 cubic units **3.** 301 cubic units **5.** 56
cubic units **7.** 0.16 L **9.** 1000 L **11.** 10
**13.** 6 **15.** 3 **17.** 15 **19.** 3 **21.** 18

PAGE 372 PROBLEMS **1.** 2,590,000 $m^3$
**3.** 14,700 $cm^3$

PAGE 372 REVIEW EXERCISES **1.** 36 m
**3.** 105 cm **5.** 51.2 km **7.** 580.8 km

PAGES 375–377 WRITTEN EXERCISES
**1.** 5700 $cm^2$ **3.** 14.56 $m^2$ **5.** 1660 $cm^2$;
2560 $cm^2$ **7.** 226 square units; 283 square
units **9.** 360 square units; 468 square units
**11.** 30 cubic units; 72 square units **13.** 5 cm
**15.** 6 **17.** 72 $m^2$; 120 $m^2$ **19.** $n - 2$

PAGE 377 REVIEW EXERCISES **1.** 6
**3.** 105 **5.** 35 **7.** 38

PAGES 379–381 WRITTEN EXERCISES
**1.** $36\pi$ square units; $36\pi$ cubic units **3.** $144\pi$
square units; $288\pi$ cubic units **5.** $400\pi$ square
units; $\frac{4000}{3}\pi$ cubic units **7.** $\frac{16}{3}\pi$ cubic units
**9.** $9.68\pi$ square units **11.** $51.84\pi$ square units;
$62.208\pi$ cubic units **13.** 9 units; $324\pi$ square
units **15.** 8 units; $\frac{2048}{3}\pi$ cubic units **17.** $324\pi$
cubic units **19.** The Earth's volume is about
49.3 times that of the moon. **21. a.** $2 : 3$
**b.** $2 : 3$ **c.** $2 : 3$

PAGE 381 REVIEW EXERCISES **1.** 2.1
**3.** 0.64 **5.** 3.5 **7.** 810

PAGE 381 CALCULATOR KEY–IN
**1.** About 126 times greater

PAGES 383–384 WRITTEN EXERCISES
**1.** 6300 **3.** 4300 **5.** 810 g **7.** 34 kg **9.** 460 t
**11.** 180 g **13.** 33.1 g **15.** 72.7 g **17.** 7460 kg

PAGES 384–385 PROBLEMS **1.** 13.3 kg
**3.** 81.1 g **5.** 3.91 t

PAGE 385 SELF-TEST B **1.** 41.8 **2.** 27,200
**3.** 9248; 9.248 **4.** 0.314 $m^3$ **5.** 1200 $m^3$
**6.** 10,800 $cm^3$ **7. a.** 98 square units **b.** 122
square units **8. a.** 94.2 square units **b.** 251
square units **9. a.** 3024 square units **b.** 3672
square units **10. a.** $1296\pi$ $cm^2$ **b.** $7776\pi$ $cm^3$
**11. a.** $144\pi$ $m^2$ **b.** $288\pi$ $m^3$ **12.** 8240 g
**13.** 181.76 g

PAGE 387 ENRICHMENT **1.** New York

**3.** Rio de Janeiro **5.** Moscow **7.** Casablanca
**9.** 40°N, 105°W **11.** 55°N, 0° **13.** 15°N,
15°W **15.** 65°N, 165°W

PAGE 388 CHAPTER REVIEW **1.** b **3.** c
**5.** d **7.** 3864 $cm^3$ **9.** 45.2 cubic units
**11.** 168; 266 **13.** 2300; 18400 **15.** 60.48 kg

PAGES 390–391 CUMULATIVE REVIEW
EXERCISES **1.** 5 **3.** 10 **5.** 12.5 **7.** $-22.5$
**9.** $\frac{37}{12}$, or $3\frac{1}{12}$ **11.** $\frac{2}{5}$ **13.** $\frac{52}{3}$, or $17\frac{1}{3}$ **15.** $-\frac{37}{4}$,
or $-9\frac{1}{4}$ **17.** All the numbers greater than 3
**19.** All the numbers less than or equal to 40
**21.** All the numbers greater than or equal to 9
**23.** 7 **25.** 21 **27.** 45°; 45° **29.** 8.1 **31.** 24%
**33.** 63,200 **35.** $y = 18 - 4x$ **37.** $y = x - 8\frac{1}{3}$
**39.** $y = \frac{20}{x}$ **41.** $196\pi$ $cm^2$ **43.** $15625\pi$ $km^2$
**45.** 3,258,025$\pi$ $cm^2$ **47.** $256\pi$ square units;
$\frac{2048}{3}\pi$ cubic units **49.** $5184\pi$ square units;
62,208$\pi$ cubic units

PAGE 391 CUMULATIVE REVIEW
PROBLEMS **1.** 1929 stories **3.** 25 m **5.** $\frac{1}{4}$
**7.** 5%

# 11 Applying Algebra to Right Triangles
PAGE 395 WRITTEN EXERCISES **1.** 6 and
7 **3.** $-4$ **5.** 1 **7.** $-6$ **9.** 7 and 8 **11.** 5
and 6 **13.** 3 and 4 **15.** 1 and 2 **17.** 3 **19.** 4
and 5 **21.** 9 **23.** $>$ **25.** $<$ **27.** $=$ **29.** $>$
**31.** 81 **33.** 11

PAGE 395 REVIEW EXERCISES **1.** 6.57
**3.** 8.67 **5.** 11.95 **7.** 0.13

PAGE 398 WRITTEN EXERCISES **1.** 3.3
**3.** 5.7 **5.** 2.8 **7.** 5.1 **9.** 7.5 **11.** 9.5 **13.** 2.3
**15.** 3.0 **17.** 2.9 **19.** 1.9 **21.** 12.3 **23.** 26.5
**25.** 3.9 **27.** 8.4 **29.** 0.6 **31.** 0.2 **33.** 1.41
**35.** 2.23

PAGE 398 REVIEW EXERCISES **1.** 58
**3.** 6.72 **5.** 3.89 **7.** 93.57

PAGE 398 CALCULATOR KEY–IN **1. a.** 1
**b.** 10 **c.** 100 **3. a.** 8.3666002 **b.** 83.666002
**c.** 836.66002 **5. a.** 0.73484692 **b.** 7.3484692
**c.** 73.484692

PAGES 400–401 WRITTEN EXERCISES
**1.** 8.06 **3.** 7.48 **5.** 9.85 **7.** 64.03 **9.** 0.87
**11.** 22.25 **13.** 26.15 **15.** 59.40 **17.** 12.57
**19.** 0.37 **21.** 0.43 **23.** 2.43 **25.** 9.03

**27.** 2.95 **29.** 7.13 **31.** 27.15 **33.** 22.36 **35.** 19.49 **37.** 58.91

PAGE 401 PROBLEMS **1.** 9.22 m **3.** $21 **5.** 3.87 cm **7.** 5.92 m

PAGE 401 REVIEW EXERCISES **1.** 0.81 **3.** 0.0036 **5.** 13.69 **7.** 590.49

PAGE 404 WRITTEN EXERCISES **1.** 289 **3.** 400 **5.** Yes **7.** Yes **9.** Yes **11.** Yes **13.** No **15.** $c = 10$ **17.** $a = 3.32$ **19.** $c = 11.40$ **21.** $b = 70$ **23.** $a = 27$, $b = 36$, $c = 45$

PAGES 404–405 PROBLEMS **1.** 108.2 km **3.** 55.7 m **5.** 7.1 cm **7.** 17.3

PAGE 405 SELF-TEST A **1.** 7 and 8 **2.** $-9$ **3.** 8 **4.** 7 **5.** 9.3 **6.** 4.2 **7.** 7.1 **8.** 3.1 **9.** 2.4 **10.** $c = 5$ **11.** $a = 28$ **12.** $b = 16$

PAGES 409–410 WRITTEN EXERCISES **1.** $RQ$ **3.** 9 **5.** 104° **7.** 25 **9.** 65°; 65° **11.** $x = 5$, $y = 9$ **13.** $x = 27$, $y = 25$ **15.** $x = 46\frac{2}{7}$, $y = 14$ **17.** 180 cm **19. a.** 18 **b.** 18 **21.** $\triangle CDA \cong \triangle BCA$; $\angle A \cong \angle A$; so $\angle B \cong \angle DCA$ **23.** Yes; angles congruent, sides proportional **25.** Not necessarily; the lengths of the sides may differ.

PAGE 410 REVIEW EXERCISES **1.** 519 **3.** 1562 **5.** 2.2454 **7.** 4.878 **9.** 3741.36

PAGES 413–414 WRITTEN EXERCISES **1.** $\frac{3\sqrt{10}}{5}$ **3.** $\sqrt{3}$ **5.** $\frac{2\sqrt{x}}{x}$ **7.** $x = 3.2$, $y = 5.5$ **9.** $x = 7.4$, $y = 8.6$ **11.** $x = 3.2$, $y = 4.5$ **13.** $2\sqrt{2}$ **15.** $5\sqrt{3}$; $10\sqrt{3}$ **17.** $AB = 5.4$ **19.** $BC = 6.0$ **21.** $BC = 1.7y$ **23.** $x = 6$, $y = 3$

PAGES 414–415 PROBLEMS **1.** 8.7 m **3.** 20 m **5.** 56.6 cm **7.** 40; 28.3 **9. a.** 0.9 r units **b.** 0.4 r² units² **11.** 5.2 m

PAGE 415 REVIEW EXERCISES **1.** 1.25 **3.** $0.\overline{2}$ **5.** 1.1 **7.** $0.1\overline{6}$

PAGES 418–419 WRITTEN EXERCISES **1.** $\sin A = \frac{3}{5}$; $\cos A = \frac{4}{5}$; $\tan A = \frac{3}{4}$; $\sin B = \frac{4}{5}$; $\cos B = \frac{3}{5}$; $\tan B = \frac{4}{3}$ **3.** $\sin A = \frac{80}{89}$; $\cos A = \frac{39}{89}$; $\tan A = \frac{80}{39}$; $\sin B = \frac{39}{89}$; $\cos B = \frac{80}{89}$; $\tan B = \frac{39}{80}$ **5.** $\sin A = \frac{a}{c}$; $\cos A = \frac{b}{c}$;

$\tan A = \frac{a}{b}$; $\sin B = \frac{b}{c}$; $\cos B = \frac{a}{c}$; $\tan B = \frac{b}{a}$ **7.** $x = 5$; $\tan A = \frac{5}{12}$ **9.** $x = \sqrt{21}$; $\sin A = \frac{\sqrt{21}}{11}$ **11.** $x = 24$; $\cos A = \frac{12}{13}$ **13. a.** $\frac{\sqrt{3}}{2}$; 0.866 **b.** $\frac{1}{2}$; 0.500 **c.** $\sqrt{3}$; 1.732 **15. a.** 6.4 **b.** 2.7 **c.** 0.4 **17. a.** 3.8 **b.** 8.2 **c.** 0.5

PAGE 419 REVIEW EXERCISES **1.** b **3.** a **5.** b **7.** b

PAGE 422 WRITTEN EXERCISES **1.** 0.4226; 0.9063; 0.4663 **3.** 0.9994; 0.0349; 28.6363 **5.** 0.6293; 0.7771; 0.8098 **7.** 0.9613; 0.2756; 3.4874 **9.** 0.6428; 0.7660; 0.8391 **11.** 0.9063; 0.4226; 2.1445 **13.** 81° **15.** 1° **17.** 76° **19.** 30° **21.** 60° **23.** 6 **25.** 58° **27.** 25° **29.** 70° **31.** $c = 9.4$, m $\angle A = 32°$, m $\angle B = 58°$ **33.** $a = 9.5$, $b = 3.1$, m $\angle B = 18°$ **35.** $a = 8.1$, m $\angle A = 64°$, m $\angle B = 26°$ **37.** $a = 41.3$, $c = 43.9$, m $\angle A = 70°$ **39.** $a = 19.2$, $c = 22.6$, m $\angle B = 32°$

PAGES 423–424 PROBLEMS **1.** 42.0 m **3.** 11° **5. a.** 42.5 m **b.** 10.7 m **7.** 39° **9.** 67°

PAGE 424 SELF-TEST B **1.** $\sim$ **2.** $\frac{TN}{TY}$ **3.** 7.5 **4.** $=$ **5.** 6 cm **6.** $4\sqrt{3}$; $8\sqrt{3}$ **7.** $\frac{p}{q}$ **8.** $\frac{r}{q}$ **9.** $\frac{p}{r}$ **10.** m $\angle P = 60°$, m $\angle R = 30°$, $p = 5.2$ **11.** $p = 23.5$, $q = 24.1$, m $\angle R = 12°$ **12.** $q = 8.7$, $r = 3.4$, m $\angle P = 67°$ **13.** $r = 20.1$, $p = 13.1$, m $\angle P = 33°$

PAGE 425 COMPUTER BYTE **1.** 3, 4, 5 **3.** 5, 12, 13 **5.** Output: THERE IS NO SUCH TRIPLE.

PAGE 427 ENRICHMENT **1.** 3 **3.** 2 **5.** 4 **7.** There is no square root because $a$ is negative and $n$ is even. **9.** $-10$ **11.** 5 **13.** $\sqrt{36} = 6$ **15.** $\sqrt[3]{64} = 4$ **17.** $\sqrt[4]{16} = 2$ **19.** $\sqrt[4]{81} = 3$    Calculator Activity **1.** 2.1544347 **3.** 2.1867241 **5.** 1.4953488 **7.** About $9.4203395\,\pi$

PAGE 428 CHAPTER REVIEW **1.** 2 **3.** 5.35 **5.** 4.899 and 5 **7.** False **9.** False **11.** F **13.** 45° **15.** $7\sqrt{2}$ **17.** A **19.** D **21.** F

PAGES 430–431 CUMULATIVE REVIEW
EXERCISES   **1.** $\frac{10}{27}$  **3.** 3300  **5.** $-3\frac{1}{3}$
**7.** $-100$  **9.** 0  **11.** $-200$  **13.** 2.5  **15.** $-1$
**17.** $>$  **19.** $<$  **21.** 38  **23.** 2.5  **25.** $7\frac{1}{3}$  **27.** 8
**29.** 12  **31.** $5\frac{3}{5}$  **33.** 11.6 ft  **35.** 4  **37.** 10
**39.**

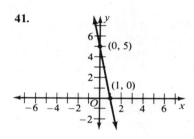

**41.**

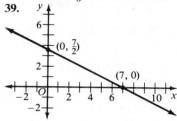

**43.** $55.5\pi$  **45.** $\frac{5\sqrt{2}}{2}$  **47.** $2\sqrt{n}$

PAGE 431 CUMULATIVE REVIEW
PROBLEMS   **1.** 40 dimes, 49 nickels, 52
quarters  **3.** \$22.50  **5.** \$2.17  **7.** 52 in. by
52 in.

## 12 Statistics and Probability
PAGES 436–437 WRITTEN EXERCISES
**1.** 100 million persons  **3.** approx. 400 million
**5.** 23%
**7.**

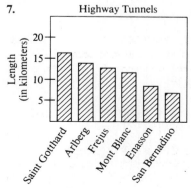

**11.**

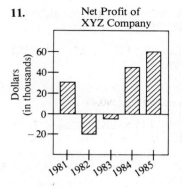

PAGE 437 REVIEW EXERCISES  **1.** 282
**3.** 354  **5.** 135  **7.** 131.4

PAGES 440–442 WRITTEN EXERCISES

**1.**

| | | |
|---|---|---|
| $\frac{1}{4}$ | 90° | |
| $\frac{3}{10}$ | 108° | |
| $\frac{1}{5}$ | 72° | |
| $\frac{1}{4}$ | 90° | |
| 240 | 1 | 360° |

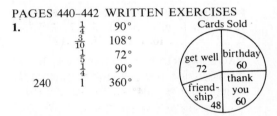

**3.** Surface of Earth

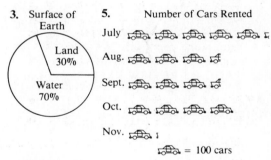

**5.** Number of Cars Rented

**9.** People Living in U.S.

*Answers* **541**

**15. a.** Distribution of Library Grant

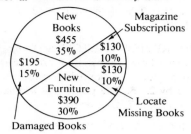

**b.**    Distribution of Library Grant

| Magazine Subscriptions | New Books | Repair Damaged Books | New Furniture | Locate Missing Books |

PAGE 442 REVIEW EXERCISES **1.** 0.84, 2.6, 3, 4, 7 **3.** $-6$, $-5$, $3\frac{1}{2}$, $4\frac{1}{4}$, 7 **5.** $-12$, $-10$, $-9$, $-6$, $-2$ **7.** 0.09, 0.9, 1.09, 1.1

PAGES 444–445 WRITTEN EXERCISES
**1.** 24; 23; 12 **3.** 16; 15; 19 **5.** 55.5; 55.5; 24 **7.** 3.2; 3.4; 0.9 **9.** $-1$; $-2$; 8 **11.** 0.3; 0.2; 2.1 **13.** $-2°$, $-1°$, 19° **15.** 16 **17.** 16 **19.** 94 **21.** It is increased by 5.

PAGE 445 REVIEW EXERCISES **1.** 6.1 **3.** 14.8 **5.** 14 **7.** 3

PAGES 447–448 WRITTEN EXERCISES
**1.** 4; 6.95; 7; 7 **3.** 20; 10; 10; 5 **5.** 6; 16.75; 17; 16 and 17

**7.**

| x | f | |
|---|---|---|
| 8 | 2 | range = 8 |
| 6 | 1 | mean = 3.9 |
| 5 | 4 | median = 4 |
| 4 | 5 | mode = 4 |
| 3 | 2 | |
| 1 | 3 | |
| 0 | 1 | |

**9.**

| x | f | |
|---|---|---|
| 20 | 6 | range = 6 |
| 19 | 7 | mean = 18.1 |
| 18 | 8 | median = 18 |
| 17 | 5 | mode = 18 |
| 16 | 2 | |
| 15 | 1 | |
| 14 | 1 | |

**11.**

| x | f | |
|---|---|---|
| 103 | 2 | range = 6 |
| 102 | 2 | mean = 100.4 |
| 101 | 6 | median = 100 |
| 100 | 9 | mode = 100 |
| 99 | 3 | |
| 97 | 1 | |

**13.** 72

PAGE 448 REVIEW EXERCISES **1.** The average of a set of data. **3.** The number from a set of data which occurs most often. **5.** An integer not evenly divisible by 2. **7.** A number greater than 1 whose only factors are 1 and itself.

PAGE 449 COMPUTER BYTE **1.** $18.\overline{3}$ **3.** $578.1\overline{6}$ **5.** approx. 4494.13042 **7.** 489,439.2

PAGES 451–452 WRITTEN EXERCISES
**1.** 15 **3.** 1 **5.** 9 **7.** 20 **9.** 34
**11.**

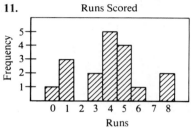

Runs Scored

**17.**

| x | f |
|---|---|
| 10–15 | 3 |
| 15–20 | 3 |
| 20–25 | 5 |
| 25–30 | 4 |
| 30–35 | 5 |
| 35–40 | 4 |
| 40–45 | 6 |

PAGE 452 SELF-TEST A
**1.**

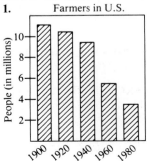

Farmers in U.S.

**2.**

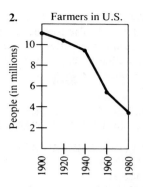

Farmers in U.S.

**3.** Farmers in United States

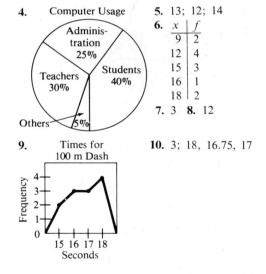

$\overset{\text{\textifamily}}{\text{\o}}$ = 1,000,000 people

**4.** Computer Usage

Administration 25%

Teachers 30%

Students 40%

Others 5%

**5.** 13; 12; 14

**6.**

| $x$ | $f$ |
|---|---|
| 9 | 2 |
| 12 | 4 |
| 15 | 3 |
| 16 | 1 |
| 18 | 2 |

**7.** 3 **8.** 12

**9.** Times for 100 m Dash

**10.** 3; 18, 16.75, 17

**PAGE 455 WRITTEN EXERCISES 1.** 120
**3.** 5040 **5.** 720 **7.** 40,320 **9.** 360 **11.** 74,046

**PAGES 455–457 PROBLEMS 1.** 6 **3.** 24
**5.** 42 **7.** 24 **9.** 12 **11.** 5040 **13.** 720
**15.** 216 **17.** 20 **19.** 6

**PAGE 457 REVIEW EXERCISES 1.** 7 **3.** 3
**5.** 12 **7.** 165

**PAGES 459–460 PROBLEMS 1.** 15 **3.** 15
**5.** 210 **7.** 252 **9. a.** 4 **b.** 6 **c.** 4 **d.** 1
**e.** 15 **11.** 10,192

**PAGE 460 REVIEW EXERCISES 1.** Any of
two or more whole numbers that are multiplied
to form a product. **3.** Any whole numbers that
is divisible by 2. **5.** The product of a number
by itself 3 times. **7.** One or less

**PAGES 464–465 WRITTEN EXERCISES**
**1.** $\frac{1}{6}$ **3.** $\frac{2}{3}$ **5.** 0 **7.** $\frac{1}{4}$ **9.** $\frac{1}{5}$ **11.** $\frac{1}{2}$ **13.** $\frac{3}{4}$
**15.** $\frac{2}{5}$ **17.** $\frac{3}{5}$ **19.** $\frac{1}{3}$ **21.** $\frac{1}{2}$ **23.** $\frac{1}{3}$ **25.** $\frac{2}{3}$ **27.** $\frac{1}{12}$
**29.** $\frac{1}{6}$ **31.** $\frac{1}{18}$ **33.** $\frac{2}{9}$ **35.** $\frac{5}{12}$ **37. a.** $\frac{1}{9}$ **b.** $\frac{5}{9}$
**c.** $\frac{4}{9}$ **d.** $\frac{8}{9}$ **e.** $\frac{4}{9}$ **39.** $\frac{1}{2}$

**PAGE 465 REVIEW EXERCISES 1.** 0 **3.** $\frac{1}{3}$
**5.** $\frac{1}{2}$ **7.** $\frac{3}{4}$

**PAGES 468–469 WRITTEN EXERCISES**
**1.** 1 to 1 **3.** 1 to 2 **5.** 5 to 1 **7.** 2 to 3 **9.** 7
to 3 **11.** 1 to 1 **13.** 3 to 1 **15.** 4 to 1 **17.** 17
to 1 **19.** 17 to 1 **21.** 13 to 5 **23. a.** 2 to 1
**b.** 1 to 2 **25.** 11 to 25 **27.** 3 to 1 **29.** $\frac{1}{9}$

**PAGE 469 REVIEW EXERCISES 1.** $\frac{23}{24}$
**3.** $\frac{11}{12}$ **5.** $\frac{25}{36}$ **7.** $\frac{71}{196}$

**PAGE 469 COMPUTER BYTE 1.** Answers
may vary; 1

**PAGES 471–472 WRITTEN EXERCISES**
**1.** $\frac{13}{15}$ **3.** 0.3 **5. a.** $\frac{1}{3}$ **b.** $\frac{1}{2}$ **c.** $\frac{5}{6}$ **7. a.** $\frac{1}{6}$ **b.** $\frac{5}{6}$
**9. a.** $\frac{1}{2}$ **b.** $\frac{1}{3}$ **c.** $\frac{5}{6}$ **11. a.** $\frac{1}{5}$ **b.** $\frac{2}{5}$ **c.** $\frac{3}{5}$
**13. a.** $\frac{1}{4}$ **b.** $\frac{3}{10}$ **c.** $\frac{11}{20}$ **15. a.** $\frac{1}{4}$ **b.** $\frac{3}{10}$ **c.** $\frac{2}{5}$
**17. a.** $\frac{1}{12}$ **b.** $\frac{11}{36}$ **c.** $\frac{11}{18}$ **19. a.** $\frac{1}{6}$ **b.** $\frac{1}{6}$ **c.** $\frac{1}{3}$
**21.** 9 to 11 **23.** 11 to 9

**PAGE 473 SELF-TEST B 1.** 24 **2.** 120 **3.** 10
**4.** 210 **5.** $\frac{1}{8}$ **6.** $\frac{1}{2}$ **7.** $\frac{1}{4}$ **8.** 1 **9.** 1 to 3
**10.** 1 to 3 **11.** 3 to 1 **12.** $\frac{7}{12}$ **13.** 0.2

**PAGE 475 ENRICHMENT 1. a.** 12 gal
**b.** −1, 1, −3, 3 **c.** 5 gal² **d.** 2.24 gal
**3. a.** 22.8 mi/gal **b.** −2.8, −3.8, −2.8, −0.8,
10.2 **c.** 26.96 mi/gal² **d.** 5.19 mi/gal

**PAGE 476 CHAPTER REVIEW 1.** Sprint
**3.** broken line **5.** 8 **7.** 68.5 **9.** 2 **11.** 3
**13.** True **15.** True

**PAGE 478 CUMULATIVE REVIEW
EXERCISES 1.** 49.5 **3.** 7.475 **5.** $m + 3n$
**7.** $1\frac{6}{35}$ **9.** $\frac{21}{32}$ **11.** $-3\frac{2}{3}$ **13.** 20 **15.** All the
numbers greater than or equal to −2 **17.** 42
**19.** 13.3 **21.** 46.8 **23.** 20% **25.** 36

*Answers* **543**

**27.** $C(-1, 4)$; $D(-7, 7)$; $m = -\frac{1}{2}$
**29.** $P = 468$ m; $A = 13308.75$ m$^2$
**31.** $P = 144$ m; $A = 756$ m$^2$ **33.** no

PAGE 479 CUMULATIVE REVIEW
PROBLEMS **1.** No **3.** 30% **5.** 9.9 cm
**7.** 180 m$^2$ **9.** $5\frac{1}{2}$ ft $\times 3\frac{1}{2}$ ft

## Skill Review

PAGE 480 ADDITION **1.** 99 **3.** 43 **5.** 121
**7.** 741 **9.** 9.2 **11.** 367.43 **13.** 325.277
**15.** 46.119 **17.** 888 **19.** 522 **21.** 20.062
**23.** 8.156 **25.** 22.098 **27.** 18.50032 **29.** 95.83
**31.** 1396 **33.** 6.88 **35.** 155.072 **37.** 1497
**39.** 98.187 **41.** 1565 **43.** 1754 **45.** 1276.208
**47.** 54.7138

PAGE 481 SUBTRACTION **1.** 13 **3.** 26
**5.** 655 **7.** 613 **9.** 5.201 **11.** 11.744
**13.** 2.137 **15.** 1.892 **17.** 435.54 **19.** 0.074
**21.** 25 **23.** 62 **25.** 44 **27.** 48 **29.** 33
**31.** 40.13 **33.** 57.03 **35.** 0.063 **37.** 3162
**39.** 3549 **41.** 6925 **43.** 21.026 **45.** 2.6205
**47.** 0.8651 **49.** 73.403 **51.** 14,548

PAGE 482 MULTIPLICATION **1.** 26
**3.** 159 **5.** 141 **7.** 360 **9.** 1008 **11.** 4615
**13.** 942 **15.** 13,897.6 **17.** 44.53 **19.** 142.576
**21.** 600 **23.** 900 **25.** 72,000 **27.** 680,000
**29.** 141,000 **31.** 15,201 **33.** 5024 **35.** 55,380
**37.** 286.165 **39.** 18.8241 **41.** 75.295
**43.** 8840 **45.** 7914.7224 **47.** 0.1162779
**49.** 1.8574 **51.** 0.00869812

PAGE 483 DIVISION **1.** 17 **3.** 46 **5.** 16
**7.** 35.2 **9.** 0.76 **11.** 15.7 **13.** 10.8 **15.** 15.8
**17.** 13.3 **19.** 548.1 **21.** 49.6 **23.** 73.29
**25.** 81 **27.** 2.75 **29.** 2.63 **31.** 7.36 **33.** 1910
**35.** 0.25 **37.** 517.06 **39.** 129,413.33
**41.** 47.91 **43.** 534.27 **45.** 307.45 **47.** 610.67
**49.** 4900.53 **51.** 18.43 **53.** 3697.69 **55.** 0.01
**57.** 99.09

## Extra Practice

PAGES 484–485 CHAPTER 1 **1.** 42.63
**3.** 0.16 **5.** 1 **7.** 24 **9.** 9 **11.** $\frac{1}{6}$ **13.** 40.8
**15.** 14 **17.** 2 **19.** 169 **21.** 1156 **23.** 16,807
**25.** 6561 **27.** 16,000 **29.** 3,456,000 **31.** 64
**33.** 1728 **35.** $(5 \times 10^2) + (1 \times 10^1) +$

$(6 \times 10^0) + \left(2 \times \frac{1}{10^1}\right) + \left(1 \times \frac{1}{10^2}\right)$

**37.** $(2 \times 10^1) + (5 \times 10^0) + \left(2 \times \frac{1}{10^1}\right)$

**39.** $(9 \times 10^1) + (1 \times 10^0) + \left(9 \times \frac{1}{10^1}\right)$

**41.** $(3 \times 10^2) + (0 + 10^1) + (7 \times 10^0) +$

$\left(0 \times \frac{1}{10^1}\right) + \left(0 \times \frac{1}{10^2}\right) + \left(9 \times \frac{1}{10^3}\right)$ **43.** 5.004
**45.** 9.0009 **47.** 0.0006 **49.** 113.9 **51.** 0.06
**53.** 45.55 **55.** 100 **57.** 1.84 **59.** 0 **61.** 8
**63.** 8 **65.** 112 **67.** 7 **69.** 11 **71.** 12 **73.** 31
**75.** 2448 **77.** 40 **79.** 52.1 s **81.** 1260 parts
**83.** 3 h

PAGES 486–487 CHAPTER 2 **1.** 3 **3.** 1
**5.** 6 **7.** 2 **9.** 9

**11.**

**13.**

**15.**

**17.** $^-8$, $^-2$, 0, 5, 6 **19.** $^-7$, $^-4$, $^-1$, 3, 5 **21.** $^-5$,
$^-4$, 0, 4, 5 **23.** $<$ **25.** 5, $^-5$ **27.** 0 **29.** 6, 5,
4, 3, 2, 1, 0, $^-1$, $^-2$, $^-3$, $^-4$, $^-5$, $^-6$
**41.** 9.9 **43.** 0 **45.** $^-404$ **47.** $^-2.9$ **49.** $^-2.3$
**51.** 20.9 **53.** 12.5 **55.** $-0.6$ **57.** 20.3
**59.** $-13.5$ **61.** $-20.3$ **63.** 5.2 **65.** 5.7
**67.** $-3.2$ **69.** $-2.5$ **71.** $-8.9$ **73.** 2.5
**75.** $-4.1$ **77.** $-610$ **79.** $-0.91$ **81.** 9100
**83.** $-0.003$ **85.** 0.5 **87.** $-4.76$ **89.** 50.02
**91.** 32.86 **93.** $-25.434$ **95.** $-13.345$
**97.** $-15.66$ **99.** 512 **101.** 128 **103.** 324
**105.** 144 **107.** $\frac{1}{16}$ **109.** $\frac{1}{1296}$ **111.** $\frac{1}{81}$
**113.** $-1$ **115.** $-\frac{1}{125}$ **117.** 10 **119.** $\frac{1}{16}$ **121.** $\frac{1}{9}$
**123.** 4

PAGES 488–489 CHAPTER 3 **1.** 1, 2, 3, 6
**3.** 1, 2, 5, 10 **5.** 1, 3, 5, 15 **7.** 1, 41 **9.** 1, 2,
4, 13, 26, 52 **11.** 1, 59 **13.** 5 **15.** 3, 9
**17.** 2, 4 **19.** 2, 3, 4, 5, 9, 10 **21.** 2, 3, 4, 5, 9,
10 **23.** none **25.** $2 \times 5$ **27.** $2 \times 3^3$
**29.** $2 \times 3 \times 5$ **31.** $\frac{1}{3}$ **33.** $1\frac{2}{3}$ **35.** $-1\frac{1}{2}$ **37.** $\frac{3}{5}$
**39.** $-1\frac{3}{4}$ **41.** $-\frac{7}{9}$ **43.** 1 **45.** $-\frac{1}{5}$ **47.** $-\frac{2}{7}$
**49.** 8 **51.** 5 **53.** 4 **55.** $\frac{17}{3}$ **57.** $-\frac{43}{10}$
**59.** $-\frac{57}{8}$ **61.** $\frac{25}{4}$ **63.** $\frac{20}{24}, \frac{9}{24}$ **65.** $\frac{12}{24}, \frac{10}{24}$
**67.** $-\frac{33}{84}, \frac{34}{84}$ **69.** $\frac{9}{13}$ **71.** $\frac{1}{3}$ **73.** $-\frac{179}{224}$
**75.** $-1\frac{17}{60}$ **77.** $3\frac{35}{48}$ **79.** $-5\frac{7}{12}$ **81.** $-\frac{3}{5}$
**83.** $-\frac{1}{2}$ **85.** $2\frac{4}{19}$ **87.** 2 **89.** $-\frac{7}{30}$ **91.** $-\frac{5}{27}$
**93.** 0.375 **95.** $0.9\overline{54}$ **97.** $0.1\overline{6}$ **99.** $-0.53$
**101.** $0.7\overline{45}$ **103.** $-2.\overline{27}$ **105.** $\frac{3}{5}$ **107.** $1\frac{17}{50}$
**109.** $-3\frac{1}{40}$ **111.** $-2\frac{1}{11}$ **113.** $8\frac{7}{33}$

PAGES 490–491 CHAPTER 4  **1.** 50  **3.** 3
**5.** 62  **7.** 0.27  **9.** 4  **11.** 72  **13.** 4  **15.** 245
**17.** $\frac{1}{3}$  **19.** 4  **21.** $\frac{25}{108}$  **23.** 0.2  **25.** 90  **27.** 5
**29.** 4  **31.** 18  **33.** $18 - t$  **35.** $\frac{40}{m} - 16$
**37.** $3y + 11$  **39.** $6n = 54$  **41.** $n + 19 = 61$
**43.** $n = 152 - (-100)$  **45.** 395.77°C  **47.** 5 h
**49.** 31 min to work; 24 min home

PAGES 492–493 CHAPTER 5
**1.** 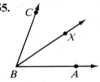  **5.**

**7.** 10,000  **9.** 900  **11.** 7,700  **13.** 1.5  **15.** 157.5
**17.** 200; 2000  **19.** right  **21.** vertex  **23.** obtuse
**25.** sides, angles  **27.** right  **29.** equilateral
**31.** True  **33.** False  **35.** False  **37.** True
**39.** 352 mm  **41.** 50.2 mm  **43.** 4.11 m
**45.** $\overline{SR}$  **47.** $\overline{NM}$  **49.** $\angle N$  **51.** $\angle L$
**53.** $\triangle CAB \cong \triangle DAB$; SSS
**55.**

PAGES 494–495 CHAPTER 6  **1.** $\frac{1}{6}$  **3.** $\frac{1}{8}$
**5.** $\frac{1}{4}$  **7.** $\frac{3}{10}$  **9.** 6.25 km/h  **11.** 24 eggs
**13.** 15¢  **15.** 9.5  **17.** 39  **19.** 72  **21.** 8
**23.** $52  **25.** $1.95  **27.** 44 km  **29.** 63.2 km
**31.** 81.6 km  **33.** 122 km  **35.** $\frac{1}{10}$  **37.** $\frac{39}{100}$
**39.** $\frac{27}{100}$  **41.** 2  **43.** $2\frac{3}{20}$  **45.** 10%  **47.** $12\frac{1}{2}$%
**49.** $43\frac{3}{4}$%  **51.** 144%  **53.** 440%  **55.** 0.71
**57.** 0.152  **59.** 0.024  **61.** 6.25  **63.** 2  **65.** 23%
**67.** 5%  **69.** 0.25%  **71.** 53%  **73.** 12.5%
**75.** 0.25; 25%  **77.** 0.625; 62.5%  **79.** 1.375;
137.5%  **81.** 8.75; 875%  **83.** 5.2; 520%
**85.** 108.9  **87.** 500  **89.** 30%  **91.** 240
**93.** 25  **95.** 3.0%  **97.** 58  **99.** 48.96

PAGES 496–497 CHAPTER 7  **1.** 46.7%
**3.** 40%  **5.** 400%  **7.** 35%  **9.** 80  **11.** 195
**13.** $1.92  **15.** 150%  **17.** $38.16  **19.** 8.7%
**21.** $138,297.87  **23.** $81\frac{1}{4}$%  **25.** $556.50
**27.** 13.8%  **29.** $296.33  **31.** 51 boxes
**33.** $1.25  **35.** $510

PAGES 498–499 CHAPTER 8  **1.** 4  **3.** $\frac{1}{2}$
**5.** $\frac{3}{7}$  **7.** $\frac{1}{2}$  **9.** $1\frac{1}{3}$  **11.** $1\frac{1}{2}$  **13.** 1  **15.** 4  **17.** 4
**19.** $-\frac{1}{2}$  **21.** 3  **23.** $-1\frac{2}{3}$  **25.** sandwich: $2.60,
milk: $0.65  **27.** 15 ft, 45 ft  **29.** $<$  **31.** $>$
**33.** $>$  **35.** $>$  **37.** $2n > 69$
**39.** $80 - 26 > 2(21)$  All the numbers:  **41.** less
than or equal to $-22$  **43.** greater than $4\frac{5}{6}$
**45.** less than 7  **47.** greater than 7  **49.** less
than $-4$  **51.** less than $-63$  **53.** greater than
or equal to $-55$  **55.** less than 70  **57.** less
than or equal to $-4\frac{2}{7}$  **59.** less than or equal
to $11\frac{1}{5}$  **61.** greater than $-5$  **63.** greater than
or equal to $-24$  **65.** less than $-9$  **67.** less
than or equal to $-3$  **69.** 3 h

PAGES 500–501 CHAPTER 9  **1.** $(-6, 2)$
**3.** $(4, 4)$  **5.** $(-2, 4)$  **7.** G  **9.** I  **11.** J
**13.** c. rectangle  **15.** c. triangle
**17.** c. right triangle
**19.** yes  **21.** no  **23.** yes  **25.** yes  **27.** yes
**29.** yes  **31. a.** $y = 3x + 4$  **b.** $(-5, -11)$,
$(0, 4)$, $(5, 19)$  **33. a.** $y = \frac{3}{2}x + 3$  **b.** $(-2, 0)$,
$(0, 3)$, $(2, 6)$  **35. a.** $y = x - 3$  **b.** $(6, 3)$,
$(-1, -4)$, $(-3, -6)$
**37.**

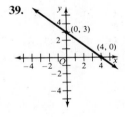

**39.**

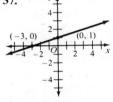

**41.**

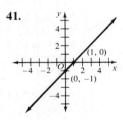

**43.**

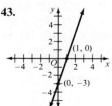

**45.**

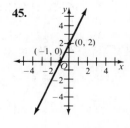

**47.**

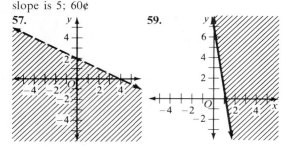

**49.** intersect

**53.** yes **55.** 6.40 **57.** 9 **59.** 9.90 **61.** 15
**63.** $x = 20, y = 40$ **65.** $x = 12, y = 12$
**67.** $x = 20, y = 10$ **69.** $\dfrac{2\sqrt{3}}{3}$ **71.** $\dfrac{x\sqrt{2a}}{2a}$

**73.** $\dfrac{x}{y}$ **75.** $\dfrac{y}{z}$ **77.** $\dfrac{x}{z}$ **79.** 0.4540, 0.8910,
0.5095 **81.** 0.9998, 0.0175, 57.2900 **83.** 26°
**85.** 53° **87.** 23°

**51.** parallel **53.** intersect **55.** $y = 5x + 25$;
slope is 5; 60¢

**57.**

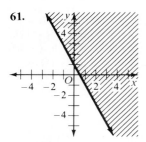

**59.**

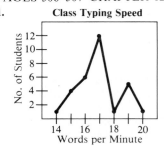

PAGES 506–507 CHAPTER 12
**1.**

Class Typing Speed

No. of Students vs Words per Minute

**61.**

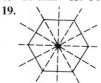

**3.** mean: 7; median: 8; range: 10 **5.** mean:
4.4; median: 3.5; range: 8 **7.** mean: 197;
median: 190; range: 95 **9.** mean: 5500;
median: 6500; range: 7000

**11.**

| $x$ | $f$ |
|---|---|
| 8 | 1 |
| 9 | 2 |
| 10 | 4 |
| 11 | 3 |
| 12 | 3 |
| 13 | 0 |
| 14 | 1 |
| 15 | 0 |
| 16 | 1 |

mean: 11; median: 11; range: 8;
mode: 10

**13.**

| $x$ | $f$ |
|---|---|
| 1 | 3 |
| 2 | 5 |
| 3 | 3 |
| 4 | 3 |
| 5 | 4 |

mean: 3; median: 3; range: 4;
mode: 2

**15.**

| $x$ | $f$ |
|---|---|
| 4.5–34.5 | 6 |
| 34.5–64.5 | 8 |
| 64.5–94.3 | 11 |

PAGES 502–503 CHAPTER 10 **1.** 5000 m²,
300 m **3.** 7625 cm², 372 cm **5.** 132 cm²
**7.** 119 km² **9.** 80.5 cm **11.** 135 m
**13.** 20 mm **15.** 380 m² **17.** 9850 m²
**19.** **21.** 865 m³

**23.** 1260 mm³ **25.** 30 cm **27.** 3536 cm³
**29. a.** 1350 cm² **b.** 1890 cm² **31. a.** 900 cm²
**b.** 1350 cm² **33.** 1296$\pi$ cm² **35.** 12 m
**37.** 2.64 kg **39.** 8530.2 g

PAGES 504–505 CHAPTER 11 **1.** 3 **3.** −6,
−5 **5.** 4, 5 **7.** 9 **9.** 7, 8 **11.** 5 **13.** 4
**15.** 9, 10 **17.** 4.9 **19.** 9.6 **21.** 7.1 **23.** 3.2
**25.** 6.1 **27.** 2.4 **29.** 2.83 **31.** 2.24 **33.** 4.90
**35.** 6.25 **37.** 5.57 **39.** 7.62 **41.** 3.97
**43.** 4.63 **45.** no **47.** no **49.** no **51.** yes

**17.** 720 **19.** 3,628,800 **21.** 120 **23.** 6 **25.** 20
**27.** 252 **29.** 1,326 **31.** $\frac{4}{7}$ **33.** 1 **35.** $\frac{5}{7}$
**37. a.** 7 to 1 **b.** 3 to 1 **c.** 3 to 1 **39. a.** $\frac{5}{18}$
**b.** $\frac{1}{36}$ **c.** $\frac{11}{36}$ **41.** $\frac{1}{2}$ **43. a.** $\frac{1}{6}$ **b.** $\frac{5}{36}$ **c.** $\frac{1}{36}$ **d.** $\frac{5}{18}$